# NUMERICAL METHODS *for* Equations and Its Applications

**Ioannis K. ARGYROS**
Cameron University
Department of Mathematical Sciences
Lawton, OK, USA

**Yeol Je CHO**
Gyeongsang National University
Department of Mathematics Education and RINS
Chinju, Korea

**Saïd HILOUT**
Poitiers University
Laboratoire de Mathematiques et Applications
Bd. Pierre et Marie Curie
Futuroscope Chasseneuil Cedex, France

CRC Press
Taylor & Francis Group
Boca Raton London New York

CRC Press is an imprint of the
Taylor & Francis Group, an **informa** business
A SCIENCE PUBLISHERS BOOK

CRC Press
Taylor & Francis Group
6000 Broken Sound Parkway NW, Suite 300
Boca Raton, FL 33487-2742



Printed in the United States of America on acid-free paper
Version Date: 20120327

International Standard Book Number: 978-1-57808-753-2 (Hardback)



**Library of Congress Cataloging-in-Publication Data**

Argyros, Ioannis K.
Numerical methods for equations and its applications / Ioannis K. Argyros, Yeol Je Cho, Said Hilout.
p. cm.
Includes bibliographical references and index.
ISBN 978-1-57808-753-2 (hardback)
1. Numerical functions. 2. Functional analysis. I. Cho, Yeol Je. II. Hilout, Said. III. Title.
QA246.A74 2012
515'.3--dc23

2011052093

**Visit the Taylor & Francis Web site at**
**http://www.taylorandfrancis.com**

**CRC Press Web site at**
**http://www.crcpress.com**

**Science Publishers Web site at**
**http://www.scipub.net**

The first author dedicates this book to his
mother *ANASTASIA* and father *KONSTANTINOS ARGYROS*

The second dedicates to his
father *Y. M. CHO* and mother *Y. S. YOON*

and

The third dedicates to his
sister *FATIMA* and brothers *MOHAMED, ABDELKADER, AL--HOUSSAIN, RACHID* and *ABDESSELAM*

# Preface

This book adopts a modern scientific approach that combines recent results in numerical methods for equations and variational inequalities with applications in various fields of engineering, optimization, economics, control theory, linear/nonlinear differential equations, partial differential equations and physics. We collect the recent results on the convergence analysis of numerical algorithms in both finite dimensional and infinite dimensional spaces and present several applications and connections with fixed point theory. The book also provides comparison between various investigations made in recent years in the field of computational sciences and combines numerical analysis, linear/nonlinear functional analysis, theory of operators and their applications. Each chapter contains several new theoretical results and important applications. The applications appear in the form of examples or study cases or they are implied since our results improve earlier one that have already been applied in concrete problems. Chapters and sections have been written as independent of each other as possible. Hence, the interested reader can go directly to a certain chapter or section and understand the material without having to go back and forth in the whole book to find related material. The book contains abundant and updated bibliography in the field of computational sciences.

Researchers in the field of computational sciences are faced with the problem of solving a variety of equations or variational inequalities. Note that in computational sciences, the practice of numerical analysis for finding such solutions is essentially connected to variants of Newton's method. For example, dynamic systems are mathematically modelled by difference or differential equations and their solutions represent usually the states of the systems. Similar equations are used in the case of discrete systems. The unknowns of engineering equations can be functions; vectors or real/complex numbers. Except special cases, the most commonly used solutions methods are iterative, when starting from one or several initial approximations a sequences is constructed, which converges to a solution of the equation. Iteration methods are applied also for solving optimization problems. In such cases the iteration sequences converge to an optimal solution of the problem in hand. Since all of these methods have the same recursive structure, they can be introduced and discussed in a general framework. On the other hand, it is well–known that, a wide variety of variational problems including linear and nonlinear complementarity problems or variational inequalities can be formulated in the form of generalized equation. In particular, generalized equations may characterize optimality or equilibrium (traffic network equilibrium, spatial price equilibrium problems, migration equilibrium problems, environmental network problems, $\cdots$ ) and then have several applications in engineering and economics (analysis of elastoplastic structures, Walrasian equilibrium, Nash equilibrium, financial equilibrium problems, $\cdots$ ). The efficient iterative methods for finding the solutions of nonlinear equations, variational inclusions and fixed points problems are the first goal of the present book. The second goal is the study of applications of these methods in several problems whose formulations are nonlinear equations or variational inequalities.

This book is intended for researchers and practitioners in applied computational sciences, engineering, optimization, mathematical economics, mathematical programming, senior undergraduate students and graduate students. The reader is assumed to have had basic courses in numerical analysis, computer programming, computational linear algebra, an introduction to real, complex and functional analysis.

**Ioannis K. Argyros**
**Yeol Je Cho**
**Saïd Hilout**

# Contents

CHAPTER 1

# Introduction

Many problems from Engineering, Optimization, Economics, Physics and other disciplines can be brought in the form of equations or variational inequalities using Mathematical Modelling.

The field of computational sciences gives a lot of opportunity to researchers to solve these problems and has seen a considerable development in mathematics.

We note that in computational sciences, the practice of numerical analysis for finding such solutions is essentially connected to variants of Newton's method. The efficient and iterative methods for finding the solutions of nonlinear equations or variational inclusions are the first goal of the present book. The second goal is the study of applications of these methods in engineering and physical problems whose formulations are nonlinear equations or variational inequalities. For example, dynamic systems are mathematically modelled by difference or differential equations, and their solutions represent usually the states of the systems. For the sake of simplicity, assume that a time–invariant systems is driven by the equation $x' = F(x)$, where $x$ is the state, then, the equilibrium states are determined by solving the equations $F(x) = 0$. Similar equations are used in the case of discrete systems. The unknowns of engineering equations can be:

- Functions (difference, differential, integral equations);
- Vectors (systems of linear or nonlinear algebraic equations);
- Real or complex numbers (single algebraic equations with single unknowns).

Except in special cases, the most commonly used solutions methods are iterative, when starting from one or several initial approximations a sequences is constructed, which converges to a solution of the equation. Iteration methods are applied also for solving optimization problems. In such cases the iteration sequences converge to an optimal solution of the problem in hand. since all of these methods have the same recursive structure, they can be introduced and discussed in a general framework.

On the other hand, it is well–known that, a wide variety of variational problems including:

- Linear and nonlinear complementarity problems;
- Variational inequalities (for example first–order necessary conditions for nonlinear programming)

can be formulated in the form of generalized equation $0 \in F(x) + G(x)$, where $F$ is a function and $G$ is a multifunction. In particular, generalized equations may characterize: Optimality or equilibrium (traffic network equilibrium, spatial price equilibrium problems, migration equilibrium problems, environmental network problems, others). Then, generalized equations have several applications in engineering and economics (analysis of elastoplastic structures, Walrasian equilibrium, Nash equilibrium, financial equilibrium problems, others).

To complicate the matter further, many of equations or generalized equations are nonlinear (i.e., $F$ is nonlinear). However, all may be formulated in terms of operators mapping a linear space into another, the solutions being sought as points in the corresponding space. Consequently, computational methods that work in this general setting for the solution of equations apply to a large number of problems, and lead directly to the development of suitable computer programs to obtain accurate approximate solutions to equations in the appropriate space.

This monograph is intended for researchers in computational sciences and as a reference book for an advanced numerical–functional analysis or computer science course. The goal is to introduce these powerful concepts and techniques at the earliest possible stage. The reader is assumed to have had basic courses in numerical analysis, computer programming, computational linear algebra and an introduction to real, complex and functional analysis.

We have divided the material into two parts. Although the monograph is of a theoretical nature, with optimization and weakening of existing hypotheses considerations each chapter contains several new theoretical results and important applications in engineering, in dynamic economic systems, in input–output systems, in the solution of nonlinear and linear differential equations and optimization problems. The applications appear in the form of Examples or Applications or study cases or they are implied since our results improve earlier one that have already been applied in concrete problems. Sections have been written as independent of each other as possible. Hence, the interested reader can go directly to a certain section and understand the material without having to go back and forth in the whole textbook to find related material.

There are four basic problems connected with iterative methods.

**Problem 1** *Show that the iterates are well defined. For Example, if the algorithm requires the evaluation of $F$ at each $x_n$, it has to be guaranteed that the iterates remain in the domain of $F$. It is, in general, impossible to find the exact set of all initial data for which a given process is well defined and we restrict ourselves to giving conditions which guarantee that an iteration sequence is well defined for certain specific initial guesses.*

**Problem 2** *Concerns the convergence of the sequences generated by a process and the question of whether their limit points are, in fact, solutions of the equation. There are several types of such convergence results. The* *first, which we call a local convergence theorem, begins with the assumption that a particular solution* $x^\star$ *exists and then asserts that there is a neighborhood* $U$ *of* $x^\star$ *such that for all initial vectors in* $U$ *the iterates generated by the process are well defined and converge to* $x^\star$*. The second type of convergence theorem, which we call semilocal, does not require knowledge of the existence of a solution, but states that, starting from initial vectors for which certain–usually stringent–conditions are satisfied, convergence to some (generally nearby) solutions* $x^\star$ *is guaranteed. Moreover, theorems of this type usually include computable (at least in principle) estimates for the error* $x_n - x^\star$*, a possibility not afforded by the local convergence theorems. Finally, the third and most elegant type of convergence result, the global theorem, asserts that starting anywhere in a linear space, or at least in a large part of it, convergence to a solution is assured.*

**Problem 3** *Concerns the economy of the entire operations and in particular, the question of how fast a given sequence will converge. Here, there are two approaches, which correspond to the local and semilocal convergence theorems. As mentioned above, the analysis which leads to the semilocal type of theorem frequently produces error estimates and these in turn, may sometimes be reinterpreted as estimates of the rate of convergence of the sequence. Unfortunately, however, these are usually overly pessimistic. The second approach deals with the behavior of the sequence* $\{x_n\}$ *when* $n$ *is large and hence when* $x_n$ *is near the solutions* $x^\star$*. This behavior may then be determined, to a first approximation, by the properties of the iteration function near* $x^\star$ *and leads to so–called asymptotic rates of convergence.*

**Problem 4** *Concerns with how to best choose a method, algorithm, or software program to solve a specific type of problem and its descriptions of when a given algorithm or method succeeds or fails.*

We have included a variety of new results dealing with Problems 1–4. Results by other authors (or us) are either explicitely stated or implicitely when compared with the corresponding ones of ours results.

This monograph is an outgrowth of research work undertaken by us and complements/updates earlier works of ours focusing on in depth treatment of convergence theory for iterative methods. Such a comprehensive study of optimal iterative procedures appears to be needed and should benefit not only those working in the field but also those interest in, or in need of, information about specific results or techniques. We have endeavored to make the main text as self contained as possible, to prove all results in full detail. In order to make the study useful as a reference source, we have complemented each section with a set of "Remarks" in which literature citations are given, other related results are discussed and various possible extensions of the results of the text are indicated. For completion, the monograph ends with a comprehensive list of references and an appendix containing useful contemporary numerical algorithms. Because we believe our readers come from diverse backgrounds and have varied interests, we provide "recommended reading" throughout the textbook. Often a long monograph summarizes knowledge in a field. This monograph, however, may be viewed as a report on work in progress. We provide a foundation for a scientific field that is rapidly changing. Therefore we list numerous conjectures and open problems as well as alternative models which need to be explored.

Isaac Newton in 1669, inaugurated his method through the use of numerical examples to solve equations, but didnot use the current iterative expression. Later, in 1690, Raphson introduced Newton's method or the also called Newton–Raphson method. Newton's method is currently and undoubtely the most popular one–point iterative procedure for generating a sequence approximating $x^\star$.

The study about convergence matter of Newton's method is usually centered on two types: semilocal and local convergence analysis. The semilocal convergence matter is, based on the information around an initial point, to give criteria ensuring the convergence of Newton's method; while the local one is, based on the information around a solution, to find estimates of the radii of convergence balls. There is a plethora of studies on the weakness and/or extension of the hypothesis made on the underlying operators. Concerning the semilocal convergence of Newton's method, one of the most important results is the celebrated Kantorovich theorem for solving nonlinear equations. This theorem provides a simple and transparent convergence criterion for operators with bounded second derivatives $F''$ or the Lipschitz continuous first derivatives. The second type analysis for numerical methods is the local convergence of Newton's method. Traub, Woźniakowsi, Rheinboldt, Rall, Argyros and other authors gave estimates of the radii of convergence balls when the Fréchet-derivatives are Lipschitz continuous around a solution.

Our goal in chapter 2 is to find weaker convergence criteria for Newton's method than in earlier studies. We present in this chapter new results on Kantorovich–Newton theory and the semilocal/local convergence analysis. It turns out that our sufficient convergence conditions are weaker and the error bounds are tighter than earlier studies in many interesting cases. In section 1 of chapter 2, we use hypotheses on the first Fréchet derivative of operator $F$. we show that Newton's method converges under weaker convergence criteria than those given in earlier studies. These new results are illustrated by several numerical examples, for which the older convergence criteria do not hold but for which our weaker convergence criteria are satisfied. In section 2, we provide semilocal result for the convergence of Newton's method using hypotheses up to the second Fréchet–derivatives and our new idea of recurrent functions. The advantages of such conditions over earlier ones in some cases are: finer bounds on the distances involved, and a better information on the location of the solution. In section 3, new sufficient semilocal convergence results for Newton's method are established on not necessarily bounded domains. Earlier results on bounded domains cannot apply. In section 4, we present a local convergence analysis for the continuous analog of Newton's method in a

Banach space setting. The radius of convergence is larger, the error bounds tighter and under the same or weaker hypotheses than earlier studies.

In section 5, we use a weaker Newton–Kantorovich theorem for solving equations to analyze interior point methods. This way our approach refines earlier work by Potra on Newton's method and interior point techniques. A numerical example is also provided in this section to show that fewer iterations than the ones suggested by Potra are needed to achieve the same error tolerance. In section 6, we use the concept of regular smoothness. This concept has been succesfully shown to be an appropriate and viable framework for the convergence of iterative procedures converging to a locally unique solution of an operator equation in a Banach space setting. Motivated by the works of Galperin and optimization considerations, we show in this section that under the same sufficient convergence conditions tighter error bounds can be obtained on the distances involved and an at least as precise information on the location of the solution. In some special cases, we also show that the sufficient convergence conditions can be weakened. Section 7 is concerned with the application of Kantorovich–type majorants for the convergence of Newton's method. The Fréchet–derivative of the operator involved satisfies only a rather weak continuity condition. Using our new idea of recurrent functions, we obtain sufficient convergence conditions as well as error estimates. In section 8, we use new concept of recurrent functions in combination with not necessarily convex majorants to provide new sufficient convergence conditions for Newton's method which compare favorably earlier works, in the sense that tighter error bounds can be obtained under weaker hypotheses. In some interesting special cases such as in the famous Kantorovich theorem for solving nonlinear equations, our sufficient convergence conditions are at least as weak. The results are also extended to include a non–differentiable term in initial nonlinear equation. Section 9 concerns ball convergence results for Newton's method. Our hypotheses in this section involve very general majorants on the Fréchet–derivatives of the operators involved. In the special case of convex majorants our results have advantages over earlier studies.

New convergence results for very general majorizing sequences of iterative methods are presented in concluding section 10 of this chapter. We unify the semilocal convergence analysis of Newton–type methods under more general Lipschitz–type conditions using our new concept of recurrent functions. We present two very general majorizing sequences and we extend the applicability of Newton–type methods.

Throughout chapter 2, we present an illustrative examples involving a nonlinear integral equation of Chandrasekhar–type and a differential equation containing a Green's kernel, to show that the different theorems are applicable in some situations in which the other are not applicable.

The Secant method uses divided differences operator and is an attractive replacement of Newton's method studied in chapter 2.

No knowledge of the derivative is required; it is a self–correcting like Newton's method and it is of high efficiency.

However the Secant method is of convergence order $1.6180339887\cdots$, whereas Newton's method is of order at most 2. We study in chapter 3 the convergence of Secant method under various conditions and space settings. In section 1 of chapter 3, we present new sufficient convergence conditions for Secant method using our new concept of recurrent functions and combining Lipschitz and center–Lipschitz conditions on the divided difference operator. Our sufficient convergence conditions expand the applicability of the Secant method in cases not covered before by several authors as Dennis, Moré, Laasonen, Potra, Schmidt and Wolfe. We propose in section 2 Secant–type iterative method for approximating a solution of nonlinear least squares problems. The radius of the local convergence is determined as well as usable error estimates. In section 3, we use nondiscrete mathematical induction inaugurated by Pták and Potra. We extend the applicability of the Secant methods. Our new approach has the following advantages over earlier works under the same information: weaker sufficient convergence conditions; tighter error bounds on the distances involved and a more information on the location of the solution. A semilocal convergence analysis for a double step Secant method is presented in section 4 using the same concept as in section 3. Special cases and numerical examples where our results apply but earlier ones fail to solve equation as well as tighter error bounds are also provided in this section. In section 5, semilocal convergence analysis for directional Secant–type methods in finite dimensional space is provided. Using weaker hypotheses than An and Bai and motivated by optimization considerations, we provide in this section a new semilocal convergence result under the same computational cost. We consider two choices of the directions used in these methods for locating $x^\star$. The first is called the directional near–gradient Secant method and the second is called the directional maximal–component–modulus. Section 6 of chapter 3 concerns the local convergence analysis for Chebyshev–Secant–type method using both the Lipschitz continuous and center–Lipshchitz continuous divided differences of order one. An estimate of the radius of the convergence ball and the error estimate matching its convergence order are established. Note that Chebyshev–Secant–type and Secant–type method are useful alternatives of Chebyshev–Newton–type method, especially in cases where the computation of $F''(x_n)$ or $F'(x_n)^{-1}$ are expensive or impossible to compute or their analytic representation is unavailable. As in chapter 2, we present in several section of chapter 3 examples and applications.

Steffensen's method is an alternative method of Newton's method studied in chapter 2 and a variant method of Secant method presented in chapter 3. We establish in chapter 4 the convergence of general Steffensen method under Lipschitz–type conditions. We also expand in some interesting special cases the applicability of this method. At the end of this chapter, we provide a numerical example, where we show that Steffensen's method can be faster than Newton's method.

Gauss–Newton method is also an alternative method for Newton's method studied in chapter 2. This method uses the well known Moore–Penrose–pseudoinverse operators. We present in chapter 5 the convergence of Gauss–Newton method under Lipschitz and average Lipschitz–type conditions in the Euclidean space.

Using our new idea of recurrent functions and a combination of center–Lipschitz, Lipschitz conditions, we provide under the same or weaker hypotheses than earlier studies, a tighter convergence analysis. The results of this chapter can be extented in the case of outer or generalized inverses.

Chapter 6 exhibits Newton–type methods and the convergence analysis using Lipschitz–type conditions. In section 1 of chapter 6, we present new semilocal convergence results for Newton–like method using outer inverses but no Lipschitz conditions. The first analysis is the Kantorovich–type approach, whereas the second uses our new concept of recurrent functions. Comparisons are given between the two techniques. Section 2 concerns with the problem for which Newton's method studied in chapter 2 is not applicable in its original form. A case of interest occurs when the derivative is not continuously invertible, as for instance, dealing with problems involving small divisors, or other important examples. Several Newton–type methods have addressed this problem. Moret unified a large class of such Newton–type methods, where, at each step, the inverse of the derivative, is replaced by a linear operator, which is obtained recursively from the previous one. We provide in this section weaker sufficient convergence conditions and finer error estimates. In section 3, we use the concept of slant differentiability of operators on a Banach space inaugurated by Chen, Nashed and Qi. We provide local as well as semilocal convergence results for the Newton–like method. In the local convergence case, our hypotheses are weaker, the radius of convergence larger and the error estimates finer. Some applications and examples are also provided in this section. We revisit in section 4 of chapter 6 a one–step intermediate Newton iterative scheme that was used by Uko and Velásquez in for the constructive solution of nonlinear splitting equations. Using weaker hypotheses of the Zabrejko–Nguen kind and a modified majorizing sequence we perform a semilocal convergence analysis which yields finer error bounds and more precise information on the location of the solution. We also give two generalizations of the well–known Kantorovich theorem on the solvability of nonlinear equations and the convergence of Newton's method.

In several cases, Newton's method presented in chapter 2 has some disadvantages. For example, it requires the exact computation of the Jacobians involved and the computational cost is great, especially when the system is large. The goal of chapter 7 is to give an answer to part of this problem using an inexact methods.

In section 1 of chapter 7, we establish the local convergence analysis of inexact Newton–type methods using a new type of residual control. We introduce the center Hölder condition on the operator involved and use it in combination with the Hölder condition to provide our new results.

In section 2, we present four local convergence results for two variants of inexact Newton–type methods using average Lipschitz–type condition.

We use in section 3 our new concept of recurrent functions to approximate the locally unique solution by an inexact two–step Newton–like algorithm. The semilocal convergence analysis of this section provides tighter error bounds and in many interesting cases, weaker sufficient convergence conditions.

Zabrejko–Zincenko–type conditions and recurrent functions concept are used in section 4 to provide convergence–existence theorems for Newton–type methods.

Some applications and numerical examples are also presented in several sections of chapter 7.

Chapters 8 and 9 contain recent results on Werner's and Halley's method, respectively. In chapter 8, we revisit Werner's method. The local convergence of Werner's method was given by several authors under Lipschitz conditions on the first and second Fréchet–derivatives given in non–affine invariant form. The two–step method uses one inverse and two function evaluations. We present a semilocal convergence analysis using our new idea of recurrent functions. Our Lipschitz hypotheses are provided in affine invariant form and our new approach can also be used on other one–step or two–step iterative methods.

In chapter 9, we provide local convergence results for Halley's method using convex majorants. Kantorovich–type and Smale–type results are also considered in this chapter as applications and special cases.

In chapter 10 we report on developments for generalized equations and variational inequalities.

In section 1 of chapter 10, we present a new approach of the convergence of Chebyshev–type iterative method in Banach space for solving variational inclusions under different assumptions used in earlier studies. We relax Lipschitz, Hölder or center–Hölder type conditions by introducing $\omega$–type–conditioned second order Fréchet derivative. Under this condition, we show that the sequence is locally superquadratically convergent if some Aubin–type continuity property is satisfied. In particular, we recover a quadratic and a cubic convergence.

In section 2, we establish a new results for the local convergence of the Newton–like method to a unique solution of nondifferentiable variational inclusions in a Banach space setting. Using the Lipschitz–like property of set–valued mappings and the concept of slant differentiability hypothesis on the operator involved, the linear convergence of the Newton–like method is also established. Our results extend the applicability of Newton–like method used for nonlinear equation to variational inclusions.

Kantorovich–type semilocal convergence analysis of the Newton–Josephy method is presented in section 3 of chapter 10. Using a combination of Lipschitz/center–Lipschitz conditions and our new concept of recurrent functions, we provide an analysis with several advantages. Numerical examples and applications are also presented in this chapter.

We present in chapter 11 the convergence analysis of two classes of two–point Newton–type methods including Özban's and the midpoint methods.

We also establish in this chapter the R–order of the semilocal convergence of these methods.

The concluding chapter 12 contains results on computing fixed points for nonlinear equations. We present the convergence analysis for successive methods to solve fixed points problems. The usual hypothesis used in several studies implies that operator is a contraction and as such the existence of $x^\star$ is guaranteed in the considered domain. We weaken in this chapter the sufficient convergence conditions provided by Cătinaş under the same computational cost. This approach extends the applicability of the method of successive approximations. Some remarks and special cases are also presented in this chapter.

# Chapter 2

# Newton's Method

Newton's method NM is undoubtedly the most popular iterative process for generating a sequence to approximate the solutions of nonlinear equations. We study in this chapter the convergence of NM under various conditions and space settings.

## 2.1 Convergence Under Fréchet Differentiability

We provide in this section new results for the semilocal convergence of NM in order to approximate a solution of a nonlinear operator equation in a Banach space setting. The first uses only the center–Lipschitz condition and the second uses both Lipschitz and center–Lipschitz conditions on the Fréchet–derivative of the involved operator. Our results extend the applicability of NM. Numerical examples are also provided in this section.

We are concerned with the problem of approximating a locally unique solution $x^\star$ of equation

$$F(x) = 0, \tag{2.1}$$

where $F$ is a Fréchet–differentiable operator defined on a closed and convex subset $\mathcal{D}$ of a Banach space $\mathcal{X}$ with values in a Banach space $\mathcal{Y}$.

Newton's method NM

$$x_{n+1} = x_n - F'(x_n)^{-1}\, F(x_n) \quad (n \geq 0,\ x_0 \in \mathcal{D}), \tag{2.2}$$

is used in this section for generating a sequence $\{x_n\}$ approximating $x^\star$. Here, $F'(x)$ $(x \in \mathcal{D})$ is the Fréchet–derivative of $F$ at $x$ (cf. [157]).

There is a plethora of local as well as semilocal convergence results on NM. We refer the reader to (cf. [157], [194], [209]) and the references therein for the history and recent results on NM, see also (cf. [170], [198], [613], [614])).

In the case of semilocal convergence Lipschitz condition

$$\| F'(x_0)^{-1}\,(F'(x) - F'(y)) \| \leq L \ \| x - y \| \quad (x, y \in \mathcal{D}) \tag{2.3}$$

has been used. Let

$$\| F'(x_0)^{-1}\, F(x_0) \| \leq \eta. \tag{2.4}$$

Then, known for its simplicity and clarity, Kantorovich hypothesis

$$h_K = 2\,L\eta \leq 1 \qquad (cf. [157],\ [194],\ [469]) \tag{2.5}$$

guarantees the convergence of NM. If strict inequality is satisfied in (2.5), the convergence is quadratic. Otherwise it is only linear (cf. [157], [469]). Hypothesis (2.5) is only a sufficient convergence condition.

As an example, let $\mathcal{X} = \mathcal{Y} = \mathbb{R}$, $x_0 = 1$, $\mathcal{D} = [a, 2-a]$, $a \in [0, .5)$ and define function $F$ on $\mathcal{D}$ by

$$F(x) = x^3 - a. \tag{2.6}$$

Then, using (2.3), (2.4) and (2.6), we get

$$\eta = \frac{1}{3}\,(1-a) \ \text{ and } \ L = 2\,(2-a).$$

Hypothesis (2.5) is violated say for $a = .45$, since

$$h_K = \frac{4}{3}\,(1-a)\,(2-a) = 1.136666 > 1.$$

Hence, there is no guarantee that NM starting at $x_0 = 1$ converges with $x^\star = \sqrt[3]{.45}$, since (2.5) is violated. However, one can see using (2.2) that $\lim\limits_{n \longrightarrow \infty} x_n = x^\star$. Note also that (2.5) is violated for all $a \in (0, .5)$. Therefore, it is very important in computational mathematics to investigate expanding the applicability of NM by weakening (2.5) under the same triplet $(x_0, L, \eta)$.

In view of (2.3) there exists $L_0$ ($L_0 > 0$) such that center–Lipschitz condition

$$\| \, F'(x_0)^{-1}\,(F'(x) - F'(x_0)) \, \| \leq L_0 \ \| \, x - x_0 \, \| \quad (x \in \mathcal{D}) \tag{2.7}$$

holds. Note that we have in general

$$L_0 \leq L \tag{2.8}$$

and $\dfrac{L}{L_0}$ can be arbitrarily large (cf. [139], [209]). In the above numerical example, we have $L_0 < L$, since $L_0 = 3 - a$ (also see Example 2.1).

Using (2.7) instead of the more expensive and less precise (2.3) for the computation of the upper bounds on $\| \, F'(x_n)^{-1}\, F'(x_0) \, \|$ $(n \geq 0)$, we provided weaker replacements of (2.5) given by

$$h_1 = (L + L_0)\,\eta \leq 1 \qquad (cf. [139],\ [136],\ [157]), \tag{2.9}$$

$$h_2 = \overline{L}\,\eta \leq 1 \qquad (cf. [198]), \tag{2.10}$$

where

$$\overline{L} = \frac{1}{4}\,(L + 4\,L_0 + (L^2 + 8\,L_0\,L)^{1/2}).$$

We have

$$\overline{L} \leq L_0 + L \leq 2\,L. \tag{2.11}$$

If $L_0 = L$, then, (2.11) holds as equality. However, if $L_0 < L$, (2.11) is a double strict inequality. Notice that for $L_0 < L$

$$h_K \leq 1 \Longrightarrow h_1 \leq 1 \Longrightarrow h_2 \leq 1, \tag{2.12}$$

but not vice versa. In particular, we get

$$\frac{h_1}{h_K} \longrightarrow \frac{1}{2}, \quad \frac{h_2}{h_K} \longrightarrow \frac{1}{4}, \quad \frac{h_2}{h_1} \longrightarrow \frac{1}{2} \quad \text{as} \quad \frac{L_0}{L} \longrightarrow 0. \tag{2.13}$$

Tighter error bounds on the distances $\| x_{n+1} - x_n \|$, $\| x_n - x^\star \|$ $(n \geq 0)$ and more precise information on the solution $x^\star$ have also been obtained under (2.9) or (2.10).

Note that these advantages have been obtained under the same triplet $(x_0, L, \eta)$, since in practice the computation of Lipschitz constant $L$ requires the computation of center Lipschitz constant $L_0$. Estimates (2.13) indicate by how many finite times (at most) the applicability of NM has been expanded using triplet $(x_0, L, \eta)$.

In this section, we weaken even further the "$h$" conditions and also provide a tighter convergence analysis. We show convergence under

$$h_3 = \overline{\overline{L}}\, \eta \leq 1, \tag{2.14}$$

where

$$\overline{\overline{L}} = \frac{1}{4}\,(4\, L_0 + (L\, L_0 + 8\, L_0^2)^{1/2} + (L_0\, L)^{1/2}).$$

Note that

$$\frac{h_3}{h_K} \longrightarrow 0, \quad \frac{h_3}{h_2} \longrightarrow 0, \quad \frac{h_3}{h_1} \longrightarrow 0 \quad \text{as} \quad \frac{L_0}{L} \longrightarrow 0. \tag{2.15}$$

Hence, the applicability of NM can be extended infinitely many times under the same triplet $(x_0, L, \eta)$. There are examples in the literature where (2.3) is not satisfied but NM converges (cf. [157], [194], [209], [348], [418]). If (2.7) only holds, then

$$h_4 = (5 + 2\sqrt{6})\, L_0\, \eta \leq 1 \tag{2.16}$$

guarantees the convergence of NM (cf. [155], [157], [348], [418]). Here, we show that under the triplet $(x_0, L_0, \eta)$, (2.16) can also be weakened.

NM requires the computation of $F'(x_n)^{-1}$ at each step which may be very expensive or impossible. That is why the Modified Newton's Method MNM

$$y_{n+1} = y_n - F'(y_0)^{-1}\; F(y_n) \quad (n \geq 0,\; y_0 = x_0 \in \mathcal{D}) \tag{2.17}$$

is used to generate a sequence $\{y_n\}$ converging linearly to $x^\star$. Here, the inverse of $F'(y_0)^{-1}$ is only required at each step. In the literature (2.5) is used as common hypothesis for the convergence of NM and MNM.

We showed in (cf. [155], [209]) that (2.5) can be replaced by weaker (for $L_0 < L$)

$$h_0 = 2\, L_0\, \eta \leq 1, \tag{2.18}$$

since (2.7) is only needed for the convergence of MNM. Note that (2.18) reduces to (2.5) for $L_0 = L$. However, if $L_0 < L$ it is weaker, the error estimates tighter and the information on the location of solution $x^\star$ more precise. Moreover, we have

$$\frac{h_0}{h_K} \longrightarrow 0 \quad \text{as} \quad \frac{L_0}{L} \longrightarrow 0. \tag{2.19}$$

Hypothesis (2.18) is the weakest of the "$h$" conditions. In case, say (2.14) (or (2.16)) is violated but (2.18) holds, we show that we can start with MNM and stop after a finite number of steps $N$, where for $x_0 = y_N$ (2.14) (or (2.16)) is satisfied. Then, we continue with faster NM. This approach was not realized before, since (2.5) was used as the sufficient convergence for both NM and MNM (cf. [155]).

We need some lemmas on majorizing sequences for NM.

**Lemma 2.1** *Let $L_0 > 0$, $L \geq L_0$ and $\eta \geq 0$ be given constants such that (2.14) holds. Set*

$$\alpha = \frac{2\,L}{L + (L^2 + 8\,L_0\,L)^{1/2}}. \tag{2.20}$$

*Then, the scalar sequence $\{t_n\}$ $(n \geq 0)$ given by*

$$t_0 = 0, \quad t_1 = \eta, \quad t_2 = \eta + \frac{L_0\,\eta^2}{2\,(1 - L_0\,\eta)}, \quad t_{n+2} = t_{n+1} + \frac{L\,(t_{n+1} - t_n)^2}{2\,(1 - L_0\,t_{n+1})} \quad (n \geq 1) \tag{2.21}$$

*is well defined, increasing, bounded from above by*

$$t^{\star\star} = \eta + \frac{L_0\,\eta^2}{2\,(1 - \alpha)\,(1 - L_0\,\eta)} \tag{2.22}$$

*and converges to its unique least upper bound $t^\star$ satisfying*

$$0 \leq t^\star \leq t^{\star\star}. \tag{2.23}$$

*Moreover, the following estimates hold for $n \geq 1$*

$$0 < t_{n+2} - t_{n+1} \leq \alpha^n\,\frac{L_0\,\eta^2}{2\,(1 - L_0\,\eta)}. \tag{2.24}$$

**Proof** Let

$$\alpha_0 = \frac{L\,(t_2 - t_1)}{2\,(1 - L_0\,t_2)}. \tag{2.25}$$

Using (2.14), it is simple algebra to show

$$\alpha_0 \leq \alpha \leq 1 - \frac{1}{2}\left(\frac{L_0\,\eta}{1 - L_0\,\eta}\right)^2 < 1 \tag{2.26}$$

and

$$L_0\,t_2 < 1. \tag{2.27}$$

We shall show using induction on $k \geq 1$

$$0 \leq \frac{L\,(t_{k+1} - t_k)}{2\,(1 - L_0\,t_{k+1})} \leq \alpha. \tag{2.28}$$

Estimate (2.28) holds true for $k = 1$ by (2.25)–(2.27). Then, we have by (2.21)

$$\begin{aligned} 0 < t_3 - t_2 \leq \alpha\,(t_2 - t_1) &\Longrightarrow t_3 \leq t_2 + \alpha\,(t_2 - t_1) \\ &\Longrightarrow t_3 \leq t_2 + (1+\alpha)\,(t_2 - t_1) - (t_2 - t_1) \\ &\Longrightarrow t_3 \leq t_1 + \frac{1-\alpha^2}{1-\alpha}\,(t_2 - t_1) \leq t^{\star\star}. \end{aligned} \tag{2.29}$$

Assume (2.28) holds, then we get

$$0 < t_{k+2} - t_{k+1} \leq \alpha^k\,(t_2 - t_1) \tag{2.30}$$

and

$$t_{k+2} \leq t_1 + \frac{1-\alpha^{k+1}}{1-\alpha}\,(t_2 - t_1). \tag{2.31}$$

Estimate (2.28) shall be true if $k$ is replaced by $k+1$ provided that

$$\frac{L}{2}\,(t_{k+2} - t_{k+1}) \leq \alpha\,(1 - L_0\,t_{k+2})$$

or

$$\frac{L}{2}\,(t_{k+2} - t_{k+1}) + \alpha\,L_0\,t_{k+2} - \alpha \leq 0$$

or

$$\frac{L}{2}\,\alpha^k\,(t_2 - t_1) + \alpha\,L_0\left(t_1 + \frac{1-\alpha^{k+1}}{1-\alpha}\,(t_2 - t_1)\right) - \alpha \leq 0. \tag{2.32}$$

Estimate (2.32) motivates us to define recurrent functions $\{f_k\}$ on $[0,1)$ by

$$f_k(s) = \frac{L}{2}\,(t_2 - t_1)\,s^k + s\,L_0\,(1 + s + s^2 + \cdots + s^k)\,(t_2 - t_1) - (1 - L_0\,t_1)\,s. \tag{2.33}$$

We need a relationship between two consecutive functions $f_k$. We get that

$$\begin{aligned} f_{k+1}(s) &= \frac{L}{2}\,(t_2 - t_1)\,s^{k+1} + s\,L_0\,(1 + s + s^2 + \cdots + s^{k+1})\,(t_2 - t_1) - (1 - L_0\,t_1)s \\ &= \frac{L}{2}\,(t_2 - t_1)\,s^{k+1} + s\,L_0\,(1 + s + s^2 + \cdots + s^{k+1})\,(t_2 - t_1) - (1 - L_0\,t_1)s \\ &\quad -\frac{L}{2}\,(t_2 - t_1)\,s^k - s\,L_0\,(1 + s + s^2 + \cdots + s^k)\,(t_2 - t_1) \\ &\quad +(1 - L_0\,t_1)\,s + f_k(s). \end{aligned}$$

Hence, we have that

$$f_{k+1}(s) = f_k(s) + \frac{1}{2}\,(2\,L_0\,s^2 + L\,s - L)\,s^k\,(t_2 - t_1). \tag{2.34}$$

Estimate (2.32) is satisfied if

$$f_k(\alpha) \leq 0. \tag{2.35}$$

But in view of (2.34) we have

$$f_{k+1}(\alpha) = f(\alpha). \tag{2.36}$$

Define function $f_\infty$ on $[0, 1)$ by

$$f_\infty(s) = \lim_{k \to \infty} f_k(s). \tag{2.37}$$

Then, we have

$$f_\infty(\alpha) = f_k(\alpha). \tag{2.38}$$

That is (2.35) is satisfied if

$$f_\infty(\alpha) \leq 0. \tag{2.39}$$

Using (2.32) we get

$$f_\infty(\alpha) = \left( \frac{L_0}{1-\alpha} (t_2 - t_1) + L_0 \, t_1 - 1 \right) \alpha. \tag{2.40}$$

It follows from (2.26) and (2.40) that (2.39) is satisfied. The induction is completed.

Hence, sequence $\{t_n\}$ is increasing, bounded from above by $t^{\star\star}$ given by (2.22) and as such that it converges to its unique least upper bound $t^\star$ which satisfies (2.23). This completes the proof of Lemma 2.1. ⊠

**Lemma 2.2** *Let $L_0 > 0$ and $\eta \geq 0$ such that (2.16) holds. Set*

$$\beta_0 = \frac{L_0 \, \eta}{2 \, (1 - L_0 \, \eta)} \tag{2.41}$$

*and*

$$\beta = \frac{1 - L_0 \, \eta - ((1 - L_0 \, \eta)^2 - 8 \, L_0 \, \eta)^{1/2}}{2}. \tag{2.42}$$

*Then, scalar sequence $\{s_n\}$ $(n \geq 0)$ given by*

$$s_0 = 0, \quad s_1 = \eta, \quad s_{n+2} = s_{n+1} + \frac{L_0 \, (2 \, s_n + \frac{1}{2} \, (s_{n+1} - s_n)) \, (s_{n+1} - s_n)}{1 - L_0 \, s_{n+1}} \quad (n \geq 0) \tag{2.43}$$

*is well defined, increasing, bounded from above by*

$$s^{\star\star} = \frac{\eta}{1 - \beta} + (\beta_0 - \beta) \, \eta \tag{2.44}$$

*and converges to its unique least upper bound $s^\star$ satisfying*

$$0 \leq s^\star \leq s^{\star\star}. \tag{2.45}$$

*Moreover, the following estimates hold* $(n \geq 0)$

$$0 < s_{n+1} - s_n \leq \beta\,(s_n - s_{n-1}) \leq \beta^n\,\eta \tag{2.46}$$

*and*

$$0 \leq s^\star - s_n \leq \frac{\beta^n\,\eta}{1-\beta}. \tag{2.47}$$

**Proof** Parameters $\beta_0$ and $\beta$ are non–negative with $\beta_0 \leq \beta$ by (2.20).
We shall show using induction

$$0 \leq \frac{L_0\,(2\,s_k + \frac{1}{2}\,(s_{k+1} - s_k))}{1 - L_0\,s_{k+1}} \leq \beta. \tag{2.48}$$

Estimate (2.48) is true for $k = 0$, since $\beta_0 \leq \beta$. It follows from (2.43) and (2.48) that

$$s_2 - s_1 \leq \beta_0\,(s_1 - s_0) \leq \beta\,(s_1 - s_0) = \beta\,\eta$$

and

$$\begin{aligned} s_2 &\leq s_1 + \beta_0\,(s_1 - s_0) \\ &\leq \eta + \beta\,\eta + (\beta_0 - \beta)\,\eta \\ &\leq \frac{1-\beta^2}{1-\beta}\,\eta + (\beta_0 - \beta)\,\eta \leq s^{\star\star}. \end{aligned}$$

Let us assume

$$0 < s_{k+1} - s_k \leq \beta^k\,\eta \tag{2.49}$$

and

$$s_k \leq \frac{1-\beta^k}{1-\beta}\,\eta + (\beta_0 - \beta)\,\eta. \tag{2.50}$$

Then, estimate (2.48) certainly holds if

$$L_0\,(2\,s_k + \frac{1}{2}\,(s_{k+1} - s_k)) \leq \beta\,(1 - L_0\,s_{k+1})$$

or

$$2\,L_0\,s_k + \frac{L_0}{2}\,(s_{k+1} - s_k) + \beta\,L_0\,s_{k+1} - \beta \leq 0$$

or

$$2\,L_0\left(\frac{1-\beta^k}{1-\beta} + (\beta_0 - \beta)\right)\eta + \frac{L_0}{2}\,\beta^k\,\eta + \beta\,L_0\left(\frac{1-\beta^{k+1}}{1-\beta} + (\beta_0 - \beta)\right)\eta - \beta \leq 0. \tag{2.51}$$

Estimate (2.51) motivates us to define recurrent functions $\{g_k\}$ on $[0, 1)$ by

$$\begin{aligned} g_k(s) = {} & \frac{L_0}{2}\,s^k\,\eta + 2\,L_0\,(1 + s + s^2 + \cdots + s^{k-1})\,\eta \\ & + s\,L_0\,(1 + s + s^2 + \cdots + s^k)\,\eta + (2 + s)\,L_0\,(\beta_0 - s)\,\eta - s. \end{aligned} \tag{2.52}$$

Estimate (2.51) is satisfied if

$$g_k(\beta) \leq 0. \tag{2.53}$$

We need a relationship between two consecutive functions $g_k$. Using (2.52) we get

$$g_{k+1}(s) = g_k(s) + \frac{1}{2}\,(2\,s^2 + s + 3)\,L_0\,s^k\,\eta. \tag{2.54}$$

But we have

$$g_{k+1}(s) \geq g_k(s), \tag{2.55}$$

since $2\,s^2 + s + 3 \geq 0$ for all $s$. Define function $g_\infty$ on $[0,1)$ by

$$g_\infty(s) = \lim_{k\to\infty} g_k(s). \tag{2.56}$$

Then, (2.53) is satisfied if

$$g_\infty(\beta) \leq 0. \tag{2.57}$$

Using (2.51) we obtain

$$g_\infty(\beta) = \frac{(2+\beta)\,L_0\,\eta}{1-\beta} + (2+\beta)\,L_0\,(\beta_0-\beta)\,\eta - \beta. \tag{2.58}$$

Define functions $p_1$, $p_2$ and $p$ on $[0,1)$ by

$$p_1(s) = s^2 - (1 - L_0\,\eta)\,s + 2\,L_0\,\eta, \tag{2.59}$$

$$p_2(s) = L_0\,(\beta_0 - s)\,\eta\,(1-s)\,(2+s) \tag{2.60}$$

and

$$p(s) = p_1(s) + p_2(s). \tag{2.61}$$

Estimate (2.57) is satisfied if

$$p(\beta) \leq 0, \tag{2.62}$$

or

$$p_1(\beta) \leq 0 \tag{2.63}$$

since $p_2(\beta) \leq 0$ and $\beta_0 \leq \beta$. This completes the induction. The rest follows as in Lemma 2.1. This completes the proof of Lemma 2.2. ⊠

**Remark 2.1** *It follows from the proof of Lemma 2.2 that (2.16) can be weakened and $\beta$ can be smaller. Indeed, let define function $q$ on $[0,1)$ by*

$$q(s) = \frac{s\,(1-s)}{2 + 2\,(1-s)\,(\beta_0 - s) + s + (1-s)\,s\,(\beta_0 - s)}. \tag{2.64}$$

*We have*

$$p(0) > 0 \quad \textit{and} \quad p(\beta) = p_1(\beta) + p_2(\beta) \leq 0.$$

*It follows from the intermediate value theorem that function p has a zero $\beta_M \in (0, \beta]$. Using Maple 13 we see*

$$q(\beta_M) \geq q(\beta) = \frac{1}{5+2\sqrt{6}} \geq L_0\,\eta. \tag{2.65}$$

*Hence, we get that*

$$L_0\,\eta \leq q(\beta_M). \tag{2.66}$$

*Parameter $\beta_M$ can replace $\beta$ in Lemma 2.2. Note that in this case error estimates and $s^{\star\star}$ are tighter.*

We also have the following generalization of Lemma 2.1.

**Lemma 2.3** *Let $L_0 > 0$, $L \geq L_0$ and $\eta \geq 0$ be given constants. Assume there exists a minimum integer $N > 1$ such that iterates $t_i$ $(i = 0, 1, \cdots, N-1)$ given by (2.21) are well defined,*

$$t_i < t_{i+1} < \frac{1}{L_0} \quad (i = 0, 1, \cdots, N-2) \tag{2.67}$$

*and*

$$t_N \leq \frac{1}{L_0}\,(1 - (1 - L_0\,t_{N-1})\,\alpha). \tag{2.68}$$

*Then, the following assertions hold*

$$L_0\,t_N < 1, \tag{2.69}$$

$$t_{N+1} \leq \frac{1}{L_0}\,(1 - (1 - L_0\,t_N)\,\alpha), \tag{2.70}$$

$$\alpha_{N-1} \leq \alpha \leq 1 - \frac{L_0\,(t_{N+1} - t_N)}{1 - L_0\,t_N}, \tag{2.71}$$

*the sequence $\{t_n\}$ $(n \geq 0)$ given by (2.21) is well defined, increasing, bounded from above by*

$$t^{\star\star} = t_{N-1} + \frac{1}{1-\alpha}\,(t_N - t_{N-1}) \tag{2.72}$$

*and converges to its unique least upper bound $t^\star$ satisfying*

$$0 \leq t^\star \leq t^{\star\star}, \tag{2.73}$$

*where $\alpha$ is given by (2.20) and*

$$\alpha_n = \frac{L\,(t_{n+2} - t_{n+1})}{2\,(1 - L_0\,t_{n+2})}. \tag{2.74}$$

*Moreover, the following estimate holds*

$$0 < t_{N+n} - t_{N+n-1} \leq \alpha^{n-1}\,(t_{N+1} - t_N) \quad (n \geq 1). \tag{2.75}$$

**Proof** Iterate $t_N$ is well defined, $L_0 \, t_N < 1$, $\alpha_{N-2} \leq \alpha$ and $t_{N-1} < t_N$ by (2.21), (2.67) and (2.68) (for $i = N-2$). Inequality (2.69) and (2.21) imply the existence of $t_{N+1}$ and $t_N < t_{N+1}$.

In order for us to show inequality (2.70) it suffices

$$t_N + \frac{L\,(t_N - t_{N-1})^2}{2\,(1 - L_0\,t_N)} \leq \frac{1}{L_0}\,(1 - (1 - L_0\,t_N)\,\alpha)$$

or

$$\frac{L\,(t_N - t_{N-1})^2}{2\,(1 - L_0\,t_N)} \leq (1-\alpha)\left(\frac{1}{L_0} - t_N\right)$$

or

$$\frac{L_0\,L}{2}\,(t_N - t_{N-1})^2 \leq (1 - L_0\,t_N)^2\,(1-\alpha)$$

or

$$(t_N - t_{N-1})^2 \leq (1 - L_0\,t_N)^2\,\frac{2\,(1-\alpha)}{L_0\,L}$$

or

$$t_N - t_{N-1} \leq \frac{2}{L}\,\alpha\,(1 - L_0\,t_N)$$

or

$$t_N \leq \frac{L\,t_{N-1} + 2\,\alpha}{L + 2\,\alpha\,L_0}.$$

We must now prove

$$\frac{L\,t_{N-1} + 2\,\alpha}{L + 2\,\alpha\,L_0} = \frac{1}{L_0}\,(1 - (1 - L_0\,t_{N-1})\,\alpha). \tag{2.76}$$

But after cross multipying and rearranging terms, we can show instead

$$(2\,\alpha^2\,L_0 + L\,\alpha - L)\,(1 - L_0\,t_N) = 0$$

which is true, since $\alpha$ solves equation

$$2\,L_0\,t^2 + L\,t - L = 0.$$

Hence, we showed (2.70). Estimate (2.71) shall be true if

$$\frac{L\,(t_{N+1} - t_N)}{2\,(1 - L_0\,t_{N+1})} \leq \alpha$$

and

$$\alpha \leq 1 - \frac{L_0\,(t_{N+1} - t_N)}{1 - L_0\,t_N}$$

or if

$$t_{N+1} \leq \frac{L\,t_N + 2\,\alpha}{L + 2\,\alpha\,L_0}$$

and

$$t_{N+1} \leq \frac{1}{L_0}\,(1 - (1 - L_0\,t_N)\,\alpha),$$

which are true by (2.76). The rest follows exactly as the last paragraph of the proof in Lemma 2.1. This completes the proof of Lemma 2.3. ⊠

**Remark 2.2** *(a) If $N = 2$, then, we must have*

$$t_2 = \eta + \frac{L_0\,\eta}{2\,(1 - L_0\,\eta)} \leq \frac{L\,\eta + 2\,\alpha}{L + 2\,\alpha\,L_0},$$

*which is (2.14). When $N > 2$, we do not have closed form inequalities (solved for $n$) any more given by*

$$c_0\,\eta \leq c_1, \tag{2.77}$$

*where $c_0$ and $c_1$ may depend on $L_0$ and $L$ (like (2.10) or (2.14)). However, the corresponding inequalities can also be verified, since only computations involving $\eta$, $L_0$ and $L$ are carried out.*

*Clearly, the sufficient convergence conditions of the form (2.68) become weaker as $N$ increases. Note also that (2.69) and (2.70) imply that $\{t_n\}$ is increasing, bounded above by $\dfrac{1}{L_0}$ and as such it converges to some limit point $t_\infty \in [0, \dfrac{1}{L_0}]$. However, we use (2.71) to arrive at $t^{\star\star}$. This way we obtain better information on the upper bounds for sequence $\{t_n\}$.*

*(b) Results similar to Lemma 2.1 or Lemma 2.3 can also be given for sequences $\{s_n\}$ given by (2.43). However, we decided to provide Lemma 2.2 instead, since in this case a direct comparison with results in (cf. [20], [155]) becomes possible (see also Remark 2.1).*

We provide the following semilocal convergence result fo NM using only center–Lipschitz condition (2.7).

**Theorem 2.1** *Let $F : \mathcal{D} \subseteq \mathcal{X} \longrightarrow \mathcal{Y}$ be a Fréchet–differentiable. Assume that there exist $x_0 \in \mathcal{D}$, $L_0 > 0$ and $\eta > 0$ such that*

$$F'(x_0)^{-1} \in \mathcal{L}(\mathcal{Y}, \mathcal{X}),$$

$$\| F'(x_0)^{-1}\,F(x_0) \| \leq \eta,$$

$$\| F'(x_0)^{-1}\,(F'(x) - F'(x_0)) \| \leq L_0\ \| x - x_0 \| \quad (x \in \mathcal{D}),$$

$$\overline{U}(x_0, s^\star) \subseteq \mathcal{D} \tag{2.78}$$

*and hypotheses of Lemma 2.2 or Remark 2.1 hold, where $s^\star$ is given in Lemma 2.2. Then, the sequence $\{x_n\}$ $(n \geq 0)$ generated by NM is well defined, remains in $\overline{U}(x_0, s^\star)$ for all $n \geq 0$ and converges to a solution $x^\star$ of equation $F(x) = 0$ in $\overline{U}(x_0, s^\star)$. Moreover, the following error estimates hold for all $n \geq 0$*

$$\| x_{n+1} - x_n \| \leq s_{n+1} - s_n \tag{2.79}$$

*and*

$$\| x_n - x^\star \| \leq s^\star - s_n, \tag{2.80}$$

*where the sequence* $\{s_n\}$ $(n \geq 0)$ *is given in Lemma 2.2. Furthermore, if there exists* $R > s^\star$ *such that*

$$U(x_0, R) \subseteq \mathcal{D} \tag{2.81}$$

*and*

$$L_0 \,(s^\star + R) \leq 2, \tag{2.82}$$

*then, the solution* $x^\star$ *is unique in* $U(x_0, R)$.

**Proof** We shall show using induction on $m \geq 0$

$$\| \, x_{m+1} - x_m \, \| \leq s_{m+1} - s_m \tag{2.83}$$

and

$$\overline{U}(x_{m+1}, s^\star - s_{m+1}) \subseteq \overline{U}(x_m, s^\star - s_m). \tag{2.84}$$

For every $z \in \overline{U}(x_1, s^\star - s_1)$,

$$\| \, z - x_0 \, \| \leq \| \, z - x_1 \, \| + \| \, x_1 - x_0 \, \| \leq s^\star - s_1 + s_1 - s_0 = s^\star - s_0$$

implies $z \in \overline{U}(x_0, s^\star - s_0)$. Since also

$$\| \, x_1 - x_0 \, \| = \| \, F'(x_0)^{-1} \, F(x_0) \, \| \leq \eta = s_1 - s_0,$$

estimates (2.83) and (2.84) hold for $m = 0$. If they hold for $n \leq m$, then

$$\| \, x_{m+1} - x_0 \, \| \leq \sum_{i=1}^{m+1} \| \, x_i - x_{i-1} \, \| \leq \sum_{i=1}^{m+1} (s_i - s_{i-1}) = s_{m+1} - s_0 = s_{m+1}$$

and

$$\| \, x_m + \theta \, (x_{m+1} - x_m) - x_0 \, \| \leq s_m + \theta \, (s_{m+1} - s_m) \leq s^\star \quad (\theta \in (0, 1)).$$

Using (2.7), Lemma 2.2 and the induction hypotheses, we get

$$\begin{aligned} \| \, F'(x_0)^{-1} \, (F'(x_{m+1}) - F'(x_0)) \, \| &\leq L_0 \, \| \, x_{m+1} - x_0 \, \| \\ &\leq L_0 \, (s_{m+1} - s_0) \leq L_0 \, s_{m+1} < 1. \end{aligned} \tag{2.85}$$

It follows from (2.85) and the Banach lemma on invertible operators (cf. [157], [209], [469]) that $F'(x_{m+1})^{-1}$ exists and

$$\| \, F'(x_{m+1})^{-1} \, F'(x_0) \, \| \leq (1 - L_0 \, \| \, x_{m+1} - x_0 \, \|)^{-1} \leq (1 - L_0 \, s_{m+1})^{-1}. \tag{2.86}$$

In view of (2.2), we obtain the approximation

$$\begin{aligned} F(x_{m+1}) &= F(x_{m+1}) - F(x_m) - F'(x_m) \, (x_{m+1} - x_m) \\ &= \int_0^1 (F'(x_m + \theta \, (x_{m+1} - x_m)) - F'(x_0)) \, (x_{m+1} - x_m) \, d\theta \\ &\quad + (F'(x_0) - F'(x_m)) \, (x_{m+1} - x_m). \end{aligned} \tag{2.87}$$

Then, by (2.7) and (2.87) we get in turn

$$
\begin{aligned}
& \| F'(x_0)^{-1} F(x_{m+1}) \| \\
& \leq \int_0^1 \| F'(x_0)^{-1} (F'(x_m + \theta (x_{m+1} - x_m)) - F'(x_0)) \| \| x_{m+1} - x_m \| \, d\theta \\
& + \| F'(x_0)^{-1} (F'(x_0) - F'(x_m)) \| \| x_{m+1} - x_m \| \\
& \leq L_0 \int_0^1 \| (x_m - x_0) + \theta (x_{m+1} - x_m) \| \| x_{m+1} - x_m \| \, d\theta \\
& + L_0 \| x_m - x_0 \| \| x_{m+1} - x_m \| \\
& \leq L_0 \left(2 s_m + \frac{1}{2} (s_{m+1} - s_m)\right) (s_{m+1} - s_m).
\end{aligned} \tag{2.88}
$$

By (2.2), (2.86), (2.88) and the induction hypotheses, we get

$$
\begin{aligned}
\| x_{m+2} - x_{m+1} \| & = \| (F'(x_{m+1})^{-1} F'(x_0)) (F'(x_0)^{-1} F(x_{m+1})) \| \\
& \leq \| F'(x_{m+1})^{-1} F'(x_0) \| \| F'(x_0)^{-1} F(x_{m+1}) \| \\
& \leq \frac{L_0 \left(2 s_m + \frac{1}{2} (s_{m+1} - s_m)\right) (s_{m+1} - s_m)}{1 - L_0 s_{m+1}} = s_{m+2} - s_{m+1},
\end{aligned} \tag{2.89}
$$

which shows (2.83) for all $m \geq 0$. Moreover, for every $z \in \overline{U}(x_{m+2}, s^\star - s_{m+2})$ we have

$$
\begin{aligned}
\| z - x_{m+1} \| & \leq \| z - x_{m+2} \| + \| x_{m+2} - x_{m+1} \| \\
& \leq s^\star - s_{m+2} + s_{m+2} - s_{m+1} = s^\star - s_{m+1}.
\end{aligned}
$$

That is $z \in \overline{U}(x_{m+1}, s^\star - s_{m+1})$. The induction for (2.83) and (2.84) is completed.

Lemma 2.2 implies that the sequence $\{s_n\}$ is a Cauchy sequence. It follows from (2.83) and (2.84) that $\{x_n\}$ $(n \geq 0)$ is also a Cauchy sequence in a Banach space $\mathcal{X}$ and as such it converges to some $x^\star \in \overline{U}(x_0, s^\star)$ (since $\overline{U}(x_0, s^\star)$ is a closed set). By letting $m \longrightarrow \infty$ in (2.88), we obtain $F(x^\star) = 0$. Estimate (2.80) is obtained from (2.79) by using standard majorization techniques (cf. [157], [209], [469]). The proof for the uniqueness part have been given in (cf. [198]). That completes the proof of Theorem 2.1. ⊠

Next, we state a semilocal convergence result for NM using both Lipschitz conditions (2.3) and (2.7). Simply use (2.14) instead of (2.9) in [136, Theorem 2.5] to obtain the proof of the following result.

**Theorem 2.2** *Let $F : \mathcal{D} \subseteq \mathcal{X} \longrightarrow \mathcal{Y}$ be a Fréchet–differentiable. Assume that there exist $x_0 \in \mathcal{D}$, $L \geq L_0 > 0$ and $\eta > 0$ such that*

$$
\begin{gathered}
F'(x_0)^{-1} \in \mathcal{L}(\mathcal{Y}, \mathcal{X}), \\
\| F'(x_0)^{-1} F(x_0) \| \leq \eta, \\
\| F'(x_0)^{-1} (F'(x) - F'(x_0)) \| \leq L_0 \| x - x_0 \| \quad (x \in \mathcal{D}), \\
\| F'(x_0)^{-1} (F'(x) - F'(y)) \| \leq L \| x - y \| \quad (x, y \in \mathcal{D}), \\
\overline{U}(x_0, t^\star) \subseteq \mathcal{D}
\end{gathered} \tag{2.90}
$$

*and hypotheses of Lemma 2.1 or Lemma 2.3 hold, where $t^\star$ is given in Lemma 2.1. Then, the sequence $\{x_n\}$ $(n \geq 0)$ generated by NM is well defined, remains in $\overline{U}(x_0, t^\star)$ for all $n \geq 0$ and converges to a solution $x^\star$ of equation $F(x) = 0$ in $\overline{U}(x_0, t^\star)$. Moreover, the following error estimates hold for all $n \geq 0$*

$$
\| x_{n+1} - x_n \| \leq t_{n+1} - t_n \tag{2.91}
$$

*and*

$$
\| x_n - x^\star \| \leq t^\star - t_n, \tag{2.92}
$$

*where the sequence $\{t_n\}$ $(n \geq 0)$ is given in Lemma 2.1. Furthermore, if there exists $R > t^\star$ such that $U(x_0, R) \subseteq \mathcal{D}$ and $L_0 (t^\star + R) \leq 2$, then, the solution $x^\star$ is unique in $U(x_0, R)$.*

**Remark 2.3** *The limit points $s^\star$, $t^\star$ can be replaced by $s^{\star\star}$ and $t^{\star\star}$, respectively, which are given in closed form.*

Next, we estimate the convergence domain of NM. The results will be presented using only hypothesis (2.9). Similar results using (2.10) or (2.14) can follow in an analogous way.

**Lemma 2.4** *Let $F : \mathcal{D} \subseteq \mathcal{X} \longrightarrow \mathcal{Y}$ be a Fréchet–differentiable. Assume that there exist $x^\star \in \mathcal{D}$, $H_\star > 0$ such that*

$$
\begin{gathered}
F'(x^\star)^{-1} \in \mathcal{L}(\mathcal{Y}, \mathcal{X}), \quad F(x^\star) = 0, \\
\| F'(x^\star)^{-1} (F'(x) - F'(x^\star)) \| \leq H_\star \| x - x^\star \| \quad \left( x \in U\left( x^\star, \frac{1}{H_\star} \right) \right), \\
U(x^\star, \frac{1}{H_\star}) \subseteq \mathcal{D}.
\end{gathered}
$$

*Then, for $x_0 \in U\left( x^\star, \dfrac{1}{H_\star} \right)$,*

$$
F'(x_0)^{-1} \text{ exists}
$$

*and*

$$
\| F'(x_0)^{-1} F(x_0) \| \leq \frac{(2 + H_\star \| x_0 - x^\star \|) \| x_0 - x^\star \|}{2 (1 - H_\star \| x_0 - x^\star \|)}. \tag{2.93}
$$

**Proof** We have that

$$\| F'(x^\star)^{-1} (F'(x_0) - F'(x^\star)) \| \leq H_\star \| x_0 - x^\star \| < 1. \tag{2.94}$$

It follows from (2.94) and the Banach lemma on invertible operators that $F'(x_0)^{-1}$ exists and

$$\| F'(x_0)^{-1} F'(x^\star) \| \leq \frac{1}{1 - H_\star \| x_0 - x^\star \|}. \tag{2.95}$$

Using the identity

$$F(x_0) = \int_0^1 ((F'(x^\star + \theta (x_0 - x^\star)) - F'(x^\star)) + F'(x^\star)) (x_0 - x^\star) \, d\theta, \tag{2.96}$$

we obtain the estimate

$$\begin{aligned} &\| F'(x^\star)^{-1} F(x_0) \| \\ &\leq \left( \int_0^1 \| F'(x^\star)^{-1} (F'(x^\star + \theta (x_0 - x^\star)) - F'(x^\star)) \| + 1 \right) \| x_0 - x^\star \| \, d\theta \\ &\leq \frac{H_\star}{2} \| x_0 - x^\star \|^2 + \| x_0 - x^\star \| . \end{aligned} \tag{2.97}$$

By combining (2.96) and (2.97) we get that

$$\begin{aligned} \| F'(x_0)^{-1} F(x_0) \| &\leq \| F'(x_0)^{-1} F'(x^\star) \| \| F'(x^\star)^{-1} F(x_0) \| \\ &\leq \frac{(2 + H_\star \| x_0 - x^\star \|) \| x_0 - x^\star \|}{2 (1 - H_\star \| x_0 - x^\star \|)}. \end{aligned}$$

That estimate shows (2.93). This completes the proof of Lemma 2.4. ⊠

**Theorem 2.3** *Let $F : \mathcal{D} \subseteq \mathcal{X} \longrightarrow \mathcal{Y}$ be a Fréchet–differentiable. For $H_\star > 0$, $H > 0$, $H_0 > 0$, set*

$$H_{\star\star} = \frac{1}{2} \min\{\frac{1}{2 H_\star}, \frac{1}{H_0 + H}\} \quad \textit{and} \quad r \in (0, H_{\star\star}].$$

*Assume there exist $x_0$, $x^\star$ in $\mathcal{D}$ such that for all $x, y \in \mathcal{D}$,*

$$F'(x^\star)^{-1} \in \mathcal{L}(\mathcal{Y}, \mathcal{X}), \quad F(x^\star) = 0;$$

$$\| F'(x^\star)^{-1} (F'(x) - F'(x_0)) \| \leq H_0 \| x - x_0 \|,$$

$$\| F'(x^\star)^{-1} (F'(x) - F'(y)) \| \leq H \| x - y \|,$$

$$\| F'(x^\star)^{-1} (F'(x) - F'(x^\star)) \| \leq H_\star \| x - x^\star \|$$

*and*

$$U(x^\star, \frac{1}{H_\star}) \subseteq \mathcal{D}.$$

*Then, for* $R^\star = \dfrac{3\,r}{1 - H_\star\, r}$*, the sequence* $\{x_n\}$ $(n \geq 0)$ *generated by NM starting at* $x_0 \in U(x^\star, r)$ *is well defined, remains in* $U(x^\star, R^\star)$ *for all* $n \geq 0$ *and converges to* $x^\star$*.*

**Proof** Let $x_0 \in U(x^\star, r)$. It follows from Lemma 2.4 that $F'(x_0)^{-1}$ exists and

$$\eta = \parallel F'(x_0)^{-1}\, F(x_0) \parallel \leq \frac{(2 + H_\star \parallel x_0 - x^\star \parallel) \parallel x_0 - x^\star \parallel}{2\,(1 - H_\star \parallel x_0 - x^\star \parallel)}. \tag{2.98}$$

Set

$$\overline{H}_0 = \frac{H_0}{1 - H_\star \parallel x_0 - x^\star \parallel}, \quad \overline{H} = \frac{H}{1 - H_\star \parallel x_0 - x^\star \parallel}$$

and

$$\overline{r} = \frac{(2 + H_\star\, r)\, r}{1 - H_\star\, r}.$$

Let $x \in U(x_0, \overline{r})$. Then, we have

$$\parallel x - x^\star \parallel \leq \parallel x - x_0 \parallel + \parallel x_0 - x^\star \parallel \leq \overline{r} + r \leq \frac{3\,r}{1 - H_\star\, r} = R^\star.$$

Using Lemma 2.4 we obtain

$$\begin{aligned} \parallel F'(x_0)^{-1}\,(F'(x) - F'(y)) \parallel &\leq \parallel F'(x_0)^{-1} F'(x^\star) \parallel \parallel F'(x^\star)^{-1}\,(F'(x) - F'(y)) \parallel \\ &\leq \frac{H \parallel x - y \parallel}{1 - H_\star \parallel x_0 - x^\star \parallel} = \overline{H} \parallel x - y \parallel \end{aligned}$$

and

$$\begin{aligned} \parallel F'(x_0)^{-1}\,(F'(x) - F'(x_0)) \parallel &\leq \parallel F'(x_0)^{-1} F'(x^\star) \parallel \parallel F'(x^\star)^{-1}\,(F'(x) - F'(x_0)) \parallel \\ &\leq \frac{H_0 \parallel x - x_0 \parallel}{1 - H_\star \parallel x_0 - x^\star \parallel} = \overline{H}_0 \parallel x - x_0 \parallel . \end{aligned}$$

Set

$$\overline{h}_1 = (\overline{H} + \overline{H}_0)\, \eta \quad \textit{and} \quad \overline{r}_1 = 2\, \eta.$$

Then, by (2.98) we get in turn

$$\overline{r}_1 = 2\eta \leq \frac{(2 + H_\star \parallel x_0 - x^\star \parallel) \parallel x_0 - x^\star \parallel}{1 - H_\star \parallel x_0 - x^\star \parallel} \leq \frac{(2 + H_\star\, r)\, r}{1 - H_\star\, r} = \overline{r}$$

and

$$\begin{aligned} \overline{h}_1 &\leq \frac{(2 + H_\star \parallel x_0 - x^\star \parallel)\,(H + H_0) \parallel x_0 - x^\star \parallel}{2\,(1 - H_\star \parallel x_0 - x^\star \parallel)^2} \\ &\leq \frac{(2 + H_\star\, r)\,(H + H_\star)\, r}{2\,(1 - H_\star\, r)^2} \leq \frac{\left(2 + \dfrac{1}{4}\right)\dfrac{1}{2}}{2\left(1 - \dfrac{1}{4}\right)^2} = 1, \end{aligned}$$

by the choice of $r$ and $H_{\star\star}$. According to [136, Theorem 1], sequence $\{x_n\}$ generated by NM converges to some $y^\star \in \overline{U}(x_0, \overline{r}_1)$ such that $F(y^\star) = 0$. Furthermore, we obtain that

$$\| x^\star - y^\star \| \leq \| x^\star - x_0 \| + \| x_0 - y^\star \| \leq r + \overline{r}_1 \leq R^\star.$$

Set $\mathcal{Q} = \int_0^1 F'(x^\star + \theta\,(y^\star - x^\star))d\theta$. Then, we have

$$\| F'(x^\star)^{-1}\,(F'(x^\star) - \mathcal{Q}) \| \leq H_\star \int_0^1 \| \theta\,(y^\star - x^\star) \| \; d\theta < \frac{1}{2}\,H_\star\,\frac{3\,r}{1 - H_\star\,r} < 1,$$

by the choice of $r$ and $H_{\star\star}$. That is $\mathcal{Q}^{-1}$ exists. Finally, in view of the identity $F(y^\star) - F(x^\star) = \mathcal{Q}\,(y^\star - x^\star)$, we deduce $x^\star = y^\star$. This completes the proof of Theorem 2.3. ⊠

**Remark 2.4** *If $H_0 = H = H_\star$, Lemma 2.4 and Theorem 2.3 reduce to earlier ones in the literature. Otherwise, they constitute an improvement, since the convergence domain of NM has been enlarged. Note also that $H_0 \leq H$, $H_\star \leq H$ and $\dfrac{H}{H_0}$, $\dfrac{H}{H_\star}$ can be arbitrarily large. We also have $R^\star = \dfrac{1}{H_\star}$ if $H_{\star\star} = r = \dfrac{1}{4\,H_\star}$ and $R^\star = \dfrac{3}{2(H_0 + H) - H_\star} \leq \dfrac{1}{H_\star}$ if $H_{\star\star} = r = \dfrac{1}{2\,(H_0 + H)}$.*

**Remark 2.5** *It turns out that convergence ball $U(x^\star, r)$, $r \in (0, H_{\star\star}]$ can be enlarged even further. Let us define parameters $\lambda$, $\mu$, $b$, $d$, $H^1_{\star\star}$, $H^2_{\star\star}$ by*

$$\lambda = \frac{2\,H_\star}{H_0 + H}, \qquad \mu = \lambda^{-1},$$

$$b = \begin{cases} \dfrac{-(2\,\lambda + 1) + \sqrt{(2\,\lambda + 1)^2 + 4\,\lambda\,(1 - \lambda)}}{2\,\lambda\,(1 - \lambda)} & \text{if} \quad \lambda \neq 1 \\ \dfrac{1}{3} & \text{if} \quad \lambda = 1 \end{cases}$$

$$d = \begin{cases} \dfrac{1 - \sqrt{\mu}}{1 + \sqrt{\mu}} & \text{if} \quad \mu \neq 1 \\ \dfrac{1}{4} & \text{if} \quad \mu = 1 \end{cases}$$

*and*

$$H^1_{\star\star} = \frac{2\,b}{H_0 + H}, \qquad H^2_{\star\star} = \frac{d}{H_\star}.$$

*Then, we have*

$$\frac{1}{4} \leq b < 1, \quad \text{if} \quad \lambda \leq 1,$$

$$\frac{1}{4} \leq d < 1, \quad \text{if} \quad \mu \leq \frac{36}{100},$$

$$d \leq \frac{1}{4}, \quad \text{if} \quad \frac{36}{100} \leq \mu \leq 1.$$

*Moreover, if* $2\,H_\star \le H_0 + H$*, then*

$$H_{\star\star} \le H^1_{\star\star} \le \frac{1}{H_\star}$$

*and*

$$\overline{h_1} \le \frac{\left(2 + \dfrac{(H_0+H)\,\lambda\,r}{2}\right)(H_0+H)\,r}{2\left(1 - \dfrac{(H_0+H)\,\lambda\,r}{2}\right)^2} \le 1.$$

*Finally, if* $H_0 + H \le 2\,H_\star$*, then*

$$H_{\star\star} \le H^2_{\star\star} \le \frac{1}{H_\star}, \quad \text{if} \quad \mu \le \frac{36}{100},$$

$$H^2_{\star\star} \le H_{\star\star} \le \frac{1}{H_\star}, \quad \text{if} \quad \frac{36}{100} \le \mu \le 1$$

*and*

$$\overline{h_1} \le \frac{(2 + H_\star\,r)\,\mu\,H_\star\,r}{(1 - H_\star\,r)^2} \le 1.$$

*Hence, we conclude that interval* $(0, H_{\star\star}]$ *can be replaced in Theorem 2.3 by the at least as large* $(0, H^1_{\star\star}]$ *if* $2\,H_\star \le H_0 + H$ *and by* $(0, H^2_{\star\star}]$ *if* $\dfrac{H_0+H}{2\,H_\star} \le \dfrac{36}{100}$.

**Remark 2.6** *In view of (2.3) and (2.7), there exist respectively constants* $M_0$, $M$ *such that*

$$\| F'(x_0)^{-1}\,(F'(x_1) - F'(x_0)) \| \le M_0 \; \| x_1 - x_0 \| \tag{2.99}$$

*and, for all* $x \in \mathcal{D}$

$$\| F'(x_0)^{-1}\,(F'(x) - F'(x_1)) \| \le M \; \| x - x_1 \|, \tag{2.100}$$

*where* $x_1 - F'(x_0)^{-1}\,F(x_0)$*. Note that in general*

$$M_0 \le L_0 \tag{2.101}$$

*and*

$$M \le L \tag{2.102}$$

*hold and* $\dfrac{L_0}{M_0}$, $\dfrac{L}{M}$ *can be arbitrarily large. It follows from the proofs of Theorems 2.1 and 2.2 that sequences* $\{v_n\}$, $\{w_n\}$ *defined by*

$$v_0 = 0,\; v_1 = \eta,\; v_2 = \eta + \frac{L_0\,\eta^2}{2\,(1 - M_0\,\eta)},$$

$$v_{n+2} = v_{n+1} + \frac{L\,(v_{n+1} - v_n)^2}{2\,(1 - M_0\,v_{n+1})} \quad (n \ge 1), \tag{2.103}$$

$$w_0 = 0,\ w_1 = \eta,\ w_2 = \eta + \frac{L_0\,\eta^2}{2\,(1 - M_0\,\eta)},\ w_3 = w_2 + \frac{M\,(w_2 - w_1)^2}{2\,(1 - L_0\,w_2)},$$

$$w_{n+2} = w_{n+1} + \frac{L\,(w_{n+1} - w_n)^2}{2\,(1 - L_0\,w_{n+1})} \quad (n \geq 2) \tag{2.104}$$

*are finer majorizing sequences than* $\{t_n\}$ *(under the hypotheses of Theorem 2.2). Using a simple inductive argument, we get*

$$w_n \leq v_n \leq t_n, \tag{2.105}$$

$$w_{n+1} - w_n \leq v_{n+1} - v_n \leq t_{n+1} - t_n \tag{2.106}$$

*and*

$$w^\star = \lim_{n\to\infty} w_n \leq v^\star = \lim_{n\to\infty} v_n \leq t^\star. \tag{2.107}$$

*Strict inequalities hold in (2.105) and (2.106) (for* $n > 2$*) if* $M_0 < L_0$ *or* $M < L$*. Moreover, it follows from the proof of Lemma 2.1 that* $\{v_n\}$, $\{w_n\}$ *converge provided that*

$$0 \leq \overline{\alpha_0} \leq \alpha \leq \frac{1 - L_0\,v_2}{1 - L_0\,v_1}, \tag{2.108}$$

$$L_0\,v_2 < 1 \tag{2.109}$$

*and*

$$0 \leq \overline{\overline{\alpha_0}} \leq \alpha \leq \frac{1 - L_0\,w_3}{1 - L_0\,w_2}, \tag{2.110}$$

$$L_0\,w_3 < 1, \tag{2.111}$$

*respectively, where*

$$\overline{\alpha_0} = \frac{L\,(v_2 - v_1)}{2\,(1 - L_0\,v_2)}, \quad \overline{\overline{\alpha_0}} = \frac{L\,(w_3 - w_2)}{2\,(1 - L_0\,w_3)}. \tag{2.112}$$

*Conditions (2.108)–(2.111) are weaker than the ones provided by Lemma 2.1, but more difficult to verify. However, these conditions involve only computations at the initial data (see also the Example 2.1).*

**Remark 2.7** *Sequence* $\{t_n\}$ *converges provided that*

$$L_0\,t_n < 1 \qquad (n \geq 1). \tag{2.113}$$

*Indeed, it follows from (2.21) and (2.113) that sequence* $\{t_n\}$ *is increasing and bounded above by* $1/L_0$ *and as such it converges to its unique least upper bound* $t^\star \in [0, 1/L_0]$*. Condition (2.14) guarantees that (2.113) holds. Next, we shall find more sufficient conditions that imply (2.113) but do not imply (2.14) (or vice versa). It turns out that this way the error bounds (2.24) can be improved.*

We need auxiliary results for zeros of functions.

**Lemma 2.5** *Let $L_0 > 0$, $L > 0$ and $\eta > 0$ be given constants. Set $\tau = L_0/L$. Let $\delta \in (\frac{L}{2}, \frac{L}{2} + L_0\,(1+\tau))$ be fixed and set $\xi = \dfrac{2\,\delta - L}{2\,L_0}$. Define a sequence of functions $p_n$ on $[0,1]$ by*

$$p_{n+1}(x) = \tau^{2^{n-1}}\,x^{2^n} + \tau^{2^{n-2}}\,x^{2^{n-1}} + \cdots + \tau^{2^0}\,x^2 + x - \xi \qquad (n \geq 1). \quad (2.114)$$

*Then, each $p_n$ $(n \geq 2)$ has a unique zero $r_n = r_n(\delta) \in (0,1)$ and*

$$r_{n+1} \leq r_n \leq \xi \qquad (n \geq 2). \quad (2.115)$$

*In particular, if*

$$p_\infty(x) = \lim_{n\to\infty} p_n(x) \quad for\ each\ fixed \quad x \in (0,1), \quad (2.116)$$

*then, the following assertion holds*

$$r_\infty \leq r_n \qquad (n \geq 2). \quad (2.117)$$

**Proof** It follows from the definition of $\delta$ that $\xi \in (0,2)$. We have $p_{n+1}(0) = -\xi < 0$ and $p_{n+1}(1) = \tau^{2^{n-1}} + \tau^{2^{n-2}} + \cdots + \tau^{2^1} + \tau + 1 - \xi > 0$ $(n \geq 0)$. The intermediate value theorem guarantees the existence of a zero $r_n$ in $(0,1)$ for each $p_n$ $(n \geq 2)$. Using (2.114), we get $p_n'(x) > 0$ $(n \geq 2)$, $(x \in [0,1])$. That is functions $p_n$ are increasing on $[0,\infty)$. Hence, functions $p_n$ cross the $x$–axis only once. That is $r_n$ is the unique zero of $p_n$ in $(0,1)$. Moreover, by (2.114) we obtain

$$p_{n+1}(x) = p_n(x) + \tau^{2^{n-1}}\,x^{2^n} \qquad (n \geq 1). \quad (2.118)$$

In particular, we have that

$$p_{n+1}(r_n) = \tau^{2^{n-1}}\,r_n^{2^n} + p_n(r_n) = \tau^{2^{n-1}}\,r_n^{2^n} > 0. \quad (2.119)$$

It follows from $p_{n+1}(0) < 0$, (2.119) and the defintion of $r_{n+1}$ that (2.115) holds true. Finally, we have $p_{n+1}(\xi) > 0$, so $r_n \leq \xi$. This completes the proof of Lemma 2.5. ⊠

The next result is shown in an analogous way.

**Lemma 2.6** *Let $L_0 > 0$, $L > 0$ and $\eta > 0$ be given constants. Set $\tau = L_0/L$. Let $\delta \in \left(\frac{L}{2} + L_0\,\tau, \frac{L}{2} + L_0\,(1+\tau)\right)$. Define a sequence of functions $q_n$ on $[0,1]$ by*

$$q_{n+1}(x) = \tau^{2^{n-1}}\,x^{2\,n} + \tau^{2^{n-2}}\,x^{2\,(n-1)} + \cdots + \tau^{2^0}\,x^2 + x - \xi \quad (n \geq 1). \quad (2.120)$$

*Then, each $q_n$ $(n \geq 2)$ has a unique zero $\overline{r_n} = \overline{r_n}(\delta) \in (0,1)$ and*

$$\overline{r_{n+1}} \leq \overline{r_n} \leq \xi \quad (n \geq 2). \quad (2.121)$$

*In particular, if*

$$q_\infty(x) = \lim_{n\to\infty} q_n(x) \quad \textit{for each fixed} \quad x \in (0,1), \tag{2.122}$$

*then, the following assertions holds for* $n \geq 2$

$$r \leq \overline{r_\infty} \leq \overline{r_n}, \tag{2.123}$$

$$\overline{r_n} \leq r_n \tag{2.124}$$

*and*

$$\overline{r_\infty} \leq r_\infty, \tag{2.125}$$

*where,* $r$ *is the unique zero of function*

$$q(x) = \frac{\tau + (1-\tau)\,(\tau\,x)^2}{1-(\tau\,x)^2} + x - \xi. \tag{2.126}$$

**Proof** We show in an analogous to Lemma 2.5 way that each $q_n$ has a unique zero $\overline{r_\infty} \in (0,1)$, so that (2.121) is satisfied. It follows from $\tau^{2^{n-1}} \leq \tau^{2(n-1)}$, $x^{2^n} \leq x^{2n}$ $(x \in [0,1])$, the definition of $q$, $q_n$ and $q_\infty$ that

$$\begin{aligned}
q_{n+1}(x) &\leq \sum_{i=1}^{n} \tau^{2\,(i-1)}\,x^{2\,i} + \tau\,x^2 + x - \xi \\
&\leq \tau^{-2}\left(\sum_{i=1}^{n} (\tau\,x)^{2\,i} + 1 - 1\right) - x^2 + \tau\,x^2 + x - \xi \\
&\leq \tau^{-2}\left(\frac{1-(\tau\,x)^{2\,(n+1)}}{1-(\tau\,x)^2} - 1\right) - (1-\tau)\,x^2 + x - \xi \\
&\leq \tau^{-2}\left(\frac{1}{1-(\tau\,x)^2} - 1\right) - (1-\tau)\,x^2 + x - \xi \\
&= \frac{\tau + (1-\tau)\,(\tau\,x)^2}{1-(\tau\,x)^2} + x - \xi = q(x).
\end{aligned}$$

Hence, we obtain that

$$q_\infty(x) \leq q(x). \tag{2.127}$$

Using (2.114) and (2.120), we get that

$$p_n(x) \leq q_n(x) \quad (x \in (0,1),\ n \geq 2). \tag{2.128}$$

In particular, we have $0 = p_n(r_n) \leq q_n(r_n)$ and $q_n(0) = -\xi > 0$, which shows (2.124). (2.125) follows from (2.124). Finally, we have $q(0) = \tau - \xi < 0$ by the choice of $\delta$ and $0 = q_\infty(\overline{r_\infty}) \leq q(\overline{r_\infty})$, which imply the existence of $r$. We also have

$$q'(x) = \frac{2\,\tau^2\,(1-\tau)\,x\,(1-(\tau\,x)^2) + 2\,\tau^2\,x\,(\tau + (1-\tau)\,(\tau\,x)^2)}{(1-(\tau\,x)^2)^2} + 1 > 0.$$

It follows that $q$ is increasing on $(0, \tau^{-1})$ and as such it crosses the $x$–axis only once. Hence, we deduce the uniqueness of $r$. This completes the proof of Lemma 2.6. ⊠

**Remark 2.8** *By Maple 13 we have the explicit form of the unique solution $r$ of $q$ on interval $(0, \tau^{-1})$ but it is very long. We just give in the following table the values of $r$ for various values of parameters $0 < \tau < 1$ and $\xi > \tau$.*

**Table 1: Values of the solution $r$**

| $\tau$ | $\xi$ | $r$ |
|---|---|---|
| .5 | .75 | 1.999899982 |
| .5 | 10 | 1.880409747 |
| .5 | 100 | 1.989822871 |
| .5 | $10^5$ | 1.999990000 |
| .7 | .71 | .009951472065 |
| .7 | .9 | .1832670750 |
| .7 | 10 | 1.346433432 |
| .7 | $10^6$ | 1.428570714 |
| .99 | .999 | .008921976324 |
| .99 | 1 | .009903856306 |
| .99 | $10^7$ | 1.010100960 |
| 1 | 1.001 | .009901941932 |

We can now show the following result for the convergence of sequence $\{t_n\}$.

**Theorem 2.4** *Let constants $L_0$, $L$, $\eta$ and $\delta$ be as given in Lemma 2.5 with $L_0 \le L$. Moreover, suppose that*

$$\delta\,\eta \le r_\infty, \tag{2.129}$$

$$\delta\,\eta \le \overline{r_\infty} \tag{2.130}$$

*or*

$$\delta\,\eta \le r. \tag{2.131}$$

*Then, the sequence $\{t_n\}$ $(n \ge 0)$ given by (2.21) is increasing, bounded above by $1/L_0$ and converges to its unique least upper bound $t^\star \in [0, 1/L_0]$. Moreover, the following error estimates hold for all $n \ge 0$*

$$t_{n+2} - t_{n+1} \le \delta\,(t_{n+1} - t_n)^2 \tag{2.132}$$

*and*

$$t^\star - t_n \le \frac{(\delta\,\eta)^{2^n}}{\delta\,(1 - \delta\,\eta)}. \tag{2.133}$$

**Proof** We shall show using induction

$$0 < \frac{L}{2\,(1 - L_0\,t_{n+1})} \le \delta \qquad (n \ge 0) \tag{2.134}$$

or

$$\delta\, t_{n+1} \le \xi \qquad (n \ge 0). \tag{2.135}$$

Note that if (2.135) holds then so does (2.110) since $\delta > L/2$. If $n = 0$, (2.135) can be rewritten as $\delta\,\eta \le \xi$, which is true by (2.129). Hence, (2.135) holds for $n = 0$. We also have by (2.21) that

$$t_2 - t_1 = \frac{L_0\,(t_1 - t_0)^2}{2\,(1 - L_0\,t_1)} \le \tau\,\frac{1}{\delta}\,(\delta\,(t_1 - t_0))^2 = \tau\,\frac{1}{\delta}\,(\delta\,\eta)^2$$

and

$$\delta\, t_2 \le \tau\,(\delta\,\eta)^2 + \delta\,\eta \le \xi, \tag{2.136}$$

since $p_2(\delta\,\eta) \le 0$ by (2.114) and (2.129). Let us assume that (2.134) holds for all $k \le n+1$. Then, we have

$$\delta\,(t_{n+2} - t_{n+1}) \le (\delta\,(t_{n+1} - t_n))^2 \le \cdots \le (\delta\,(t_1 - t_0))^{2^{n+1}} = (\delta\,\eta)^{2^{n+1}}. \tag{2.137}$$

Using the identity

$$t_{n+2} = t_1 + \sum_{i=1}^{n+1} (t_{i+1} - t_i) \tag{2.138}$$

and (2.137), we see that

$$\delta\, t_{n+2} \le \xi \tag{2.139}$$

holds if

$$\tau^{2^n}\,(\delta\,\eta)^{2^{n+1}} + \tau^{2^{n-1}}\,(\delta\,\eta)^{2^n} + \cdots + \tau^{2^0}\,(\delta\,\eta)^{2^1} + \delta\,\eta \le \xi \tag{2.140}$$

or

$$p_{n+2}(\delta\,\eta) \le 0, \tag{2.141}$$

which is true by (2.129). This completes the induction for (2.132).

Let $m \ge 1$. Then, using (2.132) and the identity

$$t_{n+m} - t_n = \sum_{i=1}^{m} (t_{n+i} - t_{n+i-1}), \tag{2.142}$$

we get in turn

$$\begin{aligned} t_{n+m} - t_n &\le \frac{1}{\delta}\left((\delta\,\eta)^{2^{n+m-1}} + (\delta\,\eta)^{2^{n+m-2}} + \cdots + (\delta\,\eta)^{2^n}\right) \\ &\le \frac{1}{\delta}\,(\delta\,\eta)^{2^n}\left((\delta\,\eta)^{m-1} + (\delta\,\eta)^{m-2} + \cdots + 1\right) \\ &\le \frac{1}{\delta}\,(\delta\,\eta)^{2^n}\,\frac{1 - (\delta\,\eta)^m}{1 - \delta\,\eta}, \end{aligned} \tag{2.143}$$

since $\delta\,\eta < 1$. By letting $m \longrightarrow \infty$ in (2.143), we obtain (2.133). This completes the proof of Theorem 2.4. ⊠

**Remark 2.9** *Zeros $r_\infty$, $\overline{r_\infty}$ or $r$ are not given in closed form (in general). Note also that another possibility for the application of Theorem 2.4 is to directly verify the validity of estimate (2.140) for all $n \geq 1$.*

We provide examples where $L_0 < L$ to compare the "$h$" conditions.

**Example 2.1** *Returning back to the example given in the Introduction, we also have in turn*

$$M_0 = \frac{1}{3}(7-a) < L_0, \quad M = \frac{2(5-2a)}{3} < L \quad (a \in [0, .5])$$

*and*

$$\alpha = \frac{2(2-a)}{2-a+(5a^2-24a+28)^{1/2}}.$$

*Using (2.41), (2.42), (2.59) and (2.60), we obtain*

$$\beta_0 = \frac{1}{2}\frac{(1-a)(3-a)}{3-(1-a)(3-a)},$$

$$\beta = \frac{1}{2}\left(1-\frac{1}{3}(1-a)(3-a)-\left(\left(1-\frac{1}{3}(1-a)(3-a)\right)^2-\frac{8}{3}(1-a)(3-a)\right)^{1/2}\right),$$

$$p_1(s) = s^2-\left(1-\frac{1}{3}(1-a)(3-a)\right)s+\frac{2}{3}(1-a)(3-a)$$

*and*

$$p_2(s) = \frac{1}{3}(1-a)(3-a)(1-s)(2+s)\left(\frac{1}{2}\frac{(1-a)(3-a)}{3-(1-a)(3-a)}-s\right).$$

*Using (2.5) and (2.16), we obtain*

$$h_K = \frac{4}{3}(1-a)(2-a) > 1$$

*and*

$$h_4 = \frac{1}{3}(5+2\sqrt{6})(3-a)(1-a) > 1 \quad \left(a \in \left[0, \frac{1}{2}\right)\right).$$

*Hence, there is no guarantee that NM starting at $x_0$ converges to $x^\star = \sqrt[3]{a}$. However, if we consider our conditions (2.9), (2.10), (2.14) and (2.18),* *we get*

$$h_0 = \frac{2}{3}(3-a)(1-a) \le 1 \quad (a \in [.418861170, .5)),$$

$$h_1 = \frac{1}{3}(7-3a)(1-a) \le 1 \quad (a \in [.464816242, .5)),$$

$$h_2 = \frac{1}{6}(8-3a+(5a^2-24a+28)^{1/2})(1-a) \le 1 \quad (a \in [.450339002, .5)),$$

$$h_3 = \frac{1}{12}(1-a)(12-4a+(84-58a+10a^2)^{1/2}+(12-10a+2a^2)^{1/2}) \le 1$$
$$(a \in [.4271907643, .5))$$

*and*

$$h_4 \le 1 \quad (a \in (.8584827889, 3.141517211)).$$

*By Remark 2.6, for all $a \in (0,1)$, we obtain*

$$(2.108) \iff a \in (.8770219758, 1),$$

$$(2.109) \iff a \in (.3136349706, 1),$$

$$(2.110) \iff a \in (.3136349706, .5)$$

*and*

$$(2.111) \iff a \in (.2862699333, .5).$$

*Then, (2.110) and (2.111) hold for $a \in (.3136349706, .5)$. Theorem 2.2 can apply to solve our problem for $a \in [.4271907643, .5)$.*

*Next, we pick three values of $a$ such that all hypotheses are satisfied, so we can compare the "h" conditions. By Maple 13, we have the following results.*

**Case** $a = .49999$

$$x^\star = .7936952346, \quad h_K = 1.000026667 > 1, \quad h_4 = 4.124673776 > 1,$$

$$h_1 = .9166899999 < 1, \quad h_2 = .8877981560 < 1, \quad h_0 = .8333533332 < 1$$

*and*

$$h_3 = .8467492049 < 1.$$

**Case** $a = .52$

$$x^\star = .8041451517, \quad h_K = .9472000000 < 1, \quad h_4 = 3.927915060 > 1,$$

$$h_1 = .8703999998 < 1, \quad h_2 = .8438043214 < 1, \quad h_0 = .7935999998 < 1$$

*and*

$$h_3 = .8059586212 < 1.$$

**Case** $a = .86$

$$x^\star = .9509685413, \quad h_K = .2128000000 < 1, \quad h_4 = .9885780846 < 1,$$

$$h_1 = .2062666666 < 1, \quad h_2 = .2040583328 < 1, \quad h_0 = .1997333334 < 1$$

*and*

$$h_3 = .2008087925 < 1.$$

*Note that for* $a = .86$, *(2.16) is satisfied. By (2.41), (2.42), (2.61) and (2.66), we have*

$$\beta_0 = .05547326320, \quad \beta = .3969002196, \quad \beta_M = .2583384304.$$

*For* $N = 3$, *by Maple 13 we obtain for* $a \in [0, 1)$

$$(2.67) \Longleftrightarrow a \in (.3394701711, 1)$$

*and*

$$(2.68) \Longleftrightarrow a \in (.1879694266, .3394701710) \cup (.4053946453, 1).$$

*Then, combining (2.14), (2.67) and (2.68), we can apply Lemma 2.3 for all* $a \in [.4271907643, .5)$.

Next, we provide more examples where the inequality is strict between the Lipschitz and the center–Lipschitz constants (i.e., $L_0 < L$).

**Example 2.2** *Define the scalar function* $F$ *by* $F(x) = d_0\, x + d_1 + d_2\, \sin e^{d_3\, x}$, $x_0 = 0$, *where* $d_i$, $i = 0, 1, 2, 3$, *are given parameters. Then, it can easily be seen that for* $d_3$ *large and* $d_2$ *sufficiently small,* $\dfrac{L}{L_0}$ *can be arbitrarily large.*

**Example 2.3** *Let* $\mathcal{X} = \mathcal{Y} = \mathcal{C}[0, 1]$, *equipped with the max–norm. Consider the following nonlinear boundary value problem (cf. [157])*

$$\begin{cases} u'' = -u^3 - \gamma\, u^2, \\ u(0) = 0, \quad u(1) = 1. \end{cases}$$

*It is well known that this problem can be formulated as the integral equation*

$$u(s) = s + \int_0^1 \mathcal{Q}(s,t)\, (u^3(t) + \gamma\, u^2(t))\, dt, \tag{2.144}$$

*where* $\mathcal{Q}$ *is the Green's function*

$$\mathcal{Q}(s,t) = \begin{cases} t\,(1-s), & t \le s \\ s\,(1-t), & s < t. \end{cases}$$

*We observe that*

$$\max_{0 \le s \le 1} \int_0^1 |\mathcal{Q}(s,t)|\, dt = \frac{1}{8}.$$

*Then, problem (2.144) is in the form (2.1), where* $F : \mathcal{D} \longrightarrow \mathcal{Y}$ *is defined as*

$$[F(x)](s) = x(s) - s - \int_0^1 \mathcal{Q}(s,t)\,(x^3(t) + \gamma\, x^2(t))\,dt.$$

*If we set* $u_0(s) = s$, $\mathcal{D} = U(u_0, R_0)$ *and since* $\| u_0 \| = 1$, *it is easy to verify that* $U(u_0, R_0) \subset U(0, R_0 + 1)$. *If* $2\,\gamma < 5$, *then, the operator* $F'$ *satisfies conditions of Theorem 2.2 with*

$$\eta = \frac{1+\gamma}{5-2\,\gamma}, \qquad L = \frac{\gamma + 6\,R_0 + 3}{4}, \qquad L_0 = \frac{2\,\gamma + 3\,R_0 + 6}{8}.$$

*Note that* $L_0 < L$ *(see also Fig. 2.1).*

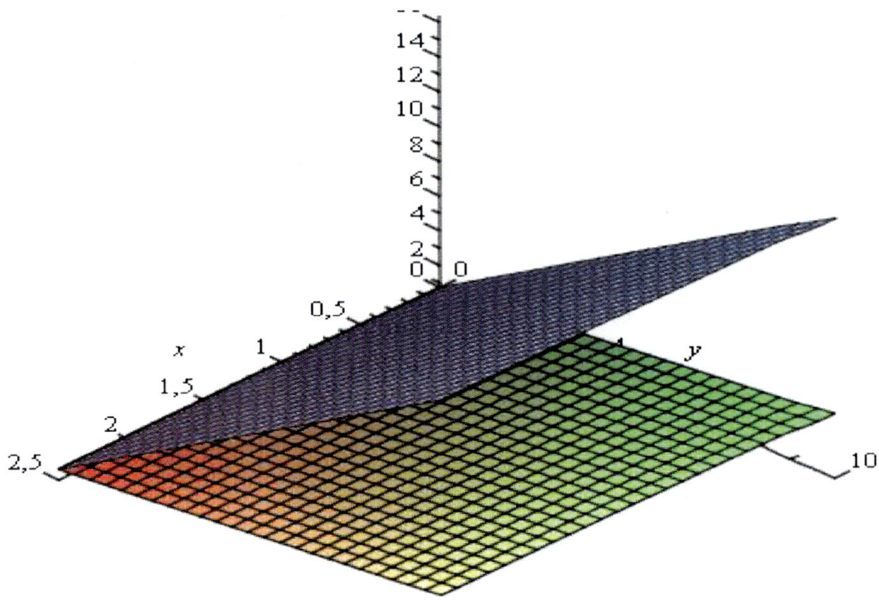


**Figure 2.1:** Functions $L_0$ and $L$ in 3d with respect to $(\gamma, R_0)$ in interval $(0, 2.5) \times (0, 10)$. $L$ is above $L_0$.

**Example 2.4** *Let* $\mathcal{X} = \mathcal{Y} = \mathcal{C}[0,1]$ *be the space of real–valued continuous functions defined on the interval* $[0,1]$ *with max–norm. Let* $\theta \in [0,1]$ *be a given parameter. Consider the "Cubic" integral equation*

$$u(s) = u^3(s) + \lambda\, u(s) \int_0^1 q(s,t)\, u(t)\, dt + y(s) - \theta. \tag{2.145}$$

*Here the kernel* $q(s,t)$ *is a continuous function of two variables defined on* $[0,1]\times[0,1]$; *the parameter* $\lambda$ *is a real number called the "albedo" for scattering;* $y(s)$ *is a given continuous function defined on* $[0,1]$ *and* $x(s)$ *is the unknown function sought in* $\mathcal{C}[0,1]$. *Equations of the form (2.145) arise in the kinetic theory of gasses (cf. [469]). For simplicity, we choose* $u_0(s) = y(s) = 1$ *and* $q(s,t) = \dfrac{s}{s+t}$ *for all* $s \in [0,1]$ *and* $t \in [0,1]$ *with* $s + t \neq 0$. *If we let* $\mathcal{D} = U(u_0, 1-\theta)$ *and define the operator* $F$ *on* $\mathcal{D}$ *by*

$$F(x)(s) = x^3(s) - x(s) + \lambda\, x(s) \int_0^1 q(s,t)\, x(t)\, dt + y(s) - \theta \quad (s \in [0,1]), \tag{2.146}$$

*for all* $s \in [0,1]$, *then, every zero of* $F$ *satisfies equation (2.145). We have the estimates*

$$\max_{0 \le s \le 1} \left| \int \frac{s}{s+t}\, dt \right| = \ln 2.$$

*Therefore, if we set* $\xi = \| F'(u_0)^{-1} \|$, *then, we have*

$$\eta = \xi\, (|\lambda| \ln 2 + 1 - \theta),$$

$$L = 2\,\xi\, (|\lambda| \ln 2 + 3\,(2-\theta)) \quad \text{and} \quad L_0 = \xi\, (2\,|\lambda| \ln 2 + 3\,(3-\theta)).$$

*It follows from our main results that if condition (2.14) holds, then problem (2.145) has a unique solution near* $u_0$. *This assumption is weaker than the one given before using the Newton–Kantorovich hypothesis. Note also that* $L_0 < L$ *for all* $\theta \in [0,1]$.

Finally, we provide examples, where $H_\star < H$.

**Example 2.5** *(cf. [157]) Let* $\mathcal{X} = \mathcal{Y} = \mathbb{R}$. *Define function* $F$ *on* $\mathcal{D} = [-1,1]$ *given by*

$$F(x) = e^x - 1. \tag{2.147}$$

*Then, for* $x^\star = 0$, *using (2.147), we have* $F(x^\star) = 0$ *and* $F'(x^\star) = e^0 = 1$. *Moreover, hypotheses of Theorem 2.3 hold for* $H = e$ *and* $H_\star = e - 1$. *Note that* $H_\star < H$.

**Example 2.6** *(cf. [157]) Let* $\mathcal{X} = \mathcal{Y} = \mathcal{C}[0,1]$, *the space of continuous functions defined on* $[0,1]$, *equipped with the max–norm and* $\mathcal{D} = \overline{U}(0,1)$. *Define function* $F$ *on* $\mathcal{D}$, *given by*

$$F(h)(x) = h(x) - 5 \int_0^1 x\, \theta\, h(\theta)^3\, d\theta. \tag{2.148}$$

*Then, we have*

$$F'(h[u])(x) = u(x) - 15 \int_0^1 x\, \theta\, h(\theta)^2\, u(\theta)\, d\theta \quad (u \in \mathcal{D}).$$

*Using (2.148), we see that hypotheses of Theorem 2.3 hold for* $x^\star(x) = 0$ $(x \in [0,1])$, $H = 15$ *and* $H_\star = 7.5$.

## 2.2 Convergence Under Twice Fréchet Differentiability

We provide in this section semilocal results for the convergence of NM given by (2.2) to a locally unique solution of an equation in a Banach space setting using hypotheses up to the second Fréchet–derivatives and our new idea of recurrent functions. The advantages of such conditions over earlier ones in some cases are: finer bounds on the distances involved and a better information on the location of the solution.

We are concerned with the problem of approximating a locally unique solution $x^\star$ of equation (2.1), where $F$ is a twice Fréchet–differentiable operator defined on a open convex subset $\mathcal{D}$ of a Banach space $\mathcal{X}$ with values in a Banach space $\mathcal{Y}$.

NM given by (2.2) has long played a central role in approximating solutions $x^\star$ of nonlinear equations and systems. Here $F'(x_n)$ denotes the Fréchet–derivative of operator $F$ evaluated at $x = x_n$ $(n \geq 0)$ (cf. [163], [469]). The geometric interpretation of NM is well known, if $F$ is a real function. In such a case $x_{n+1}$ is the point where the tangential line $y - F(x_n) = F'(x_n)\ (x - x_n)$ of function $F(x_n)$ at the point $(x_n, F(x_n))$ intersects the $x$–axis.

Local and semilocal convergence theorems for the quadratic convergence of NM to $x^\star$ have been given under several assumptions by various authors (cf. [112], [600]). For example, Lipschitz conditions have been used on the Fréchet–derivative $F'(x)$ of $F$ $(x \in \mathcal{D})$ (cf. [112], [140], [445], [600]), or center–Lipschitz conditions on the second–derivative $F''(x)$ of $F$ $(x \in \mathcal{D})$ (cf. [112], [600]). Here, we use center–Lipschitz conditions on both first and second Fréchet–derivatives of $F$ and recurrent functions. This particular combination has several advantages over the previously mentioned works. That is why we provide new semilocal convergence theorems for NM. In particular, assume the Lipschitz condition

$$\| F'(x_0)^{-1}\ (F'(x) - F'(y)) \| \leq d\ \| x - y \| \quad (x, y \in \mathcal{D}). \tag{2.149}$$

Set

$$\| F'(x_0)^{-1}\ F(x_0) \| \leq \eta.$$

Then, we arrive at the famous for its simplicity and clarity Kantorovich hypothesis for the semilocal convergence of NM (cf. [163], [195], [469])

$$h_K = d\eta \leq \frac{1}{2}. \tag{2.150}$$

In view of (2.149), there exists $b \geq 0$ such that center–Lipschitz condition

$$\| F'(x_0)^{-1}\ (F'(x) - F'(x_0)) \| \leq b\ \| x - x_0 \| \quad (x \in \mathcal{D}).$$

Clearly, $b \leq d$ holds in general and $\frac{d}{b}$ can be arbitrarily large.

Note that in practice the computation of Lipschitz constant $d$ requires that of $b$. Therefore, the introduction of the center–Lipschitz condition is not an additional hypothesis. Condition (2.149) is exclusively used in the literature to obtain upper bounds on the norms $\| F'(x_n)^{-1} F'(x_0) \|$ $(n \geq 1)$. In particular, if $x \in \mathcal{D}_0 = U(x_0, \frac{1}{d}) \subseteq \mathcal{D}$, $d \neq 0$, we obtain using (2.149)

$$\| F'(x_0)^{-1} (F'(x) - F'(x_0)) \| \leq d \ \| x - x_0 \| < 1. \tag{2.151}$$

In view of (2.151) and the Banach lemma on invertible operator, we conclude that $F'(x)^{-1}$ exists on $\mathcal{D}_0$ and

$$\| F'(x)^{-1} (F'(x_0) \| \leq \frac{1}{1 - d \ \| x - x_0 \|}. \tag{2.152}$$

The usage of estimate (2.152) and several majorizing techniques lead to condition (2.150) (cf. [163], [166]). It is clear that (2.149) is overused, when it comes to obtaining estimate (2.152). Using the needed center–Lipschitz condition, we arrive at the more precise estimate (for $b < d$)

$$\| F'(x)^{-1} F'(x_0) \| \leq \frac{1}{1 - b \ \| x - x_0 \|}. \tag{2.153}$$

Using (2.153) instead of (2.152) and our new idea of recurrent functions, we showed in section 2.1 that (2.150) can always be replaced by

$$h_A = d_0 \ \eta \leq \frac{1}{2}, \tag{2.154}$$

where

$$d_0 = \frac{1}{8} (\sqrt{d\, b} + 4\, b + \sqrt{d\, b + 8\, b^2}).$$

Note that

$$h_K \leq \frac{1}{2} \Longrightarrow h_A \leq \frac{1}{2},$$

but not necessarily vice versa, unless $b = d$.

In this case, finer errors bounds on the distances $\| x_{n+1} - x_n \|$, $\| x_n - x^\star \|$ $(n \geq 0)$ and an at least as precise information on the location of the solution than the ones provided by the Newton–Kantorovich theorem (cf. [163]) are also obtained. Hence, the applicability of NM for solving nonlinear equations has been extended and under the same computational cost. The idea of introducing recurrent functions is a logical consequence of proving the convergence of a majorizing sequence (see section 2.1).

It turns out that the idea of using the center–Lipschitz condition in combination with recurrent functions can be used, when (2.153) also replaces the corresponding estimates by Huang (cf. [445]) and Gutiérrez (cf. [399]), (see (2.222), (2.225) and Proposition 2.1). These estimates are using hypotheses on the second Fréchet–derivative $F''(x)$ $(x \in \mathcal{D})$ of operator $F$. The advantages of this approach over the works by Huang (cf. [445]) and Gutiérrez (cf. [399]) are under the same or weaker hypotheses

(a) Finer error bounds on the distances $\| x_{n+1} - x_n \|$, $\| x_n - x^\star \|$ $(n \geq 0)$;
(b) Better information on the location of the solution $x^\star$.

Let $a > 0$, $b > 0$, $c > 0$ and $\eta > 0$ be given constants. It is convenient for us to define the polynomial $g$ on $[0, +\infty)$ by

$$g(s) = 2bs^2 + cs - (c + an). \tag{2.155}$$

Assume that

$$\eta < \delta_1 = \frac{2b}{a}. \tag{2.156}$$

Then, the polynomial $g$ has a unique root $\delta/2$ in $(0,1)$. We have that

$$g(0) = -(c + a\eta) < 0 \tag{2.157}$$

and

$$g(1) = 2b - a\eta > 0 \quad (\text{by } (2.156)). \tag{2.158}$$

It follows from the Intermediate Value Theorem, (2.157) and (2.158) that there exists a root $\frac{\delta}{2} \in (0,1)$ of the polynomial $g$, given by

$$\delta = \frac{4(c + a\eta)}{c + \sqrt{c^2 + 8b(c + a\eta)}}. \tag{2.159}$$

We also have

$$g'(s) = 4bs + c > 0 \quad (s > 0). \tag{2.160}$$

That is, the polynomial $g$ crosses the positive-axis only once. Hence, $\delta$ is the only root of the polynomial $g$ on $(0,1)$. Let us define the polynomial $p$ on $[0, +\infty)$ by

$$p(s) = \frac{1}{3}\, as^2 + (c + \delta b)s - \delta. \tag{2.161}$$

The polynomial $p$ has also a unique positive root, given by

$$\delta_2 = \frac{2\delta}{c + \delta b + \sqrt{(c + \delta b)^2 + \frac{4}{3}a\delta}}. \tag{2.162}$$

Let us also define

$$\delta_3 = \frac{1}{b}\left(1 - \frac{\delta}{2}\right). \tag{2.163}$$

Set

$$\eta_0 = \min\{\delta_1, \delta_2, \delta_3\}. \tag{2.164}$$

We can show the following result on majorizing sequences for (2.2).

**Lemma 2.7** *Let $a > 0$, $b > 0$, $c > 0$ and $\eta > 0$ be given constants. Assume*

$$\eta \le \eta_0, \tag{2.165}$$

*where $\eta_0$ is defined in (2.164). Inequality (2.165) is strict if $\eta_0 = \delta_1$. Then, scalar sequence $\{t_n\}$ $(n \ge 0)$ generated by*

$$t_0 = 0, \qquad t_1 = \eta,$$
$$t_{n+2} = t_{n+1} + \frac{a\left(\dfrac{1}{3}\,(t_{n+1} - t_n) + t_n\right) + c}{2\,(1 - b\,t_{n+1})}\,(t_{n+1} - t_n)^2 \tag{2.166}$$

*is increasing, bounded from above by*

$$t^{\star\star} = \frac{2\,\eta}{2 - \delta} \tag{2.167}$$

*and converges to its unique least upper bound $t^\star$ with*

$$t^\star \in [0, t^{\star\star}]. \tag{2.168}$$

*Moreover, the following estimates hold for all $n \ge 0$*

$$0 < t_{n+2} - t_{n+1} \le \frac{\delta}{2}\,(t_{n+1} - t_n) \le \left(\frac{\delta}{2}\right)^{n+1} \eta \tag{2.169}$$

*and*

$$t^\star - t_n \le \frac{2\,\eta}{2 - \delta}\left(\frac{\delta}{2}\right)^n. \tag{2.170}$$

**Proof** We shall show using induction on $m$ that

$$\begin{aligned} 0 &< t_{m+2} - t_{m+1} \\ &= \frac{a\left(\dfrac{1}{3}\,(t_{m+1} - t_m)^2 + t_m\,(t_{m+1} - t_m)\right) + c\,(t_{m+1} - t_m)}{2\,(1 - b\,t_{m+1})}\,(t_{m+1} - t_m) \\ &\le \frac{\delta}{2}\,(t_{m+1} - t_m) \end{aligned} \tag{2.171}$$

and

$$b\,t_{m+1} < 1. \tag{2.172}$$

Estimates (2.171) and (2.172) hold for $m = 0$, since

$$\frac{a\left(\dfrac{1}{3}\,(t_1 - t_0)^2 + t_0\,(t_1 - t_0)\right) + c\,(t_1 - t_0)}{1 - b\,t_1} = \frac{\dfrac{1}{3}\,a\eta^2 + c\eta}{1 - b\eta} = \delta_0 \le \delta \tag{2.173}$$

and

$$bt_1 = b\eta < 1 \tag{2.174}$$

by the choice of $\delta$ and (2.165). Let us assume (2.171)–(2.172) for all $n \leq m+1$. Then, we get from (2.169) that

$$t_{m+2} \leq \frac{1 - \left(\dfrac{\delta}{2}\right)^{m+2}}{1 - \dfrac{\delta}{2}} \eta < \frac{2\,\eta}{2-\delta} = t^{\star\star}. \tag{2.175}$$

We shall show (2.173) and (2.174) if

$$a\left\{\frac{1}{3}\left(\left(\frac{\delta}{2}\right)^m \eta\right)^2 + \frac{1 - \left(\dfrac{\delta}{2}\right)^m}{1 - \dfrac{\delta}{2}} \left(\frac{\delta}{2}\right)^m \eta^2\right\}$$

$$+ c \left(\frac{\delta}{2}\right)^m \eta + b\,\delta\, \frac{1 - \left(\dfrac{\delta}{2}\right)^{m+1}}{1 - \dfrac{\delta}{2}} \eta - \delta \leq 0. \tag{2.176}$$

Estimates (2.176) motivates us to define polynomials $f_m$ $(m \geq 1)$ for $s = \dfrac{\delta}{2}$ given by

$$\begin{aligned} f_m(s) &= \frac{a\eta^2}{3}\, s^{2\,m-1} + a\,(1 + s + \cdots + s^{m-1})\, s^{m-1}\, \eta^2 \\ &\quad + c\, s^{m-1}\, \eta + 2b\,(1 + s + \cdots + s^m)\,\eta - 2. \end{aligned} \tag{2.177}$$

We need a relationship between two consecutive functions $f_m$. We get that

$$\begin{aligned} f_{m+1}(s) &= \frac{a\eta^2}{3}\, s^{2\,m+1} + a\,(1 + s + \cdots + s^{m-1} + s^m)\, s^m\, \eta^2 \\ &\quad + c\, s^m\, \eta + 2\,b\,(1 + s + \cdots + s^m + s^{m+1})\,\eta - 2 \\ &= \frac{a\eta^2}{3}\, s^{2\,m+1} + \frac{a\eta^2}{3}\, s^{2\,m-1} - \frac{a\eta^2}{3}\, s^{2m-1} \\ &\quad + a\,(1 + s + \cdots + s^{m-1} + s^{m-1})\, s^{m-1}\, \eta^2 \\ &\quad - a\,(1 + s + \cdots + s^{m-1} + s^{m-1})\, s^{m-1}\, \eta^2 \\ &\quad + a\,(1 + s + \cdots + s^{m-1} + s^m)\, s^m\, \eta^2 + c\, s^{m-1}\, \eta - c\, s^{m-1}\, \eta \\ &\quad + c\, s^m\, \eta + 2\,b\,(1 + s + \cdots + s^m)\,\eta + 2\,b\, s^{m+1}\, \eta - 2 \\ &= f_m(s) + g_m(s) + g(s)\, s^{m-1}\, \eta, \end{aligned} \tag{2.178}$$

where the function $g$ is given by (2.165) and

$$g_m(s) = \left(\frac{1}{3}\, s^2 + s + \frac{2}{3}\right) s^{2\,m-1}\, \eta^2\, a \geq 0 \quad (s \geq 0). \tag{2.179}$$

In view of (2.155), (2.178) and (2.179), we have that

$$f_{m+1}\left(\frac{\delta}{2}\right) \geq f_m\left(\frac{\delta}{2}\right). \tag{2.180}$$

We shall show, instead of (2.176) that

$$f_m\left(\frac{\delta}{2}\right) \leq 0 \qquad (m \geq 1). \tag{2.181}$$

Define the function $f_\infty$ on $[0,1)$ by

$$f_\infty(s) = \lim_{n\to\infty} f_n(s). \tag{2.182}$$

Then, using (2.177), we have that

$$f_\infty(s) = 2\left[\frac{b\eta}{1-s} - 1\right]. \tag{2.183}$$

It also follows from (2.180) that

$$f_\infty\left(\frac{\delta}{2}\right) \geq f_m\left(\frac{\delta}{2}\right). \tag{2.184}$$

In view of (2.181) and (2.184), it is enough to show

$$f_\infty\left(\frac{\delta}{2}\right) \leq 0, \tag{2.185}$$

which is true, since $\eta \leq \delta_3$. This completes the induction. Therefore, the sequence $\{t_n\}$ is non–decreasing, bounded above by $t^{\star\star}$ given by (2.167) and converges to its unique least upper bound $t^\star$ satisfying (2.168). Finally, estimate (2.170) follows from (2.169) by using standard majorization techniques (cf. [163], [469]). This completes the proof of Lemma 2.7. ⊠

Let us define functions $\bar{f}_m$ by

$$\begin{aligned}\overline{f}_m(s) = {}& a\left(\frac{1}{3}\, s^{2m-1} + (1+s+\cdots+s^{m-1})\, s^{m-1}\right)\eta^2 \\ &+c\, s^{m-1}\,\eta + 2c\,(1+s+\cdots+s^m)\,\eta \\ &+a\,(1+s+\cdots+s^m)^2\,\eta^2 - 2.\end{aligned} \tag{2.186}$$

Then, we have as in Lemma 2.7

$$\overline{f}_{m+1}(s) = \overline{f}_m(s) + \overline{g}_m(s) + \overline{g}(s)\, s^{m-1}\,\eta, \tag{2.187}$$

where

$$\begin{aligned}\overline{g}_m(s) = {}& \left(\frac{a}{3}\, s^{m+2} + a\eta s^{m+3} + 2a\eta(1+s+\cdots+s^m)\, s^2 \right.\\ &\left. +\frac{2}{3}\,\eta s^m + a\eta^{m+1}\right)s^{m-1}\eta > 0 \qquad (s>0)\end{aligned}$$

and

$$\overline{g}(s) = 2cs^2 + cs - (c + a\eta). \tag{2.188}$$

We also have that

$$\overline{p}(s) = \frac{a}{6}\,(2 + 3\delta)s^2 + c(1 + \delta)s - \delta \tag{2.189}$$

and

$$\overline{f}_\infty(s) = \frac{2c\eta}{1 - s} + a\left(\frac{1}{1 - s}\right)^2 \eta^2 - 2. \tag{2.190}$$

Moreover, we obtain that

$$\begin{aligned}
\overline{\delta}_1 &= \frac{2\,c}{a}, \\
\overline{\delta}_2 &= \frac{2\delta(2 + 3\delta)}{c(1 + \delta) + \sqrt{(c(1 + \delta))^2 + \frac{2}{3}a\delta(2 + 3\delta)}}, \\
\overline{\delta}_3 &= \frac{2 - \delta}{c + \sqrt{c^2 + 2a}} \quad \text{and} \\
\frac{\overline{\delta}}{2} &= \frac{2(c + a\,\eta)}{c + \sqrt{c^2 + 8c(c + a\eta)}}.
\end{aligned}$$

Set

$$\overline{\eta}_0 = \min\{\overline{\delta}_1, \overline{\delta}_2, \overline{\delta}_3\}. \tag{2.191}$$

Then, with the above changes and simply following the proof of Lemma 2.7, we can provide another majorizing sequence result for NM:

**Lemma 2.8** *Let $a > 0$, $b > 0$, $c > 0$ and $\eta > 0$ be given constants. Assume that*

$$\eta \leq \overline{\eta}_0. \tag{2.192}$$

*Inequality (2.192) is strict if $\overline{\eta}_0 = \overline{\delta}_1$. Then, the scalar sequence $\{v_n\}$ $(n \geq 0)$ given by*

$$v_0 = 0, \qquad v_1 = \eta,$$

$$v_{n+2} = v_{n+1} + \frac{a\left(\frac{1}{3}(v_{n+1} - v_n) + v_n\right) + c}{2\left(1 - cv_{n+1} - \frac{a}{2}\,v_{n+1}^2\right)}\,(v_{n+1} - v_n)^2 \tag{2.193}$$

*is increasing, bounded from above by $t^{\star\star} = \dfrac{2\eta}{2 - \overline{\delta}}$ and converges to its unique least upper bound $v^\star$ satisfying $v^\star \in [0, t^{\star\star}]$. Moreover, the following estimates hold for all $n \geq 0$*

$$0 < v_{n+2} - v_{n+1} \leq \frac{\overline{\delta}}{2}\,(v_{n+1} - v_n) \leq \left(\frac{\overline{\delta}}{2}\right)^{n+1} \eta \tag{2.194}$$

*and*

$$v^\star - v_n \leq \frac{2\eta}{2-\overline{\delta}}\left(\frac{\overline{\delta}}{2}\right)\eta. \tag{2.195}$$

Next, we present the main semilocal result involving twice Fréchet differentiable operator and center–Lipschitz conditions.

**Theorem 2.5** *Let $F : \mathcal{D} \subseteq \mathcal{X} \longrightarrow \mathcal{Y}$ be a twice Fréchet differentiable operator. Assume there exist a point $x_0 \in \mathcal{D}$, constants $\eta > 0$, $a > 0$, $b > 0$, $c > 0$ such that for all $x \in \mathcal{D}$*

$$F'(x_0)^{-1} \in \mathcal{L}(\mathcal{Y}, \mathcal{X}), \tag{2.196}$$

$$\| F'(x_0)^{-1} \; F(x_0) \| \leq \eta, \tag{2.197}$$

$$\| F'(x_0)^{-1}[F'(x) - F'(x_0)] \| \leq b \; \| x - x_0 \|, \tag{2.198}$$

$$\| F'(x_0)^{-1} \; F''(x_0) \| \leq c, \tag{2.199}$$

$$\| F'(x_0)^{-1}[F''(x) - F''(x_0)] \| \leq a \; \| x - x_0 \|, \tag{2.200}$$

$$\overline{U}(x_0, t^\star) \subseteq \mathcal{D} \tag{2.201}$$

*and hypotheses of Lemma 2.7 hold. Then, the sequence $\{x_n\}$ defined by NM (2.2) is well defined, remains in $\overline{U}(x_0, t^\star)$ for all $n \geq 0$ and converges to a unique solution $x^\star \in \overline{U}(x_0, t^\star)$ of equation $F(x) = 0$.*

*Moreover, the following estimates hold for all $n \geq 0$*

$$\| x_{n+2} - x_{n+1} \| \leq \frac{a\left(\frac{1}{6} \; \| x_{n+1} - x_n \| + \frac{1}{2} \; \| x_n - x_0 \|\right) + \frac{c}{2}}{1 - b \; \| x_{n+1} - x_0 \|} \; \| x_{n+1} - x_n \|^2$$

$$\leq \frac{a\left(\frac{1}{6} \; (t_{n+1} - t_n) + \frac{1}{2} \; (t_n - t_0)\right) + \frac{c}{2}}{1 - b \; (t_{n+1} - t_0)} \; (t_{n+1} - t_n)^2 \tag{2.202}$$

$$= t_{n+2} - t_{n+1} \tag{2.203}$$

*and*

$$\| x_n - x^\star \| \leq t^\star - t_n, \tag{2.204}$$

*where the sequence $\{t_n\}$ $(n \geq 0)$ is given by (2.166). Furthermore, if there exists $R \geq t^\star$ such that*

$$U(x_0, R) \subseteq \mathcal{D} \tag{2.205}$$

*and*

$$b \; (t^\star + R) \leq 2, \tag{2.206}$$

*then, the solution $x^\star$ is unique in $U(x_0, R)$.*

**Proof** Let us prove that

$$\| x_{k+1} - x_k \| \leq t_{k+1} - t_k \tag{2.207}$$

and

$$\overline{U}(x_{k+1}, t^\star - t_{k+1}) \subseteq \overline{U}(x_k, t^\star - t_k) \tag{2.208}$$

hold for all $k \geq 0$. For every $z \in \overline{U}(x_1, t^\star - t_1)$, we have that

$$\begin{aligned} \| z - x_0 \| &\leq \| z - x_1 \| + \| x_1 - x_0 \| \\ &\leq (t^\star - t_1) + (t_1 - t_0) = t^\star - t_0, \end{aligned}$$

implies $z \in \overline{U}(x_0, t^\star - t_0)$. Since also

$$\| x_1 - x_0 \| = \| F'(x_0)^{-1} F(x_0) \| \leq \eta = t_1 - t_0,$$

estimates (2.207) and (2.208) hold for $k = 0$. If these estimates hold for $n = 0, 1, \cdots, k$, then, we have that

$$\| x_{k+1} - x_0 \| \leq \sum_{i=1}^{k+1} \| x_i - x_{i-1} \| \leq \sum_{i=1}^{k+1} (t_i - t_{i-1}) = t_{k+1} - t_0 = t_{k+1} \tag{2.209}$$

and

$$\| x_k + \theta\,(x_{k+1} - x_k) - x_0 \| \leq t_k + \theta\,(t_{k+1} - t_k) \leq t^\star \quad (\theta \in [0,1]). \tag{2.210}$$

Using (2.2), we obtain the approximation

$$\begin{aligned} F(x_{k+1}) &= F(x_{k+1}) - F(x_k) - F'(x_k)\,(x_{k+1} - x_k) \\ &= \int_0^1 [F'(x_k + \theta\,(x_{k+1} - x_k)) - F'(x_k)]\ (x_{k+1} - x_k)\ d\theta \\ &= \int_0^1 F''(x_k + \theta\,(x_{k+1} - x_k))\ (1-\theta)\ (x_{k+1} - x_k)^2\ d\theta. \end{aligned} \tag{2.211}$$

Then, we get by (2.199), (2.200) and (2.201)

$$\begin{aligned} &\| F'(x_0)^{-1} F(x_{k+1}) \| \\ &\leq \int_0^1 (\ \| F'(x_0)^{-1}\ [F''(x_k + \theta\,(x_{k+1} - x_k)) - F''(x_0)] \| \\ &+ \| F'(x_0)^{-1}\ F''(x_0) \|\ )\ \| x_{k+1} - x_k \|^2\ (1-\theta)\ d\theta \\ &\leq \left\{ a \left( \int_0^1 \| x_k - x_0 \| + \theta\ \| x_{k+1} - x_k \|\ \ (1-\theta)\ d\theta \right) + \frac{c}{2} \right\}\ \| x_{k+1} - x_k \|^2 \\ &\leq \frac{a}{6}\ \| x_{k+1} - x_k \|^3 + \frac{a}{2}\ \| x_k - x_0 \|\ \| x_{k+1} - x_k \|^2 + \frac{c}{2}\ \| x_{k+1} - x_k \|^2 \\ &\leq \left\{ a \left( \frac{1}{6}\ (t_{k+1} - t_k) + \frac{1}{2}\ (t_k - t_0) \right) + \frac{c}{2} \right\} (t_{k+1} - t_k)^2. \end{aligned} \tag{2.212}$$

Using (2.198), we obtain that

$$\| F'(x_0)^{-1} (F'(x_{k+1}) - F'(x_0)) \| \leq b \| x_{k+1} - x_0 \| \leq b\, t_{k+1} \leq b\, t^\star < 1. \tag{2.213}$$

It follows from the Banach lemma on invertible operators and (2.213) that $F'(x_{k+1})^{-1}$ exists and

$$\| F'(x_{k+1})^{-1} F'(x_0) \| \leq (1 - b \| x_{k+1} - x_0 \|)^{-1} \leq (1 - b\, t_{k+1})^{-1}. \tag{2.214}$$

Therefore, by (2.2), (2.212) and (2.214), we obtain in turn that

$$\begin{aligned} \| x_{k+2} - x_{k+1} \| &= \| F'(x_{k+1})^{-1} F(x_{k+1}) \| \\ &\leq \| F'(x_{k+1})^{-1} F'(x_0) \| \, \| F'(x_0)^{-1} F(x_{k+1}) \| \\ &\leq t_{k+2} - t_{k+1}. \end{aligned} \tag{2.215}$$

Thus, for every $z \in \overline{U}(x_{k+2}, t^\star - t_{k+2})$, we have that

$$\begin{aligned} \| z - x_{k+1} \| &\leq \| z - x_{k+2} \| + \| x_{k+2} - x_{k+1} \| \\ &\leq t^\star - t_{k+2} + t_{k+2} - t_{k+1} = t^\star - t_{k+1}. \end{aligned} \tag{2.216}$$

That is,

$$z \in \overline{U}(x_{k+1}, t^\star - t_{k+1}). \tag{2.217}$$

Estimates (2.214) and (2.217) imply that (2.207) and (2.208) hold for $n = k + 1$. The proof of (2.207) and (2.208) is now complete. Lemma 2.7 implies that sequence $\{t_n\}$ is a Cauchy sequence. From (2.207) and (2.208), $\{x_n\}$ $(n \geq 0)$ becomes a Cauchy sequence too and as such it converges to some $x^\star \in \overline{U}(x_0, t^\star)$ (since $\overline{U}(x_0, t^\star)$ is a closed set). Estimate (2.204) follows from (2.203) by using standard majorization techniques (cf. [163], [469]). Moreover, by letting $k \to \infty$ in (2.212), we obtain $f(x^\star) = 0$.

Finally, to show uniqueness: let $y^\star$ be a solution of equation $F(x) = 0$ in $U(x_0, R)$. It follows from (2.198) for $x = y^\star + \theta\,(x^\star - y^\star)$, $\theta \in [0, 1]$, the estimate

$$\begin{aligned} &\left\| F'(x_0)^{-1} \int_0^1 (F'(y^\star + \theta\,(x^\star - y^\star)) - F'(x_0)) \right\| d\theta \\ &\leq b \int_0^1 \| y^\star + \theta\,(x^\star - y^\star) - x_0 \| \, d\theta \\ &\leq b \int_0^1 (\theta \| x^\star - x_0 \| + (1 - \theta) \| y^\star - x_0 \|) \, d\theta < \frac{b}{2} (t^\star + R) \leq 1 \quad \text{(by (2.206))} \end{aligned}$$

and the Banach lemma on invertible operators implies that the linear operator $\mathcal{M} = \int_0^1 F'(y^\star + \theta\,(x^\star - y^\star))\, d\theta$ is invertible. Using the identity $0 = F(x^\star) - F(y^\star) = \mathcal{M}\,(x^\star - y^\star)$, we deduce $x^\star = y^\star$. Similarly, we show the uniqueness in $\overline{U}(x_0, t^\star)$ using (2.206). This completes the proof of Theorem 2.5. ⊠

**Remark 2.10** *The conclusions of Theorem 2.5 hold if (2.198) is dropped from the hypotheses and Lemma 2.7, $\{t_n\}$, $t^\star$ are replaced by Lemma 2.8, $\{v_n\}$, $v^\star$, respectively. Indeed, we have the following result.*

**Theorem 2.6** *Let $F : \mathcal{D} \subseteq \mathcal{X} \longrightarrow \mathcal{Y}$ be a twice Fréchet differentiable operator. Assume hypotheses of Lemma 2.8 hold and there exist a point $x_0 \in \mathcal{D}$, a constants $\eta > 0$, $a > 0$ and $c > 0$ such that for all $x \in \mathcal{D}$*

$$F'(x_0)^{-1} \in \mathcal{L}(\mathcal{Y}, \mathcal{X}),$$

$$\| F'(x_0)^{-1} \, F(x_0) \| \leq \eta,$$

$$\| F'(x_0)^{-1} \, F''(x_0) \| \leq c,$$

$$\| F'(x_0)^{-1}[F''(x) - F''(x_0)] \| \leq a \, \| x - x_0 \|,$$

$$\overline{U}(x_0, v^\star) \subseteq \mathcal{D},$$

*where $v^\star$ is given in Lemma 2.8. Then, the sequence $\{x_n\}$ defined by NM (2.2) is well defined, remains in $\overline{U}(x_0, v^\star)$ for all $n \geq 0$ and converges to a unique solution $x^\star \in \overline{U}(x_0, v^\star)$ of equation $F(x) = 0$. Moreover, the following estimates hold for all $n \geq 0$*

$$\| x_{n+2} - x_{n+1} \| \leq v_{n+1} - v_n \tag{2.218}$$

*and*

$$\| x_n - x^\star \| \leq v_{n+2} - v_{n+1}, \tag{2.219}$$

*where the sequence $\{v_n\}$ $(n \geq 0)$ is given by (2.193).*

**Proof** We proceed as in the proof of Theorem 2.5 until (2.212). Then, we use (2.199) and (2.200) (instead of (2.198)) to obtain in turn that

$$\begin{aligned} & \| F'(x_0)^{-1} \, (F'(x_{k+1}) - F'(x_0)) \| \\ = & \left\| \int_0^1 F'(x_0)^{-1} \, (F''(x_0 + t\,(x_{k+1} - x_0)) - F''(x_0)) \, dt \, (x_{k+1} - x_0) \right. \\ & \left. + F'(x_0)^{-1} \, F''(x_0) \, (x_{k+1} - x_0) \right\| \qquad (2.220) \\ \leq & \int_0^1 a\,t\,dt \, \| x_{k+1} - x_0 \|^2 + c \, \| x_{k+1} - x_0 \| \leq \frac{a}{2} \, v_{k+1}^2 + c\,v_{k+1} < 1. \end{aligned}$$

It follows from (2.220) and the Banach lemma on invertible operators, that $F'(x_{k+1})^{-1}$ exists and

$$\| F'(x_{k+1})^{-1} \, F'(x_0) \| \leq (1 - c \, v_{k+1} - \frac{a}{2} \, v_{k+1}^2)^{-1}. \tag{2.221}$$

The rest of the proof follows as in the proof of Theorem 2.5 with (2.221) replacing (2.214) until the uniqueness part.

Let $y^\star$ be a solution of equation $F(x) = 0$ in $\overline{U}(x_0, t^\star)$. Then, since

$$\begin{aligned} y^\star & - x_{k+1} \\ & = y^\star - x_k + F'(x_k)^{-1}\, F(x_k) \\ & = -F'(x_k)^{-1}\, (F(y^\star) - F(x_k) - F'(x_k)\,(y^\star - x_k)) \\ & = \left( - F'(x_0)^{-1}\, F'(x_k)\right)^{-1} \int_0^1 F'(x_0)^{-1}\, F''(x_k + t\,(y^\star - x_k)\Big)\,(1-t)\,dt\,(y^\star - x_k)^2 \end{aligned} \tag{2.222}$$

and

$$\| \, y^\star - x_0 \, \| \leq v^\star - v_0,$$

we obtain that

$$\| \, x_k - y^\star \, \| \leq v^\star - v_k. \tag{2.223}$$

That is $\lim\limits_{k \longrightarrow \infty} x_k = y^\star$. But, we showed $\lim\limits_{k \longrightarrow \infty} x_k = x^\star$. Hence, we deduce $x^\star = y^\star$. This completes the proof of Theorem 2.6. ⊠

We can now compare majorizing sequences $\{t_n\}$ and $\{v_n\}$ $(n \geq 0)$.

**Proposition 2.1** *Assume that*

$$b < \frac{a}{2}\,\eta + c, \tag{2.224}$$

*hypotheses of Theorems 2.5 and 2.6 hold. Then, the following estimates hold for all* $n \geq 0$

$$\| \, x_{n+2} - x_{n+1} \, \| \leq t_{n+2} - t_{n+1} < v_{n+2} - v_{n+1} \tag{2.225}$$

*and*

$$\| \, x_n - x^\star \, \| \leq t^\star - t_n \leq v^\star - v_n. \tag{2.226}$$

**Proof** We only need to show using induction on the integer $k$ that

$$t_{k+2} - t_{k+1} < v_{k+2} - v_{k+1}. \tag{2.227}$$

In view of (2.177), (2.206) and (2.224), we obtain for $n = 0$ that

$$t_2 < v_2 \quad \text{and} \quad t_2 - t_1 < v_2 - v_1.$$

Let us assume that

$$t_{k+1} < v_{k+1} \tag{2.228}$$

and

$$t_{k+1} - t_k < v_{k+1} - v_k \quad (k \leq n+1). \tag{2.229}$$

Then, it follows from (2.171), (2.193), (2.228) and (2.229) that (2.227) hold for all $k \geq 0$. Estimate (2.226) follows from (2.225) by using standard majorization techniques. This completes the proof of Proposition 2.1. ⊠

**Remark 2.11** *If equality holds in (2.224), then,* $v_n = t_n$ $(n \geq 0)$*, whereas if*

$$b > \frac{a}{2}\eta + c, \tag{2.230}$$

*then, the conclusions of Proposition 2.1 hold with sequence* $\{t_n\}$, $t^\star$ *switching places with* $\{v_n\}$, $v^\star$*, respectively in (2.225) and (2.226).*

**Remark 2.12** *We can now compare our results with the ones obtained by Huang (cf. [445]) and Gutiérrez (cf. [399]). Huang (cf. [445]) used (2.196), (2.199),*

$$\| F'(x_0)^{-1}(F''(x) - F''(y)) \| \leq \alpha \| x - y \| \quad (x, y \in \mathcal{D}) \tag{2.231}$$

*and*

$$3\alpha^2\eta + 3\alpha c + c^3 \leq (c^2 + 2\alpha)^{3/2} \tag{2.232}$$

*and majorizing sequence* $\{v_n\}$ *to arrive at conclusions (2.218) and (2.219). Gutiérrez (cf. [399]) weakened Huang's conditions by using (2.196), (2.199), (2.200) (which is weaker than (2.231), since* $a \leq \alpha$*) condition*

$$3a^2\eta + 3ac + c^3 \leq (c^2 + 2a)^{3/2} \tag{2.233}$$

*and majorizing sequence* $\{v_n\}$ *to also arrive at conclusions (2.218) and (2.219).*

*Hypotheses of Lemma 2.7 use information on* $a$, $b$, $c$ *and* $\eta$*; Huang (cf. [445]) uses information on* $\alpha$, $c$ *and* $\eta$*; whereas, Gutiérrez (cf. [399]) uses* $a$, $c$ *and* $\eta$*. Therefore, a direct comparison between the sufficient convergence conditions is not possible. However, under (2.224) our majorizing sequence* $\{t_n\}$ *is finer that* $\{v_n\}$ *(See Example 2.9).*

*Note that in this section, we have simplified the sufficient convergence conditions provided by us in (cf. [112]). A favorable comparison of our approach with the corresponding one given by the Newton–Kantorovich theorem for solving nonlinear equations was also given in (cf. [112]). The same favorable comparison extends in this section. The results obtained here can be extended for* $m$ $(m \geq 2)$ *Fréchet differentiable operators (cf. [112], [140]).*

*Comparison table 1 justifies the theoretical results of Proposition 2.1, since the majorizing sequence* $\{t_n\}$ *is tighter than* $\{v_n\}$*.*

**Comparison Table 1**

| $n$ | $t_{n+1} - t_n$ | $v_{n+1} - v_n$ |
|---|---|---|
| *0* | *.21* | *.21* |
| *1* | *.0775551724* | *.0807721314* |
| *2* | *.0167871035* | *.0213332937* |
| *3* | *.0006952583* | *.0018344379* |
| *4* | *.0000025816* | *.0000139339* |
| *5* | *0* | *.0000000008* |
| *6* | *0* | *0* |

**Example 2.7** *Let $a = b = 2$, $c = 1.9$ and $\eta = .21$. Then, condition (2.233) (and (2.232)) is satisfied, since*

$$3a^2\eta + 3ac + c^3 = 20.779 < 20.99311985 = (c^2 + 2,\ a)^{3/2}$$

*However, using (2.155)–(2.164), we get $\delta_1 = 2$, $\delta_2 = .259636075$, $\delta_3 = .220086356$, $\delta = 1.119654576$ and $\eta_0 = \delta_3 > \eta = .21$. Note also that* (2.70) *holds, since*

$$b = 2 < 2.11 = \frac{a}{2}\eta + c.$$

*Note that we can also compare the error estimates using (2.166) and (2.193).*

## 2.3 Newton's Method on Unbounded Domains

We provide in this section new sufficient semilocal convergence results for NM, defined on not necessarily bounded domains. Earlier results on bounded domains cannot apply (cf. [114], [792]). As an example we solve a Chandrasekhar–type nonlinear integral equation that cannot be handled with earlier approaches (cf. [114], [792]).

We are concerned with the problem of approximating a locally unique solution $x^\star$ of equation (2.1) using (2.2), where $F$ is an $m$–times ($m \geq 3$) Fréchet–differentiable operator on a convex subset $\mathcal{D}$ of a Banach space $\mathcal{X}$ with values in a Banach space $\mathcal{Y}$.

Here, we shall provide sufficient convergence conditions for (2.2) that expand its applicability. In particular, we shall assume the following conditions.

**Condition 2.1** *Assume*

> *$F : \mathcal{D} \subseteq \mathcal{X} \longrightarrow \mathcal{Y}$ is an $m$–times ($m \geq 3$) Fréchet–differentiable operator defined on a non–empty open and convex subset $\mathcal{D}$ of a Banach space $\mathcal{X}$ with values in a Banach space $\mathcal{Y}$; there exist a point $x_0 \in \mathcal{D}$, constants $b_i \geq 0$ ($1 \leq i \leq m$), $\eta > 0$, and non–decreasing functions $v$, $w$, with $v(0) \geq 0$, $w(0) \geq 0$, such that for all $x \in \mathcal{D}$*
>
> $$F'(x_0)^{-1} \in \mathcal{L}(\mathcal{Y}, \mathcal{X}), \tag{2.234}$$
>
> $$0 < \| F'(x_0)^{-1} F(x_0) \| \leq \eta, \tag{2.235}$$
>
> $$\| F'(x_0)^{-1} (F'(x) - F'(x_0)) \| \leq v(\| x - x_0 \|), \tag{2.236}$$
>
> $$\| F'(x_0)^{-1} F^{(i)}(x_0) \| \leq b_i \quad (i = 2, 3, \cdots, m), \tag{2.237}$$
>
> $$\| F'(x_0)^{-1} (F^{(m)}(x) - F^{(m)}(x_0)) \| \leq w(\| x - x_0 \|), \tag{2.238}$$
>
> *function $g$ defined on $[0, +\infty)$ by*
>
> $$g(t) = \frac{(1 - \alpha(t) - v(t))\ \eta}{1 - 2\ \alpha(t) - v(t)} \tag{2.239}$$

*has a minimal positive fixed point, denoted by r, where for*

$$\alpha(t) = b_2\, t + \frac{1}{2!}\, b_3\, t^2 + \cdots + \frac{1}{(m-2)!}\, b_{m-1}\, t^{m-2} + \frac{b_m + w(t)}{(m-1)!}\, t^{m-1}, \tag{2.240}$$

$$\max\{v(r) = v\,,\, \alpha(r) = \alpha\} \leq \frac{1}{3} \tag{2.241}$$

*and*

$$\overline{U}(x_0, r) \subseteq \mathcal{D}. \tag{2.242}$$

The motivation for introducing Condition 2.1 is given by the following.

**Example 2.8** *Let $\mathcal{X} = \mathcal{Y} = \mathcal{D}$, where $\mathcal{D}$ is the space of real m–continuously differentiable functions on $[0, \infty)$, equipped with the max–norm. Define an operator F on $\mathcal{D}$ by*

$$F(x(s)) = \frac{1}{60}\, x^5(s) + x(s) + \Xi(x(s)) - \eta, \tag{2.243}$$

*where $\Xi$ is any linear operator on $\mathcal{X}$, and $\eta$ is a non–negative parameter. Choose $x_0(s) = 0$. The set of conditions used in the literature are:*

($\mathcal{C}_1$) *Newton–Kantorovich (cf. [469], [165])*

$$\| F'(x_0)^{-1}\, (F'(x) - F'(y)) \| \leq L\ \| x - y \| \quad (x, y \in \mathcal{D}),$$

$$h_K = L\, \eta \leq \frac{1}{2}.$$

($\mathcal{C}_2$) *Argyros (cf. [165], [215])*

$$\| F'(x_0)^{-1}\, (F'(x) - F'(y)) \| \leq L\ \| x - y \|$$

$$\| F'(x_0)^{-1}\, (F'(x) - F'(x_0)) \| \leq L_0\ \| x - x_0 \| \quad (x, y \in \mathcal{D}),$$

$$h_K = \overline{L}\, \eta \leq \frac{1}{2},$$

*where*

$$\overline{L} = \frac{1}{8}\, (L + 4\, L_0 + \sqrt{L^2 + 8\, L_0\, L}).$$

($\mathcal{C}_3$) *Huang (cf. [445])*

$$\| F'(x_0)^{-1}\, F''(x_0) \| \leq b_2,$$

$$\| F'(x_0)^{-1}\, (F''(x) - F''(y)) \| \leq B_2\ \| x - y \| \quad (x, y \in \mathcal{D})$$

*and*

$$3\, B_2^2\, \eta + 3\, B_2\, b_2 + b_2^3 \leq (b_2^2 + 2\, B_2)^{\frac{3}{2}}.$$

($\mathcal{C}_4$) *Gutiérrez (cf. [399])*

$$\| F'(x_0)^{-1} F''(x_0) \| \leq b_2,$$

$$\| F'(x_0)^{-1} (F''(x) - F''(x_0)) \| \leq C_2 \| x - x_0 \| \quad (x \in \mathcal{D}),$$

*and*

$$3C_2^2 \eta + 3 C_2 b_2 + b_2^3 \leq (b_2^2 + 2C_2)^{\frac{3}{2}}.$$

($\mathcal{C}_5$) *Argyros (cf. [139])*

$$\| F'(x_0)^{-1} F^{(i)}(x_0) \| \leq b_i \ (i = 2, 3, \cdots, m),$$

$$\| F'(x_0)^{-1} (F^{(m)}(x) - F^{(m)}(x_0)) \| \leq b_{m+1} \| x - x_0 \|$$

$$\| F'(x_0)^{-1} (F'(x) - F'(x_0)) \| \leq q_0(\| x - x_0 \|) + 1 < 1 \quad (x \in \mathcal{D}),$$

$$p(t) \leq 0 \quad (t \in [0, t_0], \ t_0 > \eta)$$

*and*

$$q_0(t) \leq p'(t) \qquad (t \in [0, t_0]),$$

*where,*

$$p(t) = \frac{b_{m+1}}{(m+1)!} t^{m+1} + \cdots + \frac{b_2}{2!} t^2 - t + \eta,$$

*$q_0$ is some continuous, monotonically increasing real function and $t_0$ is a uniquely determined point.*

*More related conditions can be found in (cf. [157], [165], [313], [326], [600], [792]). Conditions ($\mathcal{C}_1$)–($\mathcal{C}_5$) cannot be used, since e.g. for $m = 3$, constants $L_0$, $L$, $B_2$, $C_2$, $b_4$ do not exist. It turn out that under Condition 2.1, this problem can be resolved (see also Example 2.9).*

We need an auxiliary result.

**Lemma 2.9** *Under Condition 2.1, let*

$$f(t) = \frac{1}{1 - v(t)}, \qquad v(t) \neq 1.$$

*Then, the following estimates hold*

$$\left(1 + \sum_{j=1}^{n} 2^{j-1} (\alpha f(\alpha))^j \right) \eta < \left(1 + \frac{\alpha f(\alpha)}{1 - 2 \alpha f(\alpha)}\right) \eta = r, \tag{2.244}$$

$$F'(x)^{-1} \in \mathcal{L}(\mathcal{Y}, \mathcal{X}) \quad (x \in \overline{U}(x_0, r)), \tag{2.245}$$

$$\| F'(x)^{-1} F'(x_0) \| \leq f(\| x - x_0 \|) \quad (x \in \overline{U}(x_0, r)), \tag{2.246}$$

$$\| F'(x)^{-1} F(x_n) \| \leq 2 \alpha \| x_n - x_{n-1} \| \quad (n \geq 2), \tag{2.247}$$

$$\| x_{n+1} - x_n \| \leq 2 \alpha f(\alpha) \| x_n - x_{n-1} \| \quad (n \geq 2), \tag{2.248}$$

$$\| x_{n+1} - x_0 \| \leq \left(1 + \alpha\ f(\alpha) + \cdots + 2^{n-1}\ (\alpha\ f(\alpha))^n\right)\ \eta < r \qquad (n \geq 1) \tag{2.249}$$

*and*

$$\| x_{n+m} - x_n \| \leq \frac{1}{2}\ \eta \sum_{j=n}^{n+m-1} (2\ \alpha\ f(\alpha))^j \quad (m = 0, 1, 2, \cdots). \tag{2.250}$$

**Proof** By hypothesis (2.241), we have

$$f(\alpha) = \frac{1}{1 - v(\alpha)} < \frac{3}{2} \quad \text{and} \quad 0 < \alpha\ f(\alpha) < 1,$$

which imply in turn that

$$\sum_{j=1}^{n} 2^{j-1}\ (\alpha\ f(\alpha))^j < \frac{\alpha\ f(\alpha)}{1 - 2\ \alpha\ f(\alpha)},$$

$$\left(1 + \sum_{j=1}^{n} 2^{j-1}\ (\alpha\ f(\alpha))^j\right)\ \eta < \left(1 + \frac{\alpha\ f(\alpha)}{1 - 2\ \alpha\ f(\alpha)}\right)\ \eta = g(\alpha) = r.$$

That shows (2.244). Let $x \in \overline{U}(x_0, r)$. Then, using (2.236), and (2.241), we obtain

$$\| F'(x_0)^{-1}\ (F'(x) - F'(x_0)) \| \leq v(\| x - x_0 \|)\ \leq v(r) = v < \frac{1}{3} < 1. \tag{2.251}$$

In view of (2.251) and the Banach lemma on invertible operators, we get that $F'(x) \in \mathcal{L}(\mathcal{Y}, \mathcal{X})$ and (2.246) holds. By hypothesis (2.235) and the definition of $r$, we have that

$$\| x_1 - x_0 \| = \| F'(x_0)^{-1}\ F(x_0) \| \leq \eta < r.$$

Hence, we get $x_1 \in \overline{U}(x_0, r)$. That is, $\| F'(x_1)^{-1}\ F'(x_0) \| \leq f(\alpha)$.

Using (2.237), (2.238), (2.240) and Taylor's formula (cf. [114], [165]), as in (cf. [139]), we obtain in turn for $\theta_1, \theta_2 \in [0, 1]$ that

$$\| F'(x_0)^{-1}\, F(x_1) \| = \left\| \int_0^1 F'(x_0)^{-1}(F'(x_0 + \theta_1\,(x_1 - x_0)) - F'(x_0))\ d\theta_1 (x_1 - x_0) \right\|$$

$$\leq \int_0^1 \| F'(x_0)^{-1}(F''(x_0)\,\theta_1 (x_1 - x_0) + \cdots + \frac{F^{(m-1)}(x_0)}{(m-2)!}(\theta_1 (x_1 - x_0))^{m-2}$$

$$+ \frac{1}{(m-2)!} \int_0^1 F^{(m)}(x_0 + \theta_2 \theta_1 (x_1 - x_0))(1 - \theta_2)^{m-2} d\theta_2 (\theta_1 (x_1 - x_0))^{m-1} \|\ d\theta_1\ \| x_1 - x_0 \|$$

$$\leq \int_0^1 \left(b_2 r + \frac{1}{2!} b_3 r^2 + \cdots + \frac{1}{(m-2)!} b_{m-1} r^{m-2} + \frac{b_m + w(r)}{(m-1)!} r^{m-1}\right) d\theta_1\ \| x_1 - x_0 \|$$

$$= \alpha\ \| x_1 - x_0 \|,$$

$$\| x_2 - x_1 \| = \| F'(x_1)^{-1}\, F(x_1) \|$$

$$\leq \| F'(x_1)^{-1}\, F'(x_0) \|\, \| F'(x_0)^{-1}\, F(x_1) \| \leq \alpha\, f(\alpha)\ \| x_1 - x_0 \|$$

and

$$\| x_2 - x_0 \| \leq \| x_2 - x_1 \| + \| x_1 - x_0 \| \leq (1 + \alpha\, f(\alpha))\, \eta < r.$$

Hence, $x_2 \in \overline{U}(x_0, r)$, $F'(x_2)^{-1}$ exists and (2.246) holds for $x = x_2$. We also obtain, as above for $\ell_2 = \theta_1\,(x_2 - x_0) + (1 - \theta_1)\,(x_1 - x_0)$ and $\| \ell_2 \| \leq \theta_1 \| x_2 - x_0 \| + (1 - \theta_1) \; \| x_1 - x_0 \| < r$ that

$$\| F'(x_0)^{-1} F(x_2) \| = \left\| \int_0^1 F'(x_0)^{-1}(F'(x_1 + \theta_1(x_2 - x_1)) - F'(x_1)) d\theta_1 (x_2 - x_1) \right\|$$

$$\leq \int_0^1 \| F'(x_0)^{-1}(F''(x_0)\ell_2 + \cdots + \frac{F^{(m-1)}(x_0)}{(m-2)!} \ell_2^{m-2}$$

$$+ \frac{1}{(m-2)!} \int_0^1 F^{(m)}(x_0 + \theta_2(\ell_2 - x_0))(1-\theta_2)^{m-2} d\theta_2 \ell_2^{m-1}$$

$$-F''(x_0)(x_1 - x_0) - \cdots - \frac{F^{(m-1)}(x_0)}{(m-2)!}(x_1 - x_0)^{m-2}$$

$$-\frac{1}{(m-2)!} \int_0^1 F^{(m)}(x_0 + \theta_3(x_1 - x_0))(1-\theta_3)^{m-2} d\theta_3 (x_1 - x_0)^{m-1} \| d\theta_1 \| x_2 - x_1 \|$$

$$\leq 2 \int_0^1 \left( b_2 r + \frac{1}{2!} b_3 r^2 + \cdots + \frac{1}{(m-2)!} b_{m-1} r^{m-2} + \frac{b_m + w(r)}{(m-1)!} r^{m-1} \right) d\theta_1 \| x_2 - x_1 \|$$

$$= 2\; \alpha \| x_2 - x_1 \| .$$

Then, we get

$$\begin{aligned} \| x_3 - x_2 \| &= \| F'(x_2)^{-1}\, F(x_2) \| \\ &\leq \| F'(x_2)^{-1}\, F'(x_0) \| \| F'(x_0)^{-1}\, F(x_2) \| \\ &\leq 2\alpha\, f(\alpha) \; \| x_2 - x_1 \| \end{aligned}$$

and

$$\| x_3 - x_0 \| \leq \| x_3 - x_2 \| + \; \| x_2 - x_0 \| \leq (1 + \alpha\, f(\alpha) + 2\, (\alpha \; f(\alpha))^2)\, \eta < r.$$

Assume that assertions of lemma hold for all $k \leq n - 1$. Using Taylor's formula, we obtain for $\ell_n = \theta_1\,(x_n - x_0) + (1 - \theta_1)\,(x_{n-1} - x_0)$ and

$$\| \ell_n \| \leq \theta_1 \; \| x_n - x_0 \| + (1 - \theta_1) \; \| x_{n-1} - x_0 \| < r :$$

$$\| F'(x_0)^{-1} F(x_n) \|$$

$$= \left\| \int_0^1 F'(x_0)^{-1}(F'(x_{n-1} + \theta_1(x_n - x_{n-1})) - F'(x_{n-1})) d\theta_1 (x_n - x_{n-1}) \right\|$$

$$\leq \int_0^1 \| F'(x_0)^{-1}(F''(x_0)\ell_n + \cdots + \frac{F^{(m-1)}(x_0)}{(m-2)!} \ell_n^{m-2}$$

$$+\frac{1}{(m-2)!}\int_0^1 F^{(m)}(x_0+\theta_2(\ell_n-x_0))(1-\theta_2)^{m-2}d\theta_2\ell_n^{m-1}$$

$$-F''(x_0)\,(x_{n-1}-x_0)-\cdots-\frac{F^{(m-1)}(x_0)}{(m-2)!}(x_{n-1}-x_0)^{m-2}$$

$$-\frac{1}{(m-2)!}\int_0^1 F^{(m)}(x_0+\theta_3(x_{n-1}-x_0))(1-\theta_3)^{m-2}d\theta_3(x_{n-1}-x_0)^{m-1}\parallel d\theta_1$$

$$\parallel x_n-x_{n-1}\parallel$$

$$\leq 2\int_0^1\left(b_2 r+\frac{1}{2!}b_3 r^2+\cdots+\frac{1}{(m-2)!}b_{m-1}r^{m-2}+\frac{b_m+w(r)}{(m-1)!}r^{m-1}\right)d\theta_1\parallel x_n-x_{n-1}\parallel$$

$$=2\,\alpha\parallel x_n-x_{n-1}\parallel.$$

That shows (2.247). Hence, we have that

$$\begin{aligned}\parallel x_{n+1}-x_n\parallel &= \parallel F'(x_n)^{-1}F(x_n)\parallel\\ &\leq \parallel F'(x_n)^{-1}F'(x_0)\parallel\parallel F'(x_0)^{-1}F(x_n)\parallel\\ &\leq 2\,\alpha\, f(\alpha)\ \parallel x_n-x_{n-1}\parallel,\end{aligned}$$

$$\begin{aligned}\parallel x_{n+1}-x_0\parallel &\leq \parallel x_{n+1}-x_n\parallel+\parallel x_n-x_0\parallel\\ &\leq 2\ \alpha\, f(\alpha)\ \parallel x_n-x_{n-1}\parallel+(1+\alpha\, f(\alpha)+\cdots+2^{n-2}\,(\alpha\, f(\alpha))^{n-1})\,\eta\\ &\leq (1+\alpha\, f(\alpha)+\cdots+2^{n-1}\,(\alpha\, f(\alpha))^n)\,\eta<r\end{aligned}$$

and for all $m\geq 0$ small

$$\begin{aligned}\parallel x_{n+m}-x_n\parallel &\leq \parallel x_{n+m}-x_{n+m-1}\parallel+\cdots+\parallel x_{n+1}-x_n\parallel\\ &\leq \parallel F'(x_{n+m-1})^{-1}F(x_{n+m-1})\parallel+\cdots+\parallel F'(x_n)^{-1}F(x_n)\parallel\\ &\leq \parallel F'(x_{n+m-1})^{-1}F'(x_0)\parallel\parallel F'(x_0)^{-1}F(x_{n+m-1})\parallel+\cdots+\\ &\quad\parallel F'(x_n)^{-1}F'(x_0)\parallel\parallel F'(x_0)^{-1}F(x_n)\parallel\\ &\leq f(\alpha)\sum_{j=n}^{n+m-1}\parallel F'(x_0)^{-1}F(x_j)\parallel\\ &\leq \frac{1}{2}\eta\sum_{j=n}^{n+m-1}(2\,\alpha\ f(\alpha))^j.\end{aligned}$$

That is show (2.248), (2.249) and (2.250) are true. This completes the proof of Lemma 2.9. ⊠

Hence, we arrived at the main semilocal convergence result for (2.2).

**Theorem 2.7** *Assume Condition 2.1 holds. Then, sequence* $\{x_n\}$ $(n \geq 0)$ *generated by (2.2) is well defined, remains in* $\overline{U}(x_0, r)$ *for all* $n \geq 0$ *and converges to a solution* $x^\star \in \overline{U}(x_0, r)$ *of equation* $F(x) = 0$. *Moreover, if there exists* $R \geq r$ *such that*

$$\overline{U}(x_0, R) \subseteq \mathcal{D} \tag{2.252}$$

*and*

$$\int_0^1 v((1-\theta)\, r + \theta r)\, d\theta < 1, \tag{2.253}$$

*then,* $x^\star$ *is the unique solution of equation* $F(x) = 0$ *in* $\overline{U}(x_0, R)$.

**Proof.** It follows from (2.250) that sequence $\{x_n\}$ is a Cauchy sequence in a Banach space $\mathcal{X}$ and as such it converges to some $x^\star \in \overline{U}(x_0, r)$ (since $\overline{U}(x_0, r)$ is a closed set). By letting $n \longrightarrow \infty$ in (2.247), we obtain $F(x^\star) = 0$.

Finally, to show the uniqueness, let $x^\star$, $y^\star \in \overline{U}(x_0, R)$ be such that $F(x^\star) = F(y^\star)$ and set

$$\mathcal{M} = \int_0^1 F'(x^\star + \theta\, (y^\star - x^\star))\, d\theta. \tag{2.254}$$

Using (2.251)–(2.254), we have in turn that

$$\begin{aligned} \| F'(x_0)^{-1}\, (\mathcal{M} - F'(x_0)) \| &\leq \int_0^1 v(\| x^\star + \theta\, (y^\star - x^\star) - x_0 \|)\, d\theta \\ &\leq \int_0^1 v((1-\theta)\ \ \| x^\star - x_0 \| + \theta\ \ \| y^\star - x_0 \|)\, d\theta \\ &\leq \int_0^1 v((1-\theta)\, r + \theta\, r)\, d\theta < 1. \end{aligned} \tag{2.255}$$

In view of (2.255) and the Banach lemma on invertible operators we get that $\mathcal{M}^{-1}$ exists. It follows from the estimate $0 = F(x^\star) - F(y^\star) = \mathcal{M}\, (x^\star - y^\star)$ that $x^\star = y^\star$. This completes the proof of Theorem 2.7. ⊠

**Remark 2.13** *We present two choices of function* $v$.

**Case 1.** *A possible choice for function* $v$ *is given by*

$$v(t) = b_2\, t + \frac{1}{2}\, b_3\, t^2 + \cdots + \frac{1}{(m-2)!}\, b_{m-1}\, t^{m-2} + \frac{b_m + w(t)}{(m-1)!}\, t^{m-1}. \tag{2.256}$$

*If we set* $v(r) = \alpha$, *condition (2.241) becomes*

$$\alpha < \frac{1}{3}. \tag{2.257}$$

*The choice is justified for $p = x - x_0$ from the estimate*

$$\| F'(x_0)^{-1} (F(x) - F(x_0)) \| = \left\| F'(x_0)^{-1} \int_0^1 F''(x_0 + \theta_1 p)\, d\theta_1\, p \right\|$$

$$\leq b_2 r + \left\| F'(x_0)^{-1} \int_0^1 \int_0^1 F'''(x_0 + \theta_2\theta_1 p)\theta_1 d\theta_2 d\theta_1 p^2 \right\|$$

$$\leq \sum_{j=0}^{m-3} \frac{b_{j+2}}{(j+1)!} r^{j+1} + \left\| F'(x_0)^{-1} \underbrace{\int_0^1 \cdots \int_0^1}_{(m-2)\ times} \left( F^{(m)}(x_0 + \theta_{m-1} \cdots \theta_1 p) \right.\right.$$

$$\left.\left. - F^{(m)}(x_0) \right) \theta_1^{m-2}\, \theta_2^{m-3} \cdots \theta_{m-2}\, d\theta_{m-1} \cdots d\theta_1\, p^{m-1} \right\|$$

$$\leq \sum_{j=0}^{m-3} \frac{b_{j+2}}{(j+1)!}\, r^{j+1} + \frac{b_m + w(r)}{(m-1)!}\, r^{m-1} = \alpha < \frac{1}{3} < 1. \tag{2.258}$$

*Condition (2.253) can be replaced by (2.257). Indeed, we have for $\gamma = x^\star + \theta\, (y^\star - x^\star) - x_0$ by Taylor's theorem that*

$$\left\| \int_0^1 F'(x_0)^{-1} \left( F'(x^\star + \theta\, (y^\star - x^\star)) - F'(x_0) \right) d\theta \right\|$$

$$= \left\| \int_0^1 F'(x_0)^{-1} \left( F''(x_0)\, \gamma + \cdots + \frac{F^{(m-1)}(x_0)}{(m-2)!}\, \gamma^{m-2} \right.\right.$$

$$\left.\left. + \frac{1}{(m-2)!} \int_0^1 F^{(m)}(x_0 + \theta_1\, (\gamma - x_0))\, (1 - \theta_1)^{m-2}\, d\theta_1\, \gamma^{m-1} \right) d\theta \right\|$$

$$\leq \int_0^1 \left( \sum_{j=0}^{m-3} \frac{b_{j+2}}{(j+1)!}\, r^{j+1} + \frac{b_m + w(r)}{(m-1)!}\, r^{m-1} \right) d\theta = \alpha < \frac{1}{3} < 1. \tag{2.259}$$

*Concerning the uniqueness hypothesis (2.253) for $r = R$, we must have that*

$$v(r) < 1. \tag{2.260}$$

**Case 2** *Another possible choice for function $v$ is given by*

$$v(t) = \beta\, t \qquad (\beta > 0,\ t \in [0, +\infty)). \tag{2.261}$$

*Denote by $r_\beta$ the corresponding fixed point of $g$.*

*If*

$$v(r) < \alpha(r), \tag{2.262}$$

*then, the estimates (2.244)–(2.250) are finer (smaller) than in case 1 and*

$$r_\beta < r. \tag{2.263}$$

*Note that these advantages hold, even if function $v$ is not necessarily given by (2.251) as long as (2.262) is satisfied.*

**Example 2.9** *Using Example 2.8, $\eta = 253/1016$, consider first*

$\underline{v(t) = \alpha(t).}$ *Then, we have $b_2 = b_3 = 0$, $\alpha(t) = t^4$ and $w(t) = 2t^2$. It can easily be seen that the fixed point $r$ of function $g$ given in (2.239) is*

$$r = \frac{1}{4}.$$

*Condition (2.241) is satisfied, since*

$$\alpha = v = \alpha(\frac{1}{4}) = \left(\frac{1}{4}\right)^4 < \frac{1}{3}.$$

$\underline{v(t) \neq \alpha(t).}$ *Using (2.236), we have that*

$$v(t) = \frac{1}{12}\, t^4 < t^4 = \alpha(t).$$

*In this case the fixed point of $g$ is given by*

$$r_0 = .247 < r.$$

*In view of (2.253), we can set*

$$R = 1.614.$$

*Operator $\Xi$ in (2.243) is given by*

$$\Xi(x(s)) = \int_0^1 \frac{s}{s+y}\; x(t)\; dt.$$

*In this case, $F(x(s)) = 0$ is a Chandrasekhar–type nonlinear integral equation. Equations of this type appear in radiative transfer in connection with the problem of determination of the angular distribution of the radiant flux emerging from a plane radiation field (cf. [165], [313], [469]).*

## 2.4 Continuous Analog of Newton's Method

A local convergence result is provided in this section for the continuous analog of NM in a Banach space setting. The radius of convergence is larger, the error bounds tighter and under the same or weaker hypotheses than before (cf. [624]). In this section we are concerned with the problem of approximating a locally unique solution $x^\star$ of equation (2.1), where $F$ is a Fréchet–differentiable operator defined on a closed subset $\mathcal{D}$ of a real Hilbert space $\mathcal{X}$ with values in a Hilbert space $\mathcal{Y}$.

Let $x_0 \in \mathcal{D}$ be given. Let $U(x_0, r)$ for some $r > 0$. Assume that for $x \in \mathcal{D}$, $F'(x)^{-1} \in \mathcal{L}(\mathcal{Y}, \mathcal{X})$.

$$U(x_0, r) \subseteq \mathcal{D}, \tag{2.264}$$

$$\sup_{x \in U(x_0, r)} \| F'(x)^{-1} \; F'(x_0) \| \leq m(r), \tag{2.265}$$

$$\sup_{x \in U(x_0, r)} \| F'(x_0)^{-1} \; F^{(i)}(x) \| \leq M_i \quad (i = 0, 1), \tag{2.266}$$

there exists a unique solution $x^\star$ of equation (2.1) in $U(x_0, r)$.

Let us now consider the Newton–type continuous method NTCM for solving (2.1) by

$$\dot{x} = -\alpha \, F'(x)^{-1} \, F(x), \quad x(0) = x_0, \tag{2.267}$$

where $\alpha$ is a given positive constant.

The main local convergence result of this section is:

**Theorem 2.8** *If the above assumptions hold and*

$$m(r) \; \| F'(x_0)^{-1} \, F(x_0) \| \leq r, \tag{2.268}$$

*then*

*(i) Equation (2.1) has a unique global solution $x(t) \in U(x^\star, r)$ for any $x_0 \in U(x^\star, r)$;*

*(ii) $x_\infty$ exists, $F(x_\infty) = 0$, $x_\infty = x^\star$, so that*

$$\| x(t) - x^\star \| \leq r \; e^{-\alpha \, t} \quad (t > 0). \tag{2.269}$$

**Remark 2.14** *Theorem 2.8 improves the corresponding one given by the elegant study in (cf. [624]). Some special cases and applications are also provided in this section. Other related work can be found in (cf. [3], [623], [625], [628]).*

**Proof of Theorem 2.8** As in (cf. [624]) we let

$$Z(t) = x(t) - x^\star, \quad g(t) = \| Z(t) \|,$$

$$q(t) = \| F'(x_0)^{-1} \; F(x(t)) \|, \dot{q} = \frac{dq}{dt}.$$

In view of (2.267), we get that

$$q \, \dot{q} = -\alpha \, (F'(x)^{-1} \, F'(x) \, F'(x)^{-1} \, F'(x) \, F'(x)^{-1} \, F \; F'(x)^{-1} \, F) = -\alpha \, q^2. \tag{2.270}$$

Then, we have that

$$(2.270) \Longrightarrow \dot{q} = -\alpha \, q \Longrightarrow q(t) = q(0) \, e^{-\alpha \, t} = \| F'(x_0)^{-1} \, F(x_0) \| \; e^{-\alpha \, t}. \tag{2.271}$$

It follows from (2.265), (2.267) and (2.271) that

$$\begin{aligned} \| \dot{x} \| &\leq \alpha \ \| F'(x)^{-1} F'(x_0) \| \ \| F'(x_0)^{-1} F(x) \| \\ &\leq \alpha\, m(r) \ \| F'(x_0)^{-1} F(x_0) \| \ e^{-\alpha\, t}. \end{aligned} \tag{2.272}$$

Hence, we deduce $x_\infty$ exists, $\| F'(x_0)^{-1} \ F(x_\infty) \| = 0$, which imply

$$\| F(x_\infty) \| = 0.$$

Moreover, by the uniqueness hypotheses, we obtain $x^\star = x_\infty$. Furthemore, in view of (2.265) and (2.272), we obtain in turn that

$$\begin{aligned} \| x(t) - x_\infty \| &\leq \| \int_t^\infty \dot{x}(s)\, ds \| \\ &= \alpha \left\| \int_t^\infty (F'(x(s))^{-1} F'(x_0))(F'(x_0)^{-1} F(x(s)))\, ds \right\| \\ &\leq \alpha\, m(r) \ \| F'(x_0)^{-1} F(x_0) \| \lim_{p \longrightarrow \infty} \int_t^p e^{-\alpha\, s}\, ds \\ &= m(r) \ \| F'(x_0)^{-1} F(x_0) \| \lim_{p \longrightarrow \infty} (e^{-\alpha\, p} - e^{-\alpha\, t}) \\ &= m(r) \ \| F'(x_0)^{-1} F(x_0) \| \ e^{-\alpha\, t} \\ &\leq \ \varkappa e^{-\alpha\, t} \leq r, \end{aligned} \tag{2.273}$$

which implies (2.269) and $x(t) \in U(x^\star, r)$ $(t > 0)$. Moreover, as in (2.273), we get that

$$\| x(t) - x_0 \| = \left\| \int_0^1 \dot{x}(t)\, dt \right\| \leq m(r) \ \| F'(x_0)^{-1} F(x_0) \| \ (1 - e^{-\alpha\, t}) \leq r, \tag{2.274}$$

which implies $x(t) \in U(x_0, r)$ $(t > 0)$. Hence, we conclude $x(t) \in U(x_0, r) \cap U(x^\star, r)$. This completes the proof of Theorem 2.8. ⊠

**Remark 2.15** *The results in Theorem 2.8 reduce to the corresponding ones in (cf. [624]) if $\mathcal{X} = \mathcal{Y} = \mathcal{D}$ and $F'(x_0)^{-1} \ F$ is replaced by $F$ (non–affine invariant form). Otherwise, they extend the applicability of (2.267). Note also that the advantages of providing convergence results in affine invariant instead of non–affine invariant form are well known in the literature and have been given in (cf. [139], [157]). Theorem 2.8 can be weakened in some cases.*

*(i) Hypotheses (2.264)–(2.266) can be replaced by*

$$x_0 \in U(x^\star, r) \subseteq \mathcal{D}, \tag{2.275}$$

$$\| F'(x^\star)^{-1} (F'(x) - F'(x^\star)) \| \leq L \ \| x - x^\star \| \qquad (L > 0, \ x \in \mathcal{D}). \tag{2.276}$$

*If* $x \in U^\circ\left(x^\star, \frac{1}{L}\right) \subseteq \mathcal{D}$, *then, we have*

$$\| F'(x^\star)^{-1} (F'(x) - F'(x^\star)) \| \leq L \| x - x^\star \| < 1. \tag{2.277}$$

*In view of (2.277), it follows from the Banach lemma on invertible operators that* $F'(x)^{-1}$ *exists and*

$$\| F'(x)^{-1} F'(x^\star) \| \leq (1 - L \| x - x^\star \|)^{-1}. \tag{2.278}$$

*Hence, we can set*

$$m(r) = \frac{1}{1 - L\, r} \qquad \left(r \in \left(0, \frac{1}{L}\right)\right). \tag{2.279}$$

*In this case, (2.268) is replaced by*

$$h = 4\, L \| F'(x_0)^{-1} F(x_0) \| \leq 1 \tag{2.280}$$

*and*

$$r \in [r_1, r_2], \tag{2.281}$$

*for* $h \neq 0$, *if* $r = r_2$, *where,* $r_1$, $r_2$ *are the real zeros of quadratic polynomial*

$$g(s) = Ls^2 - s + \| F'(x_0)^{-1} F(x_0) \| \tag{2.282}$$

*given by*

$$r_1 = \frac{1 - \sqrt{1-h}}{2\, L} \quad \text{and} \quad r_2 = \frac{1 + \sqrt{1-h}}{2\, L}. \tag{2.283}$$

*(ii) Estimates (2.280) and (2.281) can be replaced as follows*

*Assume instead of (2.266) that*

$$\| F'(x^\star)^{-1} (F'(x_0) - F'(x^\star)) \| \leq m_1 \| x_0 - x^\star \| . \tag{2.284}$$

*Then, we get that*

$$\begin{aligned} \| F'(x_0)^{-1} F(x_0) \| &\leq \| F'(x_0)^{-1} F(x^\star) \| \| F'(x^\star)^{-1} (F'(x_0) - F'(x^\star)) \| \\ &\leq \frac{m_1\, r}{1 - L\, r}. \end{aligned} \tag{2.285}$$

*In view of (2.273), we should have that*

$$m(r) \| F'(x_0)^{-1} F(x_0) \| \leq r \tag{2.286}$$

*or, by (2.279) and (2.285),*

$$\frac{1}{1 - L\, r} \frac{m_1\, r}{1 - L\, r} \leq r \tag{2.287}$$

*or*

$$r^\star = \frac{1 - \sqrt{m_1}}{L}, \tag{2.288}$$

*provided that*

$$m_1 \in [0, 1). \tag{2.289}$$

We refer the reader to (cf. [136], [157]), where, in the more general setting of a Banach space, we have provided a larger radius of convergence for many interesting examples by using the theoretical approach given in this section. The same examples can be used here by introducing the norms using standard inner products and replacing $\mathcal{C}[a,b]$ by $L^2[a,b]$ in the appropriate places.

## 2.5 Interior Point Techniques

We use in this section a weaker Newton–Kantorovich theorem for solving equations, introduced in (cf. [165]) to analyze interior point methods. This way our approach refines earlier work in (cf. [602]) on NM and interior point techniques. We are concerned with the problem of approximating a locally unique solution $x^\star$ of equation (2.1) where, $F$ is a differentiable operator defined on a convex domain $\mathcal{D}$ of $\mathbb{R}^i$ ($i$ an integer) with values in $\mathbb{R}^i$.

The famous Newton–Kantorovich theorem (cf. [469]) has been used extensively to solve equation (2.1). A survey of such results can be found in (cf. [139]) and the references there. Recently (cf. [139], [165]) we improved the Newton–Kantorovich theorem. Here, we use this development to show that the results obtained in the elegant work in (cf. [602]) in connection with interior point methods can be improved if our convergence conditions simply replace the stronger ones given there. Finally, a numerical example is provided to show that fewer iterations than the ones suggested in (cf. [602]) are needed to achieve the same error tolerance.

Let $\|\cdot\|$ be a given norm on $\mathbb{R}^i$ and $x_0$ be a point of $\mathcal{D}$ such that $\overline{U}(x_0,r)\subseteq \mathcal{D}$. We assume that the Jacobian $F'(x_0)$ is nonsingular and that the following affine-invariant Lipschitz condition is satisfied: for some $\omega\geq 0$ and for all $x,y\in\overline{U}(x_0,r)$,

$$\|F'(x_0)^{-1}[F'(x)-F'(y)]\|\leq \omega\,\|x-y\|. \tag{2.290}$$

The famous Newton–Kantorovich Theorem (cf. [469]) states that, if the quantity

$$\alpha := \|F'(x_0)^{-1}\,F(x_0)\| \tag{2.291}$$

together with $\omega$ satisfy

$$k=\alpha\omega\leq\frac{1}{2}, \tag{2.292}$$

then, there exists $x^\star\in\overline{U}(x_0,r)$ with $F(x^\star)=0$. Moreover, the sequences produced by NM given by (2.2) and by the modified Newton method

$$y_{n+1}=y_n-F'(y_0)^{-1}F(y_n),\quad y_0=x_0\quad (n\geq 0) \tag{2.293}$$

are well defined and converge to $x^\star$. In (cf. [139], [165]) we introduced the center–Lipschitz condition

$$\|F'(x_0)^{-1}[F'(x)-F'(x_0)]\|\leq\omega_0\,\|x-x_0\| \tag{2.294}$$

for some $\omega_0 \geq 0$ and for all $x \in \overline{U}(x_0, r)$. This way we provided a finer local and semilocal convergence analysis of method (2.2) by using the combination of conditions (2.290) and (2.294) given by

$$k^0 = \alpha \overline{\omega} \leq \frac{1}{2}, \tag{2.295}$$

where

$$\overline{\omega} = \frac{1}{8}\left(4\,\omega_0 + \sqrt{\omega\,\omega_0 + 8\,\omega_0^2} + \sqrt{\omega\,\omega_0}\right) \quad (cf.[165]). \tag{2.296}$$

In general

$$\omega_0 \leq \overline{\omega} \leq \omega \tag{2.297}$$

holds and $\frac{\omega}{\omega_0}$, $\frac{\overline{\omega}}{\omega_0}$, $\frac{\omega}{\overline{\omega}}$ can be arbitrarily large. Note also that

$$k \leq \frac{1}{2} \Rightarrow k^0 \leq \frac{1}{2}, \tag{2.298}$$

but not vice versa unless if $\omega_0 = \omega$. Examples where weaker condition (2.295) holds but (2.292) fails have been given in (cf. [139], [165]).

Similarly, by simply replacing $\omega$ with $\omega_0$ (since (2.294) instead of (2.290) is actually needed in the proof) and condition (2.292) by the weaker

$$k^1 = \alpha \omega_0 \leq \frac{1}{2} \tag{2.299}$$

in the proof of Theorem 1 in (cf. [602]), we showed that method (2.293) also converges to $x^\star$ and the improved bounds

$$\|y_n - x^\star\| \leq \frac{2\,\beta_0\,\lambda_0^2}{1-\lambda_0^2}\,\xi_0^{n-1} \quad (n \geq 1), \tag{2.300}$$

where

$$\beta_0 = \frac{\sqrt{1-2k^1}}{k^1}, \quad \lambda_0 = \frac{1-\sqrt{1-2k^1}-h^1}{k^1} \quad \text{and} \quad \xi_0 = 1-\sqrt{1-2k^1}$$

hold. In case $\omega_0 = \omega$, (2.300) reduces to (2.9) in (cf. [602]). Otherwise, our error bounds are finer. Note also that

$$k \leq \frac{1}{2} \Rightarrow k^1 \leq \frac{1}{2},$$

but not vice versa unless if $\omega_0 = \omega$. Let us provide an example to show that (2.299) holds but (2.292) fails.

**Example 2.10** *Let $i = 1$, $x_0 = 1$, $\mathcal{D} = [p, 2-p]$, $p \in \left[0, \frac{1}{2}\right)$ and define functions $F$ on $D$ by*

$$F(x) = x^3 - p. \tag{2.301}$$

*Using (2.290), (2.291), (2.294) and (2.301), we obtain*

$$\alpha = \frac{1}{3}(1-p), \quad \omega = 2(2-p) \quad \textit{and} \quad \omega_0 = 3-p, \tag{2.302}$$

*which imply that*

$$k = \frac{2}{3}(1-p)(2-p) > \frac{1}{2} \quad \left(p \in \left[0, \frac{1}{2}\right)\right). \tag{2.303}$$

*That is, there is no guarantee that NM (2.2) converges to* $x^\star = \sqrt[3]{p}$, *since the Newton–Kantorovich hypothesis (2.292) is violated. However, condition (2.299) holds for all* $p \in \left[\frac{4-\sqrt{10}}{2}, \frac{1}{2}\right)$ *(see also section 2.1).*

The above suggest that all results on interior point methods obtained in (cf. [602]) for NM using (2.292) can now be rewritten using only (2.295). The same holds true for the modified Newton's method, where (2.299) also replaces (2.292). It has already been shown in (cf. [560]) that the Newton–Kantorovich theorem can be used to construct and analyze optimal–complexity path following algorithms for linear complementary problems. Potra has chosen to apply this theorem to linear complementary problems because such problems provide a convenient framework for analyzing primal–dual interior point methods. Theoretical and experimental work conducted over the past decade has shown that primal–dual path following algorithms are among the best solution methods for LP, quadratic programming QP and linear complementary problems (LCP) (see for example (cf. [609], [773])). Primal–dual path following algorithms are the basis of the best general–purpose practical methods and they have important theoretical properties (cf. [662], [773], [795]). Potra, using (2.292), in particular showed how to construct path–following algorithms for LCP that have $O(\sqrt{n}L)$ iteration complexity (cf. [602]).

Given a point $x$ that approximates a point $x(\tau)$ on the central path of the LCP with complementarity gap $\tau$, the algorithms compute a parameter $\theta \in (0,1)$ so that $x$ satisfies the Newton–Kantorovich hypothesis (2.292) for the equation defining $x\left((1-\theta)\tau\right)$. It is proven that $\theta$ is bounded below by a multiple of $n^{-1/2}$. Since (2.292) is satisfied, the sequence generated by NM or by the modified Newton method (take $F'(x_n) = F'(x_0)$, $n \geq 0$) with starting $x$, will converge to $x\left((1-\theta)\tau\right)$. He showed that the number of steps required to obtain an acceptable approximation of $x\left((1-\theta)\tau\right)$ is bounded above by a number independent of $n$. Therefore, a point with complementarity less than $\varepsilon$ can be obtained in at most $O\big(\sqrt{n}\log\left(\frac{\varepsilon}{\varepsilon_0}\right)\big)$ steps (for both methods), where $\varepsilon_0$ is the complementary gap of the starting point. For linear complementarity problems with rational input data of bit length $L$, this implies that an exact solution can be obtained in at most $O\left(\sqrt{n}L\right)$ iterations plus a rounding procedure including $O\left(n^3\right)$ arithmetic operations (cf. [773]) (see also (cf. [636])).

We also refer the reader to the excellent monograph of Nesterov and Nemirovskii (cf. [560]) for an analysis of the construction of interior point methods for a larger class of problems than that considered in (cf. [602]) (see also (cf. [636])). We can now describe the linear complementarity problem as follows: Given two matrices $Q, R \in \mathbb{R}^{n\times n}$ $(n \geq 2)$ and a vector $b \in \mathbb{R}^n$, the horizontal linear complementarity problem (HLCP) consists of approximating a pair of vectors $(w, s)$ such that

$$\begin{aligned} ws &= 0, \\ Q(w) + R(s) &= b, \\ w, s &\geq 0. \end{aligned} \tag{2.304}$$

The monotone linear complementarity problem LCP is obtained by taking $R = -I$ and $Q$ positive semidefinite. Moreover, the linear programming problem LP and the quadratic programming problem QP can be formulated as HLCPs. That is, HLCP is a suitable way for studying interior point methods.

We assume HLCP (2.304) is monotone in the sense that

$$Q(u) + R(v) = 0 \text{ implies } u^t v \geq 0 \text{ for all } u, v \in \mathbb{R}^n. \tag{2.305}$$

Condition (2.305) holds if the HLCP is a reformulation of a QP. If the HLCP is a reformulation of a LP, then, the following stronger condition holds

$$Q(u) + R(v) = 0 \text{ implies } u^t v = 0 \text{ for all } u, v \in \mathbb{R}^n. \tag{2.306}$$

Then, we say in this case that the HLCP is skew–symmetric.

If the HLCP has an interior point, i.e., there is $(w, s) \in \mathbb{R}^n_{++} \times \mathbb{R}^n_{++}$ satisfying $Q(w) + R(s) = b$, then, for any parameter $\tau > 0$, the nonlinear system

$$\begin{aligned} ws &= \tau e, \\ Q(w) + R(s) &= b, \\ w, s &\geq 0 \end{aligned} \tag{2.307}$$

has a unique positive solution $x(\tau) = \left[w(\tau)^t, s(\tau)^t\right]^t$. The set of all such solutions defines the central path $C$ of the HLCP. It can be proved that $(w(\tau), s(\tau))$ converges to a solution of the HLCP as $\tau \to 0$. Such an approach for solving the HLCP is called the path following algorithm.

At a basic step of a path following algorithm, an approximation $(w, s)$ of $(w(\tau),\ s(\tau))$ has already been computed for some $\tau > 0$. The algorithm determines the smaller value of the central path parameter $\tau_+ = (1 - \theta)\,\tau$, where the value $\theta \in (0, 1)$ is computed in some unspecified way. The approximation $(w^t, s^t)$ of $(w(\tau_+), s(\tau_+))$ is computed. The procedure is then, repeated with $(w^+, s^+, \tau^+)$ in place of $(w, s.\tau)$.

In order for us to relate the path following algorithm and the the Newton–Kantorovich theorem, we introduce the notations

$$x = \begin{bmatrix} w \\ s \end{bmatrix},\quad x(\tau) = \begin{bmatrix} w(\tau) \\ s(\tau) \end{bmatrix},$$

$$x^+ = \begin{bmatrix} w^+ \\ s^+ \end{bmatrix},\quad x(\tau_+) = \begin{bmatrix} w(\tau_+) \\ s(\tau_+) \end{bmatrix}, \text{ etc.}$$

Then, for any $\theta > 0$, we define the nonlinear operator

$$F_\sigma(x) = \begin{bmatrix} ws - \sigma e \\ Q(w) + R(s) - b \end{bmatrix}. \tag{2.308}$$

Then, system (2.307) defining $x(\tau)$ becomes

$$F_\sigma(x) = 0, \tag{2.309}$$

whereas the system defining $x(\tau_+)$ is given by

$$F_{(1-\theta)\tau}(x) = 0. \tag{2.310}$$

We assume that the initial guess $x$ belongs in the interior of the feasible set of the HLCP

$$F^0 = \left\{x = (w^t, s^t)^t \in \mathbb{R}^{2n}_{++} : Q(w) + R(s) = b\right\}. \tag{2.311}$$

In order to verify the Newton–Kantorovich hypothesis for equation (2.309), we introduce the quantity

$$\eta = \eta(x,\tau) = \left\|F'(x)^{-1}F_\tau(x)\right\|, \tag{2.312}$$

the measure of proximity

$$\begin{aligned} k &= k(x,\tau) = \eta\ell,\ \ell = \ell(x), \\ k^0 &= k^0(x,\tau) = \eta\bar{\ell},\ \bar{\ell} = \bar{\ell}(x), \\ k^1 &= k^1(x,\tau) = \eta\ell_0,\ \ell_0 = \ell_0(x) \end{aligned} \tag{2.313}$$

and the normalized primal–dual gap

$$\mu = \mu(x) = \frac{w^t s}{\eta}. \tag{2.314}$$

If, for a given interior point $x$ and a given parameter $\tau$, we have $k^0(x,\tau) \leq .5$ for the Newton–Kantorovich method or $k^1(x,\tau) \leq .5$ for the modified Newton–Kantorovich method, then, corresponding sequences starting from $x$ will converge to the point $x(\tau)$ on the central path. We can now describe our algorithm which is a weaker version of the one given in (cf. [602]).

**Algorithm 2.1** *(using Newton–Kantorovich method).*
*Given* $0 < k_1^0 < k_2^0 < .5$, $\varepsilon > 0$ *and* $x_0 \in F^0$ *satisfying* $k^0(x_0, \mu(x_0)) \leq k_1^0$,

(a) *Set* $k^0 \leftarrow 0$ *and* $\tau_0 \leftarrow \mu(x_0)$;

(b) *repeat (outer iteration)*

*Set* $(x,\tau) \leftarrow (x_k, \tau_k)$, $\bar{x} \leftarrow x_k$,

*Determine the largest* $\theta \in (0,1)$ *such that* $k^0(x, (1-\theta)\tau) \leq k_2^0$,

*Set* $\tau \leftarrow (1-\theta)\tau$,

*(c) repeat (inner iteration)*

$$\textit{Set } x \leftarrow x - F'(x)^{-1}F_\tau(x), \ \textit{until } k^0(x,\mu) \le k_1^0, \tag{2.315}$$

*Set* $(x_{k+1}, \tau_{k+1}) \leftarrow (x, \tau)$,

*Set* $k \leftarrow k+1$, *until* $\left(w^k\right)^t s^k \le \varepsilon$.

For the modified the Newton–Kantorovich algorithm $k_1^0, k_2^0, k^0$ should be replaced by $k_1^1, k_2^1, k^1$ and (2.315) by

$$\text{Set } x \leftarrow x - F'(\bar{x})^{-1}F_\tau(x),$$

respectively. In order to obtain Algorithm 1 in (cf. [602]) we need to replace $k_1^0, k_2^0, k^0$ by $k_1, k_2, k$, respectively.

The above suggest that all results on interior point methods obtained in (cf. [602]) using (2.292) can now be rewritten using only the weaker (2.295) (or (2.299)). We only state those results for which we will provide applications.

Let us introduce the notation

$$\Psi_i^a = \begin{cases} 1 + \theta_i^a + \sqrt{2\theta_i^a + r_i^a}, & \text{if HLCP is monotone,} \\ 1 + q_{ia} + \sqrt{2q_{ia} + q_{ia}^2}, & \text{if HLCP is skew–symmetric,} \end{cases} \tag{2.316}$$

where $\sqrt{r_i^a} = \theta_i^a$, $\sqrt{t_i^a} = k_i^a$, $a = 0, 1$,

$$\theta_i^a = t_i\left[1 + \frac{t_i^a}{1 - t_i^a}\right], \quad q_{ia} = \frac{t_i^a}{2} \ (i = 1, 2). \tag{2.317}$$

Then, by simply replacing $k, k_1, k_2$ by $k^0, k_1^0, k_2^0$, respectively, in the corresponding results in (cf. [602]), we obtain the following improvements.

**Theorem 2.9** *The parameter $\theta$ determined at each outer iteration of Algorithm 2.1 satisfies*

$$\theta \ge \frac{\chi^a}{\sqrt{n}} = \lambda^a,$$

*where*

$$\chi^a = \begin{cases} \dfrac{\sqrt{2}\,(k_2^a - k_1^a)}{\sqrt{2 + p^2 t_i \sqrt{\psi_1^a}}}, & \textit{if HLCP is skew–symmetric or if no simplified Newton–Kanorovich steps are performed,} \\[2ex] \dfrac{\sqrt{2}\,(k_2^a - k_1^a)}{(\sqrt{2} + pk_1^a)\sqrt{\psi_1^a}}, & \textit{otherwise,} \end{cases} \tag{2.318}$$

*where*

$$p = \begin{cases} \sqrt{2}, & \textit{if HLCP is monotone,} \\ 1, & \textit{if HLCP is skew–symmetric.} \end{cases} \tag{2.319}$$

*Clearly, the lower bound on $\lambda^a$ on $\theta$ is an improvement over the corresponding one in [602, Corollary 4].*

In the next result, a bound on the number of steps of the inner iteration that depends only on $k_1^0$ and $k_2^0$ is provided.

**Theorem 2.10** *If Newton–Kantorovich method is used in Algorithm 2.1, then, each inner iteration terminates in at most* $N^0\left(k_1^0, k_2^0\right)$ *steps, where*

$$N^0\left(k_1^0, k_2^0\right) = \text{integer}\, part\left[\log_2\left(\frac{\log_2\left(x_{N^0}\right)}{\log_2\left[\left(1-\sqrt{1-2k_2^0}-k_2^0\right)/k_2^0\right]}\right)\right] \tag{2.320}$$

*and*

$$x_{N^0} = \frac{\left(1-pk_2^0/\sqrt{2}\right)\left[t_2^0-\left(1-\sqrt{1-2k_2^0}\right)^2\right]k_1^0}{2\sqrt{2}t_2^0\sqrt{1-2k_2^0}\left[\sqrt{\psi_2^0}+1-\sqrt{1-2k_2^0}\right]\left(1+k_1^0\right)}. \tag{2.321}$$

*If the modified the Newton–Kantorovich method is used in Algorithm 2.1, then, each iteration terminates in at most* $S^0\left(k_1, k_2\right)$ *steps, where*

$$S^1\left(k_1^1, k_2^1\right) = \text{integer}\, part\left[\frac{\log_2\left(x_{S^1}\right)}{\log_2\left(1-\sqrt{1-\sqrt{1-2k_2^1}}\right)}\right]+1 \tag{2.322}$$

*and*

$$x_{S^1} = \frac{\left(1-pk_2^1/\sqrt{2}\right)\left[t_2^1-\left(1-\sqrt{1-2k_2^1}-k_2^1\right)^2\right]k_1^0}{2\sqrt{2}\sqrt{1-2k_2^1}\left(1-\sqrt{1-2k_2^1}-k_2^1\right)^2\left(\sqrt{\psi_2^1}+1-\sqrt{1-2k_2^1}\right)\left(1+k_1^1\right)}.$$

**Remark 2.16** *Clearly, if* $k_1^1 = k_1^0 = k_1$, $k_2^1 = k_2^0 = k_2$, $k^1 = k^0 = k$, *Theorem 2.9 reduces to the corresponding Theorem 2 in (cf. [602]). Otherwise, the following improvement holds*

$$N^0\left(k_1^0, k_2^0\right) < N\left(k_1, k_2\right),\ N^0 < N,\ S^1\left(k_1^1, k_2^1\right) < S\left(k_1, k_2\right) \text{ and } S^1 < S.$$

*Since* $\frac{k_1}{k_1^0}, \frac{k_2}{k_2^0}, \frac{k_1}{k_1^1}, \frac{k_2}{k_2^1}$ *can be arbitrarily large for a given triplet* $\eta$, $\ell$ *and* $\ell_0$, *the choices*

$$k_1^0 = k_1^1 = .12,\quad k_2^0 = k_2^1 = .24,\quad \text{when } k_1 = .21 \text{ and } k_2 = .42$$

*and*

$$k_1^0 = k_1^1 = .24,\quad k_2^0 = k_2^1 = .48,\quad \text{when } k_1 = .245 \text{ and } k_2 = .49$$

*are possible.*

Then, using formulas (2.319), (2.320) and (2.322), we obtain the following tables:

(a) If the HLCP is monotone and only Newton directions are performed, then, we have that

| Potra (2.318) | Argyros (2.318) |
|---|---|
| $\chi(.21, .42) > .17$ | $\chi(.12, .24) > .1$ |
| $\chi(.245, .49) > .199$ | $\chi(.24, .48) > .196$ |

| Potra (2.320) | Argyros (2.320) |
|---|---|
| $N(.21, .42) = 2$ | $N(.12, .24) = 1$ |
| $N(.245, .49) = 4$ | $N(.24, .48) = 3$ |

(b) If the HLCP is monotone and Modified Newton directions are performed, then, we have that

| Potra (2.318) | Argyros (2.318) |
|---|---|
| $\chi(.21, .42) > .149$ | $\chi(.12, .24) > .098$ |
| $\chi(.245, .49) > .164$ | $\chi(.24, .48) > .162$ |

| Potra (2.322) | Argyros (2.322) |
|---|---|
| $S(.21, .42) = 5$ | $S(.12, .24) = 1$ |
| $S(.245, .49) = 18$ | $S(.24, .48) = 12$ |

All the above improvements are obtained under weaker hypotheses and the same computational cost (in the case of NM) or less computational cost (in the case of the modified Newton method) since in practice the computation of $\omega$ requires that of $\omega_0$ and in general the computation of $\omega_0$ is less expensive than that of $\omega$.

## 2.6 Regular Smoothness

The concept of regular smoothness has been succesfully shown to be an appropriate and viable framework for the convergence of iterative procedures converging to a locally unique solution of an operator equation in a Banach space setting (cf. [379], [383]).

Motivated by these works and optimization considerations, we show that under the same sufficient convergence conditions tighter error bounds can be obtained on the distances involved and an at least as precise information on the location of the solution. In some special cases, we also show that the sufficient convergence conditions can be weakened.

In this section we are concerned with the problem of approximating a locally unique solution $x_\infty$ of nonlinear equation (2.1), where $F$ is a Fréchet–differentiable operator defined on a open convex subset $\mathcal{D}$ of a Banach space $\mathcal{X}$ with values in a Banach space $\mathcal{Y}$.

The most popular iterative procedures for generating a sequence approximating $x_\infty$ are NM and the Secant method:

$$x_{n+1} = x_n - [x_n, x_{n-1}; F]^{-1}\, F(x_n) \quad (x_{-1},\ x_0 \in \mathcal{D},\ n \geq 0). \tag{2.323}$$

Here, $F'(x) \in \mathcal{L}(\mathcal{X}, \mathcal{Y})$ denotes the Fréchet–derivative of operator $F$ and $[x, y; F] \in \mathcal{L}(\mathcal{X}, \mathcal{Y})$ the divided difference of order one at $x \in \mathcal{D}$, $y \in \mathcal{D}$ (cf. [163], [194], [195], [469]). NM requires one function evaluation and the computation of one inverse at each step. It is sellf–correcting, and has a quadratic convergence under natural conditions (cf. [139], [195], [288], [383], [469], [485]).

The Secant method has some attractive properties: it is also self–correcting, it exhibits superlinear convergence and no knowledge of the derivatives of the operators involved is required. These facts not only make the Secant iteration much cheaper than Newton's, but also makes this method applicable to equations with nondifferentiable operators (cf. [139], [195], [426], [494], [673]).

In the one dimensional case, the Secant method is of higher efficiency than the corresponding NM. A convergence analysis for both methods has been provided under various assumptions by many authors. In particular, in the excellent works by Galperin, the concept of regular smoothness was introduced, which became a viable framework for the study of the convergence of iterative procedures such as NM and Secant method. This way, the applicability of these methods was extended and in the case of NM tighter than before error bounds on the distances involved were found (cf. [379], [383]).

In some interesting special cases, we show that the sufficient convergence conditions can be weakened, indicating that the regular smoothness approach and the majorant systems introduced by Galperin do not necessarily lead to weakest possible or usable sufficient convergence conditions.

Let $\mathcal{T}$ denote the class of nondecreasing continuous functions $\omega : [0, \infty) \longrightarrow [0, \infty)$ that have convex subgraphs $\{(s, t) : s \geq 0 \,\text{and}\, t \leq \omega(s)\}$ and vanish at zero, i.e., they are concave (cf. [163], [194], [195]). The functions of $\mathcal{T}$ have left– and right–hand derivatives at each $s > 0$ (being monotone) and they coincide everywhere except, perhaps, for a countable number of points.

**Definition 2.1** *(cf. [383]) Let $F : \mathcal{D} \subseteq \mathcal{X} \longrightarrow \mathcal{Y}$ be a smooth operator. Denote by $\underline{h}(F)$ the $\inf\limits_{x \in \mathcal{D}} \parallel F'(x) \parallel$. Given an $\omega \in \mathcal{T}$, we say that $F$ is $\omega$–regularly smooth on $\mathcal{D}$ or, equivalently, $\omega$ is a regular smoothness modulus of $F$ on $\mathcal{D}$ if there exists an $\underline{h} \in [0, \underline{h}(F)]$ such that the inequality*

$$\omega^{-1}(h_F(x, y) + \parallel F'(y) - F'(x) \parallel) - \omega^{-1}(h_F(x, y)) \leq \parallel y - x \parallel, \tag{2.324}$$

*where*

$$h_F(x, y) = \min\{\parallel F'(x) \parallel, \parallel F'(y) \parallel\} - \underline{h} \quad (x, y \in \mathcal{D}) \tag{2.325}$$

*holds.*

*The operator is regularly smooth on $\mathcal{D}$ if it is $\omega$–regularly smooth there for some $\omega \in \mathcal{T}$.*

The notation $\omega^{-1}$ is for a function whose closed epigraph $cl\{(s,t) : s \geq 0 \text{ and } t \geq \omega^{-1}(s)\}$ is symmetrical to closure of the subgraph of $\omega$ with respect to the axis $t = s$. Clearly, $\omega^{-1}$ is a convex function on $[0, \infty)$ vanishing at zero, continuously increasing in $[0, \omega(\infty))$ and equal to $\infty$ for all $s > \omega(\infty)$. In view of the convexity of $\omega^{-1}$, each $\omega$–regularly smooth operator is also $\omega$–smooth, in the sense that

$$\| F'(x) - F'(y) \| \leq \omega(\| x - y \|) \quad (x, y \in \mathcal{D}). \tag{2.326}$$

However, the converse is not true (cf. [383]).

**Remark 2.17** *In view of (2.324), for $x_0 \in \mathcal{D}$ fixed, there exists $\omega_0 \in \mathcal{T}$, which is a regular smoothness modulus of $F$ on $\mathcal{D}$ at $x_0$, such that*

$$\omega_0^{-1}(h_F(x, x_0) + \| F'(x_0) - F'(x) \|) - \omega_0^{-1}(h_F(x, x_0)) \leq \| x_0 - x \| \quad (x \in \mathcal{D}) \tag{2.327}$$

*and*

$$\| F'(x_0) - F'(x) \| \leq \omega_0(\| x_0 - x \|) \quad (x \in \mathcal{D}). \tag{2.328}$$

*Note that, in general*

$$\omega_0(t) \leq \omega(t) \qquad (t \geq 0) \tag{2.329}$$

*holds and $\dfrac{\omega(t)}{\omega_0(t)}$ can be arbitrarily large.*

*It turns out that when upper bounds on the norms $\| F'(x)^{-1} \|$ are to be found the stronger condition (2.324) (or (2.326)) is used in the literature (cf. [288], [383], [469], [485]), instead of the needed condition (2.327) (or (2.328)), which is weaker and tighter.*

*In the case $\omega_0 < \omega$, this modification creates a tighter majorizing sequence than before with consequences (under the same hypotheses and computational cost) as stated in the introduction of this section. It is also expected that the corresponding sufficient convergence conditions must be at least as weak.*

*We shall show that this is indeed the case for NM. It is convenient for us to adopt some of the now standard notation in (cf. [383]), so we can make the comparison between the two works easier to carry out.*

Let $\omega \in \mathcal{T}$ and $\Omega(t) = \displaystyle\int_0^t \omega(\tau)\, d\tau$. Denote by

$$\theta(\sigma, \tau) := \min\{\tau,\, \sigma - \tau\}, \quad m(u, v, \tau) = \min\{u,\, (u - \theta(u - v, \tau))^+\}$$

and

$$\psi(u, v, w) = \int_0^w (\omega(m(u, v, \tau) + \tau) - \omega(m(u, v, \tau)))\, d\tau \qquad (u, w > 0), \tag{2.330}$$

where

$$a^+ = \max\{a,\, 0\}. \tag{2.331}$$

Denote also by $\psi_0$ the function given in (2.330) with $\omega_0$ replacing $\omega$.

Set

$$F_0 = F'(x_0)^{-1}\ F \tag{2.332}$$

for some $x_0 \in \mathcal{D}$ such that $F'(x_0)^{-1} \in \mathcal{L}(\mathcal{Y}, \mathcal{X})$. Clearly, NM for $F$ and $F_0$ are identical.

Let $\underline{h}$ be a lower bound for $\underline{h}(F_0)$

$$0 \leq \underline{h} \leq \underline{h}(F_0) \tag{2.333}$$

and let $\omega, \omega_0 \in \mathcal{T}$ satisfy (2.324), (2.327), respectively, with $F_0$ replacing $F$. Define the constant

$$\kappa = \omega_0^{-1}(1 - \underline{h}) \tag{2.334}$$

and consider an upper bound $a$ for the norm

$$\| \ F_0(x_0) \ \| \leq a. \tag{2.335}$$

Moreover, define the sequence of triplets $(\alpha_n, \gamma_n, \delta_n)$ $(n \geq 0)$ by

$$\begin{aligned}
&\alpha_0 = \kappa,\ \ \gamma_0 = 1,\ \ \delta_0 = a,\\
&\alpha_n = (\alpha_{n-1} - \delta_{n-1})^+,\\
&\gamma_n = 1 - \omega_0(\alpha_n + t_n) + \omega_0(\alpha_n),\\
&\delta_n = \gamma_n^{-1}\ \psi(\alpha_{n-1}, \alpha_{n-1} - \delta_{n-1}, \delta_{n-1}),
\end{aligned} \tag{2.336}$$

where,

$$t_n = \sum_{i=0}^{n-1} \delta_i.$$

If $\omega(t) = c\ t$ and $\omega_0(t) = c_0\ t$, then, we have the sequence

$$t_{n+1} = t_n + \frac{c\ (t_n - t_{n-1})^2}{2\ (1 - c_0\ t_n)} \tag{2.337}$$

and the triplet is given by

$$\alpha_n = \kappa - t_n, \quad \gamma_n = 1 - c_0\ t_n, \quad \delta_n = t_{n+1} - t_n. \tag{2.338}$$

Sequence $\{t_n\}$ further reduces to the one considered by Kantorovich for $c = c_0$

$$s_{n+1} = s_n + \frac{c\ (s_n - s_{n-1})^2}{2\ (1 - c\ s_n)} = \frac{.5\ c\ s_n^2 - s_n + a}{1 - c\ s_n}. \tag{2.339}$$

The analog of the Kantorovich's majorant function $t \longrightarrow .5\ c\ t^2 - t + a$ is given by

$$\phi_{\underline{h}}(t) = a - t + \psi(\kappa, (\kappa - t)^+ - t, t) \quad (t > 0). \tag{2.340}$$

Denote also by $\phi^0_{\underline{h}}$ function $\phi_{\underline{h}}$ with $\psi$ replaced by $\psi_0$. The triple $(\alpha_n, \gamma_n, \delta_n)$ given in (2.336) is well defined, if $\gamma_n > 0$. This is true if

$$t_n < \omega_0^{-1}(1) \qquad (n \geq 0). \tag{2.341}$$

In this case, $\delta_n \geq 0$ and $\alpha_n = (\kappa - t_n)^+$.

As shown in (cf. [383]) $\phi_{\underline{h}}$ is decreasing on $[0, \omega^{-1}(1)]$ and is increasing on $[\omega^{-1}(1), +\infty)$. Denote by $\phi^{-1}_{\underline{h},1}$, $\phi^{-1}_{\underline{h},2}$ the inverse of the restrictions of $\phi_{\underline{h}}$ to $[0, \omega^{-1}(1)]$ and to $[\omega^{-1}(1), \infty)$, respectively. Similarly, for the inverses of $\phi^0_{\underline{h},1}$ and $\phi^0_{\underline{h},2}$, which are defined on $[0, \omega_0^{-1}(1)]$, $[\omega_0^{-1}(1), \infty)$, respectively.

We can show the main semilocal convergence result for NM.

**Theorem 2.11** *Let $F_0 : \mathcal{D} \subseteq \mathcal{X} \longrightarrow \mathcal{Y}$ be a $\omega_0$ and $\omega$–regularly smooth. Assume that Condition (2.341) and*

$$\overline{U}(x_0, t_\infty) \subseteq \mathcal{D} \tag{2.342}$$

*hold, where*

$$t_\infty = \lim_{n \to \infty} t_n. \tag{2.343}$$

*Then, the sequence $\{x_n\}$ $(n \geq 0)$ generated by (2.2) is well defined, remains in $\overline{U}(x_0, t_\infty)$ for all $n \geq 0$ and converges to a solution $x_\infty$ of equation $F(x) = 0$. Moreover, the following estimates hold*

$$\| F_0'(x_n) \| \geq \omega_0(\alpha_n) + \underline{h}, \tag{2.344}$$

$$\overline{\gamma_n}^{-1} \leq \gamma_n^{-1}, \tag{2.345}$$

$$\overline{\delta_n} \leq \gamma_n^{-1}\, \psi(\alpha_{n-1}, \alpha_n - \delta_{n-1}, \delta_{n-1}), \tag{2.346}$$

$$\Delta_{n+1} = \| x_{n+1} - x_\infty \| \leq \overline{\gamma_n}^{-1}\ \psi(\overline{\alpha_n}, \omega_0^{-1}(\| F_0'(x_\infty) \| - \underline{h}) - \Delta_n, \Delta_n) \tag{2.347}$$

*and*

$$\| x_n - x_\infty \| \leq t_\infty - t_n, \tag{2.348}$$

*where, $\overline{\gamma_n}$ and $\overline{\delta_n}$ are defined in (2.349). Furthermore, if $a$ is such that $t_\infty \leq \kappa$, then, the solution $x_\infty$ is unique in $\overline{U}(x_0, \phi^{0,-1}_{\underline{h},2}(0))$.*

**Proof** It is convenient for us to set

$$\overline{\alpha_n} = \omega_0^{-1}(\| F_0'(x_n) \| - \underline{h}), \quad \overline{\gamma_n} = \| F_0'(x_n)^{-1} \|^{-1}, \quad \overline{\delta_n} = \| x_{n+1} - x_n \|. \tag{2.349}$$

These numbers are well defined and the relations:

$$\overline{\alpha_0} = \kappa = \alpha_0, \quad \overline{\gamma_0} = 1 = \gamma_0, \quad \overline{\delta_0} \leq a = \delta_0 \tag{2.350}$$

hold. Suppose that $n \geq 1$ and, for all $k \leq n-1$, the following statements hold $F_0'(x_k)$ exists, $F_0'(x_k)^{-1} \in \mathcal{L}(\mathcal{Y}, \mathcal{X})$ and

$$\overline{\alpha_k} \geq \alpha_k, \quad \overline{\gamma_k} \geq \gamma_k, \quad \overline{\delta_k} \leq \delta_k. \tag{2.351}$$

Then, we have that

$$\| x_n - x_0 \| \le \sum_{k=0}^{n-1} \overline{\delta_k} \le \sum_{k=0}^{n-1} \delta_k = t_n. \tag{2.352}$$

Hence, $F_0'(x_n)$ exists. Moreover, due to $\omega_0$–regular smoothness of $F_0$ on $U(x_0, t_\infty)$, we have in turn that

$$\begin{aligned} \| F_0'(x_n) - F_0'(x_0) \| &\le \omega_0(\omega_0^{-1}(\min\{\| F_0'(x_0) \|, \| F_0'(x_n) \|\} - \underline{h}) + \| x_n - x_0 \|) \\ &\quad - \min\{\| F_0'(x_0) \|, \| F_0'(x_n) \|\} + \underline{h} \\ &= \omega_0(\min\{\kappa, \overline{\alpha_n}\} + \| x_n - x_0 \|) - \omega_0(\min\{\kappa, \overline{\alpha_n}\}) \\ &\le \omega_0(\min\{\kappa, \alpha_n\} + \| x_n - x_0 \|) - \omega_0(\min\{\kappa, \alpha_n\}) \\ &\le \omega_0(\alpha_n + t_n) - \omega_0(\alpha_n). \end{aligned} \tag{2.353}$$

We also have $\alpha_n \le \alpha_0 = \kappa$ and $\sum_{k=0}^{n-1} \overline{\delta_k} \le \delta_k$, which lead to

$$\begin{aligned} \overline{\gamma_n} = d(F_0'(x_n)) &\ge d(F_0'(x_0)) - \| F_0'(x_n) - F_0'(x_0) \| \\ &\ge 1 - \omega_0(\alpha_n + t_n) + \omega_0(\alpha_n) \qquad (2.354) \\ &\ge 1 - \omega_0(t_n) > 0 \qquad \text{(by (2.341))}. \end{aligned}$$

It follows from (2.354) and the Banach lemma of invertible operators that $F_0'(x_n)^{-1} \in \mathcal{L}(\mathcal{Y}, \mathcal{X})$, so that (2.344) and (2.345) hold. Note that in (cf. [383]), looser estimates were obtained with $\omega$ replacing $\omega_0$ in estimates (2.353) and (2.354). Using the identity

$$x_{n+1} - x_n = F_0'(x_n)^{-1} \left(F_0(x_n) - F_0(x_{n-1}) - F_0'(x_{n-1}) \left(x_n - x_{n-1}\right)\right) \tag{2.355}$$

and (2.345), we obtain that

$$\begin{aligned} \overline{\delta_n} &\le \overline{\gamma_n}^{-1} \ \psi(\overline{\alpha_{n-1}}, \overline{\alpha_n} - \overline{\delta_{n-1}}, \overline{\delta_{n-1}}) \\ &\le \psi(\alpha_{n-1}, \alpha_n - \delta_{n-1}, \delta_{n-1}) = \delta_n, \end{aligned} \tag{2.356}$$

which completes the induction and also show (2.346). In view of the estimate

$$\| x_{n+m} - x_n \| \le \sum_{k=n}^{n+m-1} \overline{\delta_k} \le \sum_{k=n}^{n+m-1} \delta_k < \sum_{k=n}^{\infty} \overline{\delta_k} = t_\infty - t_n, \tag{2.357}$$

we deduce that sequence $\{x_n\}$ is a Cauchy sequence in a Banach space $\mathcal{X}$ and as such it converges to some $x_\infty \in \overline{U}(x_0, t_\infty)$ (since $\overline{U}(x_0, t_\infty)$ is a closed set). We also have that

$$\| F_0(x_n) \| \le \psi(\alpha_{n-1}, \alpha_n - \delta_{n-1}, \delta_{n-1}) \longrightarrow 0 \quad \text{as } n \to \infty. \tag{2.358}$$

That is $F(x_\infty) = 0$. The rest of the proof follows by simply replacing $\omega$ by $\omega_0$ in the corresponding proof in (cf. [383]). This completes the proof of Theorem 2.11. ⊠

**Remark 2.18** *If $\omega_0(s) = \omega(s)$ $(s > 0)$, then our Theorem 2.11 reduces to Theorem 4.3 in (cf. [383]). Otherwise, it constitutes an improvement with advantages as noted in the discussion after (2.328). Note also that a posteriori bounds given in (cf. [383]) are also becoming tighter with our approach and in Application 2.1, we show how to replace delicate condition (2.341).*

We need an analogous definition of $\omega$–regularly continuity for the Secant method (see also Definition 2.1).

**Definition 2.2** *(cf. [379]) The dd $[x, y; F]$ is said to be $\omega$–regularly continuous on $\mathcal{D}$ if there exist an $\omega \in \mathcal{T}$ (call it regularity modulus and a constant $\underline{h} \in [0, \underline{h}([x, y; F])]$ such that for all $x, y, u, v \in \mathcal{D}$,*

$$\omega^{-1}(\min\{\| [x, y; F] \|, \| [u, v; F] \|\} - \underline{h} + \| [x, y; F] - [u, v; F] \|)$$

$$- \omega^{-1}(\min\{\| [x, y; F] \|, \| [u, v; F] \|\} - \underline{h}) \leq \| x - u \| + \| y - v \|, \quad (2.359)$$

*where*

$$\underline{h}([x, y; F]) = \inf_{x,y \in \mathcal{D}^2} \{\| [x, y; F] \|\}. \quad (2.360)$$

As in (2.324), we have that (2.359) implies that

$$\| [x, y; F] - [u, v; F] \| \leq \omega(\| x - u \| + \| y - v \|) \quad (x, y, u, v \in \mathcal{D}). \quad (2.361)$$

In view of (2.359) and, for fixed $x_{-1}, x_0 \in \mathcal{D}$, there exists $\omega_0 \in \mathcal{T}$ such that for all $x, y \in \mathcal{D}$

$$\omega_0^{-1}(\min\{\| [x, y; F] \|, \| [x_0, x_{-1}; F] \|\} - \underline{h} + \| [x, y; F] - [x_0, x_{-1}; F] \|)$$

$$- \omega_0^{-1}(\min\{\| [x, y; F] \|, \| [x_0, x_{-1}; F] \|\} - \underline{h}) \leq \| x - x_0 \| + \| y - x_{-1} \|, \quad (2.362)$$

which implies

$$\| [x, y; F] - [u, v; F] \| \leq \omega_0(\| x - x_0 \| + \| y - x_{-1} \|). \quad (2.363)$$

We shall now define a more precise majorant generator than the one in (cf. [379]). Let

$$\bar{t} : = \| x - x_0 \|, \; \overline{\alpha} : = \omega_0^{-1}(\| [x, x_-; F] \| - \underline{h}), \; \overline{\beta} : = \| x - x_- \|,$$

$$\overline{\delta}: =\| x_+ - x \|,\ [x_0, x_{-1}; F] = \mathcal{I},$$

$$\alpha_0: = \omega_0^{-1}(1-\underline{h}),\quad \overline{\beta}_0 : =\| x_0 - x_{-1} \|,\quad \overline{a} := \alpha_0 - \overline{\beta}_0,$$

$$\psi(s,t) = \omega(s+t) - \omega(s),\quad \psi_0(s,t) = \omega_0(s+t) - \omega_0(s),$$

$$\overline{d} = \psi_0((\overline{a} - 2\,\overline{t} - \overline{\delta})_+, 2\,\overline{t} + \overline{\delta} + \overline{\beta_0}),\quad a = \alpha_0 - \beta_0 = \omega_0^{-1}(1-\underline{h}) - \beta_0,$$

$$\Delta_n = \| x_n - x_\infty \| .$$

Our majorant generator is defined for $g : Q \subseteq \mathbb{R}^4 \longrightarrow \mathbb{R}^4$ $q = (t, \alpha, \beta, \delta)$ where

$Q = \{q : g_0((\alpha_0 - 2t - \delta - \beta_0)_+, \beta_0 + 2t + \delta) < 1$ for $t \geq 0, \alpha \geq 0, \beta \geq 0$, and $\delta \geq 0\}$

into $q_+ = (t_+, \alpha_+, \beta_+, \delta_+)$ by

$$t_+ := t + \delta,\quad \alpha_+ := a - 2\,t - \delta,\quad \beta_+ := \delta,$$

$$\delta_+ := \delta\left(\frac{\omega(a - 2\,t + \beta) - \omega(a - 2\,t - \delta)}{\omega_0(a - 2\,t - \delta)}\right). \tag{2.364}$$

**Remark 2.19** *Note that, if $\omega_0 = \omega$, majorant generator (2.364) coincides with the corresponding one in (cf. [383]). Otherwise, (i.e. if $\omega_0 < \omega$) it is more precise. We say that the quadruple $q' = (t', \alpha', \beta', \delta')$ is majorizing $q = (t, \alpha, \beta, \delta)$ ($q \prec q'$) if $t \leq t'$, $\alpha \geq \alpha'$, $\beta \leq \beta'$ and $\delta \leq \delta'$.*

Let $q_0 = (t_0, \alpha_0, \beta_0, \delta_0) \in Q$, then, we have the iteration

$$q_{n+1} = g(q_n) \qquad (n \geq 0). \tag{2.365}$$

If $d_n = \psi_0(\alpha_{n+1}, \beta_0 + 2\,t_n + \delta_n) < 1$, iteration $\{q_n\}$ is well defined. In particular, if

$$\overline{d} = \psi_0((\overline{a} - 2\,\overline{t} - \overline{\delta})_+, 2\,\overline{t} + \overline{\delta} + \overline{\beta_0}) < 1, \tag{2.366}$$

then, dd $[x_+, x; F]$ is boundedly invertible and

$$\| [x_+, x; F]^{-1} \| \leq (1 - \overline{d})^{-1}. \tag{2.367}$$

Indeed, we have in view of (2.362) that

$$\begin{aligned}
&\| \mathcal{I} - [x_+, x; F] \| \\
&= \| [x_0, x_{-1}; F] - [x_+, x; F] \| \\
&\leq \omega_0\big(\omega_0^{-1}(\min\{1, \| [x_+, x; F] \|\} - \underline{h}) + \| x_+ - x_0 \| + \| x - x_{-1} \| \big) \\
&\quad - \min\{1, \| [x_+, x; F] \|\} + \underline{h} \\
&\leq \omega_0(\min\{\alpha_0, \overline{\alpha_+}\} + \overline{t_+} + \overline{t} + \overline{\beta_0}) - \omega_0(\min\{\alpha_0, \overline{\alpha_+}\}) \\
&\leq \psi_0(\min\{\alpha_0, \overline{\alpha_+}\}, 2\,\overline{t} + \overline{\delta} + \overline{\beta_0}) = \overline{d} < 1,
\end{aligned}$$

which together with the Banach lemma on invertible operators implies (2.367). Estimates (2.367) is tighter than the corresponding one in (cf. [379]) using $\omega$ instead of $\omega_0$. This substitution in the proofs of the results in (cf. [379]) produces the advantages as already stated in the introduction of this section. Hence, we have the following result.

**Theorem 2.12** *We have that for all $n \in \mathbb{N}$, we have that $d_n < 1 \Longleftrightarrow 2\,t_n + \delta_n < a$;*
*If*

$$d_n < 1 \quad (n \geq 0) \tag{2.368}$$

*then, sequence $q_n$ generated by (2.365) is well defined and converges to $(t_\infty, a - 2\,t_\infty, 0, 0)$, where $t_\infty \leq .5\,a$,*

$$\overline{q_0} \prec q_0 \Longrightarrow \overline{q_n} \prec q_n,$$

*sequence $\{x_n\}$ generated by (2.323) is well defined, remains in $U(x_0, t_\infty - t_0)$ and converges to a solution $x_\infty$ of $F(x) = 0$. Moreover, the following estimates hold for all $n \geq 0$*

$$\Delta_{n+1} \leq \Delta_n \,\frac{\omega(a - t_{n-1} - t_n) - \omega(a - t_{n-1} - t_n - \Delta_{n-1})}{\omega_0(a - t_n - t_{n+1})} < \frac{\Delta_n\,\omega(\Delta_{n-1})}{\omega_0(a - 2\,t_\infty)}$$

*and*

$$\parallel x_n - x_\infty \parallel \leq t_\infty - t_n.$$

*Furthermore, $x_\infty$ is the only solution in $U(x_0, a - t_\infty)$.*

We would like to known how small $q_0 \in Q$ should be to force the series $\sum \delta_n$ or, equivalently, the sequence $\{t_n\}$ to converge. That is, we need to find the convergence domain $Q_c$ of generator (2.364). Let

$$Q(t) = \{(t_0, \beta_0, \delta_0)\,:\, t_\infty = t_\infty(t_0, \beta_0, \delta_0) = t\}$$

denotes the attraction bassin of the fixed point $(t, 0, 0)$ of generator (2.364). Clearly, we have that

$$Q_c = \bigcup_{0 \leq t \leq .5\ a} Q(t). \tag{2.369}$$

For all triples $(t', t, \beta)$ with $0 \leq t \leq t' \leq .5\ a$ and $\beta \geq 0$, the equation

$$t_\infty(t, \beta, \delta) = t' \tag{2.370}$$

is uniquely solvable for $\delta$ $\delta = X_{t'}(t, \beta)$. The function $X_{t'}(t, \beta)$ uniquely solves the system

$$X(t, \beta)\,\frac{\omega(a - 2\,t + \beta) - \omega(a - 2\,t - X(t, \beta))}{\omega_0(a - 2\,t - X(t, \beta))} = X(t + X(t, \beta), X(t, \beta)) \tag{2.371}$$

and

$$X(t', \beta) = 0.$$

**Application 2.1** *If $\omega(s) = c\ s$ and $\omega_0(s) = c_0\ s$, the generator (2.364) reduces to*

$$t_+ = t + \delta, \quad \beta_+ = \delta, \quad \delta_+ = \lambda\ \delta\ \frac{\beta_+\ \delta}{c_0^{-1} - \beta_0 - 2\ t - \delta}, \quad \lambda = \frac{c}{c_0}. \tag{2.372}$$

*The change of variables $(u, v, w) = \dfrac{1}{c_0^{-1} - \beta_0}\ (t, \beta, \delta)$ leads to*

$$u_+ = u + w, \quad v_+ = w, \quad w_+ = \lambda\ w\ \frac{v + w}{1 - 2\ u - w}. \tag{2.373}$$

*In view of (2.373), the system (2.371) reduces to*

$$\begin{aligned} &\lambda\ X(u,v)\ \frac{v + X(u,v)}{1 - 2\ u - X(u,v)} = X(u + X(u,v), X(u,v)) \\ &X(u',v) = 0 \qquad (u' \in [0, .5]) \end{aligned} \tag{2.374}$$

(1) *Case $\lambda = 1$. The solution of the system is given by*

$$X_{u'}(u,v) = \frac{(.5 - u)^2 - (.5 - u')^2}{2\ (.5 - u) + v}. \tag{2.375}$$

*In view of (2.375), we have that*

$$\delta_n = X_{t'}(t_n, \beta_n) = \frac{r_n^2 - (.5\ a - t')^2}{2\ r_n + \beta_n}, \tag{2.376}$$

*where $r_n = .5\ (c^{-1} - \beta_0) - t_n$. Hence, we get that*

$$X_{.5\ a}(0, \beta_0) = \frac{(1 - c\ \beta_0)^2}{4\ c}. \tag{2.377}$$

*Condition (2.368) in Theorem 2.12 for starters $x_{-1}$, $x_0$ can be replaced by*

$$\|\ x_{-1} - x_0\ \| \le \beta_0 \le c^{-1}, \quad \|\ F(x_0)\ \| \le \frac{(1 - c\ \beta_0)^2}{4\ c}, \tag{2.378}$$

*whereas in the general case by*

$$\|\ x_{-1} - x_0\ \| \le \beta_0 \le \omega^{-1}(1 - \underline{h}), \quad \|\ F(x_0)\ \| \le \delta_0 \le X_{.5\ a}(0, \beta_0), \tag{2.379}$$

*where $X_{t'}(t, \beta)$ is the only solution of the system (2.374) and $a = \omega_0^{-1}(1 - \underline{h}) - \beta_0$.*

*Note that, for $\beta_0 = 0$, (2.378), (2.379) can replace condition (2.341) for NM, provided that the Lipschitz constant $\ell = 2c$ ($\ell_0 = 2c_0$). That is, we have for NM:*

$$\|\ F_0(x_0)\ \| \le \delta_0 \le \frac{1}{2\ \ell} \tag{2.380}$$

*or*

$$\| F_0(x_0) \| \leq \delta_0 \leq X_{.5\ a}(0,0), \tag{2.381}$$

*where* $X_{t'}(t,\beta)$ *is the only solution of the system (2.374) and* $a = \omega^{-1}(1 - \underline{h})$.

*Condition (2.380) is the famous for its simplicity and clarity Newton–Kantorovich hypothesis for solving nonlinear equations in the Lipschitz case (cf. [469]).*

(2) *Case* $\lambda \neq 1$. *The solution* $X_{t'}(t,\beta)$ *is given implicitly, and is of only theoretical use. A more direct approach however leads to a weaker sufficient convergence for NM than (2.380) given by*

$$\| F_0(x_0) \| \leq \delta_0 \leq \frac{1}{2\,L}, \tag{2.382}$$

*where*

$$L = \frac{1}{8}\,(\sqrt{\ell\,\ell_0} + 4\,\ell_0 + \sqrt{\ell_0\,\ell + 8\,\ell_0^2}) \tag{2.383}$$

*(cf. [163], [194], [195]) (see also (2.337) and section 2.1). Note that*

$$L < \ell, \tag{2.384}$$

*which improves (2.380).*

*In the case* $\omega_0(t) = c_0\, t^p$, $\omega(t) = c\, t^p$, $p \in (0,1)$, *the analogous to condition (2.382) is given by*

$$\| F_0(x_0) \| \leq \delta_0 \leq h(c_1)^{\frac{1}{p}}, \tag{2.385}$$

*where*

$$c_1 = \frac{\ell + \sqrt{\ell^2 + 4\,\ell_0\,\ell\,(1+p)^p\,p^{1-p}}}{2\,\ell} \tag{2.386}$$

*and*

$$h(c_1) = \left(1 - \frac{1}{t}\right)^p \frac{1+p}{\left((\ell_0\,(1+p))^{\frac{1}{1-p}} + (\ell\,t\,(t-1))^{\frac{1}{1-p}}\right)^{1-p}} \tag{2.387}$$

*(cf. [163], [195], [288]).*

Other applications and examples including the solution of nonlinear Chandrasekhar–type integral equations appearing in radiative transfer can be found in section 2.1.

## 2.7 $\omega$–convergence

This section is concerned with the application of Kantorovich–type majorants for the convergence of NM to a locally unique solution of a nonlinear equation in a Banach space setting. The Fréchet–derivative of the operator involved satisfies only a rather weak continuity condition. Using our new idea of recurrent functions, we obtain sufficient convergence conditions as well as error estimates. The results compare favorably to earlier ones (cf. [355], [614]).

We are concerned with the problem of approximating a locally unique solution $x^\star$ of perturbed equation of (2.1) in the form

$$F(x) + G(x) = 0, \tag{2.388}$$

where $F$ is defined on an open convex subset $\mathcal{D}$ of a Banach space $\mathcal{X}$ with values in a Banach space $\mathcal{Y}$ and $G : \mathcal{D} \longrightarrow \mathcal{Y}$ is a continuous operator.

We shall use the Newton–type method NTM given by

$$x_{n+1} = x_n - F'(x_n)^{-1}\ (F(x_n) + G(x)) \quad (x_0 \in \mathcal{D},\ n \geq 0), \tag{2.389}$$

to generate a sequence $\{x_n\}$ approximating $x^\star$. Here, $F'(x) \in \mathcal{L}(\mathcal{X}, \mathcal{Y})$, $(x \in \mathcal{D})$ denotes the Fréchet–derivative of operator $F$ (cf. [163]).

If $G(x) = 0$ $(x \in \mathcal{D})$, then, we obtain the popular NM studied the above sections. Zinčenko (cf. [812]), Zabrejko and Nguen (cf. [806]), Chen and Yamamoto (cf. [281]), Appell et. al. (cf. [31]), (cf. Deuflhard [325], [326]), Yamamoto (cf. [792]), Rheinboldt (cf. [637]), Dennis (cf. [313]), Cătinaş (cf. [261]), Hernández and Ezquerro (cf. [355]), Potra (cf. [596]–[600]), Proinov (cf. [613], [614]), and Argyros (cf. [88], [139], [170]) have provided a convergence analysis of NTM under various conditions. A survey of such results can be found in (cf. [163], [194], [195]) and the references there.

Let $x_0 \in \mathcal{D}$ and $R > 0$ be such that

$$U(x_0, R) \subseteq \mathcal{D}. \tag{2.390}$$

In this section, we are motivated by the elegant works by Hernández and Ezquerro (cf. [355]), Proinov (cf. [613], [614]), and optimization considerations. They provided a semilocal convergence analysis using the following conditions:

$(\mathcal{C})$:

$F'(x_0)^{-1} \in \mathcal{L}(\mathcal{Y}, \mathcal{X})$ and, for any $x, y \in \overline{U}(x_0, r)$ $(0 < r \leq R)$

$$\begin{gathered} \| F'(x_0)^{-1}\ (F'(x) - F'(y)) \| \leq \omega(\| x - y \|), \\ \omega(t\, r) \leq h(t)\, \omega(r), \quad t \in [0,1], \quad r \in [0, R], \\ \| F'(x_0)^{-1}\ (G(x) - G(y)) \| \leq \omega_1(\| x - y \|)\ \| x - y \|, \\ \| F'(x_0)^{-1}\ (F(x_0) + G(x_0)) \| \leq \eta \end{gathered} \tag{2.391}$$

and scalar iteration $\{v_n\}$ $(n \geq 0)$ given by

$$v_0 = 0, \quad v_1 = \eta,$$

$$v_{n+1} = v_n + \frac{H\ \omega(v_n - v_{n-1}) + \omega_1(v_n - v_{n-1})}{1 - \omega(v_n)}\ (v_n - v_{n-1}) \quad (n \geq 1),$$

as the majorizing sequence for $\{x_n\}$. Here, $\omega$, $\omega_1$ are non–decreasing, non–negative function defined on interval $[0, R]$. $h$ is a function on $[0, 1]$ and

$$H = \int_0^1 h(t)\,dt.$$

Condition (2.391) has been succesfully used to sharpen the error bounds obtained for particular expressions (cf. [355]) (see also [614, section 7]). Note that such a function $h$ always exists. Indeed, if $\omega$ is a nonzero function on $\mathcal{J} = [0, +\infty)$, then one can define function $h : [0, 1] \longrightarrow \mathbb{R}$ by

$$h(s) = \sup \left\{ \frac{\omega(s\,t)}{\omega(t)} \; : \; t \in [0, \infty) \quad with\ \omega(t) > 0 \right\}.$$

Clearly, function $h$ so defined satisfies (2.391) and has the following properties (cf. [614])

- $h(0) = 0$, $h(1) = 1$ provided that $\omega(0) = 0$;
- $h$ is nondecreasing on $[0, 1]$ provided that $\omega$ is nondecreasing on $\mathcal{J}$;
- $h$ is continuous on $[0, 1]$ provided that $\omega$ is nondecreasing on $\mathcal{J}$;
- $h$ is identical to 1 on $[0, 1]$ if $\omega$ is non–decreasing on $\mathcal{J}$ and $\omega(0) > 0$.

Several choices of function $h$ can be found in (cf. [614]). Recently (cf. [203]), we provided a finer convergence than in (cf. [355], [614]), under the same computational cost using the following set of conditions:

$(\mathcal{H})$:

$F'(x_0)^{-1} \in \mathcal{L}(\mathcal{Y}, \mathcal{X})$ and, for any $x, y \in \overline{U}(x_0, r)$ $(0 < r \leq R)$

$$\begin{aligned}
&\| F'(x_0)^{-1} \, (F'(x) - F'(y)) \| \leq \omega(\| x - y \|), \\
&\quad \omega(t\,r) \leq h(t)\,\omega(r) \quad (t \in [0, 1], \; r \in [0, R]), \\
&\| F'(x_0)^{-1} \, (F'(x) - F'(x_0)) \| \leq \omega_0(\| x - x_0 \|), \qquad (2.392) \\
&\| F'(x_0)^{-1} \, (G(x) - G(y)) \| \leq \omega_1(\| x - y \|) \; \| x - y \|, \\
&\| F'(x_0)^{-1} \, (F(x_0) + G(x_0)) \| \leq \eta,
\end{aligned}$$

and scalar iteration $\{t_n\}$ $(n \geq 0)$ given by

$t_0 = 0, \quad t_1 = \eta$

$$t_{n+1} = t_n + \frac{H\;\omega(t_n - t_{n-1}) + \omega_1(t_n - t_{n-1})}{1 - \omega_0(t_n)} \; (t_n - t_{n-1}) \quad (n \geq 1), \quad (2.393)$$

as the majorizing sequence for $\{x_n\}$. Here, $\omega_0$ is a non–decreasing, non–negative function on $[0, R]$. Condition (2.392) is not an additional hypothesis, since in practice, the computation of function $\omega$ requires that of $\omega_0$. Note also that

$$\omega_0(r) \leq \omega(r) \qquad (r \in [0, R]) \qquad (2.394)$$

holds in general and $\dfrac{\omega}{\omega_0}$ can be arbitrarily large.

In this section, we still use the set of hypotheses $(\mathcal{H})$, but we generate new convergence conditions for majorizing sequence $\{t_n\}$. This way we provide an even finer analysis than in the previously stated approaches.

We need to define some sequences and functions.

**Definition 2.3** *Let constants $\eta$, $H$ and functions $h_0$, $\omega_0$, $\omega$ and $\omega_1$ be as in the introduction of this section. Let scalar sequence $\{t_n\}$ given by (2.393). Define functions $f_n$, $g_n$, $p_n$ on $[0,1)$ and $q$ on $I_q = [0,1) \times [0,\eta] \times \left[\eta, \frac{\eta}{1-s}\right]^3$ by*

$$f_n(s) = H\,\omega(s^{n-1}\,\eta) + \omega_1(s^{n-1}\,\eta) + s\,(\omega_0((1+s+\cdots+s^{n-1})\,\eta) - 1), \tag{2.395}$$

$$g_n(s) = H\,(\omega(s^n\,\eta) - \omega(s^{n-1}\,\eta)) + \omega_1(s^n\,\eta) - \omega_1(s^{n-1}\,\eta) \\ +s\,(\omega_0((1+s+\cdots+s^n)\,\eta) - \omega_0((1+s+\cdots+s^{n-1})\,\eta)), \tag{2.396}$$

$$p_n(s) = H\,(\omega(s^{n+1}\,\eta) + \omega(s^{n-1}\,\eta) - 2\,\omega(s^n\,\eta)) \\ +\omega_1(s^{n+1}\,\eta) + \omega_1(s^{n-1}\,\eta) - 2\,\omega_1(s^n\,\eta) \\ +s\,(\omega_0((1+s+\cdots+s^{n+1})\,\eta) + \omega_0((1+s+\cdots+s^{n-1})\,\eta) \\ -2\,\omega_0((1+s+\cdots+s^n)\,\eta)) \tag{2.397}$$

*and*

$$q(s,\lambda,\beta_0,\beta,\gamma_0) = H\,(\omega(s^2\,\lambda) + \omega(\lambda) - 2\,\omega(s\,\lambda)) \\ +\,\omega_1(s^2\,\lambda) + \omega_1(\lambda) - 2\,\omega_1(s\,\lambda) \\ +\,s\,(\omega_0(\beta_0+\beta+\gamma_0) + \omega_0(\beta) - 2\,\omega_0(\beta_0+\beta)). \tag{2.398}$$

*Define function $f_\infty$ on $[0,1)$ by*

$$f_\infty(s) = \lim_{n\longrightarrow\infty} f_n(s). \tag{2.399}$$

*It follows from (2.395) and (2.399) that*

$$f_\infty(s) = \omega_0\left(\frac{\eta}{1-s}\right) - 1. \tag{2.400}$$

*It follows from (2.395)–(2.398) that*

$$f_{n+1}(s) = f_n(s) + g_n(s), \tag{2.401}$$

$$h_{n+1}(s) = g_n(s) + p_n(s) \tag{2.402}$$

*and, for*

$$\lambda = s^{n-1}\,\eta, \quad \beta_0 = s^{n+1}\,\eta, \quad \beta = \eta\sum_{i=0}^{n-1} s^i, \quad \gamma_0 = s^n\,\eta, \tag{2.403}$$

*we have*

$$q(s, \lambda, \beta_0, \beta, \gamma_0) = p_n(s). \tag{2.404}$$

We need the following result on majorizing sequences for NTM.

**Lemma 2.10** *Let iteration $\{t_n\}$ given by (2.393) and functions $f_n$, $g_n$, $p_n$, $q$ be as given in Defintion 2.3. Assume there exists $\alpha \in (0, 1)$ such that*

$$\omega_0(\eta) < 1, \tag{2.405}$$

$$\frac{H\,\omega(\eta) + \omega_1(\eta)}{1 - \omega_0(\eta)} \leq \alpha, \tag{2.406}$$

$$q(s, \lambda, \beta_0, \beta, \gamma_0) \geq 0 \quad \text{on} \quad I_q, \tag{2.407}$$

$$g_1(\alpha) \geq 0 \tag{2.408}$$

*and*

$$f_\infty(\alpha) \leq 0. \tag{2.409}$$

*Then, scalar iteration $\{t_n\}$ $(n \geq 0)$, is well defined, non–decreasing, bounded from above by*

$$t^{\star\star} = \frac{\eta}{1 - \alpha} \tag{2.410}$$

*and converges to its unique least upper bound $t^\star$ satisfying*

$$t^\star \in [0, t^{\star\star}]. \tag{2.411}$$

*Moreover, the following estimates hold for all $n \geq 0$*

$$0 \leq t_{n+1} - t_n \leq \alpha\,(t_n - t_{n-1}) \leq \alpha^n\,\eta \tag{2.412}$$

*and*

$$0 \leq t^\star - t_n \leq \frac{\eta}{1 - \alpha}\,\alpha^n. \tag{2.413}$$

**Proof** Estimate (2.412) is true if

$$0 \leq \frac{H\,\omega(t_n - t_{n-1}) + \omega_1(t_n - t_{n-1})}{1 - \omega_0(t_n)} \leq \alpha \tag{2.414}$$

hold for all $n \geq 1$. In view of (2.393), (2.405) and (2.406), estimate (2.414) holds for $n = 1$. We also have by (2.414) that $0 \leq t_2 - t_1 \leq \alpha\,(t_1 - t_0)$. That is (2.412) holds for $n = 1$. Let us assume that (2.412) and (2.414) hold for all $k \leq n$. Then, we have by (2.412) that

$$t_n \leq \frac{1 - \alpha^n}{1 - \alpha}\,\eta. \tag{2.415}$$

Using the induction hypotheses and (2.415), we see that (2.414) certainly holds if

$$H\,\omega(\alpha^{n-1}\,\eta)+\omega_1(\alpha^{n-1}\,\eta)\leq\alpha\left(1-\omega_0\left(\frac{1-\alpha^n}{1-\alpha}\,\eta\right)\right) \tag{2.416}$$

or

$$H\,\omega(\alpha^{n-1}\,\eta)+\omega_1(\alpha^{n-1}\,\eta)+\alpha\,\omega_0\left(\frac{1-\alpha^n}{1-\alpha}\,\eta\right)-\alpha\leq 0.$$

Let $s=\alpha$. Estimates (2.416) motivates us to define function $f_n$ given by (2.395) and show instead of (2.416)

$$f_n(\alpha)\leq 0. \tag{2.417}$$

We have by (2.401)–(2.404) (for $s=\alpha$) and (2.408) that

$$f_{n+1}(\alpha)\geq f_n(\alpha)\qquad (n\geq 1). \tag{2.418}$$

In view of (2.399), (2.400) and (2.418), estimate (2.417) holds if (2.409) is true, since

$$f_n(\alpha)\leq f_\infty(\alpha)\qquad (n\geq 1). \tag{2.419}$$

The induction is completed. It follows that iteration $\{t_n\}$ is non–decreasing, bounded from above by $t^{\star\star}$ (given by (2.410)) and as such it converges to its unique least upper bound $t^\star$ satisfying (2.411). Finally, estimate (2.413) follows from (2.412) by using standard majorization techniques (cf. [163], [194], [195]). This completes the proof of Lemma 2.10. ⊠

Hypotheses $(\mathcal{H})$ and those of Lemma 2.395 will be called from now $(\mathcal{A})$. We provide the main semilocal convergence result for NTM.

**Theorem 2.13** *Assume hypotheses $(\mathcal{A})$ hold. Then, sequence $\{x_n\}$ $(n\geq 0)$ generated by NTM is well defined, remains in $\overline{U}(x_0,t^\star)$ for all $n\geq 0$ and converges to a solution $x^\star\in\overline{U}(x_0,t^\star)$ of equation $F(x)+G(x)=0$. Moreover, the following estimates hold for all $n\geq 0$*

$$\|\, x_n-x^\star\,\|\leq t^\star-t_n. \tag{2.420}$$

*Furthermore, if there exists*

$$R_0\in[t^\star,R] \tag{2.421}$$

*such that*

$$H\,\omega(R_0)+\omega_1(R_0)+\omega_0(t^\star)<1, \tag{2.422}$$

*then, $x^\star$ is the unique solution of equation $F(x)+G(x)=0$ in $\overline{U}(x_0,R_0)$.*

**Proof** We shall show using indution on $k$ the following assertions

$$\|\, x_k-x_{k-1}\,\|\leq t_k-t_{k-1} \tag{2.423}$$

and

$$\| x_k - x_0 \| \le t_k. \tag{2.424}$$

Estimate (2.423) holds by (2.393), whereas (2.424) is true for $k = 1$ as equality. Let us assume (2.423) and (2.424) hold for all $m \le k$. Using hypotheses (2.392) and (2.405), we get

$$\| F'(x_0)^{-1} \, (F'(x_1) - F'(x_0)) \| \le \omega_0(\| x_1 - x_0 \|) \; \le \omega_0(\eta) < 1. \tag{2.425}$$

It follows from (2.425) and the Banach lemma on invertible operators (cf. [163], [194], [195]) that $F'(x_0)^{-1}$ exists and

$$\| F'(x_1)^{-1} \, F'(x_0) \| \le \frac{1}{1 - \omega_0(\| x_1 - x_0 \|)} \le \frac{1}{1 - \omega_0(\eta)}. \tag{2.426}$$

We also showed in Lemma 2.10 that

$$\omega_0(t_k) < 1. \tag{2.427}$$

It follows as in (2.425) with $t_k$, $x_k$ replacing $t_1$, $x_1$ that $F'(x_k)^{-1}$ exists and

$$\| F'(x_k)^{-1} \, F'(x_0) \| \le \frac{1}{1 - \omega_0(\| x_k - x_0 \|)} \le \frac{1}{1 - \omega_0(t_k)}. \tag{2.428}$$

Using (2.389), hypotheses $(\mathcal{H})$ and (2.428), we get in turn that

$$\begin{aligned}
\| x_{k+1} - x_k \| &= \| F'(x_k)^{-1} \, (F(x_k) + G(x_k)) \| \\
&\le \| F'(x_k)^{-1} \, F'(x_0) \| \, \| F'(x_0) \, \big(F(x_k) + G(x_k) \\
&- F'(x_{k-1}) \, (x_k - x_{k-1}) - F(x_{k-1}) - G(x_{k-1})\big) \| \\
&\le \frac{1}{1 - \omega_0(t_k)} \left( \int_0^1 \| F'(x_0)^{-1} \, (F'(x_{k-1} + t \, (x_k - x_{k-1})) \right. \\
&- F'(x_{k-1})) \| \, \| x_k - x_{k-1} \| \, dt \\
&\left. + \| F'(x_0)^{-1} \, (G(x_k) - G(x_{k-1})) \| \right) \qquad (2.429) \\
&\le \frac{1}{1 - \omega_0(t_k)} \left( \int_0^1 \omega(t \, \| x_k - x_{k-1} \|) \, dt \right. \\
&\left. + \omega_1(\| x_k - x_{k-1} \| \right) \| x_k - x_{k-1} \| \\
&\le \frac{H \, \omega(t_k - t_{k-1}) + \omega_1(t_k - t_{k-1})}{1 - \omega_0(t_k)} \, (t_k - t_{k-1}) = t_{k+1} - t_k,
\end{aligned}$$

which shows (2.423) for all $k \geq 1$. Moreover, we have that

$$\begin{aligned} \| x_{k+1} - x_0 \| &\leq \| x_{k+1} - x_k \| + \| x_k - x_0 \| \\ &\leq t_{k+1} - t_k + t_k = t_{k+1} \leq t^\star. \end{aligned} \tag{2.430}$$

The induction for (2.423) and (2.424) is completed.

In view of Lemma 2.10, (2.423) and (2.424), sequence $\{x_n\}$ is a Cauchy sequence in a Banach space $\mathcal{X}$ and as such it converges to some $x^\star \in \overline{U}(x_0, t^\star)$ (since $\overline{U}(x_0, t^\star)$ is a closed set). Using (2.429), we get that

$$\| F'(x_0)^{-1} \, (F(x_k) + G(x_k)) \| \leq (H \, \omega(t_k - t_{k-1}) + \omega_1(t_k - t_{k-1})) \, \omega(t_k - t_{k-1}). \tag{2.431}$$

By letting $k \longrightarrow \infty$ in (2.431), we obtain $F(x^\star) + G(x^\star) = 0$. Estimate (2.420) follows from (2.423) by using standard majorization techniques (cf. [163], [194], [195]).

Finally, to show the uniqueness of $x^\star$ in $\overline{U}(x_0, R_0)$, let us assume $y^\star$ is a solution in $\overline{U}(x_0, R_0)$. Using the approximation

$$x^\star - x_{k+1} = x^\star - x_k + F'(x_k)^{-1} \, (F(x_k) + G(x_k)) - F'(x_k)^{-1} \, (F(y^\star) + G(y^\star)) \tag{2.432}$$

as in (2.429), we obtain in turn that

$$\begin{aligned} \| y^\star - x_{k+1} \| &\leq \frac{1}{1 - \omega_0(t_k)} \left( \int_0^1 \| F'(x_0)^{-1} \, (F'(x_{k-1} + t \, (x^\star - x_k)) \right. \\ &\quad \left. - F'(x_k)) \| \| x^\star - x_k \| \, dt + \| F'(x_0)^{-1} \, (G(y^\star) - G(x_k)) \| \right) \\ &\leq \frac{1}{1 - \omega_0(t_k)} \left( \int_0^1 \omega(t \, \| y^\star - x_k \|) \, dt + \omega_1(\| y^\star - x_k \|) \right) \| y^\star - x_k \| \\ &\leq \frac{H \, \omega(R_0) + \omega_1(R_0)}{1 - \omega_0(t^\star)} \, \| y^\star - x_k \| \\ &< \| y^\star - x_k \|, \quad \text{(by (2.422))}, \end{aligned} \tag{2.433}$$

which implies $\lim_{k \to \infty} x_k = y^\star$. But we showed $\lim_{k \to \infty} x_k = x^\star$. Hence, we deduce that $x^\star = y^\star$. This completes the proof of Theorem 2.13. ⊠

**Remark 2.20** *(a) The limit point $t^\star$ can be replaced in Theorem 2.13 by $t^{\star\star}$ given in the closed form by (2.410).*

*(b) Note that more general conditions than the ones given in (cf. [614]) and ours (introduced in this section) were provided in (cf. [139]) to show the local (and semilocal) convergence of two–point Newton–like methods (see also (cf. [163], [194], [468])).*

Let us consider the case of NM. That is set $G(x) = 0$ for all $x \in \mathcal{D}$ and $\omega_1(s) = 0$ for all $s \in [0, 1]$. In this case, we note that

$$\omega_0(s) \leq \omega(s) \quad (s \in [0, 1]), \tag{2.434}$$

holds and $\dfrac{\omega}{\omega_0}$ can be arbitrarily large. Comparison with a result by Ezquerro and Hernández (cf. [355]) (see also Proinov [614, Theorem 7.3]) If one reproduces these results in affine invariant form, then, the corresponding to $\{t_n\}$ majorizing sequence is essentially given by

$$v_0 = 0, \quad v_1 = \eta,$$

$$v_{n+1} = v_n + \frac{H\ \omega(v_n - v_{n-1}) + \omega_2(v_{n-1})}{1 - \omega(v_n)}\ (v_n - v_{n-1}) \quad (n \geq 1). \tag{2.435}$$

Then, under the hypotheses of our Theorem 2.13 and the corresponding ones in (cf. [355], [614]), we note the following advantages.

(i) Majorizing sequence $\{t_n\}$ is tighter than $\{v_n\}$, since an inductive argument shows:

$$t_n < v_n \quad (n \geq 2) \tag{2.436}$$

and

$$t_{n+1} - t_n < v_{n+1} - v_n \quad (n \geq 2). \tag{2.437}$$

(ii) The information on the location of the solution is at least as precise, since

$$t^\star = \lim_{n\to\infty} t_n \leq \lim_{n\to\infty} v_n = v^\star. \tag{2.438}$$

It turns out our sufficient convergence conditions are weaker. Indeed, for simplicity, let us consider the case, when

$$\omega(s) = L\ s, \quad \omega_0(s) = L_0\ s \text{ and } \quad h(t) = \frac{1}{2}. \tag{2.439}$$

Then, iterations $\{t_n\}$, $\{v_n\}$ become

$$t_0 = 0, \quad t_1 = \eta, \quad t_{n+1} = t_n + \frac{L\ (t_n - t_{n-1})^2}{2\ (1 - L_0\ t_n)} \quad (n \geq 1), \tag{2.440}$$

$$v_0 = 0, \quad v_1 = \eta, \quad v_{n+1} = v_n + \frac{L\ (v_n - v_{n-1})^2}{2\ (1 - L\ v_n)} \quad (n \geq 1). \tag{2.441}$$

Note that iteration (2.441) converges if the famous for its simplicity and clarity Newton–Kantorovich hypothesis

$$K = L\ \eta \leq \frac{1}{2} \tag{2.442}$$

holds (cf. [468]). However, iteration (2.440) converges under weaker conditions provided that $L_0 < L$.

We need the following result for the convergence of majorizing sequence $\{t_n\}$.

**Lemma 2.11** *(cf. [203]) Assume there exist constants $L_0 \geq 0$, $L \geq 0$ with $L_0 \leq L$ and $\eta \geq 0$ such that*

$$q_0 = \overline{L}\,\eta \leq \frac{1}{2}, \tag{2.443}$$

*where*

$$\overline{L} = \frac{1}{8}\left(L + 4\,L_0 + \sqrt{L^2 + 8\,L_0\,L}\right). \tag{2.444}$$

*Then, sequence $\{t_k\}$ $(k \geq 0)$ given by (2.440) is well defined, nondecreasing, bounded above by $t^{\star\star}$ and converges to its unique least upper bound $t^\star \in [0, t^{\star\star}]$, where*

$$t^{\star\star} = \frac{2\,\eta}{2-\delta}, \tag{2.445}$$

$$1 \leq \delta = \frac{4\,L}{L + \sqrt{L^2 + 8\,L_0\,L}} < 2 \quad (L_0 \neq 0). \tag{2.446}$$

*Moreover, the following estimates hold*

$$L_0\,t^\star \leq 1, \tag{2.447}$$

$$0 \leq t_{k+1} - t_k \leq \frac{\delta}{2}\,(t_k - t_{k-1}) \leq \cdots \leq \left(\frac{\delta}{2}\right)^k \eta \quad (k \geq 1), \tag{2.448}$$

$$t_{k+1} - t_k \leq \left(\frac{\delta}{2}\right)^k (2\,q_0)^{2^k - 1}\,\eta \quad (k \geq 0), \tag{2.449}$$

$$0 \leq t^\star - t_k \leq \left(\frac{\delta}{2}\right)^k \frac{(2\,q_0)^{2^k-1}\,\eta}{1 - (2\,q_0)^{2^k}} \quad (2\,q_0 < 1, \quad k \geq 0). \tag{2.450}$$

**Remark 2.21** *If $L_0 = L$, Lemma 2.11 provides the usual error bounds appearing essentially in the Newton–Kantorovich theorem (cf. [468]). However, if $L_0 < L$, then, our sufficient convergence condition (2.443) is weaker than (2.442).*

*Finally, our ratio $2\,q_0$ is also smaller than $2\,K$. Hence, the $\omega$–function approach (cf. [355], [614]) does not necessarily produce the weakest sufficient convergence conditions not even in the simplest case of NM under the simple Lipschitz conditions (2.439). Moreover, in the general case, our approach takes advantage of estimate (2.434). Note that the more precise than $\omega$, function $\omega_0$ does not appear in the convergence analysis of NTM (cf. [355], [614]). Furthemore, the ration $\alpha$ of linear convergence is given for iteration $\{t_n\}$, but not for $\{v_n\}$. A weaker than (2.443) condition was given in section 2.1.*

Finally, we note that relevant results have been given by us for more general Newton–type method in (cf. [194]) under different sufficient convergence conditions. In the special case of NTM, our conditions are different from (cf. [355], [614]). Moreover, we provide the ratio of linear convergence, not given with the approach in (cf. [355], [614]).

## 2.8 Semilocal Convergence and Convex Majorants

We provide in this section new sufficient convergence conditions for NM using majorants and recurrent functions. The advantages of this approach over earlier works (cf. [162], [167], [367], [751]) (under the same computational cost) are tighter error bounds under the same or weaker hypotheses.

We use NM given by (2.2) for approximating a locally unique solution $x^\star$ of equation (2.1), where $F$ is a Fréchet–differentiable operator defined on a convex subset $\mathcal{D}$ of a Banach space $\mathcal{X}$ with values in a Banach space $\mathcal{Y}$. Recently, Argyros (cf. [162]) provided some results using convex majorants which improved and extended the corresponding ones given by Ferreira, Svaiter (cf. [367]) and Wang (cf. [751]). Here, we use new concept of recurrent functions in combination with not necessarily convex majorants to provide new sufficient convergence conditions for NM which compare favorably with the works already mentioned above, in the sense that tighter error bounds can be obtained under weaker hypotheses. Moreover, in some interesting special cases such as in the famous Kantorovich theorem for solving nonlinear equations, our sufficient convergence conditions are at least as weak. The results are also extended to include a non–differentiable term in (2.1).

We need to define some parameters, functions and scalar sequences.

**Definition 2.4** *Let $x_0 \in \mathcal{D}$ and $c \in [0,1]$. Let $f, g : [0,R) \longrightarrow (-\infty,+\infty)$ be a two continuously differentiable functions such that $f(0) > 0$, $g(0) > 0$ and $f'$, $g'$, $f'(r+t) - f'(r)$, $g'(r) - g'(0)$ ($r \geq t \geq 0$ and $0 \leq r+t < R$) are non–decreasing. Define parameter $\eta$, sequence $\{t_n\}$, functions $f_n$, $h_n$, $p_n$ on $[0,1)$ and $q$ on $\mathcal{I}_q = [0,1)^2 \times [1, \frac{1}{1-s}]$ ($s \in [0,1)$) by*

$$\eta \geq \| F'(x_0)^{-1} F(x_0) \| > 0, \tag{2.451}$$

$$\begin{cases} t_0 = 0, \quad t_1 = \eta, \\ t_{n+2} = t_{n+1} + \dfrac{\displaystyle\int_0^1 (f'(t_n + \theta\,(t_{n+1} - t_n)) - f'(t_n))\,d\theta\,(t_{n+1} - t_n)}{1 - c - g'(t_{n+1})}, \end{cases} \tag{2.452}$$

$$\begin{aligned} f_n(s) = & \int_0^1 \left( f'\!\left(\frac{1-s^n}{1-s}\,\eta + \eta\,\theta\,s^n\right) - f'\!\left(\frac{1-s^n}{1-s}\,\eta\right)\right) d\theta \\ & + s\,g'\!\left(\frac{1-s^{n+1}}{1-s}\,\eta\right) - s\,(1-c), \end{aligned} \tag{2.453}$$

$$h_n(s) = \int_0^1 \left( f'\!\left(\frac{1-s^{n+1}}{1-s}\,\eta + \eta\theta\,s^{n+1}\right) - f'\!\left(\frac{1-s^n}{1-s}\,\eta + \eta\,\theta\,s^n\right)\right.$$

$$+f'\left(\frac{1-s^n}{1-s}\eta\right)-f'\left(\frac{1-s^{n+1}}{1-s}\eta\right)\Bigg)\,d\theta$$

$$+\,s\left(g'\left(\frac{1-s^{n+2}}{1-s}\eta\right)-g'\left(\frac{1-s^{n+1}}{1-s}\eta\right)\right), \tag{2.454}$$

$$p_n(s)=\int_0^1\Bigg(f'\left(\frac{1-s^{n+2}}{1-s}\eta+\eta\,\theta\,s^{n+2}\right)+f'\left(\frac{1-s^n}{1-s}\eta+\eta\theta\,s^n\right)$$

$$-\,2\,f'\left(\frac{1-s^{n+1}}{1-s}\eta+\eta\theta\,s^{n+1}\right)+2\,f'\left(\frac{1-s^{n+1}}{1-s}\eta\right) \tag{2.455}$$

$$-\,f'\left(\frac{1-s^{n+2}}{1-s}\eta\right)-f'\left(\frac{1-s^n}{1-s}\eta\right)\Bigg)\,d\theta+s\,\Bigg(g'\left(\frac{1-s^{n+3}}{1-s}\eta\right)$$

$$+\,g'\left(\frac{1-s^{n+1}}{1-s}\eta\right)-2\,g'\left(\frac{1-s^{n+2}}{1-s}\eta\right)\Bigg)$$

*and*

$$q(s,b,a)=\int_0^1 (f'((a+b+b\,s)\,\eta+\eta\theta b\,s^2)+f'(a\eta+\eta\theta s^n)$$

$$-\,2f'((a+b)\eta+\eta\theta bs)+2\,f'((a+b)\eta)-f'((a+b+b\,s)\eta) \tag{2.456}$$

$$-\,f'(a\eta))d\theta+s\,(g'((a+(1+s+s^2)b)\eta)+g'((a+b)\eta)$$

$$-\,2g'((a+b+bs)\eta)).$$

*Moreover, define function $f_\infty$ on $[0,1)$ by*

$$f_\infty(s)=\lim_{n\longrightarrow\infty} f_n(s). \tag{2.457}$$

**Remark 2.22** *It follows from (2.452) and (2.457) that*

$$f_\infty(s)=s\left(g'\left(\frac{\eta}{1-s}\right)-(1-c)\right). \tag{2.458}$$

*Using (2.452), (2.453), (2.456) and (2.457), we obtain the following identities*

$$f_{n+1}(s)=f_n(s)+h_n(s), \tag{2.459}$$

$$h_{n+1}(s)=h_n(s)+p_n(s) \tag{2.460}$$

*and*

$$p_n(s)=q(s,s^n\,\eta,(1+s+\cdots+s^n)\,\eta). \tag{2.461}$$

We can show the following result on majorizing sequences for NM.

**Lemma 2.12** *Let $\eta$, $c$, $f_n$, $h_n$, $p_n$, $q$ and $f_\infty$ be as in Definition 2.4. Assume there exists $\alpha \in (0,1)$ such that*

$$\int_0^1 (f'(\eta\theta) - f'(0))\, d\theta + \alpha\, g'(\eta) - \alpha\,(1-c) \leq 0, \tag{2.462}$$

$$q(s,b,a) \geq 0 \quad \text{on} \quad \mathcal{I}_q, \tag{2.463}$$

$$h_1(\alpha) \geq 0, \tag{2.464}$$

$$f_\infty(\alpha) \leq 0. \tag{2.465}$$

*Then, scalar sequence $\{t_n\}$ given by (2.452) is non–decreasing, bounded from above by*

$$t^{\star\star} = \frac{\eta}{1-\alpha} \tag{2.466}$$

*and converges to its unique least upper bound $t^\star$ satisfying*

$$t^\star \in [0, t^{\star\star}]. \tag{2.467}$$

*Moreover, the following estimates hold for all $n \geq 0$*

$$0 \leq t_{n+1} - t_n \leq \alpha\,(t_n - t_{n-1}) \leq \alpha^n\, \eta \tag{2.468}$$

*and*

$$0 \leq t^\star - t_n \leq \frac{\alpha^n\, \eta}{1-\alpha}. \tag{2.469}$$

**Proof** Estimate (2.468) is true if

$$\int_0^1 (f'(t_n + \theta\,(t_{n+1} - t_n)) - f'(t_n))\, d\theta + \alpha\, g'(t_{n+1}) - \alpha\,(1-c) \leq 0 \tag{2.470}$$

holds for all $n \geq 0$. Note also that (2.470) implies that the denominator in (2.452) is positive. In view of (2.462), estimate (2.470) holds for $n = 0$. We also have that (2.468) holds for $n = 0$. Let us assume (2.468) and (2.470) hold for all $k \leq n$. Then, we have

$$t_n \leq \frac{1-\alpha^n}{1-\alpha}\, \eta < t^{\star\star}. \tag{2.471}$$

Using the induction hypotheses and (2.473), we see that (2.470) holds if

$$\int_0^1 \left( f'\left(\frac{1-\alpha^n}{1-\alpha}\, \eta + \alpha^n\, \eta\theta\right) - f'\left(\frac{1-\alpha^n}{1-\alpha}\, \eta\right)\right) d\theta + \alpha\, g'\left(\frac{1-\alpha^{n+1}}{1-\alpha}\, \eta\right) - \alpha\,(1-c) \leq 0. \tag{2.472}$$

Estimate (2.472) (for $s = \alpha$) motivates us to introduce function $f_k$ given by (2.452) and show instead of (2.472)

$$f_n(\alpha) \leq 0 \quad (n \geq 1). \tag{2.473}$$

We have by (2.459)–(2.461) (for $s = \alpha$) and (2.464) that

$$f_{n+1}(\alpha) \geq f_n(\alpha) \quad (n \geq 1). \tag{2.474}$$

Using (2.457), (2.458) and (2.474), we see that estimate (2.473) shall holds if (2.465) is true, since

$$f_n(\alpha) \leq f_\infty(\alpha) \quad (n \geq 1) \tag{2.475}$$

and the induction is completed.

It follows from (2.468) and (2.471) that sequence $\{t_n\}$ is non–decreasing, bounded from above by $t^{\star\star}$ and as such it converges to $t^\star$.

Finally, estimate (2.469) follows from (2.468) by using standard majorization techniques. This completes the proof of Lemma 2.12. ⊠

We state the main semilocal convergence for NM.

**Theorem 2.14** *Let $F : \mathcal{D} \subseteq \mathcal{X} \longrightarrow \mathcal{Y}$ be a Fréchet–differentiable operator and let $x_0 \in \mathcal{D}$ be such that $F'(x_0)^{-1} \in \mathcal{L}(\mathcal{Y}, \mathcal{X})$. Suppose there exist $R > 0$, $c \in [0, 1)$, $c_1 \in [0, 1)$ and continuously differentiable function $f, g : [0, R) \longrightarrow (-\infty, +\infty)$ such that*

$$U(x_0, R) \subset \mathcal{D}, \tag{2.476}$$

$$\| F'(x_0)^{-1} (F'(x) - F'(x_0)) \| \leq g'(\| x - x_0 \|) - g'(0), \tag{2.477}$$

$$\| F'(x_0)^{-1} (F'(x) - F'(y)) \| \leq f'(\| y - x \| + \| x - x_0 \|) - f'(\| x - x_0 \|), \tag{2.478}$$

$$\| x - x_0 \| + \| y - x \| < R \quad (x, y \in U(x_0, R))$$

$$f(0) > 0, \quad g(0) > 0, \quad f'(0) = -c_1, \quad g'(0) = -c,$$

*$f'$, $g'$, $f'(r + t) - f'(r)$, $g'(r) - g'(0)$ ($r \geq t \geq 0$ and $0 \leq r + t < R$) are non–decreasing and hypotheses of Lemma 2.12 hold with*

$$t^\star < R. \tag{2.479}$$

*Then, sequence $\{x_n\}$ generated by NM is well defined, remains in $\overline{U}(x_0, t^\star)$ for all $n \geq 0$ and converges to a zero $x^\star$ of (2.1) in $\overline{U}(x_0, t^\star)$. Moreover, the following estimates hold ($n \geq 1$)*

$$\| x_{n+1} - x_n \| \leq e_n \leq t_{n+1} - t_n \tag{2.480}$$

*and*

$$\| x_n - x^\star \| \leq t^\star - t_n, \tag{2.481}$$

*where*

$$e_n = \frac{\int_0^1 (f'(\| x_{n-1} - x_0 \| + \theta \| x_n - x_{n-1} \|) - f'(\| x_{n-1} - x_0 \|)) \, d\theta \, \| x_n - x_{n-1} \|^2}{(1 - c - g'(\| x_n - x_0 \|)) \, (t_n - t_{n-1})}.$$

*Furthemore, if there exists*

$$R_0 \in [t^\star, R] \tag{2.482}$$

*such that*

$$\int_0^1 (g'((1-\theta)\,R_0+\theta\,t^\star)-g'(0))\,d\theta<1, \tag{2.483}$$

*then, the solution $x^\star$ is unique in $U(x_0,R_0)$.*

**Proof** We shall show using induction on $k$ the following

$$\|\, x_k-x_{k-1}\,\| \le e_k \le t_k-t_{k-1} \tag{2.484}$$

and

$$\|\, x_k-x_0\,\| \le t_k. \tag{2.485}$$

Estimates (2.484) and (2.485) hold for $k=1$ by (2.451) and (2.452). Let $x\in \overline{U}(x_0,t^\star)$. Using (2.470), (2.476), (2.477) and hypotheses on functions $g$, $g'$, we have

$$\begin{aligned}\|\, F'(x_0)^{-1}\,(F'(x_1)-F'(x_0))\,\| &\le g'(\|\, x_1-x_0\,\|)-g'(0)\\ &\le g'(t_1-t_0)+c<1\end{aligned} \tag{2.486}$$

by Lemma 2.12. It follows from (2.486) and the Banach Lemma on invertible operators (cf. [157], [469]) that $F'(x_1)^{-1}$ is invertible with

$$\|\, F'(x_1)^{-1}\,F'(x_0)\,\| \le \frac{1}{1-c-g'(t_1)}. \tag{2.487}$$

Using NM for $n=1$, we get

$$\begin{aligned}F(x_1)&=F(x_1)-F(x_0)-F'(x_0)\,(x_1-x_0)\\ &=\int_0^1 (F'(x_0+\theta\,(x_1-x_0))-F'(x_0))\,(x_1-x_0)\,d\theta.\end{aligned} \tag{2.488}$$

It follows from NM, (2.452), (2.478), (2.487) and (2.488) that

$$\begin{aligned}\|\, x_2-x_1\,\| &= \|\,(F'(x_1)^{-1}\,F'(x_0))\,(F'(x_0)^{-1}\,F(x_1))\,\|\\ &\le e_1\\ &\le \frac{\int_0^1 (f'(\theta\,(t_1-t_0))-f'(t_0))\,d\theta\,(t_1-t_0)}{1-c-g'(t_1-t_0)}=t_2-t_1.\end{aligned} \tag{2.489}$$

Then, we get that

$$\|\, x_2-x_0\,\| \le \|\, x_2-x_1\,\|+\|\, x_1-x_0\,\| \le t_2-t_1+t_1-t_0=t_2\le t^\star, \tag{2.490}$$

which imply estimates (2.484) and (2.485) hold for $k=2$.

Let us assume (2.484) and (2.485) hold all $n\le k$. As in (2.486), we have (by Lemma 2.12)

$$\begin{aligned}\|\, F'(x_0)^{-1}\,(F'(x_k)-F'(x_0))\,\| &\le g'(\|\, x_k-x_0\,\|)-g'(0)\\ &\le g'(t_k-t_0)+c<1\end{aligned} \tag{2.491}$$

and

$$\| F'(x_k)^{-1} F'(x_0) \| \leq \frac{1}{1 - c - g'(t_k)}. \tag{2.492}$$

Using NM, we have the identity

$$\begin{aligned} F(x_k) &= F(x_k) - F(x_{k-1}) - F'(x_{k-1})(x_k - x_{k-1}) \\ &= \int_0^1 (F'(x_{k-1} + \theta(x_k - x_{k-1})) - F'(x_{k-1}))(x_k - x_{k-1})\, d\theta. \end{aligned} \tag{2.493}$$

Then, we get that

$$\begin{aligned} &\| F'(x_0)^{-1} F(x_k) \| \\ &= \left\| \int_0^1 F'(x_0)^{-1} (F'(x_{k-1} + \theta(x_k - x_{k-1})) - F'(x_{k-1}))(x_k - x_{k-1})\, d\theta \right\| \\ &\leq \int_0^1 (f'(\| x_{k-1} - x_0 \| + \theta \| x_k - x_{k-1} \|) - f'(\| x_{k-1} - x_0 \|))\, d\theta \ \| x_k - x_{k-1} \| \\ &\leq \int_0^1 (f'(t_{k-1} + \theta(t_k - t_{k-1})) - f'(t_{k-1}))\, d\theta\, (t_k - t_{k-1}), \end{aligned} \tag{2.494}$$

which implies that

$$\begin{aligned} \| x_{k+1} - x_k \| &\leq \| F'(x_k)^{-1} F'(x_0) \| \, \| F'(x_0)^{-1} F(x_k) \| \leq e_k \\ &\leq \frac{\int_0^1 (f'(t_{k-1} + \theta(t_k - t_{k-1})) - f'(t_{k-1}))\, d\theta\, (t_k - t_{k-1})}{1 - c - g'(t_k)} \\ &= t_{k+1} - t_k \end{aligned} \tag{2.495}$$

and

$$\| x_{k+1} - x_0 \| \leq \| x_{k+1} - x_k \| + \| x_k - x_0 \| \leq t_{k+1} - t_k + t_k - t_0 = t_{k+1} \leq t^\star. \tag{2.496}$$

The induction for (2.484) and (2.485) is complete. It follows from Lemma 2.12, ((2.484), and (2.485) that sequence $\{x_n\}$ $(n \geq 0)$ is a Cauchy sequence in a Banach space $\mathcal{X}$ and as such it converges to some $x^\star \in \overline{U}(x_0, t^\star)$ (since $\overline{U}(x_0, t^\star)$ is a closed set). By letting $k \longrightarrow \infty$ in (2.494), we obtain $F(x^\star) = 0$. Furthermore, estimate (2.481) is obtained from (2.480) by using standard majorization techniques.

Finally, to show that $x^\star$ is the unique solution of equation (2.1) in $\overline{U}(x_0, R_0)$, let $y^\star \in \overline{U}(x_0, R_0)$ with $F(y^\star) = 0$.

Set $\mathcal{M} = \int_0^1 F'(y^\star + \theta\,(x^\star - y^\star))\,d\theta$. Using (2.477), we have

$$\begin{aligned}\| F'(x_0)^{-1}\,(\mathcal{M} - F(x_0)) \| &\le \int_0^1 (g'(\| y^\star + \theta\,(x^\star - y^\star) - x_0 \|) - g'(0))\,d\theta \\ &\le \int_0^1 (g'((1-\theta)\ \| y^\star - x_0 \| + \theta\ \| x^\star - x_0 \|) - g'(0))\,d\theta \\ &\le \int_0^1 (g'((1-\theta)\,R_0 + \theta\,t^\star) - g'(0))\,d\theta < 1. \end{aligned} \tag{2.497}$$

It follows from (2.497) that $\mathcal{M}^{-1}$ exists. In view of the identity $0 = F(x^\star) - F(y^\star) = \mathcal{M}(x^\star - y^\star)$, we deduce $x^\star = y^\star$. This completes the proof of Theorem 2.14. ⊠

**Remark 2.23** *(a) Hypothesis (2.478) was used by Ferreira and Svaiter (cf. [367]) to provide a semilocal convergence analysis for NM. Argyros extended the applicability of NM in (cf. [162], [194]) by using a combination of (2.477) and (2.478). Note that (2.478) was also used by Wang (cf. [751]) in the form*

$$\| F'(x_0)^{-1}\,(F'(x) - F'(y)) \| \le \int_{\|x-x_0\|}^{\|x-x_0\|+\|x-y\|} L(t)\,dt,$$

*where $L$ is a non–negative, non–decreasing function on $[0, R)$.*

*(b) The limit point $t^\star$ can be replaced in (2.479), (2.481) and (2.482) by $t^{\star\star}$ (given in closed form by (2.466)).*

*(c) It follows from (2.489) that sequence $\{\overline{t}_n\}$ given by*

$$\begin{cases} \overline{t}_0 = 0, \quad \overline{t}_1 = \eta, \\ \overline{t}_{n+2} = \overline{t}_{n+1} + \dfrac{\displaystyle\int_0^1 (g'(\overline{t}_n + \theta\,(\overline{t}_{n+1} - \overline{t}_n)) - g'(\overline{t}_n))\,d\theta\,(\overline{t}_{n+1} - \overline{t}_n)}{1 - c - g'(\overline{t}_{n+1})} \end{cases}$$

*is a tighter majorizing sequence for $\{x_n\}$ than $\{t_n\}$ with*

$$\overline{t}_n \le t_n,$$

$$\overline{t}_{n+1} - \overline{t}_n \le t_{n+1} - t_n$$

*and*

$$\overline{t^\star} = \lim_{n \longrightarrow \infty} \overline{t}_n = t^\star.$$

*(d) In view of (2.478), estimate (2.477) always holds if we set $g' = f'$ on $[0, R)$ and $c = c_1$. Moreover, if $c = 1$, our result reduces to the one in (cf. [367]). However, in general, we have that*

$$g'(t) \le f'(t) \quad (t \in [0, R)) \tag{2.498}$$

*and* $\dfrac{f'(t)}{g'(t)}$ *can arbitrarily large. Hence, under the same hypotheses (with the exception of the convexity of* $f'$, $g'$*), our error estimates are tighter (see also (2.507)–(2.510)). Let us choose*

$$f(t) = \eta - (1 - c_1)\, t + \frac{L}{2}\, t^2, \quad L > 0, \tag{2.499}$$

$$g(t) = \eta - (1 - c)\, t + \frac{L_0}{2}\, t^2, \quad L_0 > 0, \tag{2.500}$$

*with*

$$L_0 \leq L \tag{2.501}$$

*and*

$$c \leq c_1, \tag{2.502}$$

*then, estimate (2.498) holds.*

(e) *The integrals in (2.483), (2.489), (2.495) and (2.497) can easily be computed using the Newton–Leibnitz theorem. However, we decided to leave them as uncluttered (short) as possible. Note also that the convexity of* $f'$ *and* $g'$*, which is a usual hypothesis (cf. [367], [751]) is not used in this section.*

(f) *The results obtained in this section can be extended to hold for the equation*

$$F(x) + G(x) = 0, \tag{2.503}$$

*where* $F$ *is as in the introduction and* $G : \mathcal{D} \longrightarrow \mathcal{Y}$ *is a continuous operator satisfying*

$$\| F'(x_0)^{-1}\, (G(x) - G(y)) \| \leq c_2\, \| x - y \| \quad (x, y \in U(x_0, R)). \tag{2.504}$$

*Indeed, simply replace* $f_n(s)$, $f_\infty(s)$ *by*

$$\overline{f_n}(s) = f_n(s) + c_2$$

*and*

$$\overline{f_\infty}(s) = f_\infty(s) + c_2,$$

*respectively, in Defintion 2.4 and Lemma 2.12. Moreover, define majorizing sequence* $\{t_n\}$ *by*

$$\begin{cases} t_0 = 0, \quad t_1 = \eta, \\ t_{n+2} = t_{n+1} + \dfrac{\left( \displaystyle\int_0^1 (f'(t_n + \theta\,(t_{n+1} - t_n)) - f'(t_n))\, d\theta + c_2 \right) (t_{n+1} - t_n)}{1 - c - g'(t_{n+1})}. \end{cases} \tag{2.505}$$

*The new iteration corresponding to (2.503) is given by*

$$x_{n+1} = x_n - F'(x_n)^{-1}\, (F(x_n) + G(x_n)) \quad (n \geq 0,\ x_0 \in \mathcal{D}). \tag{2.506}$$

*Then, we have the following semilocal result for (2.506).*

**Theorem 2.15** *Let $F : \mathcal{D} \subseteq \mathcal{X} \longrightarrow \mathcal{Y}$ be a Fréchet–differentiable operator, and $G : \mathcal{D} \longrightarrow \mathcal{Y}$ a continuous operator. Let $x_0 \in \mathcal{D}$ be such that $F'(x_0)^{-1} \in \mathcal{L}(\mathcal{Y}, \mathcal{X})$. Suppose there exist $R > 0$, $c \in [0,1)$, $c_1, c_2 \in [0,1)$ and continuously differentiable function $f, g : [0,R) \longrightarrow (-\infty, +\infty)$ such that (2.476)–(2.478), (2.504) hold,*

$$f(0) > 0, \quad g(0) > 0, \quad f'(0) = -c_1, \quad g'(0) = -c,$$

*$f'$, $g'$, $f'(r+t) - f'(r)$, $g'(r) - g'(0)$ ($r \geq t \geq 0$ and $0 \leq r + t < R$) are non–decreasing, hypotheses of Lemma 2.12 hold with*

$$\overline{f_n}(s) = f_n(s) + c_2, \quad \overline{f_\infty}(s) = f_\infty(s) + c_2$$

*replacing functions $f_n(s)$ and $f_\infty(s)$, respectively, and*

$$t^\star < R.$$

*Then, the sequence $\{x_n\}$ generated by (2.506) is well defined, remains in $\overline{U}(x_0, t^\star)$ for all $n \geq 0$ and converges to a zero $x^\star$ of (2.503) in $\overline{U}(x_0, t^\star)$. Moreover, the following estimates hold ($n \geq 1$)*

$$\| x_{n+1} - x_n \| \leq \overline{e_n} \leq t_{n+1} - t_n$$

*and*

$$\| x_n - x^\star \| \leq t^\star - t_n,$$

*where*

$$\overline{e_n} = \frac{e_n + c_2 \| x_n - x_{n-1} \|}{1 - c - g'(\| x_n - x_0 \|)},$$

*$\{t_n\}$ is given by (2.505) and $t^\star$ is its limit. Furthermore, if*

$$\int_0^1 f'((1-\theta) \| x^\star - x_0 \| + \theta\, t^\star)\, d\theta - f'(t^\star) + g'(t^\star) + c_2 + c < 1,$$

*the solution $x^\star$ is unique in $\overline{U}(x_0, t^\star)$.*

**Proof** It follows from (2.506) that

$$F(x_k) + G(x_k) = F(x_k) - F(x_{k-1}) - F'(x_{k-1})\,(x_k - x_{k-1}) + G(x_k) - G(x_{k-1}).$$

Then, using the above approximation instead of (2.493), as in Theorem 2.14, we arrive at (2.480), (2.481) with scalar $\{t_n\}$ given by (2.505).

Finally, to show uniqueness, let $y^\star \in \overline{U}(x_0, t^\star)$ with $F(y^\star) + G(y^\star) = 0$. We first obtain the identity

$$\begin{aligned} x_{k+1} - y^\star &= x_k - F'(x_k)^{-1}\,(F(x_k) + G(x_k)) - y^\star \\ &= -F'(x_k)^{-1}\,(F(x_k) - F(y^\star) - F'(x_k)\,(x_k - y^\star) + G(x_k) - G(x^\star)) \\ &= -F'(x_k)^{-1} \left( \int_0^1 (F'(y^\star + \theta\,(x_k - y^\star)) - F'(x_k))\, d\theta\,(x_k - y^\star) + G(x_k) - G(y^\star) \right). \end{aligned}$$

Then, we get

$$\begin{aligned}
&\| x_{k+1} - y^\star \| \\
&\leq \frac{\displaystyle\int_0^1 (f'((1-\theta)\ \| y^\star - x_0 \| + \theta\ \| x_k - y^\star \|) - f'(\| x_k - x_0 \|) + c_2)\, d\theta\ \ \| x_k - y^\star \|}{1 - c - g'(t_k)} \\
&\leq \frac{\displaystyle\int_0^1 (f'((1-\theta)\ \| y^\star - x_0 \| + \theta\, t^\star) - f'(t^\star) + c_2)\, d\theta\ \ \| x_k - y^\star \|}{1 - c - g'(t^\star)} \\
&< \| x_k - y^\star \|,
\end{aligned}$$

which implies $\lim\limits_{k \longrightarrow \infty} x_k = y^\star$. But we have $\lim\limits_{k \longrightarrow \infty} x_k = x^\star$. Hence, we deduce $x^\star = y^\star$. This completes the proof of Theorem 2.15. ⊠

The results in (cf. [367], [751]) used the majorizing sequence $\{u_n\}$ given by

$$\left\{ \begin{array}{l} u_0 = 0, \quad u_1 = \eta, \\ u_{n+2} = u_{n+1} - \dfrac{\displaystyle\int_0^1 (f'(u_n + \theta\,(u_{n+1} - u_n)) - f'(u_n))\, d\theta\, (u_{n+1} - u_n)}{f'(u_{n+1})}. \end{array} \right. \tag{2.507}$$

Under certain conditions, they showed the quadratic convergence of majorizing sequence $\{u_n\}$. If we let $c = 1$, then, a simple induction argument shows that

$$t_n \leq u_n, \tag{2.508}$$

$$t_{n+1} - t_n \leq u_{n+1} - u_n \tag{2.509}$$

and

$$t^\star \leq u^\star, \tag{2.510}$$

where $u^\star = \sup\limits_n u_n \in (0, u^{\star\star}]$, provided that there exists $u^{\star\star}$ with

$$f'(u^{\star\star}) < 0. \tag{2.511}$$

If strict inequality hold in (2.498), then, so does in (2.508) and (2.509) for $n > 1$. Moreover, hypotheses in (cf. [367], [751]) differ from ours here (and also differ from the ones in (cf. [162], [194])). However, as we showed in (2.508)–(2.510), our bounds are tighter. Hence, our sequence $\{t_n\}$ is tighter than $\{u_n\}$ and quadratically convergent. It turns out that our sufficient convergence conditions can also be weaker.

**NM and Lipschitz condition.** Let $c_1 = c = 0$ in (2.499) and (2.500), respectively. Then, the conditions in (cf. [367], [751]) reduce to the classical and famous for its simplicity and clarity Kantorovich hypothesis for solving nonlinear equations (cf. [157], [469])

$$h_K = L\,\eta \leq \frac{1}{2} \quad \text{and} \quad t^{\star\star} = \frac{1 - \sqrt{1 - 2\,h_K}}{L}. \tag{2.512}$$

The hypotheses of Lemma 2.12 and Theorem 2.14 reduce to

$$h_{AH} = \overline{L}\,\eta \leq \frac{1}{2} \quad \text{and} \quad t^{\star\star} = \frac{\eta}{1-b}, \tag{2.513}$$

where

$$\overline{L} = \frac{1}{8}\left(L + 4\,L_0 + \sqrt{L^2 + 8\,L_0\,L},\right) \tag{2.514}$$

and

$$b = \frac{4\,L}{L + \sqrt{L^2 + 8\,L_0\,L}}. \tag{2.515}$$

Note that

$$h_K \leq \frac{1}{2} \Longrightarrow h_{AH} \leq \frac{1}{2}, \tag{2.516}$$

but not necessarily vice versa unless if $L_0 = L$. Note that under (2.513), the convergence is quadratic with ratio $2\,\overline{L}\,\eta$, which is smaller than $2\,L\,\eta$ given in the famous Kantorovich theorem (cf. [157], [469]).

**Example 2.11** *Let $\mathcal{X} = \mathcal{Y} = \mathbb{R}^2$ be equipped with the max–norm, $\mathcal{D} = [\beta, 2-\beta]^2$, $\beta \in [0,1)$ and define function $F$ on $\mathcal{D}$ by*

$$F(x) = (\xi_1^3 - \beta, \xi_2^3 - \beta)^T \qquad (x = (\xi_1, \xi_2)^T). \tag{2.517}$$

*The Fréchet–derivative of operator $F$ is given by*

$$F'(x) = \begin{bmatrix} 3\,\xi_1^2 & 0 \\ 0 & 3\,\xi_2^2 \end{bmatrix}.$$

*(a) Let $x_0 = (1,1)^T$. Using hypotheses of Theorem 2.14, we get*

$$\eta = \frac{1}{3}\,(1-\beta), \quad L_0 = 3-\beta, \quad L = 2\,\,(2-\beta).$$

*The Kantorovich condition is violated, since*

$$2\,h_K = \frac{4}{3}\,(1-\beta)\,(2-\beta) > 1 \quad (\beta \in [0,.5)).$$

*Hence, there is no guarantee that NM converges to $x^\star = (\sqrt[3]{\beta}, \sqrt[3]{\beta})^T$ starting at $x_0$. However, our condition (2.513) is true for all $\beta \in I = \left[.450339002, \frac{1}{2}\right)$. That is, the conclusions of our Theorem 2.14 can apply to solve equation (2.517) for all $\beta \in I$.*

*(b) Let $x_0 = (.9,.9)^T$ and $\beta = .7$. Using hypotheses of Theorem 2.14, we get*

$$\eta = .1, \quad L = 2.6, \quad L_0 = 2.3, \quad \overline{L} = 2.39864766,$$

$$h_K = .26, \quad h_{AH} = .239864766.$$

*Then, (2.512) and (2.513) are satisfied.*

*We have also that*

$$F'(.9, .9)^{-1} = .4115226337\ \mathcal{I},$$

*where $\mathcal{I}$ is the identity $2 \times 2$ matrix.*

*The hypotheses of our Theorem 2.14 and the Kantorovich theorem are satisfied. Then, NM converges to $x^\star = (.8879040017, .8879040017)^T$ starting at $x_0$. We also can provide the comparison table using the software Maple 13.*

**Comparison table.**

| | NM | (2.452) | (2.507) |
|---|---|---|---|
| $k$ | $\parallel x_{k+1} - x_k \parallel$ | $t_{k+1} - t_k$ | $u_{k+1} - u_k$ |
| 0 | .01193415638 | .1 | .1 |
| 1 | .0001618123748 | .01468926554 | .03513513514 |
| 2 | 2.946995467e-8 | .0003231246464 | .004948198199 |
| 3 | 4.228114294e-11 | 1.564213506e-7 | .0001001286551 |
| 4 | $\sim$ | 3.665624833e-14 | 4.101652297e-8 |
| 5 | $\sim$ | 2.013038029e-27 | 6.882715732e-15 |
| 6 | $\sim$ | 6.070995489e-54 | 1.938035887e-28 |
| 7 | $\sim$ | 5.521737690e-107 | 1.536617512e-55 |
| 8 | $\sim$ | 4.567804358e-213 | 9.659923912e-110 |
| 9 | $\sim$ | 3.125870200e-425 | 3.817592426e-218 |
| 10 | $\sim$ | 1.463854229e-849 | 5.962402218e-435 |

*The table shows that our error bounds (2.452) are tighter than (2.507) proposed in (cf. [367], [751]) in the special case of Kantorovich's Theorem.*

*Even weaker conditions and tighter error estimates have been given in section 2.1 for this special case.*

## 2.9 Local Convergence and Convex Majorants

We provide in this section ball convergence results for NM in order to approximate a locally unique solution of a nonlinear operator equation in a Banach space setting. Our hypotheses involve very general majorants on the Fréchet-derivatives of the operators involved. In the special case of convex majorants our results have advantages over earlier ones (cf. [364], [469], [638], [751]) These advantages are at least as: large radius of convergence; tight error bounds on the distances involved and precise information on the uniqueness of the solution.

NM converges quadratically to $x^\star$ (under certain conditions (cf. [139], [157], [469])) one function evaluation and requires one inverse at each step.

Therefore the efficiency index $E = p^{\frac{1}{q}}$ ($p$ is the order of convergence and $q$ the number of function evaluations per iterative step) is given by $E = \sqrt{2}$ (cf. [157], [751]).

The convergence for Newton–type methods under very general conditions has been studied by several authors. A survey of recent results can be found in (cf. [157], [469] and the references therein), see also (cf. [139], [683], [751]).

We suppose that the nonlinear operator equation (2.1) has a solution $x^\star$. An interesting problem is to estimate the radius of convergence ball of NM defined by (2.2). An open ball $U(x^\star, r) \subseteq \mathcal{D}$ with center $x^\star$ and radius $r$ is called a convergence ball of an iterative method if the sequence generated by this iterative method starting from any initial values in it converges. The convergence ball of an iterative method is very important in computational mathematics, because it shows the extent of difficulty of choosing initial points. Ball convergence theorems can be immediately obtained by specializing results on Newton-type methods in the case of Newton's method (cf. [31], [170], [402], [638]). However, a more direct approach is expected to generate more exact results.

In particular, we are motivated by optimization considerations and the work of Ferreira (cf. [364]), where it was claimed that the best possible radius of convergence can be obtained for Newton's method using the information $(F, F')$ under convex majorants.

We introduce even more general majorant conditions that do not necessarily imply the convexity of the functions involved. We provide a ball convergence result for NM (2.2). If we specialize our theorem in the case of convex majorants, we obtain a result with the following advantages over earlier ones (cf. [364], [638], [751]) (under the same hypotheses and computational cost) at least as

(a) large radius of convergence,

(b) tight error estimates on the distance $\|x_n - x^\star\| \quad (n \geq 0)$, and

(c) precise information on the uniqueness of the solution $x^\star$.

Advantages (a)–(c) are very important in computational mathematics, since they allow a wider choice of initial guesses $x_0$ and the computation of less iterative steps to obtain the same desired error tolerance. Numerical examples further validating the theoretical results are also provided in this section.

We shall show the main local convergence result for (2.2).

**Theorem 2.16** *Let $\mathcal{X}$ and $\mathcal{Y}$ be Banach spaces, $\mathcal{D} \subseteq \mathcal{X}$ be an open set and $F : \mathcal{D} \longrightarrow \mathcal{Y}$ be a continuously Fréchet-differentiable operator. Let $x^\star \in \mathcal{D}$, $R > 0$ and $\alpha = \sup\{t \in [0, R) : U(x^\star, t\} \subset \mathcal{D}$. Suppose that $F(x^\star) = 0$, $F'(x^\star)$ is invertible and there exist functions $f, f_0 : [0, R) \to \mathbb{R}$ continuously differentiable such that, for all $\theta \in [0, 1]$ and $x \in U(x^\star, \alpha)$*

**(H1)** $\| F'(x^\star)^{-1} (F'(x) - F'(x^\star + \theta(x - x^\star))) \| \leq f'(\| x - x^\star \|) - f'(\theta \| x - x^\star \|)$;

**(H2)** $\| F'(x^\star)^{-1} (F'(x) - F'(x^\star)) \| \leq f_0'(\| x - x^\star \|) - f_0'(0)$;

**(H3)** *let* $\beta = \sup\{t \in [0, R) : f_0'(t) < 0\}$, $\gamma = \sup\{t \in (0, \beta) : r_{f_0,f} \leq 1\}$, $\delta = \min\{\alpha, \gamma\}$,

*where function* $r_{f_0,f} : [0, \delta)^3 \to \mathbb{R}$ *is given for* $0 \leq t < v < \delta$, $x \in U(x^\star, \delta)$ *by*

$$r_{f_0,f} = r_{f_0,f}(t, v, \|x - x^\star\|) = -\frac{e(t, v)}{f_0'(\|x - x^\star\|)}$$

*and*

$$e(t, v) = f(v) - f(t) - f'(v)(v - t);$$

**(H4)** *functions* $f'(t) - f'(\theta t)$, $f_0'(t) - f_0'(0)$ *are increasing for all* $t \in [0, \infty)$ *and*$\theta \in [0, 1]$;

**(H5)** $f_0(t) \leq f(t)$, $f_0'(t) \leq f'(t)$ *for all* $t \in [0, R)$;

**(H6)** $f_0'(0) \geq -1$.

*Then, sequence* $\{x_n\}$ *generated by (2.2) for* $x_0 \in U(x^\star, \delta) - \{x^\star\}$ *is well defined, remains in* $U(x^\star, \delta)$ *for all* $n \geq 0$ *and converges to* $x^\star$*, which is the unique solution of equation* $F(x) = 0$ *in* $U(x^\star, \alpha)$.

From now, one we assume hypotheses of Theorem 2.16 hold. We shall show Theorem 2.16 through a series of lemmas.

**Lemma 2.13** *If* $x \in \overline{U}(x^\star, t)$ *and* $t \in [0, \beta_0)$, *where,* $\beta_0 = \min\{\alpha, \beta\}$, *then*

$$F'(x)^{-1} \in L(\mathcal{Y}, \mathcal{X})$$

*and*

$$\| F'(x)^{-1} F'(x^\star) \| \leq -\frac{1}{f_0'(\| x - x^\star \|)} \leq -\frac{1}{f_0'(t)} \leq -\frac{1}{f'(t)}. \tag{2.518}$$

**Proof** Let $x \in \overline{U}(x^\star, t)$ and $t \in [0, \beta_0)$. Using hypotheses $(H2) - (H6)$, we obtain in turn

$$\begin{aligned} \| F'(x^\star)^{-1} (F'(x) - F'(x^\star)) \| &\leq f_0'(\| x - x^\star \|) - f_0'(0) \\ &\leq f_0'(\| x - x^\star \|) + 1 \qquad (2.519) \\ &\leq f_0'(t) + 1 < 1. \end{aligned}$$

It follows from (2.519) and the Banach Lemma on invertible operators that $F'(x)^{-1} \in L(\mathcal{Y}, \mathcal{X})$. Hence, (2.518) holds true. This completes the proof of Lemma 2.13. ⊠

**Lemma 2.14** *Let* $x \in U(x^\star, t)$ *and* $0 \leq t < v < \beta_0$. *Then, for*

$$E(x^\star, x) = F(x) - F(x^\star) - F'(x)(x - x^\star) = \int_0^1 [F'(x^\star + \theta(x - x^\star)) - F'(x)](x - x^\star) d\theta,$$

*we have the following estimates*

$$\| F'(x^\star)^{-1} \; E(x^\star, x) \| \leq e(t, v) \; \| x^\star - x \| \tag{2.520}$$

*and*

$$r_{f_0, f} \leq 1, \tag{2.521}$$

*where functions* $e(t, v)$ *and* $r_{f_0, f}$ *are given in (H3).*

**Proof** Using hypothesis $(H1)$, $(H4)$ and the definition of operator $E$, we obtain in turn that

$$\begin{aligned} \| F'(x^\star)^{-1} E(x^\star, x) \| &\leq \int_0^1 \| F'(x^\star)^{-1} \left(F'(x^\star + \theta \, (x - x^\star)) - F'(x)\right) \| \, \| x - x^\star \| \; d\theta \\ &\leq \int_0^1 \left(f'(\| x - x^\star \|) - f'(\theta \, \| x - x^\star \|)\right) \; \| x - x^\star \| \; d\theta, \end{aligned} \tag{2.522}$$

which implies estimate (2.520). In view of hypothesis $(H5)$, we get that

$$\frac{\displaystyle\int_0^1 \left(f'(t) - f'(\theta t)\right) d\theta}{-f_0'(t)} \leq 1, \tag{2.523}$$

which together with $(H3)$ imply (2.521). This completes the proof of Lemma 2.14. ⊠

**Proof of Theorem 2.16.** According to Lemmas 2.13 and 2.14, it is left to show $x_n \in U(x^\star, \delta)$ $(n \geq 1)$, $\lim\limits_{n \to \infty} x_n = x^\star$ and the uniqueness part. By hypothesis $x_0 \in U(x^\star, \delta)$. Let us assume $x_k \in U(x^\star, \delta)$ for all $k \leq n$. We shall show $x_{k+1} \in U(x^\star, \delta)$. Using (2.2) and Lemma 2.14 for $x = x_n$, we get

$$\| x_{k+1} - x^\star \| < \| x_k - x^\star \| < \delta, \tag{2.524}$$

which show $x_{k+1} \in U(x^\star, \delta)$ and $\lim\limits_{k \to \infty} x_k = x^\star$.

Finally, to show uniqueness in $U(x^\star, \alpha)$, let $y^\star$ be a solution of equation $F(x) = 0$ in $U(x^\star, \alpha)$. Define linear operator $\mathcal{M}$ by

$$\mathcal{M} = \int_0^1 F'(x^\star + \theta \, (y^\star - x^\star)) \, d\theta.$$

Using (2.518) for $x^\star + \theta \, (y^\star - x^\star) \in U(x^\star, \alpha)$ and replacing $x$, we conclude $\mathcal{M}^{-1}$ exists. It follows from the identity $F(y^\star) - F(x^\star) = \mathcal{M} \, (y^\star - x^\star)$, that $x^\star = y^\star$. This completes the proof of Theorem 2.16. ⊠

If functions $f_0, f$ satisfy certain convexity conditions, then Theorem 2.16 specializes to Theorem 2.1 in (cf. [364, p. 748]). The proofs are omitted, since they follow from the corresponding ones in (cf. [364]), where the estimate

$$\|F'(x)^{-1} F'(x^\star)\| \leq - \frac{1}{f'(\|x - x^\star\|)} \tag{2.525}$$

was used instead of the at least, as tight

$$\|F'(x)^{-1}F'(x^\star)\| \leq -\frac{1}{f_0'(\|x - x^\star\|)} \tag{2.526}$$

(see Lemma 2.13).

**Proposition 2.2** *Let $\mathcal{X}$ and $\mathcal{Y}$ be Banach spaces, $\mathcal{D} \subseteq \mathcal{X}$ be an open set and $F : \mathcal{D} \to \mathcal{Y}$ be a continuously Fréchet-differentiable operator. Let $x^\star \in \mathcal{D}$, $R > 0$ and $\alpha = \sup\{t \in [0,R) : U(x^\star,t) \subset \mathcal{D}\}$. Suppose that $F(x^\star) = 0$, $F'(x^\star)$ is invertible and there exist functions $f, f_0 : [0,R) \to \mathbb{R}$ twice continuously differentiable such that, for all $x \in U(x^\star,\alpha)$ and $\theta \in [0,1]$*
**(H1)** $\| F'(x^\star)^{-1} (F'(x) - F'(x^\star + \theta(x - x^\star))) \| \leq f'(\| x - x^\star \|) - f'(\theta \| x - x^\star \|)$;

**(H2)** $\| F'(x^\star)^{-1} (F'(x) - F'(x^\star)) \| \leq f_0'(\| x - x^\star \|) - f_0'(0)$;

**($\overline{\mathbf{H}}$3)** *let* $\beta = \sup\{t \in [0,R) : f_0'(t) < 0\}$, $\gamma = \sup\{t \in (0,\beta) : \overline{r}_{f_0,f} \leq 1\}$, $\delta = \min\{\alpha,\gamma\}$,

*where function $\overline{r}_{f_0,f} : [0,\delta)^3 \to \mathbb{R}$ is given for $0 \leq t < v < \delta$, $x \in U(x^\star,\delta)$ by*

$$\overline{r}_{f_0,f} = \overline{r}_{f_0,f}(t,v,\|x - x^\star\|) = -\frac{e(t,v)\|x - x^\star\|}{(v-t)^2 f_0'(\|x - x^\star\|)}$$

*and*

$$e(t,v) = f(v) - f(t) - f'(v)(v-t);$$

**($\overline{\mathbf{H}}$4)** *$f_0'$, $f'$ are convex and strictly increasing functions on $[0,R)$;*

**(H5)** $f_0(t) \leq f(t)$, $f_0'(t) \leq f'(t)$ *for all* $t \in [0,R)$;

**($\overline{\mathbf{H}}$6)** $f_0(0) = f(0) = 0$, *and* $f_0'(0) = f'(0) = -1$. *Then, sequence $\{x_n\}$ generated by (2.1) is well defined, remains in $U(x^\star,\alpha)$ for all $n \geq 0$ and converges to a unique solution $x^\star$ in $U(x^\star,\lambda)$ of equation $F(x) = 0$, where*

$$\lambda = \sup\{0 < t < \alpha : f_0(t) < 0\}. \tag{2.527}$$

*Moreover, the scalar sequence $\{t_n\}$ given for $t_0 = \|x^\star - x_0\|$ by*

$$t_{n+1} = \left| t_n - \frac{f(t_n)}{f'(t_n)} \right| \left| \frac{f'(t_n)}{f_0'(t_n)} \right| \qquad (n \geq 0) \tag{2.528}$$

*is well defined, remains in $(0,\alpha)$, strictly decreasing and converges to zero. Furthermore, $\{\frac{t_{n+1}}{t_n^2}\}$ is strictly decreasing,*

$$\|x^\star - x_{n+1}\| \leq \frac{t_{n+1}}{t_n^2}\|x^\star - x_n\|^2 \qquad (n \geq 0), \tag{2.529}$$

$$\|x^\star - x_n\| \leq t_0 \left(\frac{t_1}{t_0}\right)^{2^n - 1} \qquad (n \geq 0) \tag{2.530}$$

*and*

$$\frac{t_{n+1}}{t_n^2} \leq \frac{f''(t_0)}{2|f_0'(t_0)|} \qquad (n \geq 0). \tag{2.531}$$

*If, additionally, $\overline{r}_{f_0,f}(\gamma,\gamma,\gamma) = 1$ and $\gamma < \alpha$, then, $\alpha = \gamma$ is the best possible convergence radius.*

**Proof** As noted above, the proof follows as in Theorem 2.1 in (cf. [364]) (simply replace (2.525) by (2.526) in the computation of the upper bounds of the norm $\|F'(x)^{-1}F'(x^\star)\|$) with the exception of the uniqueness part. We have that

$$y^\star - x^\star = -\int_0^1 F'(x^\star)^{-1}[F'(x^\star + \theta(y^\star - x^\star)) - F'(x^\star)](y^\star - x^\star)d\theta. \tag{2.532}$$

Using $(H2)$, $(\overline{H}4)$, $(\overline{H}6)$ and (2.532) with $x = x^\star + \theta(y^\star - x^\star)$ and $\theta = 0$, we get

$$\begin{aligned} \|y^\star - x^\star\| &\leq \int_0^1 [f_0'(\theta\|y^\star - x^\star\|) - f_0'(0)]\|y^\star - x^\star\|d\theta \\ &= f_0(\|y^\star - x^\star\|) - f_0(0) - f_0'(0)\|y^\star - x^\star\|, \end{aligned} \tag{2.533}$$

which implies

$$f_0(\|y^\star - x^\star\|) \geq 0. \tag{2.534}$$

Function $f_0$ is strictly convex and $f(t) < 0$ in $[0, \alpha)$. That is, 0 is the unique solution of equation $f_0(t)$ in $[0, \alpha)$. Hence, estimate (2.534) implies $\|y^\star - x^\star\| = 0$. Hence, we conclude that $x^\star = y^\star$. This completes the proof of Proposition 2.2. ⊠

**Remark 2.24** *If $f_0' = f'$ on $[0, R)$, then, Proposition 2.2 reduces to Theorem 2.1 in (cf. [364]). Otherwise, i.e, if*

$$f_0'(t) < f'(t) \qquad (t \in [0, R)), \tag{2.535}$$

*then, our results are finer with advantages as stated in the abstract of this section.*

*Note that $(H2)$ is not an additional to $(H1)$ hypothesis, since $f_0$ always exists in this case. Hence, through Theorem 2.16 we have studied the local convergence of Newton's method under very general majorants (not necessarily convex). Moreover, using Proposition 2.2 we expanded the applicability of Theorem 2.16 using the same hypotheses $(F, F')$ and convex majorants.*

We provide now numerical examples where (2.535) holds.

**Example 2.12** *Let us assume there exist $L > 0$ such that the Lipschitz condition*

$$\| F'(x^\star)^{-1}(F'(y) - F'(x)) \| \leq L \| x - y \| \tag{2.536}$$

*for all* $x, y \in \overline{U}(x_0, R) \subseteq \mathcal{D}$. *Define a scalar majorant function* $f : [0, R] \longrightarrow (-\infty, +\infty)$ *by*

$$f(t) = \frac{L}{2}\, t^2 - t \tag{2.537}$$

*and set*

$$f_0(t) = f(t) \qquad (t \in [0, R]). \tag{2.538}$$

*It follows from Proposition 2.2 (cf. [364, Theorem 2.1]) that we can set*

$$r_{RTW} = \frac{2}{3\,L}, \tag{2.539}$$

*which is the radius of convergence obtained by Rheinboldt (cf. [638], [157]) and Traub (cf. [751]).*

*It follows from (2.536) that there exists* $L_0 > 0$ *such that*

$$\| \, F'(x^\star)^{-1} \, (F'(x) - F'(x^\star)) \, \| \leq L_0 \, \| \, x - x^\star \, \| \qquad (x \in \overline{U}(x_0, R)). \tag{2.540}$$

*Clearly,*

$$L_0 \leq L \tag{2.541}$$

*holds and* $\dfrac{L}{L_0}$ *can be arbitrarily large. Let us define a function* $f_0$ *by*

$$f_0(t) = \frac{L_0}{2}\, t^2 - t. \tag{2.542}$$

*It follows from Proposition 2.2 that we can set*

$$r_{AH} = \frac{2}{2\,L_0 + L}. \tag{2.543}$$

*By comparing (2.539) with (2.543), we conclude*

$$r_{RTW} \leq r_{AH}. \tag{2.544}$$

*If strict inequality holds in (2.541), then, so does in (2.544). Note also that*

$$\frac{r_{RTW}}{r_{AH}} = \frac{2 \frac{L_0}{L} + 1}{3} \to \frac{1}{3} \quad as \quad \frac{L_0}{L} \to 0.$$

*Hence, our approach triples (at most) the radius of convergence given by (2.539) (cf. [139], [157], [364], [469]).*

**Example 2.13** *Let* $f : [0, R) \longrightarrow (-\infty, +\infty)$ *be a twice continuously differentiable function with* $f'$ *convex. Then,* $F$ *satisfies (2.519) if and only if*

$$\| \, F'(x^\star)^{-1} \; F''(x) \, \| \leq f''(\| \, x - x^\star \, \|) \tag{2.545}$$

*for all $x \in \mathcal{D}$ such that $x \in U(x^\star, R)$ (see (cf. [364, Lemma 14] or [751])). Let us define a function $f$ on $[0, R)$ by*

$$f(t) = \frac{\gamma\, t^2}{1 - \gamma\, t} - t, \tag{2.546}$$

*where $R < \dfrac{1}{\gamma}$ for some $\gamma > 0$. If, for example, $F$ is an analytic operator, then,(2.545) is satisfied for*

$$\gamma^\star = \sup_{k \geq 2} \left\| \frac{F'(x^\star)^{-1} F^{(k)}(x^\star)}{k!} \right\|^{\frac{1}{k-1}}. \tag{2.547}$$

*Smale (cf. [683]) and Wang (cf. [751]) have used (2.546) to provided a convergence analysis for (2.2). In particular, Wang (cf. [751]) showed convergence for $F$ being only twice Fréchet continuously differentiable for $\gamma$ satisfying*

$$\gamma^\star \leq \gamma. \tag{2.548}$$

*We have also used (2.546) to provide a convergence analysis for the Secant method (cf. [157]).*

*Let us also define function $f_0$ by*

$$f_0(t) = f(t) \qquad (t \in [0, R)). \tag{2.549}$$

*We obtain for analytic operators $F$ that Smale's radius of convergence (cf. [683]) is given by*

$$t_S^\star = \frac{5 - \sqrt{13}}{6\, \gamma^\star} \tag{2.550}$$

*and for twice Fréchet continuously differentiable operator $F$, Wang's radius (cf. [751]) is given by*

$$t_W^\star = \frac{5 - \sqrt{13}}{6\, \gamma}. \tag{2.551}$$

*In what follows we shall show that we can enlarge radii given by (2.550) and (2.551). We can see that, for function $f$ given by (2.546), condition (2.545) or, equivalently, (2.519) imply that there exists $\gamma_0 > 0$ satisfying*

$$\gamma_0 \leq \gamma, \tag{2.552}$$

*so that function $f_0 : \left[0, \dfrac{1}{\gamma_0}\right) \longrightarrow (-\infty, +\infty)$ satisfies condition (2.518) for $R \in \left[0, \dfrac{1}{\gamma_0}\right)$.*

*Note also that $\dfrac{\gamma}{\gamma_0}$ can be arbitrarily large. It follows from (2.552) that there exists $a \in [0, 1]$ such that*

$$\gamma_0 = a\, \gamma. \tag{2.553}$$

*Set*

$$b = 1 - a \tag{2.554}$$

*and define scalar polynomial* $P_a$ *by*

$$P_a(t) = 3\,a^2\,t^3 + a\,(6\,b - a)\,t^2 + (3\,b^2 - 2\,a\,b - 1)\,t - b^2. \tag{2.555}$$

*By the definition of polynomial* $P_a$ *and for fixed* $a$*, we get*

$$P_a(0) = -b^2 \leq 0, \qquad P_a(1) = 1. \tag{2.556}$$

*Using (2.556) and the intermediate value theorem, we conclude that there exists* $t_a \in [0, 1)$ *such that* $P_a(t_a) = 0$*. Let us denote by* $t_a$ *the minimal number in* $[0, 1)$ *satisfying* $P_a(t_a) = 0$*. Define*

$$t_a^\star = \frac{1 - t_a}{\gamma}. \tag{2.557}$$

*In particular, for* $a = 1$*,* $t_1 = \dfrac{1 + \sqrt{13}}{6}$ *and, consequently,*

$$t_a^\star = \frac{5 - \sqrt{13}}{6\,\gamma} = t_W^\star. \tag{2.558}$$

*It is simple algebra to show that, for all* $a \in [0, 1]$*,* $P_a(t_1) \geq 0$*, which implies that*

$$t_a \leq t_1 \tag{2.559}$$

*and*

$$t_1^\star \leq t_a^\star. \tag{2.560}$$

*We also note that strict inequality holds in (2.559) and (2.560) for* $a \neq 1$*.*

*As an example, let* $a = \dfrac{1}{2}$*. Then, we obtain that*

$$t_{1/2} = .65185 < t_1 = \frac{1 + \sqrt{13}}{6} = .76759$$

*and*

$$t_1^\star = \frac{.23241}{\gamma} < \frac{.34815}{\gamma} = t_{1/2}^\star. \tag{2.561}$$

*Finally, note that clearly if strict inequality holds in (2.521), i.e., in (2.541) or (2.552), then our estimates on* $\| \, x_{n+1} - x^\star \, \|$ $(n \geq 0)$ *are finer (more precise) than the corresponding ones in (cf. [31], [364], [469], [638], [751]) (see e.g. (2.526)). These results are also obtained under the same computational cost since in practice the evaluation of* $L$ *(or* $\gamma$*) requires that* $L_0$ *(or* $\gamma_0$*).*

**Remark 2.25** *As noted in (cf. [136], [170], [560], [751], [801]), the local results obtained here can be used for projection method such us Arnold's, the generalized minimum residual method (GMRES), the generalized conjugate residual method (GCR) for combined Newton/finite projection methods and in connection with the mesh independence principle to develop the cheapest and most efficient mesh refinement strategies.*

**Remark 2.26** *The local results obtained can also be used to solve equation of the form $F(x) = 0$, where $F'$ satisfies the autonomous differential equation (cf. [157], [469])*

$$F'(x) = T(F(x)), \tag{2.562}$$

*where $T : \mathcal{Y} \longrightarrow \mathcal{X}$ is a known continuous operator. Since $F'(x^\star) = T(F(x^\star)) = T(0)$, we can apply our results without actually knowing the solution of $x^\star$ of equation $F(x) = 0$.*

Examples where $L_0 < L$ have been given in section 2.1.

## 2.10 Majorizing Sequences

We provide in this section convergence results for very general majorizing sequences of iterative methods. Using our new concept of recurrent functions, we unify the semilocal convergence analysis of Newton–type methods NTM under more general Lipschitz–type conditions. We present two very general majorizing sequences and we extend the applicability of NTM using the same information before (cf. [281], [325], [469], [524], [525], [637]). Applications, special cases and examples are also provided in this study to justify the theoretical results of our new approach.

We are concerned with the problem of approximating a locally unique solution $x^\star$ of (2.388). We use Newton–type methods NTM

$$x_{n+1} = x_n - A_n^{-1}\,(F(x_n) + G(x_n)) \quad (n \geq 0,\ x_0 \in \mathcal{D}). \tag{2.563}$$

Here, $A(x) \in \mathcal{L}(\mathcal{X}, \mathcal{Y})$ the space of bounded linear operators from $\mathcal{X}$ into $\mathcal{Y}$ is an approximation to Fréchet–derivative $F$. The most popular choice for operators $A$ and $G$ is $A(x) = F'(x)$ and $G(x) = 0$ $(x \in \mathcal{D})$. Then, we obtain Newton method NM given by (2.2). Another interesting choice of operator $A$ is

$$A(x) = \mathcal{I} - \frac{1}{2}\,F'(x)^{-1}\,F''(x)\,F'(x)^{-1}\,F(x),$$

then, if $G(x) = 0$ $(x \in \mathcal{D})$, (2.563) becomes Halley's method

$$x_{n+1} = x_n - \left(\mathcal{I} - \frac{1}{2}\,F'(x_n)^{-1}\,F''(x_n)\,F'(x_n)^{-1}\,F(x_n)\right)^{-1} F(x_n) \quad (x_0 \in \mathcal{D},\ n \geq 0). \tag{2.564}$$

In this section, we unify the semilocal study of iterative methods under more general Lipschitz–type conditions and majorizing sequences than before (cf. [281], [325], [469], [524], [525], [637]). In particular, we show that it is possible to expand the applicability of iterative methods and improve the error bound involved using the same information as before. Majorizing sequences play important role in the study of iterative methods. We define a very general majorizing sequence. Let $\eta \geq 0$ and $\alpha_n \geq 0$ be given. Define scalar sequence $\{t_n\}$ $(n \geq 0)$ by

$$t_0 = 0, \quad t_1 = \eta, \quad t_{n+2} = t_{n+1} + \alpha_n\,(t_{n+1} - t_n). \tag{2.565}$$

If there exists $\alpha$ such that

$$\alpha_n \leq \alpha < 1 \qquad (n \geq 1), \tag{2.566}$$

then, sequence $\{t_n\}$ $(n \geq 0)$ is well defined, non–decreasing, bounded above by

$$t^{\star\star} = \frac{\eta}{1-\alpha} \tag{2.567}$$

and converges to its unique least upper bound $t^\star$ satisfying

$$0 \leq t^\star \leq t^{\star\star}. \tag{2.568}$$

Moreover, the following estimates hold for all $n \geq 1$

$$0 \leq t_{n+2} - t_{n+1} \leq \alpha\ (t_{n+1} - t_n) \leq \alpha^{n+1}\ \eta \tag{2.569}$$

and

$$t^\star - t_n \leq \frac{\eta}{1-\alpha}\ \alpha^n \leq t^{\star\star}. \tag{2.570}$$

For example, many majorizing sequences of NTM (2.563) are special cases of (2.565) (cf. [157]). Therefore, choosing sequence $\{\alpha_n\}$ and also veryfing (2.566) is very important.

In this study, we provide conditions implying (2.566), when for $\beta_n \geq 0$, $\gamma_n \geq 0$ $(n \geq 0)$

$$\alpha_n = \frac{\beta_n}{1 - \gamma_n\, t_{n+1}}. \tag{2.571}$$

Note that (2.566) is obviously such a condition. Sequence (2.565) for $\alpha_n$ given by (2.571) becomes

$$t_0 = 0, \quad t_1 = \eta, \quad t_{n+2} = t_{n+1} + \frac{\beta_n\ (t_{n+1} - t_n)}{1 - \gamma_n\ t_{n+1}}. \tag{2.572}$$

We also examine the special case of (2.571), when for $\delta_n \geq 0$

$$\beta_n = \delta_n\, (t_{n+1} - t_n). \tag{2.573}$$

In this case, (2.572) can be written as

$$t_0 = 0, \quad t_1 = \eta, \quad t_{n+2} = t_{n+1} + \frac{\delta_n\ (t_{n+1} - t_n)^2}{1 - \gamma_n\ t_{n+1}}. \tag{2.574}$$

We can show the following convergence result for (2.572).

**Lemma 2.15** *Let $\eta > 0$, $\gamma > 0$, $\{\beta_n\} > 0$ and $\gamma_n > 0$. Assume that together with*

$$\gamma_0 \eta < 1 \tag{2.575}$$

*and*

$$\gamma_n \leq \gamma, \tag{2.576}$$

*one of the following set of hypotheses holds*

*(I) There exists parameter $\alpha$ such that*

$$\lim_{n \longrightarrow \infty} \beta_n = 0, \tag{2.577}$$

$$g_n(\alpha) = \beta_{n+1} - \beta_n + \alpha \left( \gamma_{n+1} \frac{1-\alpha^{n+2}}{1-\alpha} - \gamma_n \frac{1-\alpha^{n+1}}{1-\alpha} \right) \eta \geq 0 \tag{2.578}$$

*and*

$$\frac{\beta_0}{1 - \gamma_0 \, \eta} \leq \alpha \leq 1 - \gamma \, \eta. \tag{2.579}$$

*(II) There exists parameter $\alpha$ such that*

$$g_n(\alpha) \leq 0, \tag{2.580}$$

$$f_1(\alpha) \leq 0 \tag{2.581}$$

*and*

$$\frac{\beta_0}{1 - \gamma_0 \, \eta} \leq \alpha \leq 1 - \gamma \, \eta,$$

*where*

$$f_1(s) = \beta_1 + s \, \gamma_1 \, (1+s) \, \eta - s. \tag{2.582}$$

*Then, scalar sequence defined by (2.565) satisfies (2.567)–(2.570).*

**Proof**

**Case (I)** We shall use induction to show (2.569). We must have that

$$0 \leq \frac{\beta_n}{1 - \gamma_n \, t_{n+1}} \leq \alpha. \tag{2.583}$$

Estimate (2.583) holds for $n = 0$ by (2.579). Then, we have by (2.572) (for $n = 0$) that

$$t_2 - t_1 \leq \alpha \, (t_1 - t_0) = \alpha \, \eta \quad \text{and} \quad t_2 \leq \eta + \alpha \eta = \frac{1-\alpha^2}{1-\alpha} \eta.$$

Assume (2.583) holds for all $k \leq n$. Then, we have again by (2.572) that

$$t_{k+1} - t_k \leq \alpha \, (t_k - t_{k-1}) \leq \alpha^k \, \eta$$

and

$$\begin{aligned} t_{k+1} &\leq t_k + \alpha^k \, \eta \leq t_{k-1} + (\alpha^{k-1} + \alpha^k) \, \eta \\ &\leq t_1 + (\alpha + \cdots + \alpha^{k-1} + \alpha^k) \, \eta \\ &= \frac{1-\alpha^{k+1}}{1-\alpha} \, \eta. \end{aligned}$$

Using this expression for $t_{k+1}$, we can show instead of (2.583) that

$$\beta_n + \alpha\,\gamma_n\,\frac{1-\alpha^{n+1}}{1-\alpha}\,\eta - \alpha \leq 0. \tag{2.584}$$

Estimate (2.584) motivates us to introduce recurrent functions $f_n$ on $[0,\infty)$ by

$$f_n(s) = \beta_n + s\,\gamma_n\,(1+s+\cdots+s^n)\,\eta - s \leq 0. \tag{2.585}$$

We need a relationship between two consecutive recurrent functions $f_n$

$$\begin{aligned} f_{n+1}(s) &= \beta_{n+1} + s\,\gamma_{n+1}\,(1+s+\cdots+s^{n+1})\,\eta - s - \beta_n \\ &\quad - s\,\gamma_n\,(1+s+\cdots+s^n)\,\eta + s + f_n(s). \end{aligned}$$

Hence, we have that

$$\begin{aligned} f_{n+1}(s) &= f_n(s) + \beta_{n+1} - \beta_n \\ &\quad + s\,(\gamma_{n+1}\,(1+s+\cdots+s^{n+1}) - \gamma_n\,(1+s+\cdots+s^n))\,\eta. \end{aligned} \tag{2.586}$$

Estimate (2.584) certainly holds if

$$f_n(\alpha) \leq 0. \tag{2.587}$$

Using hypotheses (2.578) and (2.586), we have that

$$f_n(\alpha) \leq f_{n+1}(\alpha). \tag{2.588}$$

Define a function $f_\infty$ on $[0,1)$ by

$$f_\infty(s) = \lim_{n\longrightarrow\infty} f_n(s). \tag{2.589}$$

In view of (2.576), (2.585) and (2.589), we get that

$$f_\infty(\alpha) \leq \left(\frac{\gamma\,\eta}{1-\alpha} - 1\right)\alpha \tag{2.590}$$

and

$$f_n(\alpha) \leq f_\infty(\alpha). \tag{2.591}$$

Hence, (2.587) is satisfied if

$$\frac{\gamma\,\eta}{1-\alpha} - 1 \leq 0, \tag{2.592}$$

which is true by (2.579). The induction for (2.569) is completed.

**Case (II).** We must again show (2.587). This time by (2.580), (2.581) and (2.586), we have that

$$f_2(\alpha) = f_1(\alpha) + g_1(\alpha) \leq 0. \tag{2.593}$$

Assume $f_k(\alpha) \leq 0$ for all $k \leq n$. Then, again by (2.580) and (2.586), we get that

$$f_{n+1}(\alpha) = f_n(\alpha) + g_n(\alpha) \leq 0. \tag{2.594}$$

We also have by (2.587) and (2.589)

$$f_\infty(\alpha) = \lim_{n \longrightarrow \infty} f_n(\alpha) \leq 0.$$

That completes the induction.

Therefore, the sequence $\{t_n\}$ is non–decreasing, bounded above by $t^{\star\star}$ and as such it converges to its unique least bound $t^\star$ satisfying (2.568).

Finally, estimate (2.570) follows from (2.569) by using standard majorization techniques. This completes the proof of Lemma 2.15. ⊠

**Lemma 2.16** *Let $\gamma_n > 0$, $\delta_n > 0$ and $\eta > 0$ be given parameters. Define functions $\{f_n\}$ $(n \geq 1)$, $\{g_n\}$ $(n \geq 0)$ on $[0, \infty)$ by*

$$f_n(s) = \delta_n\, s^{n-1}\, \eta + \gamma_n\, (1 + s + \cdots + s^n)\, \eta - 1 \tag{2.595}$$

*and*

$$g_n(s) = \gamma_{n+1}\, s^{n+1} + (\delta_{n+1} + \gamma_{n+1} - \gamma_n)\, s^n + (\gamma_{n+1} - \gamma_n - \delta_n)\, s^{n-1}$$
$$+ (\gamma_{n+1} - \gamma_n)\, s^{n-2} + \cdots + (\gamma_{n+1} - \gamma_n). \tag{2.596}$$

*Assume there exist parameters $\alpha$, $\gamma$ such that*

$$\gamma_0 \eta < 1, \quad \gamma_n \leq \gamma \tag{2.597}$$

*and*

$$\frac{\delta_0}{1 - \gamma_0 \eta} \leq \alpha \leq 1 - \gamma \eta. \tag{2.598}$$

*We also suppose that the following assumptions (I) or (II) hold*

*(I) There exists parameter $\alpha$ such that*

$$g_n(\alpha) \geq 0 \tag{2.599}$$

*and*

$$f_\infty(\alpha) \leq 0, \tag{2.600}$$

*where*

$$f_\infty(s) = \lim_{n \longrightarrow \infty} f_n(s). \tag{2.601}$$

*(II)*

$$f_1(\alpha) \leq 0 \tag{2.602}$$

*and*

$$g_n(\alpha) \leq 0. \tag{2.603}$$

*Then, scalar sequence defined by (2.574) is well defined, non–decreasing, bounded from above by*

$$t^{\star\star} = \frac{\eta}{1-\alpha} \tag{2.604}$$

*and converges to its unique least upper bound $t^\star$ satisfying*

$$0 \leq t^\star \leq t^{\star\star}. \tag{2.605}$$

*Moreover, the following estimates hold for all $n \geq 1$*

$$0 \leq t_{n+2} - t_{n+1} \leq \alpha\,(t_{n+1} - t_n) \leq \alpha^{n+1}\,\eta \tag{2.606}$$

*and*

$$0 \leq t^\star - t_n \leq \frac{\eta}{1-\alpha}\,\alpha^n \leq t^{\star\star}. \tag{2.607}$$

**Proof** We shall show using induction that

$$0 < \frac{\delta_n\,(t_{n+1} - t_n)}{1 - \gamma_n\,t_{n+1}} \leq \alpha \quad (n \geq 0). \tag{2.608}$$

Then, estimate (2.606) follows from (2.574) and (2.608). Inequality (2.608) holds for $n = 0$ by (2.598) and (2.574). That is (2.606) holds for $n = 0$. Assume (2.608) holds for all $k \leq n$. Replacing the expression of $t_k$ in (2.583), we obtain that estimate (2.608) certainly holds if

$$\delta_n\,\alpha^k\,\eta + \alpha\,\gamma_k\,\frac{1-\alpha^{k+1}}{1-\alpha}\,\eta - \alpha \leq 0 \tag{2.609}$$

or

$$\delta_n\,\alpha^{k-1}\,\eta + \gamma_k\,(1 + \alpha + \cdots + \alpha^k)\,\eta - 1 \leq 0 \leq 0. \tag{2.610}$$

Estimate (2.610) motivates us to define recurrent functions $f_k$ $(k \geq 0)$ given by (2.595). Estimate (2.608) is true if

$$f_k(\alpha) \leq 0 \quad (k \geq 1). \tag{2.611}$$

We need a relationship between two consecutive functions $f_k$. We have that

$$\begin{aligned} f_{k+1}(s) &= \delta_{k+1}\,s^k\,\eta + \gamma_{k+1}\,(1 + s + \cdots + s^{k+1})\,\eta - 1 \\ &\quad - \delta_k\,s^{k-1}\,\eta - \gamma_k\,(1 + s + \cdots + s^k)\,\eta + 1 + f_k(s). \end{aligned}$$

Hence, we get

$$f_{k+1}(s) = f_k(s) + g_k(s)\,\eta, \tag{2.612}$$

where $g_k$ is given by (2.596).

**Case of (I)** It follows from (2.599), (2.601) and (2.612) that

$$f_\infty(s) \geq f_{k+1}(\alpha) \geq f_k(\alpha) \quad (k \geq 1). \tag{2.613}$$

Hence, (2.611) holds by (2.600) and (2.613).

**Case of (II)** Using (2.602), (2.603) and (2.612), we get

$$f_2(\alpha) = f_1(\alpha) + g_1(\alpha)\eta \leq 0. \tag{2.614}$$

That is, (2.611) holds for $k = 1$.

Assume $f_k(\alpha) \leq 0$ for all $k \leq n$, then again by (2.603) and (2.612)

$$f_{k+1}(\alpha) = f_k(\alpha) + g_k(\alpha)\eta \leq 0,$$

which shows (2.612) for all $k \geq 1$. Note that

$$f_\infty(\alpha) = \lim_{n \longrightarrow \infty} f_n(\alpha) \leq 0.$$

The induction for (2.612) is completed. Hence, sequence $\{t_n\}$ is non–decreasing, bounded above by $t^{\star\star}$ and as such it converges to $t^\star$ satisfying (2.603).

Finally, estimate (2.607) follows from (2.587) by using standard majorization techniques. This completes the proof of Lemma 2.16. ⊠

We need the following assumptions.

$(\mathcal{C}_1)$ There exist $a$, $b$, $\alpha$ and a unique positive zero denoted by $s_n$ for each function $g_n$ such that

$$\gamma_n \leq a,$$

$$\delta_n \leq b$$

and

$$\max\left\{\frac{\delta_0\eta}{1-\gamma_0\eta},\ s^\star\right\} \leq \alpha \leq 1 - a\,\eta,$$

where $s^\star = \sup\limits_{n\geq 0} s_n$.

$(\mathcal{C}_2)$ There exist $a$, $b$, $\alpha$ such that

$$\gamma_{n+1} < \gamma_n \leq a,$$

$$\delta_n \leq b$$

and

$$\max\left\{\frac{\delta_0\,\eta}{1-\gamma_0\,\eta},\ s^\star\right\} \leq \alpha \leq 1 - a\eta.$$

Note that, by (2.596) and $\gamma_{n+1} < \gamma_n$, we have $g_n(0) = \gamma_{n+1} - \gamma_n < 0$. Moreover, $\lim\limits_{s \longrightarrow \infty} g_n(s) = \infty$. It follows by the intermediate value theorem that each function $g_n$ has at least one positive zero $s_n$. By Descarte's rule of signs, $s_n$ is the unique positive zero of $g_n$.

$(\mathcal{C}_3)$ There exist $a$, $b$, $\alpha$ such that

$$\gamma_n \leq \gamma_{n+1} \leq a,$$
$$\delta_n \leq b$$

and

$$\max\left\{\frac{\delta_0\,\eta}{1-\gamma_0\,\eta},\ r^\star\right\} \leq \alpha \leq 1 - a\,\eta,$$

where

$$r_n = \frac{2\,\delta_n}{\delta_{n+1} + \sqrt{\delta_{n+1}^2 + 4\,\gamma_{n+1}\,\delta_n}} \quad \text{and} \quad r^\star = \sup r_n.$$

$(\mathcal{C}_{11})$ Each function $g_n$ has a minimal zero $s_n$ in $(0, 1)$ and there exists $\alpha$ such that

$$\frac{\delta_0\,\eta}{1-\gamma_0\,\eta} \leq \alpha \leq \min\{s^\star,\ s_0\} < 1,$$

where

$$s^\star = \inf s_n \quad \text{and} \quad s_0 = \frac{1-(\gamma_1+\delta_1)\,\eta}{\gamma_1\,\eta}$$

Note that $f_1(s_0) = 0$.

$(\mathcal{C}_{22})$ $\gamma_{n+1} < \gamma_n$ and there exists $\alpha$ such that

$$\frac{\delta_0\,\eta}{1-\gamma_0\,\eta} \leq \alpha \leq \min\{s^\star,\ s_0\} < 1.$$

See $(\mathcal{C}_2)$ for the existence and uniqueness of zeros $s_n$.

$(\mathcal{C}_{33})$ $\gamma_{n+1} \leq \gamma_n$ and there exists $\alpha$ such that

$$\frac{\delta_0\,\eta}{1-\gamma_0\,\eta} \leq \alpha \leq \min\{r^\star,\ s_0\} < 1,$$

where

$$r^\star = \inf r_n.$$

With the notation of Lemma 2.16, we have the following result.

**Corollary 2.1** *Assume together with (2.597) that any of conditions* $(\mathcal{C}_i)$, $(\mathcal{C}_{ii})$ *(*$i = 1, 2, 3$*) holds. Then, the conclusions of Lemma 2.16 hold.*

**Proof** We note that

$$g_n(s) = \overline{g}_n(s) + (\gamma_{n+1} - \gamma_n)\,(1 + s + \cdots + s^n) \tag{2.615}$$

and $r_n$ are the unique positive roots of polynomials $\overline{g}_n$.

**Case** $(\mathcal{C}_1)$ We have $g_n(\alpha) \geq 0$ since $\alpha \geq s^\star$ and for $s \in (0,1)$

$$\begin{aligned} f_\infty(s) &= \lim_{n \longrightarrow \infty} (\delta_n\, s^{n-1}\, \eta + \gamma_n\, (1 + s + \cdots + s^n)\, \eta - 1) \\ &\leq b \lim_{n \longrightarrow \infty} s^{n-1}\, \eta + a \lim_{n \longrightarrow \infty} \frac{1 - s^{n+1}}{1 - s}\, \eta - 1 \\ &= \frac{a\, \eta}{1 - s} - 1. \end{aligned}$$

In particular, $f_\infty(1 - a\, \eta) = 0$, so by the choice of $\alpha$, we conclude $f_\infty(\alpha) \leq 0$.

**Case** $(\mathcal{C}_2)$ This part follows from $(\mathcal{C}_1)$ and the remark after $(\mathcal{C}_1)$.

**Case** $(\mathcal{C}_3)$ It follows from (2.615), $\gamma_{n+1} \leq \gamma_n$ and the choice of $\alpha$ that (2.599) holds. Inequality (2.602) holds in all case $(\mathcal{C}_{ii})$ $(i = 1, 2, 3)$, since $\alpha \leq s_0$. Moreover, (2.603) holds by $\alpha \leq s^\star$. This completes the proof of Corollary 2.1. ⊠

**Remark 2.27** *(1) If $a = b$, we obtain the famous for its simplicity and clarity Newton–Kantorovich hypothesis for solving nonlinear equations*

$$h_K = a\eta \leq \frac{1}{2}. \tag{2.616}$$

*(2) Scalar sequences $\{\beta_n\}$, $\{\gamma_n\}$ and $\{\delta_n\}$ are functions of $t_n$ (see Application 2.2). Then, in the case of $r^\star$, we shall choose an upper bound $\overline{r}^\star$ (or a lower bound $\overline{s}^\star$ in case of $s^\star$) independent of $n$ to replace $t_n$ with such as $\eta$ or $\dfrac{\eta}{1-\alpha}$.*

We provide the main semilocal convergence result of NTM.

**Theorem 2.17** *Let $F : \mathcal{D} \subseteq X \longrightarrow \mathcal{Y}$ be a Fréchet–differentiable operator. Let $G : \mathcal{D} \longrightarrow \mathcal{Y}$ be a continuous operator and $A(x) \in \mathcal{L}(\mathcal{X}, \mathcal{Y})$ be an approximation of $F'(x)$ $(x \in \mathcal{D})$. Assume that there exist $x_0 \in \mathcal{D}$, $\eta > 0$, a bounded inverse $A(x_0)^{-1}$ of $A(x_0)$, functionals $v : \mathcal{D}^2 \longrightarrow [0, \infty)$, $w : \mathcal{D} \longrightarrow [0, 1)$, $z : \mathcal{D} \longrightarrow [0, \infty)$ and $p : \mathcal{D}^2 \longrightarrow [0, \infty)$ such that, for all $x$, $y \in \mathcal{D}$*

$$\| A(x_0)^{-1}\, (F(y) - F(x) - F'(x)\, (y - x))\, \| \leq v(x, y)\, \| y - x \|, \tag{2.617}$$

$$\| A(x_0)^{-1}\, (A(x) - A(x_0))\, \| \leq w(x), \tag{2.618}$$

$$\| A(x_0)^{-1}\, (F'(x) - A(x))\, \| \leq z(x), \tag{2.619}$$

$$\| A(x_0)^{-1}\, (G(x) - G(y))\, \| \leq p(x, y)\, \| x - y \|, \tag{2.620}$$

$$\eta \geq \| A(x_0)^{-1}\, (F(x_0) + G(x_0))\, \|, \tag{2.621}$$

*there exists $\alpha$ such that*

$$0 < \alpha_n = \frac{\beta_n}{1 - \lambda_n} \leq \alpha < 1 \qquad (n \geq 0) \tag{2.622}$$

*and*

$$\overline{U}(x_0, t^\star) \subseteq \mathcal{D}, \tag{2.623}$$

*where*

$$\lambda_n = w(x_{n+1}) \quad \text{and} \quad \beta_n = v(x_n, x_{n+1}) + p(x_n, x_{n+1}) + z(x_n).$$

*Then*

(a) *Scalar sequence $\{t_n\}$ generated by (2.565) is non–decreasing and convergent to $t^\star$, so that estimates (2.567)–(2.570) hold.*

(b) *Sequence $\{x_n\}$ $(n \geq 0)$ generated by NTM is well defined, remains in $\overline{U}(x_0, t^\star)$ for all $n \geq 0$ and converges to a zero $x^\star \in \overline{U}(x_0, t^\star)$ of equation (2.388). Moreover, the following estimates bounds hold for all $n \geq 1$*

$$\| x_{n+1} - x_n \| \leq t_{n+1} - t_n \leq \alpha^n \eta \tag{2.624}$$

*and*

$$\| x_n - x^\star \| \leq t^\star - t_n \leq \frac{\alpha^n \eta}{1 - \alpha}. \tag{2.625}$$

*Furthermore, if there exists $\alpha^\star$ such that*

$$0 < \alpha_n^\star = \frac{\beta_n^\star}{1 - \lambda_n} \leq \alpha^\star < 1 \qquad (n \geq 0), \tag{2.626}$$

*then, the limit point $x^\star$ is unique zero of (2.388) in $U(x_0, t^\star)$, where*

$$\beta_n^\star = v(x_n, x^\star) + p(x_n, x^\star) + z(x_n).$$

**Proof**

(a) This part can be found in the introduction of this study.

(b) We shall show using induction that sequence $\{x_k\}$ is well defined,

$$\| x_{k+1} - x_k \| \leq t_{k+1} - t_k \tag{2.627}$$

and

$$\overline{U}(x_{k+1}, t^\star - t_{k+1}) \subseteq \overline{U}(x_k, t^\star - t_k). \tag{2.628}$$

Let $q \in \overline{U}(x_1, t^\star - t_1)$. Then, estimate

$$\| q - x_0 \| \leq \| q - x_1 \| + \| x_1 - x_0 \| \leq t^\star - t_1 + t_1 - t_0 = t^\star - t_0 \tag{2.629}$$

implies $q \in \overline{U}(x_0, t^\star - t_0)$. We also have

$$\| x_1 - x_0 \| = \| A_0^{-1} (F(x_0) + G(x_0)) \| \leq \eta = t_1 - t_0, \tag{2.630}$$

which together with (2.629) show (2.627) and (2.628) for $k = 0$.

Let assume $\{x_k\}$ is well defined for all $k \leq n$. Using the induction hypotheses, we get

$$\| x_k - x_0 \| \leq \sum_{i=1}^{k} \| x_i - x_{i-1} \| \leq \sum_{i=1}^{k} (t_i - t_{i-1}) = t_k - t_0 = t_k < t^\star.$$

That is, $x_k \in \overline{U}(x_0, t^\star)$. Using (2.563), we obtain the approximation

$$\begin{aligned} & F(x_{k+1}) + G(x_{k+1}) \\ & \quad = F(x_{k+1}) + G(x_{k+1}) - F(x_k) - G(x_k) - A(x_k)\,(x_{k+1} - x_k) \\ & \quad = (F(x_{k+1}) - F(x_k) - F'(x_k)\,(x_{k+1} - x_k)) \\ & \quad + (F'(x_k) - A(x_k))\,(x_{k+1} - x_k) + (G(x_{k+1}) - G(x_k)). \qquad (2.631) \end{aligned}$$

It follows from (2.617), (2.619), (2.620) and (2.631), the definition of $\beta_k$ and the induction hypotheses that

$$\begin{aligned} & \|A(x_0)^{-1}\,(F(x_{k+1}) + G(x_{k+1}))\,\| \\ & \leq (v(x_k, x_{k+1}) + p(x_k, x_{k+1}) + z(x_{k+1}))\ \| x_{k+1} - x_k \| \\ & \leq \beta_k\,(t_{k+1} - t_k). \qquad (2.632) \end{aligned}$$

Using (2.618) for $x = x_{k+1}$ and the Banach lemma on invertible operators, we have $A(x_{k+1})^{-1} \in \mathcal{L}(\mathcal{Y}, \mathcal{X})$ so that

$$\| A(x_{k+1})^{-1}\ A(x_0) \| \leq \frac{1}{1 - w(x_{k+1})} = \frac{1}{1 - \lambda_k}. \qquad (2.633)$$

Then, by (2.563), (2.565), (2.632) and (2.633), we get

$$\begin{aligned} \| x_{k+2} - x_{k+1} \| & \leq \| A(x_{k+1})^{-1}\ A(x_0) \| \| A(x_0)^{-1}\,(F(x_{k+1}) + G(x_{k+1})) \| \\ & \leq \frac{\beta_k\,(t_{k+1} - t_k)}{1 - \lambda_k} = \alpha_k\,(t_{k+1} - t_k) = t_{k+2} - t_{k+1}, \quad (2.634) \end{aligned}$$

which completes the induction for (2.627). Let $q_0 \in \overline{U}(x_{k+1}, t^\star - t_{k+1})$, then, we get

$$\begin{aligned} \| q_0 - x_k \| & \leq \| q_0 - x_{k+1} \| + \| x_{k+1} - x_k \| \\ & \leq t^\star - t_{k+1} + t_{k+1} - t_k = t^\star - t_k, \qquad (2.635) \end{aligned}$$

which shows (2.628) for all $k \geq 1$. The induction for (2.627) and (2.628) is completed. In view of (2.627) and (2.628), sequence $\{x_k\}$ $(k \geq 0)$ is a Cauchy sequence ( since $\{t_n\}$ is a Cauchy sequence) in a Banach space $\mathcal{X}$ and as such it converges to some $x^\star \in \overline{U}(x_0, t^\star)$ (since $\overline{U}(x_0, t^\star)$ is a closed set). By letting $k \longrightarrow \infty$ in (2.632), we get $F(x^\star) + G(x^\star) = 0$ (since $A(x_0)^{-1}$, $\beta_k$

are bounded and $\{t_k\}$ is a Cauchy sequence). Estimate (2.625) follows from (2.627) by using standard majorization techniques.

Finally, to show the uniqueness part, let $y^\star$ be a zero of $F+G$ in $\overline{U}(x_0, t^\star)$. Using (2.563), we get the approximation

$$x_{k+1} - y^\star = A(x_0)^{-1}\left(F(y^\star) - F(x_k) - F'(x_k)\,(y^\star - x_k)\right) \\ + A(x_0)^{-1}\left(F'(x_k) - A(x_k)\right)(y^\star - x_k) + A(x_0)^{-1}\left(G(y^\star) - G(x_k)\right). \tag{2.636}$$

Using (2.564), (2.617)–(2.620), (2.633) and (2.636), we get

$$\| x_{k+1} - y^\star \| \leq \frac{\beta_n^\star}{1-\lambda_n} \| x_k - y^\star \| \\ = \alpha_n^\star \| x_k - y^\star \| \leq \alpha^\star \| x_k - y^\star \| < \| x_k - y^\star \|, \tag{2.637}$$

from which we deduce $x^\star = y^\star$.

This completes the proof of Theorem 2.617. ⊠

Note that $t^{\star\star}$ given in closed form by (2.567) can replace $t^\star$ in Theorem 2.17.In this section, we show how to choose sequences and functions introduced in the previous sections for some interesting cases.

**Application 2.2** *Newton's method (2.2). Let $G(x) = 0$ and $A(x) = F'(x)$ ($x \in \mathcal{D}$).*

*(i) Assume the Lipschitz conditions (cf. [157], [194])*

$$\| F'(x_0)^{-1}\left(F'(x) - F'(y)\right) \| \leq b \ \| x - y \|,$$

$$\| F'(x_0)^{-1}\left(F'(x) - F'(x_0)\right) \| \leq a \ \| x - x_0 \| \quad (x, y \in U(x_0, t^\star) \subseteq \mathcal{D}).$$

*Choose*

$$\beta_n = b, \quad \gamma_n = a \qquad (n \geq 0),$$

$$v(x,y) = \| \int_0^1 \left(F'(x + t\,(y-x)) - F'(x)\right) dt \|, \quad z(x) = 0,$$

$$w(x) = a \ \| x - x_0 \|, \quad p(x,y) = 0 \quad (x, y \in \mathcal{D}).$$

*Then, delicate conditions (2.622) and (2.626) reduce to (2.513) and $a\,t^\star \leq 1$ (or $(a+b)\,t^\star < 1$), respectively. The advantages of this approach over the one given in the Newton–Kantorovich theorem for solving nonlinear equations (cf. [157], [194]) have already been given in Corollary 2.1 with ($\mathcal{C}_3$) and and Remark 2.27.*

*According to Corollary 2.1 with ($\mathcal{C}_3$) and Remark 2.27, our condition (2.513) can be weakened if we can find upper bounds on $\gamma_n$ and $\delta_n$ tighter than $a$ and $b$, respectively. Below we present such cases.*

*(ii) Assume (cf. [88], [157], [194]) that*

$$\| F'(x_0)^{-1} F(x_0) \| \leq \eta,$$

$$\| F'(x_0)^{-1} F''(x_0) \| \leq b_0,$$

$$\| F'(x_0)^{-1} (F''(x) - F''(x_0)) \| \leq b_1 \ \| x - x_0 \|$$

*and*

$$\| F'(x_0)^{-1} (F'(x) - F'(x_0)) \| \leq a \ \| x - x_0 \| \quad (x \in U(x_0, r) \subseteq \mathcal{D}).$$

*Define a polynomial p on $[0, \infty)$ by*

$$p(s) = \frac{b_1}{6} s^3 + \frac{b_0}{2} s^2 - s + \eta.$$

*Note that we must have $p(r) = 0$. In view of the estimate*

$$F(y) - F(x) - F'(x)(y - x) = \int_0^1 F''(x + t(y - x))(1 - t)(y - x)^2 \, dt,$$

*we can choose*

$$v(x, y) = \| \int_0^1 F''(x + t(y - x))(1 - t)(y - x) \, dt \|, \quad w(x) = a \ \| x - x_0 \|,$$

$$z(x) = 0, \quad p(x, y) = 0, \quad \gamma_n = a, \quad \delta_n = \frac{1}{2} \left( \frac{b_1}{3} (t_{n+1} - t_n) + b_0 \right),$$

$$b_2 = \frac{1}{2} \left( \frac{b_1 \eta}{3(1 - \alpha)} + b_0 \right), \quad \overline{b}_2 = \frac{1}{2} \left( \frac{b_1 \eta}{3} + b_0 \right)$$

*and*

$$r^\star \geq \overline{r}^\star = \frac{2 b_2}{\overline{b}_2 + \sqrt{\overline{b}_2^2 + 4 a \overline{b}_2}} \quad (x, y \in \mathcal{D}).$$

*If $b_2 < b$ (used in (i) above), then hypothesis (2.513) with $b_2$ replacing $b$ is weaker than in case (i).*

*The sufficient convergence condition already in the literature (cf. [88], [157], [194]) is*

$$\eta \leq \frac{(b_0^2 + 2 b_1)^{3/2} - 3 b_0 b_1 - b_0^3}{3 b_1^2},$$

*which is different from (2.513) with b being $b_2$ or not.*

*(iii) Assume (cf. [88], [157], [194]) that*

$$\| F'(x_0)^{-1} F(x_0) \| \leq \eta,$$

$$\| F'(x_0)^{-1} F''(x) \| \leq p_1''(\| x - x_0 \|) = \frac{2 \gamma}{(1 - \gamma \ \| x - x_0 \|)^3}$$

*and*

$$\| F'(x_0)^{-1} (F'(x) - F'(x_0)) \| \leq a \| x - x_0 \| \quad \left( x \in U\left( x_0, \frac{1}{\gamma} \left( 1 - \frac{1}{\sqrt{2}} \right) \right) \subseteq \mathcal{D} \right),$$

*where*

$$p_1(s) = \frac{\gamma\, s^2}{1 - \gamma\, s} - s + \eta.$$

*We can choose*

$$v(x, y) = \left\| \int_0^1 F''(x + t\,(y - x))\,(1 - t)\,(y - x)\,dt \right\|, w(x) = a \| x - x_0 \|,$$

$$z(x) = 0, \quad p(x, y) = 0, \quad \gamma_n = a,$$

$$\delta_n = \int_0^1 p_1''(t_n + t\,(t_{n+1} - t_n))\,(1 - t)\,(t_{n+1} - t_n)\,dt$$

*and*

$$b_3 = \frac{1}{2}\, p_1''\left( \frac{\eta}{1 - \alpha} \right).$$

*If $b_3 < b$ ($b$ as used in (i) above), then, hypothesis (2.513) with $b_3$ replacing $b$ in weaker than in case (i).*

*The sufficient convergence condition (cf. [88], [157], [194]) is*

$$\eta \leq \frac{1}{\gamma}\,(3 - 2\sqrt{2}).$$

*The three cases in Application 2.2 constitute an incomplete list of possibilities. Note that, in practice, we will test these conditions to see which one is verified (if any). If more than one set of conditions is verified, then, we will use the one providing the tighter error bounds on the distances $\| x_{n+1} - x_n \|$, $\| x_n - x^\star \|$ and the more precise information on the location of the solution $x^\star$.*

**Application 2.3** *Newton–type method (2.563).*
*Assume the following conditions (cf. [157], [194])*

$$\| A(x_0)^{-1} (F(x_0) + G(x_0)) \| \leq \eta,$$

$$\| A'(x_0)^{-1} (F'(x) - F'(y)) \| \leq c_1 \| x - y \|,$$

$$\| A'(x_0)^{-1} (F'(x) - A(x)) \| \leq c_2 \| x - x_0 \| + c_3,$$

$$\| A'(x_0)^{-1} (A(x) - A(x_0)) \| \leq c_4 \| x - x_0 \| + c_5,$$

$$\| A'(x_0)^{-1} (G(x) - G(y)) \| \leq c_6 \| x - y \| \; (x, y \in U(x_0, r) \subseteq \mathcal{D}),$$

*where*

$$p_2(s) = \frac{\sigma}{2}\, s^2 - (1 - \sigma_0)\, s + \eta, \quad \sigma_0 = c_3 + c_5,$$

$$\sigma = \max\{c_1,\, c_2 + c_4\} \quad \text{and} \quad p_2(r) = 0.$$

*We can choose for* $x, y \in \mathcal{D}$ *the following functions*

$$v(x,y) = \| \int_0^1 (F'(x + t\,(y-x)) - F'(x))\,dt\,(x-y) \,\|,$$

$$w(x) = c_4 \, \| \, x - x_0 \, \| + c_5, \quad p(x,y) = c_6, \quad z(x) = c_2 \, \| \, x - x_0 \, \| + c_3,$$

*constant* $\alpha$ *(used in (2.622)) by*

$$\alpha = \frac{c_7}{c_8},$$

*where*

$$c_7 = \frac{c_4\,\eta}{1-\alpha} \quad \text{and} \quad c_8 = 1 - \left( \frac{1}{2}\,\frac{c_1\,\eta}{1-\alpha} + \frac{c_2\,\eta}{1-\alpha} + c_3 + c_6 \right).$$

*The hypothesis in (cf. [157], [194])*

$$\eta \leq \frac{1}{2\,\sigma}\,(1-\sigma_0)^2.$$

*More general than the above choices of functions* $v$*,* $w$ *and* $z$ *can be found in (cf. [139]).*

Applications and examples can be found in (cf. [163], [209]), see also section 2.1.

## 2.11 Upper Bounds for Newton's Method

We present in this section upper bounds on the limit points of majorizing sequences for Newton's method (2.2). Numerical examples are provided, where our new results compare favorably to earlier ones such as (cf. [139], [198], [209], [569], [608]).

Note that if the Kantorovich condition

$$h_\star = L\eta \leq \frac{1}{2} \tag{2.638}$$

holds, where $L$ and $\eta$ are given by (2.3) and (2.4), respectively, then, Newton's method (2.2) converges to a solution $x^\star$ of equation (2.1) (cf. [163], [469]). Using the majorizing sequence $\{u_n\}$ given by

$$u_0 = 0, \quad u_1 = \eta, \quad u_{n+2} = u_{n+1} + \frac{L\,(u_{n+1}-u_n)^2}{2\,(1 - L\,u_{n+1})} \quad (n \geq 0), \tag{2.639}$$

a plethora of error estimates on the distances $\| \, x_{n+1} - x_n \, \|$, $\| \, x_{n+1} - x^\star \, \|$ $(n \geq 0)$ have been given by several authors (cf. [163], [209] and the references therein). For examples, we have for all $n \geq 0$

$$\| \, x_{n+1} - x_n \, \| \leq u_{n+1} - u_n \tag{2.640}$$

and

$$\| x_{n+1} - x^\star \| \leq \varrho^\star - u_n, \tag{2.641}$$

where,

$$\varrho^\star = \lim_{n\to\infty} u_n = \frac{1 - \sqrt{1 - 2\,L\,\eta}}{L}. \tag{2.642}$$

However, (2.638) is only sufficient to guarantee the convergence of Newton's method (2.2). There are examples in the literature where (2.638) is not satisfied but $\{x_n\}$ generated by (2.2) converges to $x^\star$.

Therefore, it is possible and very important to find weaker convergence criteria and tighter error estimates. We first state what has already been done in this direction. It follows from (2.3) that there exists $L_0 > 0$ such that center-Lipschitz condition (2.7) is satisfied. Using the more precise that $\{u_n\}$ majorizing sequence $\{t_n\}$ given by

$$t_0 = 0, \quad t_1 = \eta, \quad t_{n+2} = t_{n+1} + \frac{L\,(t_{n+1} - t_n)^2}{2\,(1 - L_0\,t_{n+1})} \quad (n \geq 0), \tag{2.643}$$

we provided tighter error estimates and weaker sufficient convergence conditions for Newton's method (cf. [139], [198], [209], [569], [608]). These conditions are given by

$$h_1 = L_1\,\eta \leq \frac{1}{2}, \tag{2.644}$$

$$h_2 = L_2\,\eta \leq \frac{1}{2}, \tag{2.645}$$

where,

$$L_1 = \frac{L_0 + L}{2} \quad \text{and} \quad L_2 = \frac{1}{8}\,(L + 4\,L_0 + (L^2 + 8\,L_0\,L)^{1/2}).$$

We also showed that

$$t^\star = \lim_{n\to\infty} t_n \leq r_1 := 2\,\eta \leq r_2 := \frac{\eta}{1-\alpha}, \tag{2.646}$$

where,

$$\alpha = \frac{2\,L}{L + (L^2 + 8\,L_0\,L)^{1/2}}. \tag{2.647}$$

Recently, we used in (cf. [210]) majorizing sequence $\{s_n\}$ given by

$$\begin{aligned} s_0 = 0, \quad s_1 = \eta, \quad s_2 = \eta + \frac{L_0\,\eta^2}{2\,(1 - L_0\,\eta)} \\ s_{n+2} = s_{n+1} + \frac{L\,(s_{n+1} - s_n)^2}{2\,(1 - L_0\,s_{n+1})} \quad (n \geq 1). \end{aligned} \tag{2.648}$$

We showed convergence provided that

$$h_3 = L_3\,\eta \leq \frac{1}{2} \tag{2.649}$$

and

$$s^\star = \lim_{n\to\infty} s_n \le r_3 := \left(1 + \frac{L_0\,\eta}{2\,(1-\alpha)\,(1-L_0\,\eta)}\right)\eta \tag{2.650}$$

where,

$$L_3 = \frac{1}{8}\,((L_0\,L)^{1/2} + 4\,L_0 + (8\,L_0^2 + L_0\,L)^{1/2}).$$

Note that

$$r_3 \begin{cases} \le r_2 & \text{if} \quad L_0 \le L \\ < r_2 & \text{if} \quad L_0 < L \end{cases} \quad \text{and} \quad r_3 \begin{cases} \le r_1 & \text{if} \quad L_0 \le L \\ < r_1 & \text{if} \quad L_0 < L. \end{cases} \tag{2.651}$$

Moreover, $r_2$ can be smaller than $t^\star$ for sufficiently small $L_0$. It follows from (2.638), (2.644), (2.645) and (2.649)

$$h_\star \le \frac{1}{2} \Longrightarrow h_1 \le \frac{1}{2} \Longrightarrow h_2 \le \frac{1}{2} \Longrightarrow h_3 \le \frac{1}{2}, \tag{2.652}$$

but not necessarily vice versa unless if $L_0 = L$. Moreover, we have

$$\left.\begin{array}{lll} \dfrac{h_1}{h_\star} \longrightarrow \dfrac{1}{2}, & \dfrac{h_2}{h_\star} \longrightarrow \dfrac{1}{4}, & \dfrac{h_2}{h_1} \longrightarrow \dfrac{1}{2} \\ \dfrac{h_3}{h_\star} \longrightarrow 0, & \dfrac{h_3}{h_2} \longrightarrow 0, & \dfrac{h_3}{h_1} \longrightarrow 0 \end{array}\right\} \text{as} \quad \frac{L_0}{L} \longrightarrow 0. \tag{2.653}$$

Clearly, $\{s_n\}$ is tighter than $\{t_n\}$ which is tighter than $\{u_n\}$ provided that $L_0 < L$ and $s^\star \le t^\star \le \varrho^\star$.

We present a family of new sufficient convergence conditions for Newton's method which can be weaker than all the above "$h$" conditions. These conditions lead to tighter upper bounds on the limit points of majorizing sequences.

We need auxiliary results concerning the convergence of majorizing sequence $\{t_n\}$.

**Lemma 2.17** *Let $a$, $b$, $c$ be real constants, $L_0 > 0$ and $\lambda \in [0, 1/L_0]$. Define quadratic polynomials $p$, $q$ on $(-\infty, +\infty)$ and functions $f$, $g$ on $[0, 1/L_0)$ by*

$$p(t) = at^2 + bt + c, \tag{2.654}$$

$$f(t) = \frac{p(t)}{1 - L_0\,t},$$

$$g(t) = t + f(t) \tag{2.655}$$

*and*

$$q(t) = (1 - L_0\,t)^2\,g'(t) = L_0\,(L_0 - a)\,t^2 - 2\,(L_0 - a)\,t + 1 + b + L_0\,c. \tag{2.656}$$

*Suppose that the polynomial $p$ has a unique root $\varrho$ in the interval $[\lambda, 1/L_0]$ ,*

$$p(\lambda) \ge 0, \tag{2.657}$$

$$p(1/L_0) \leq 0 \tag{2.658}$$

*and*

$$q(t) \geq 0 \quad for\ each \quad t \in [\lambda, \varrho]. \tag{2.659}$$

*Then, function $g$ is nondecreasing and bounded above by $\varrho$ for each $t \in [\lambda, \varrho]$. The root $\varrho$ is given by*

$$\varrho = \frac{-b \pm (b^2 - 4ac)^{1/2}}{2\,a}. \tag{2.660}$$

**Proof** The expression under the radical in (2.660) is nonnegative, since $p$ has a unique root $\varrho$ in $[\lambda, 1/L_0]$. Polynomial $p$ is obviously nonnegative on the interval $[\lambda, \varrho)$, since $\varrho$ is the only root of $p$ in $[\lambda, \varrho]$. Moreover, the function $f$ is nonnegative on $[\lambda, \varrho]$ with the only exception when $\varrho = 1/L_0$; but the L'Hospital theorem implies that $f$ admits a continuous extension on the interval $[\lambda, \varrho]$. We must prove that function $g$ is nondecreasing on the interval $[\lambda, \varrho]$. In fact, its derivative is given by

$$g'(t) = 1 + f'(t) = \frac{q(t)}{(1 - L_0\, t)^2}. \tag{2.661}$$

It follows from (2.659) and (2.661) that $g'(t) \geq 0$ for each $t \in [\lambda, \varrho]$. Therefore, we have for each $t \in [\lambda, \varrho]$ that

$$g(t) = t + f(t) \leq \varrho + f(\varrho) = \varrho. \tag{2.662}$$

That completes the proof of Lemma 2.17. ⊠

**Lemma 2.18** *Let $L > 0$ be a constant. Suppose hypotheses of Lemma 2.17. Define functions $\phi$ and $\psi$ on $\mathcal{I} = [\lambda, t] \times [s, \varrho]$ for each $t \in [\lambda, \varrho]$ by*

$$\phi(s,t) = t + \frac{\frac{L}{2}\,(t-s)^2}{1 - L_0\, t} \tag{2.663}$$

*and*

$$\psi(s,t) = \begin{cases} \frac{L}{2}\,(t-s)^2 - p(t) & if \quad t \neq \varrho \\ 0 & if \quad t = \varrho. \end{cases} \tag{2.664}$$

*Moreover, suppose that*

$$\psi(s,t) \leq 0 \quad for\ each \quad (s,t) \in [\lambda, t] \times [s, \varrho]. \tag{2.665}$$

*Then, the following assertion holds*

$$\phi(s,t) \leq g(t) \quad for\ each \quad (s,t) \in [\lambda, t] \times [s, \varrho]. \tag{2.666}$$

**Proof** The inequality (2.665) follows immediately from the definition of functions $g$, $\phi$, $\psi$ and hypothesis (2.664). That completes the proof of Lemma 2.18. ⊠

**Lemma 2.19** *Let $N = 0, 1, 2, \cdots$ be fixed. Under the hypotheses of Lemma 2.18 with $\lambda = t_N$, further suppose that*

$$t_1 \leq t_2 \leq \cdots \leq t_N \leq t_{N+1} \leq \varrho \tag{2.667}$$

*and*

$$p(t_{N+1}) \geq 0. \tag{2.668}$$

*Then, sequence $\{t_n\}$ generated by (2.643) is nondecreasing, bounded by $\varrho$ and converges to its unique least upper bound $t^\star$ which satisfies $t^\star \in [t_N, \varrho]$.*

**Proof** We can write

$$t_{n+1} = \phi(t_{n-1}, t_n). \tag{2.669}$$

Using (2.662), (2.666) and (2.669), we get

$$t_{N+2} = \phi(t_N, t_{N+1}) \leq g(t_{N+1}) \leq \varrho. \tag{2.670}$$

Moreover, we get by (2.643), (2.667) and (2.668) that $t_{N+1} \leq t_{N+2}$. That completes the proof of Lemma 2.19. ⊠

**Remark 2.28** *(a) The hypotheses of Lemma 2.19 are satisfied in the Kantorovich case. Indeed, let $N = 0$, $\lambda = 0$, $a = L/2$, $b = -1$, $c = \eta$ and $L_0 = L$. Suppose (2.638) holds. Then, all hypotheses of Lemma 2.19 are satisfied (cf. [569, 12.6.2, p. 422]). Moreover, the sequence $\{t_n\}$ reduces to $\{u_n\}$ in this case. Furthemore, if $L_0 \neq L$, sequence $\{t_n\}$ is tighter than $\{u_n\}$ and converges to $t^\star$ provided that (2.638) or (2.644) or (2.645) hold.*

*In a more general setting, if $t_1 \leq t_2 \leq \cdots \leq t_N \leq t_{N+1} < 1/L$ for some $N = 0, 1, 2, \cdots$, iteration $\{t_n\}$ can be written as*

$$t_{N+n+1} = t_{N+n} - \frac{p_N(t_{N+n})}{p'_N(t_{N+n})} \quad \textit{for each} \quad n = N, N+1, \cdots,$$

*where,*

$$p_N(t) = \frac{L}{2}\, t^2 - t + t_{N+1} - t_N.$$

*Then, the Kantorovich hypothesis is given by*

$$h_\star^N = L\,(t_{N+1} - t_N) \leq \frac{1}{2}.$$

*We also have*

$$\lambda = t_N, \quad a = \frac{L}{2}, \quad b = -1 \quad \textit{and} \quad c = t_{N+1} - t_N.$$

*Note that for $N = 0$, we obtain the usual Kantorovich iteration and $h_\star^0 = h_\star$.*

(b) *Let us define function* $\psi_N$ *on* $[\lambda, \varrho]$ *by*

$$\psi_N(t) = (\frac{L}{2} - a)\, t^2 - (b + L\, t_N)\, t + \frac{L}{2}\, t_N^2 - c.$$

*Then, since* $0 \le t - s \le t - t_N$ *for each* $s \ge t_N$, $t \ge s$ *and for* $t_N = \lambda$, *function* $\psi_N$ *can replace* $\psi$ *in hypothesis (2.665).*

(c) *Cartesian product* $\mathcal{I}$ *can be replaced by the more practical* $\mathcal{J} = [\lambda, \varrho]^2$ *in Lemma 2.18.*

(d) *The conclusions of Lemma 2.17 hold if hypotheses above (2.659) are replaced by the hypotheses "Suppose that polynomial* $p$ *has a root in the interval* $[\lambda, 1/L_0]$. *Denote by* $\varrho$ *the smaller root of* $p$ *in the interval* $[\lambda, 1/L_0]$*".*

(e) *We have chosen* $p$ *to be a quadratic polynomial, since it can have a unique root* $\varrho$ *inside the interval* $[0, 1/L_0]$ *given in closed form by (2.660). However, the results obtained in Lemmas 2.17–2.19 hold if* $p$ *is a function with a unique preferably in closed form zero* $\varrho$ *on* $[0, 1/L_0]$. *Note that for example function* $q$ *will then be given by*

$$q(t) = (1 - L_0\, t)^2 + p'(t)\,(1 - L_0\, t) + L_0\, p(t).$$

(f) *It follows from (2.655) that (2.664) should only hold for* $t = g(s)$.

**Remark 2.29** *We have already given in Remark 2.28 a combination of constants* $a$, $b$, $c$, $L_0$, $L$, $\eta$ *and functions* $p$, $q$, $g$, $\psi$, $\phi$ *such that the hypotheses of Lemma 2.19 are satisfied. Many other choices exist. Let us consider the case* $a = 2\, L_0$. *Then, (2.665) is certainly satisfied provided that*

$$-(b + L\, s)\, t + \frac{L}{2}\, s^2 - c \le 0$$

*or if*

$$t \le \frac{\frac{L}{2}\, s^2 - c}{b + L\, s} \quad \text{and} \quad b + L\, s < 0$$

*or if*

$$\frac{L}{2}\, t^2 - c < 0, \quad b + L\, t < 0 \quad \text{and} \quad \frac{\frac{L}{2}\, t^2 - c}{b + L\, t} \ge t_N.$$

*Then, all hypotheses of Lemma 2.19 except (2.659) are satisfied if*

$$b < 0, \quad c > 0, \quad b^2 - 4\, a\, c \ge 0, \quad \varrho < \varrho_N^0 \quad \text{and} \quad t_N \le \lambda_0,$$

*where,*

$$\varrho_N^0 = \frac{1}{L}\, \min\{(2\, L\, c)^{1/2},\, L\, t_N + (L\,(L\, t_N^2 + 2\, b\, t_N + 2\, c)^{1/2},\, -b,\, 1\}$$

*and*

$$\lambda_0 = \min\{\varrho, -\frac{c}{b}\}.$$

*There are many possibilities under which the remaining hypothesis (2.659) is satisfied. Choose for example* $b = -1$. *Suppose* $1 + b + L_0\, c > 0$ *and* $L_0 < 2\, L$. *Then, quadratic polynomial* $q$ *has a positive root* $q^\star$ *given by*

$$q^\star = \frac{1 + b + L_0\, c}{L_0 - 2\, L + ((L_0 - 2\, L)^2 - L_0\, (L_0 - 2\, L)\, (1 + b + L_0\, c))^{1/2}}$$

*and a negative root. Then, (2.659) is satisfied if* $t \in [0, q^\star]$, *so we must have*

$$t_N \leq q^\star \quad \text{and} \quad \varrho \leq q^\star.$$

*Hence, we conclude that hypotheses of Lemma 2.19 are satisfied for*

$$L_0 < 2\, L, \quad a = 2\, L, \quad 8\, L\, c \leq b^2, \quad 1 + b + L_0\, c > 0,$$

$$\varrho < \varrho_N^\star$$

*and*

$$t_N \leq \lambda^\star$$

*where,*

$$\varrho_N^\star = \min\{\varrho_N^0, q^\star\} \quad \text{and} \quad \lambda^\star = \min\{\lambda_0, q^\star\}.$$

**Remark 2.30** *(a) Another interesting choice but not necessarily the best possible is given by the following scheme. Choose*

$$a = \frac{L_0}{2},\ b = -1,\ c = \frac{L_i}{L_0}\, (t_N - t_{N-1}) \quad \textit{and} \quad \lambda = t_N \ \textit{for each}\ N = 1, 2, \cdots.$$

*Suppose that*

$$h_i^N = L_i\, (t_N - t_{N-1}) \leq \frac{1}{2} \quad \textit{for some} \quad i = 1, 2, 3.$$

*Set*

$$\varrho_i^N = \frac{2\, (t_N - t_{N-1})\, L_i}{L_0\, (1 + (1 - 2\, L_i\, (t_N - t_{N-1}))^{1/2})},\ i = 1, 2, 3.$$

*Moreover, suppose*

$$\psi(t_N, t) \leq 0 \quad \textit{for each} \quad t = g(t_N) \in [t_N, \varrho_i^N].$$

*Then,* $\varrho_i^N$ *are well defined. Elementary computations now show that all hypotheses of Lemma 2.19 are satisfied for* $\varrho = \varrho_i^N$.

*Moreover, the following estimates hold for each* $i = 1, 2, 3$

$$\frac{2\, (t_N - t_{N-1})}{1 + (1 - 2\, L_i\, (t_N - t_{N-1}))^{1/2})} \leq \varrho_i^N \leq \frac{1}{L_0},$$

$$\frac{L_i}{L_0} \geq 1$$

*and*

$$\varrho_i^N \leq \varrho_{i-1}^N.$$

*Furthemore, if*

$$(L_0^2 + L\,L_i)\,(t_N - t_{N-1}) \leq L_0\,(1 - (1 - 2\,h_i^N)^{1/2}\,(1 - 2\,h_\star)^{1/2})$$

*and*

$$h_\star \leq \frac{1}{2}$$

*then, we obtain that*

$$\varrho_i^N \leq \varrho^\star = \frac{2\,(t_N - t_{N-1})}{1 + (1 - 2\,h_\star)^{1/2}}.$$

*We also have that for sufficiently small* $t_N - t_{N-1}$, $\varrho_i^N$ *are smaller than* $r_i$. *Indeed, for example, we have that if*

$$L < 3\,L_0 \quad \text{and} \quad h_1 \leq \frac{1}{2}\left(1 - \left(\frac{L - L_0}{2\,L_0}\right)^2\right)$$

*then*

$$\varrho_1^1 \leq r_1.$$

*Similar results hold for the rest of the* $\varrho_i^N$ *and* $r_i$ *for* $i = 2, 3$.

*(b) Constant* $L$ *can replace* $L_0$ *in the definition of* $c$. *Then, we have*

$$\frac{L_i}{L_0} \leq 1 \quad \text{and} \quad \varrho_i^N \leq \varrho_{i-1}^N \leq \varrho^\star \leq 1/L_0 \quad \text{for} \quad i = 1, 2, 3.$$

Next we state a semilocal convergence result for Newton's method using Lipschitz conditions (2.3) and (2.7). The result is obtained by simply using hypotheses of Lemma 2.19 instead of (2.644) in the proof of Theorem 2.5 in (cf. [136]).

**Theorem 2.18** *Let* $F : \mathcal{D} \subseteq \mathcal{X} \longrightarrow \mathcal{Y}$ *be Fréchet-differentiable. Suppose there exist* $x_0 \in \mathcal{D}$, $L \geq L_0 > 0$ *and* $\eta > 0$ *such that* $F'(x_0)^{-1} \in \mathcal{L}(\mathcal{Y}, \mathcal{X})$, *(2.3), (2.4), (2.7) hold,* $\overline{U}(x_0, t^\star) \subseteq \mathcal{D}$ *and hypotheses of Lemma 2.19 hold, where* $t^\star$ *is given in Lemma 2.19. Then, sequence* $\{x_n\}$ *generated by Newton's method is well defined, remains in* $\overline{U}(x_0, t^\star)$ *and converges to a solution* $x^\star \in \overline{U}(x_0, t^\star)$ *of equation* $F(x) = 0$. *Moreover, the following estimates hold*

$$\| x_{n+1} - x_n \| \leq t_{n+1} - t_n$$

*and*

$$\| x_n - x^\star \| \leq t^\star - t_n \leq \varrho - t_n \quad \text{for each} \quad n = 0, 1, 2, \cdots,$$

*where sequence* $\{t_n\}$ *and* $\varrho$ *are given by (2.643) and (2.660), respectively. Furthermore, if there exists* $R > t^\star$ *such that*

$$U(x_0, R) \subseteq \mathcal{D} \quad \text{and} \quad L_0\,(t^\star + R) \leq 2,$$

*then, the solution* $x^\star$ *of equation* $F(x) = 0$ *is unique in* $U(x_0, R)$.

**Remark 2.31** *(a) If* $L = L_0$ *Theorem 2.18 reduces to Kantorovich theorem (cf. [469]).*

*(b) The limit points* $s^\star$, $t^\star$ *can be replaced by* $\varrho$ *given in closed form (see (2.660)).*

*(c) Clearly, if* $N = 2, 3, \cdots$, *sequence* $\{t_n\}$ *can be replaced by tighter* $\{s_n\}$ *in all results involving* $\{t_n\}$ *from Lemma 2.19 until Theorem 2.18.*

We provide examples where $L_0 < L$ so we can compare the "$h$" conditions.

**Example 2.14** *We consider our conditions (2.644), (2.645) and (2.649) in the example (see (2.6) with* $a = d$*). Then, we obtain that*

$$h_1 = \frac{1}{6}\,(7 - 3\,d)\,(1 - d) \leq .5 \quad for\ all \quad d \in [.4648162415, .5),$$

$$h_2 = \frac{1}{12}\,(8 - 3\,d + (5\,d^2 - 24\,d + 28)^{1/2})\,(1 - d) \leq .5 \quad for\ all \quad d \in [.450339002, .5)$$

*and*

$$h_3 = \frac{1}{24}\,(1 - d)\,(12 - 4\,d + (84 - 58\,d + 10\,d^2)^{1/2} + (12 - 10\,d + 2\,d^2)^{1/2}) \leq .5$$
$$for\ all \quad d \in [.4271907643, .5).$$

*Next, we pick some values of d such that all hypotheses are satisfied, so we can compare the "h" conditions. Using Maple 13, we can compare conditions (2.644), (2.645) and (2.649) (see Table 1).*

**Table 1**

| $d$ | $x^\star$ | $h_\star$ | $h_1$ | $h_2$ | $h_3$ |
|---|---|---|---|---|---|
| .49999 | .7936952346 | .5000133334 | .4583450000 | .4438990778 | .4233746025 |
| .52 | .8041451517 | .4736000000 | .4352000000 | .4219021608 | .4029793108 |
| .68 | .8793659344 | .2816000000 | .2645333333 | .2586823517 | .2502395182 |
| .7555 | .9107758147 | .2028535000 | .1928901250 | .1894931527 | .1845538775 |
| .86 | .9509685413 | .1064000000 | .1031333333 | .1020291664 | .1004043962 |
| .95 | .9830475725 | .03500000000 | .03458333333 | .03444370170 | .03423578560 |

*In Fig. 2.2, we compare the "h" conditions for $d \in (0, .9)$.*

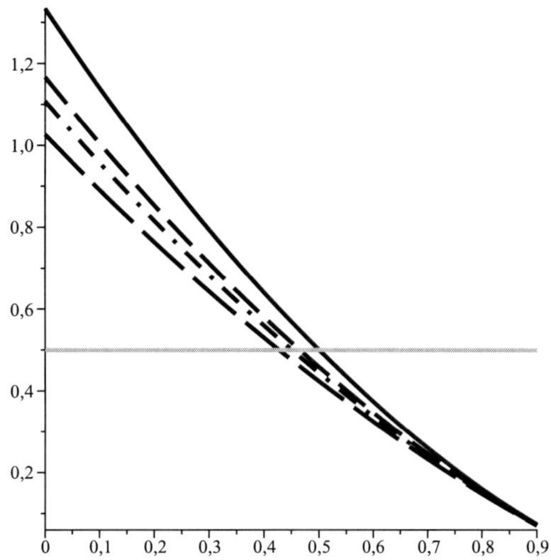


**Figure 2.2:** Functions $h_\star$, $h_1$, $h_2$ and $h_3$ (from top to bottom) with respect to $d$ in interval $(0, .9)$, respectively. The horizontal red line is of equation $y = .5$.

**Example 2.15** *We also consider Example 2.14 in one more case. Let $d = .74137931$. Then, $L = 2.51724138$, $L_0 = 2.25862069$ and $\eta = .0862068966$. Using (2.646), (2.647), (2.650), (2.660) and hypotheses of Lemma 2.17 for $b = -1$, $c = \eta$, $a = 2.4$ we obtain*

$$\varrho = .1218276252, \quad \alpha = .5181703378, \quad 1/L_0 = .4427480915,$$

$$r_1 = .1724137932, \quad r_2 = .1789157110 \quad \text{and} \quad r_3 = .1078366107.$$

*Using (2.643), we get*

$$t_2 = .09782207463, \quad t_3 = .09804003491, \quad t_4 = .09804011171$$

*and for all $n \geq 5$*

$$t_n = t_4 = .09804011171 = t^\star < \varrho^\star = .09839144902 < .4427480915 = 1/L_0.$$

*Moreover, if we use the choices in Remark 2.30 (a) with $N = 1$, we get*

$$L_1 = 2.387931037, \quad L_2 = .343795815, \quad L_3 = 2.279717419,$$

$$\varrho_1^1 = .1031606656, \quad \varrho_2^1 = .100971435 \quad \text{and} \quad \varrho_3^1 = .09781769768.$$

*We also have by (2.664) that*

(i) <u>*For $\varrho := \varrho_1^1$*</u>

$$\psi(\eta, t) = .1293103450t^2 + .7829964326t - .09357445995 \quad \text{for} \quad \eta \leq t < \varrho_1^1;$$

(ii) For $\varrho := \varrho_2^1$

$$\psi(\eta,t) = .1293103450t^2 + .7829964326t - .09167207975 \quad for \quad \eta \le t < \varrho_2^1;$$

(iii) For $\varrho := \varrho_3^1$

$$\psi(\eta,t) = .1293103450t^2 + .7829964326t - .08891007985 \quad for \quad \eta \le t < \varrho_3^1.$$

*Hence, hypotheses of Lemma 2.19 are satisfied for* $\lambda = \eta$ *and* $\varrho := \varrho_i^1$ *for* $i = 1,2,3$. *Finally, note that we have*

$$r_3 \le r_1 \le r_2, \quad \varrho_3^1 \le \varrho_2^1 \le \varrho_1^1 \quad and \quad r_i \ge \varrho_i^1 \quad for \quad i = 1,2,3.$$

# Chapter 3

# Secant Method

Secant method SM is an alternative method of Newton's method NM studied in chapter 2. We study in this chapter the convergence of SM under various conditions and space settings.

## 3.1 Convergence

We establish in this section new sufficient convergence conditions for Secant method to a locally unique solution of a nonlinear equation in a Banach space. Using our new concept of recurrent functions and combining Lipschitz and center–Lipschitz conditions on the divided difference operator, we obtain a new semilocal convergence analysis of the Secant method. Moreover, our sufficient convergence conditions expand the applicability of the Secant method in cases not covered before (cf. [313], [427], [496], [596], [600], [668], [770]).

We are concerned with the problem of approximating a locally unique solution $x^\star$ of equation (2.1) using Secant method SM

$$x_{n+1} = x_n - \delta F(x_{n-1}, x_n)^{-1}\ F(x_n) \quad (n \geq 0,\ x_{-1},\ x_0 \in \mathcal{D}), \tag{3.1}$$

where $\delta F(x, y) \in \mathcal{L}(\mathcal{X}, \mathcal{Y})$ $(x, y \in \mathcal{D})$ is a consistent approximation of the Fréchet–derivative of $F$. The Banach spaces $\mathcal{X}$ and $\mathcal{Y}$ are equipped with the norm $\| \, . \, \|$. SM is an alternative method of NM given by (2.2).

Bosarge and Falb (cf. [238]), Potra (cf. [596], [600]), Hernández et al. (cf. [427]), Dennis (cf. [313]), Argyros (cf. [163]) and others (cf. [426], [569], [792]) have provided sufficient convergence conditions for SM based on Lipschitz–type conditions on $\delta F$ (see also relevant works in (cf. [149], [139], [194], [496], [613], [614], [668], [770])).

The conditions usually associated with the semilocal convergence of Secant method (3.1) are:

($\mathcal{H}_1$) $F$ is a nonlinear operator defined on a convex subset $\mathcal{D}$ of a Banach space $\mathcal{X}$ with values in a Banach space $\mathcal{Y}$;

($\mathcal{H}_2$) $x_{-1}$ and $x_0$ are two points belonging to the interior $\mathcal{D}^0$ of $\mathcal{D}$ and satisfying the inequality

$$\| x_0 - x_{-1} \| \leq c;$$

($\mathcal{H}_3$) $F$ is Fréchet–differentiable on $\mathcal{D}^0$ and there exists an operator $\delta F : \mathcal{D}^0 \times \mathcal{D}^0 \to \mathcal{L}(\mathcal{X}, \mathcal{Y})$ such that the linear operator $A = \delta F(x_{-1}, x_0)$ is invertible, its inverse $A^{-1}$ is bounded,

$$\| A^{-1} F(x_0) \| \leq \eta,$$

$$\| A^{-1} (\delta F(x,y) - F'(z)) \| \leq \ell(\| x - z \| + \| y - z \|) \quad (x, y, z \in \mathcal{D}),$$

$$\overline{U}(x_0, r) \subseteq \mathcal{D}^0$$

for some $r > 0$ depending on $\ell$, $c$ and $\eta$

and

$$\ell c + 2\sqrt{\ell \eta} \leq 1. \tag{3.2}$$

The sufficient convergence condition (3.2) is easily violated (see Example 3.2). Hence, there is no guarantee that equation (2.1) under the information $(\ell, c, \eta)$ has a solution that can be found using SM. Here, we are motived by the above observation and optimization consideration.

It is convenient for us to introduce some constants, functions and sequences.

**Definition 3.1** *Let $\ell_0 > 0$, $\ell > 0$, $c \geq 0$ and $\eta \in (0, c]$ be given constants such that*

$$(\ell + 2\,\ell_0)\,\eta + \ell_0\, c < 1. \tag{3.3}$$

*Define constants $\delta_0$, $\delta_1$ and $b$ by*

$$\delta_0 = \frac{\ell\,(c+\eta)}{1 - \ell_0\,(c+\eta)},$$

$$\delta_1 = \frac{2\,(1 - ((\ell + 2\,\ell_0)\,\eta + \ell_0\,c))}{(\ell + 2\,\ell_0)\,\eta + \sqrt{((\ell + 2\,\ell_0)\,\eta)^2 + 4\,\ell_0\,\eta\,(1 - ((\ell + 2\,\ell_0)\,\eta + \ell_0\,c))}},$$

*and*

$$b = \frac{\ell + 4\,\ell_0 + \sqrt{\ell^2 + 8\,\ell_0\,\ell}}{4}.$$

*Define a function $f$ on $(0,1)$ by*

$$f(t) = \ell_0\, t^3 + (\ell_0 + \ell)\, t^2 - \ell \tag{3.4}$$

*and $\delta$ denotes the unique positive root of polynomial $f$ in $(0,1)$ given by*

$$\delta = \frac{1}{2}\,\frac{-\ell + \sqrt{\ell^2 + 4\,\ell\,\ell_0}}{\ell_0}.$$

*The existence and the uniqueness of $\delta$ follow from that the other roots of $f$ are negative:*

$$-1 \quad \text{and} \quad \frac{-1}{2}\,\frac{\ell + \sqrt{\ell^2 + 4\,\ell\,\ell_0}}{\ell_0}.$$

*Moreover, in view of (3.3), we have* $\delta_0 \in [0,1)$. *Define functions* $\phi$, $\varphi$ *and* $\psi$ *on* $\mathbb{R}^3$ *by*

$$\phi(x,y,z) = z + \frac{\ell\,(z-x)\,(z-y)}{1-\ell_0\,(z-c+y)}, \tag{3.5}$$

$$\varphi(x,y,z) = z + \frac{\ell\,(z-x)\,(z-y)}{1-\ell\,(z-c+y)} \tag{3.6}$$

*and*

$$\psi(x,y,z) = z + \frac{\ell_0\,(z-x)\,(z-y)}{1-\ell_0\,(z-c+y)}, \tag{3.7}$$

*respectively, for* $(x,y,z) \in \mathbb{R}^3$. *Finally, define a function* $\Theta$ *on* $\mathbb{R}^2$ *by*

$$\Theta(x,y) = y + \frac{\ell\,(y-x)^2}{2\,(1-\ell_0\,y)} \quad ((x,y) \in \mathbb{R}^2). \tag{3.8}$$

We need the following result on majorizing sequences for SM.

**Lemma 3.1** *Let* $\ell_0 > 0$, $\ell > 0$, $c \geq 0$, $\eta \in (0,c]$, $\delta$, $\delta_0$, $\delta_1$, $f$ *and* $\phi$ *as given in Definition 3.1. Assume*

$$\delta_0 \leq \delta \leq \delta_1. \tag{3.9}$$

*Then, scalar sequence* $\{t_n\}$ $(n \geq -1)$ *given by*

$$t_{-1} = 0,\ t_0 = c,\ t_1 = c+\eta,\ t_{n+2} = \phi(t_{n+1}, t_n, t_{n-1}) \quad (n \geq 0) \tag{3.10}$$

*is non–decreasing, bounded from above by*

$$t^{\star\star} = \frac{\eta}{1-\delta} + c \tag{3.11}$$

*and converges to its unique least upper bound* $t^\star$ *such that*

$$0 \leq t^\star \leq t^{\star\star}. \tag{3.12}$$

*Moreover, the following estimates hold for all* $n \geq 0$

$$0 \leq t_{n+1} - t_n \leq \delta\ (t_n - t_{n-1}) \leq \delta^n\ \eta. \tag{3.13}$$

**Proof** We shall show using mathematical induction on $n$ that

$$0 \leq t_{n+1} - t_n \leq \delta\ (t_n - t_{n-1}). \tag{3.14}$$

By (3.10) for $n = 0$, we must show that

$$0 \leq \frac{\ell\ (t_1 - t_{-1})}{1 - \ell_0\ t_1} \leq \delta \quad \text{or} \quad 0 \leq \frac{\ell\ (c+\eta)}{1-\ell_0\ (c+\eta)} \leq \delta,$$

which is true from (3.3) and the choice of $\delta \geq \delta_0$. Let assume that (3.14) holds for $k \leq n$. Then, by the induction hypothesis, we get

$$\begin{aligned} t_{n+1} &\leq t_n + \delta\,(t_n - t_{n-1}) \\ &\leq t_{n-1} + \delta\,(t_{n-1} - t_{n-2}) + \delta\,(t_n - t_{n-1}) \\ &\leq t_1 + \delta\,(t_1 - t_0) + \cdots + \delta\,(t_n - t_{n-1}) \\ &\leq c + \eta + \delta\,\eta + \cdots + \delta^n\,\eta \\ &= c + \frac{1-\delta^{n+1}}{1-\delta}\,\eta < \frac{\eta}{1-\delta} + c = t^{\star\star}. \end{aligned} \tag{3.15}$$

Moreover, we have that

$$\begin{aligned} &\ell\,(t_{n+1} - t_{n-1}) + \delta\,\ell_0\,(t_{n+1} - t_0 + t_n) \\ &\leq \ell\,\big((t_{n+1} - t_n) + (t_n - t_{n-1})\big) + \delta\,\ell_0 \left(\frac{1-\delta^{n+1}}{1-\delta} + \frac{1-\delta^n}{1-\delta}\right)\eta + \delta\,\ell_0\,c \\ &\leq \ell\,(\delta^{n-1} + \delta^n)\,\eta + \frac{\delta\,\ell_0}{1-\delta}\,(2 - \delta^n - \delta^{n+1})\,\eta + \delta\,\ell_0\,c. \end{aligned} \tag{3.16}$$

We prove now (3.14). By (3.10), we have the estimate

$$\ell\,(\delta^{n-1} + \delta^n)\,\eta + \frac{\delta\,\ell_0}{1-\delta}\,(2 - \delta^n - \delta^{n+1})\,\eta + \delta\,\ell_0\,c \leq \delta \tag{3.17}$$

or

$$\ell\,(\delta^{n-2} + \delta^{n-1})\,\eta + \ell_0\,\big((1+\delta+\cdots+\delta^{n-1}) + (1+\delta+\cdots+\delta^{n+1})\big)\,\eta + \ell_0\,c - 1 \leq 0. \tag{3.18}$$

In view of (3.18), we are motivated to define (for $\delta = s$) the functions $f_n$ on $[0,1]$ for $n \geq 2$

$$f_{n-1}(s) = \ell\,(s^{n-2} + s^{n-1})\,\eta + \ell_0\,(2\,(1+s+\cdots+s^{n-1}) + s^n)\,\eta + \ell_0\,c - 1. \tag{3.19}$$

We need the relationship between two consecutive functions $f_n$. Using (3.19), we obtain that

$$\begin{aligned} f_n(s) &= \ell\,(s^{n-1} + s^n)\,\eta + \ell_0\,\big(2\,(1+s+\cdots+s^n) + s^{n+1}\big)\,\eta + \ell_0\,c - 1 \\ &= f_{n-1}(s) + \ell\,(s^n - s^{n-2})\,\eta + \ell_0\,(s^n + s^{n+1})\,\eta \\ &= f(s)\,s^{n-2}\,\eta + f_{n-1}(s). \end{aligned} \tag{3.20}$$

Note that $\delta$ is the unique positive root of polynomial $f_1$. We can show instead of (3.18) that

$$f_n(\delta) \leq 0 \qquad (n \geq 2). \tag{3.21}$$

However, using (3.20), we get

$$f_n(\delta) = f_2(\delta) \qquad (n \geq 2). \tag{3.22}$$

Hence, in view of (3.21) and (3.22), it suffices to only show $f_2(\delta) \leq 0$, which is true, since $\delta \leq \delta_1$. Moreover, define function $f_\infty$ on $[0,1)$ by

$$f_\infty(s) = \lim_{k \longrightarrow \infty} f_k(s). \tag{3.23}$$

By (3.21) and (3.23), we obtain

$$f_\infty(\delta) = \lim_{k \longrightarrow \infty} f_k(\delta) \leq 0. \tag{3.24}$$

The induction is completed. Hence, we showed sequence $\{t_n\}$ $(n \geq -1)$ is non–decreasing and bounded above from by $t^{\star\star}$, so that (3.13) holds. It follows that there exists $t^\star \in [0, t^{\star\star}]$, so that $\lim_{n \longrightarrow \infty} t_n = t^\star$. This completes the proof of Lemma 3.1. ⊠

We shall study SM for triplets $(F, x_{-1}, x_0)$ belonging to the class $\mathcal{C}(\ell, \ell_0, \eta, c, \delta)$ defined as follows.

**Definition 3.2** *Let $\ell$, $\ell_0$, $\eta$, $c$, $\delta$ be non–negative constants satisfying the hypotheses of Lemma 3.1. A triplet $(F, x_{-1}, x_0)$ belongs to the class $\mathcal{C}(\ell, \ell_0, \eta, c, \delta)$ if*

$(\mathcal{A}_1)$ *$F$ is a nonlinear operator defined on a convex subset $\mathcal{D}$ of a Banach space $\mathcal{X}$ with values in a Banach space $\mathcal{Y}$;*

$(\mathcal{A}_2)$ *$x_{-1}$ and $x_0$ are two points belonging to the interior $\mathcal{D}^0$ of $\mathcal{D}$ and satisfying the inequality*

$$\| x_0 - x_{-1} \| \leq c;$$

$(\mathcal{A}_3)$ *$F$ is Fréchet–differentiable on $\mathcal{D}^0$ and there exists an operator $\delta F : \mathcal{D}^0 \times \mathcal{D}^0 \to \mathcal{L}(\mathcal{X}, \mathcal{Y})$ such that $A^{-1} = \delta F(x_{-1}, x_0)^{-1} \in \mathcal{L}(\mathcal{Y}, \mathcal{X})$ and for all $x, y, z \in \mathcal{D}$, the following hold*

$$\| A^{-1} F(x_0) \| \leq \eta,$$

$$\| A^{-1} (\delta F(x,y) - F'(z)) \| \leq \ell \, (\| x - z \| + \| y - z \|),$$

*and*

$$\| A^{-1} (\delta F(x,y) - F'(x_0)) \| \leq \ell_0 \, (\| x - x_0 \| + \| y - x_0 \|);$$

$(\mathcal{A}_4)$

$$\overline{U}(x_0, t^\star) \subseteq \mathcal{D}_c = \{x \in \mathcal{D} \, : \, F \text{ is continuous at } x\} \subseteq \mathcal{D},$$

*where $t^\star$ is given in Lemma 3.1;*

The semilocal convergence theorem for SM is as follows.

**Theorem 3.1** *If $(F, x_{-1}, x_0) \in \mathcal{C}(\ell, \ell_0, \eta, c, \delta)$, then, sequence $\{x_n\}$ $(n \geq -1)$ generated by SM is well defined, remains in $\overline{U}(x_0, t^\star)$ for all $n \geq 0$ and converges to a unique solution $x^\star \in \overline{U}(x_0, t^\star)$ of (2.1). Moreover, the following estimates hold for all $n \geq 0$*

$$\| x_n - x_{n-1} \| \leq t_n - t_{n-1} \tag{3.25}$$

*and*

$$\| x_n - x^\star \| \leq t^\star - t_n, \tag{3.26}$$

*where* $\{t_n\}$ $(n \geq 0)$ *is given by (3.10). Furthermore, if there exists* $R > 0$ *such that*

$$t^\star \leq t_0 + R, \quad U(x_0, R) \subseteq \mathcal{D} \quad \text{and} \quad \ell_0 \, (t^\star + R) \leq 1, \tag{3.27}$$

*then, the solution* $x^\star$ *is unique in* $U(x_0, R)$.

**Proof** First, we show that $L = \delta F(x_k, x_{k+1})$ is invertible for $x_k, x_{k+1} \in \overline{U}(x_0, t^\star)$. By (3.10), (3.11), $(\mathcal{A}_2)$ and $(\mathcal{A}_3)$, we have that

$$\begin{aligned}
\| I - A^{-1}\, L \| = \| A^{-1}\, (L - A) \| &\leq \| A^{-1}(L - F'(x_0)) \| + \| A^{-1}(F'(x_0) - A) \| \\
&\leq \ell_0 \, (\| x_k - x_0 \| + \| x_{k+1} - x_0 \| + \| x_0 - x_{-1} \|) \\
&\leq \ell_0 \, (t_k - t_0 + t_{k+1} - t_0 + c) \\
&\leq \ell_0 \, (t^\star - t_0 + t^\star - t_0 + c) \\
&\leq \ell_0 \left( 2 \left( \frac{\eta}{1 - \delta} + c \right) - c \right) < 1
\end{aligned} \tag{3.28}$$

by (3.18) and the choice of $\delta$. Using the Banach Lemma on invertible operators (cf. [163], [469]) and (3.28), $L$ is invertible and

$$\| L^{-1}\, A \| \leq \big(1 - \ell_0 \, (\| x_k - x_0 \| + \| x_{k+1} - x_0 \| + c)\big)^{-1}. \tag{3.29}$$

By $(\mathcal{A}_3)$, we have

$$\| A^{-1}\, (F'(u) - F'(v)) \| \leq 2\, \ell \; \| u - v \| \quad (u, v \in \mathcal{D}^0). \tag{3.30}$$

We can write the identity

$$F(x) - F(y) = \int_0^1 F'(y + t(x - y))\, dt\, (x - y) \tag{3.31}$$

and then, for all $x, y, u, v \in \mathcal{D}^0$, we obtain

$$\begin{aligned}
&\| A_0^{-1}\, (F(x) - F(y) - F'(u)(x - y)) \| \\
&\leq \ell \; (\| x - u \| + \| y - u \|) \; \| x - y \|
\end{aligned} \tag{3.32}$$

and

$$\begin{aligned}
&\| A_0^{-1}\, (F(x) - F(y) - \delta F(u, v)\, (x - y)) \| \\
&\leq \ell \, (\| x - v \| + \| y - v \| + \| u - v \|) \, \| x - y \| \, .
\end{aligned} \tag{3.33}$$

By a continuity argument (3.30)–(3.33) remain valid if $x$ and/or $y$ belong to $\mathcal{D}_c$.

Now, we show (3.25). If (3.25) holds for all $n \leq k$ and $\{x_n\}$ $(n \geq 0)$ is well defined for $n = 0, 1, 2, \cdots, k$, then

$$\| x_n - x_0 \| \leq t_n - t_0 < t^\star - t_0, \quad n \leq k. \tag{3.34}$$

That is, (3.1) is well defined for $n = k+1$. For $n = -1$ and $n = 0$, (3.25) reduces to $\| x_{-1} - x_0 \| \le c$ and $\| x_0 - x_1 \| \le \eta$. Suppose (3.25) holds for $n = -1, 0, 1, \cdots, k$ $(k \ge 0)$. By (3.29), (3.33) and

$$F(x_{k+1}) = F(x_{k+1}) - F(x_k) - \delta F(x_{k-1}, x_k)\,(x_{k+1} - x_k), \tag{3.35}$$

we obtain the following estimate

$$\begin{aligned}
\| x_{k+2} - x_{k+1} \| &= \| \delta F(x_k, x_{k+1})^{-1}\, F(x_{k+1}) \| \\
&\le \| \delta F(x_k, x_{k+1})^{-1}\, A \| \, \| A^{-1}\, F(x_{k+1}) \| \\
&\le \frac{\ell\,(\| x_{k+1} - x_k \| + \| x_k - x_{k-1} \|)}{1 - \ell_0\,(\| x_{k+1} - x_0 \| + \| x_k - x_0 \| + c)} \, \| x_{k+1} - x_k \| \\
&\le \frac{\ell\,(t_{k+1} - t_k + t_k - t_{k-1})}{1 - \ell_0\,(t_{k+1} - t_0 + t_k - t_0 + t_0 - t_{-1})}\,(t_{k+1} - t_k) \\
&= t_{k+2} - t_{k+1}
\end{aligned} \tag{3.36}$$

and the induction for (3.25) is completed. It follows from (3.25) and Lemma 3.1 that $\{x_n\}$ $(n \ge -1)$ is a Cauchy sequence in a Banach space $\mathcal{X}$ and as such it converges to some $x^\star \in \overline{U}(x_0, t^\star)$ (since $\overline{U}(x_0, t^\star)$ is a closed set). By letting $k \to \infty$ in (3.36), we obtain $F(x^\star) = 0$. Estimate (3.26) follows from (3.25) by using standard majoration techniques.

Finally, for showing the uniqueness in $U(x_0, R)$, let $y^\star \in U(x_0, R)$ be a solution (2.1). Set

$$\mathcal{M} = \int_0^1 F'(y^\star + t\,(y^\star - x^\star))\,dt.$$

It follows from $(\mathcal{A}_3)$ that

$$\begin{aligned}
\| A^{-1}\,(A - \mathcal{M}) \| &= \ell_0\,(\| y^\star - x_0 \| + \| x^\star - x_0 \| + \| x_0 - x_{-1} \|) \\
&\le \ell_0\,(R + t^\star - t_0 + t_0) = \ell_0\,(R + t^\star) \le 1.
\end{aligned} \tag{3.37}$$

It follows from (3.37) and the Banach lemma on invertible operators that $\mathcal{M}^{-1}$ exists on $U(x_0, R)$. Using the identity:

$$F(x^\star) - F(y^\star) = \mathcal{M}\,(x^\star - y^\star) \tag{3.38}$$

we deduce $x^\star = y^\star$. This completes the proof of Theorem 3.1. ⊠

**Remark 3.1** *The point $t^{\star\star}$ given in closed form by (3.11) can replace $t^\star$ in $(\mathcal{A}_4)$ and (3.27).*

**Remark 3.2** *Let us define the majoring sequence $\{w_n\}$ used in (cf. [313], [427], [496], [596], [600], [668], [770]) (under condition (3.2))*

$$w_{-1} = 0,\ w_0 = c,\ w_1 = c + \eta,\ w_{n+2} = \varphi(w_{n+1}, w_n, w_{n-1}) \quad (n \ge 0), \tag{3.39}$$

*where $\varphi$ is given by (3.6). Note that the sufficient convergence condition for (3.39) is given by (3.2). Note that in general,*

$$\ell_0 \leq \ell \tag{3.40}$$

*holds and $\dfrac{\ell}{\ell_0}$ can be arbitrarily large (cf. [139], [163]). In the case $\ell_0 = \ell$, then, $t_n = w_n$ $(n \geq -1)$. Otherwise, a simple inductive argument shows*

$$t_n < w_n, \quad t_{n+1} - t_n < w_{n+1} - w_n \quad (n \geq 2) \tag{3.41}$$

*and*

$$0 \leq t^\star - t_n \leq w^\star - w_n \quad (n \geq 0), \quad w^\star = \lim_{n \longrightarrow \infty} w_n. \tag{3.42}$$

*The proof of (3.41) and (3.42) can also be found in (cf. [163]). Note that the only difference in the proofs is that the conditions of Lemma 3.1 are used here, instead of the ones in (cf. [146]). However, this makes no difference in the proofs.*

**Remark 3.3** *Let $\phi$ and $\psi$ are given, respectively, by (3.5) and (3.7). It turns out from the proof of Theorem 3.1 that $\{v_n\}$ given by*

$$v_{-1} = 0, \; v_0 = c, \; v_1 = c + \eta, \; v_2 = \psi(v_1, v_0, v_{-1})$$

$$v_{n+2} = \phi(v_{n+1}, v_n, v_{n-1}) \qquad (n \geq 1) \tag{3.43}$$

*is a majorizing sequence for $\{x_n\}$, which is tighter than $\{t_n\}$ and such that for all $n \geq 0$*

$$v_n \leq t_n, \tag{3.44}$$

$$\| x_{n+1} - x_n \| \leq v_{n+1} - v_n \leq t_{n+1} - t_n, \tag{3.45}$$

$$\| x_{n+1} - x^\star \| \leq v^\star - v_n \leq t^\star - t_n, \quad v^\star = \lim_{n \longrightarrow \infty} v_n \tag{3.46}$$

*and*

$$v^\star \leq t^\star. \tag{3.47}$$

*Moreover, if $\ell_0 < \ell$, then, we have for $(n \geq 2)$ that*

$$v_n < t_n \tag{3.48}$$

*and*

$$v_{n+1} - v_n < t_{n+1} - t_n. \tag{3.49}$$

We shall consider NM given by (2.2) as a special case of the Secant method (3.1). We need a result similar to Lemma 3.1 for (2.2).

**Lemma 3.2** *Let $\ell_0 > 0$, $\ell > 0$, $\eta > 0$ be given parameters. Let $\Theta$ and $b$ be as given in Definition 3.1. Assume*

$$2h_0 = b\eta \leq 1. \tag{3.50}$$

*Then, scalar sequence* $\{t_n\}$ $(n \geq 0)$ *given by*

$$t_0 = 0, \quad t_1 = \eta, \quad t_{n+2} = \Theta(t_{n+1}, t_n) \quad (n \geq 0), \tag{3.51}$$

*is non–decreasing, bounded above by*

$$t^{\star\star} = \frac{2\eta}{2-\delta}, \tag{3.52}$$

*where,*

$$\delta = \frac{4\,\ell}{\ell + \sqrt{\ell^2 + 8\,\ell_0\,\ell}}, \tag{3.53}$$

*and converges to some* $t^\star \in [0, t^{\star\star}]$. *Moreover, the following estimates hold for all* $n \geq 0$

$$0 < t_{n+2} - t_{n+1} \leq \frac{\delta}{2}\,(t_{n+1} - t_n) \leq \cdots \leq \left(\frac{\delta}{2}\right)^{n+1} \eta. \tag{3.54}$$

**Proof** We follow the proof of Lemma 3.1. The estimate corresponding to (3.16) is given by

$$\left(\ell \left(1 - \frac{\delta}{2}\right) \left(\frac{\delta}{2}\right)^k + \delta\,\ell_0 \left(1 - \left(\frac{\delta}{2}\right)^{k+1}\right)\right) \eta \leq \delta \left(1 - \frac{\delta}{2}\right). \tag{3.55}$$

which leads to the definition of functions

$$f(s) = 2\,\ell_0\,s^2 + \ell_0\,s - \ell, \tag{3.56}$$

$$f_k(s) = \left(\ell\,s^{k-1} + 2\,\ell_0\,(1 + s + s^2 + \cdots + s^k)\right) \eta - 2 \quad (k \geq 1). \tag{3.57}$$

Hence, we get that

$$f_{k+1}(s) = f(s)\,s^{k-1}\,\eta + f_k(s). \tag{3.58}$$

It follows from (3.50) and (3.58) that

$$f_{k+1}\left(\frac{\delta}{2}\right) = f_1\left(\frac{\delta}{2}\right) \leq 0,$$

where $\dfrac{\delta}{2}$ is the only positive root of $f$. We also have that

$$\delta_0 = \frac{\ell\,\eta}{1 - \ell_0\,\eta} \leq \delta \quad \text{by} \quad (3.50).$$

This completes the proof of Lemma 3.2. ⊠

In the next result, we provide more estimates on the distances $t_{n+1} - t_n$ and $t^\star - t_n$ $(n \geq 0)$

**Proposition 3.1** *(cf. [202]) Under the hypotheses of Lemma 3.2, the following estimates hold for all* $n \geq 0$

$$t_{n+1} - t_n \leq \left(\frac{\delta}{2}\right)^n (2\,h_0)^{2^n-1}\,\eta \tag{3.59}$$

*and*

$$t^\star - t_n \leq \left(\frac{\delta}{2}\right)^n \frac{(2\,h_0)^{2^n-1}\,\eta}{1-(2\,h_0)^{2^n}} \quad (2\,h_0 < 1). \tag{3.60}$$

We shall study (2.2) for couples $(F, x_0)$ belonging to a class $\mathcal{C}(\ell, \ell_0, \eta, \delta)$ defined as follows (see also corresponding Definition 3.2)

**Definition 3.3** *Let* $\ell$, $\ell_0$, $\eta$, $\delta$ *be non–negative parameters satisfying the hypotheses of Lemma 3.2. A triplet* $(F, x_{-1}, x_0)$ *belongs to the class* $\mathcal{C}(\ell, \ell_0, \eta, \delta)$ *if:*

$(\mathcal{K}_1)$ *F is a nonlinear operator as defined in* $(\mathcal{A}_1)$*;*

$(\mathcal{K}_2)$ *F is Fréchet–differentiable on the interior* $\mathcal{D}^0$ *of* $\mathcal{D}$ *and there exists* $x_0 \in \mathcal{D}$ *such that* $A = F'(x_0)$ *is invertible, its inverse* $A^{-1}$ *is bounded and for all* $x, y \in \mathcal{D}$*, the following hold*

$$\| A^{-1}\, F(x_0) \| \leq \eta,$$

$$\| A^{-1}\, [F'(x) - F'(x_0)] \| \leq \ell_0 \;\; \| x - x_0 \|$$

*and*

$$\| A^{-1}\, [F'(x) - F'(y)] \| \leq \ell \;\; \| x - y \|;$$

$(\mathcal{K}_3) \equiv (\mathcal{A}_4)$*, where* $t^\star$ *is given in Lemma 3.2.*

We present the semilocal convergence theorem for (2.2).

**Theorem 3.2** *If* $(F, x_0) \in \mathcal{C}(\ell, \ell_0, \eta, \delta)$*, then, sequence* $\{x_n\}$ $(n \geq 0)$ *generated by (2.2) is well defined, remains in* $\overline{U}(x_0, t^\star)$ *for all* $n \geq 0$ *and converges to a unique solution* $x^\star \in \overline{U}(x_0, t^\star)$ *of equation* $F(x) = 0$*. Moreover, the following estimates hold for all* $n \geq 0$

$$\| x_{n+1} - x_n \| \leq t_{n+1} - t_n$$

*and*

$$\| x_n - x^\star \| \leq t^\star - t_n$$

*where, the sequence* $\{t_n\}$ $(n \geq 0)$ *given by (3.51). Furthermore, if there exists* $R > t^\star$ *such that* $\overline{U}(x_0, R) \subseteq \mathcal{D}$ *and* $\ell_0\,(t^\star + R) \leq 2$*, then, the solution* $x^\star$ *is unique in* $\overline{U}(x_0, R)$*.*

**Proof** The proof as identical to Theorem 1 in (cf. [196]) is omitted. Note that, in (cf. [196]) we simply used sufficient convergence conditions different than the ones in Lemma 3.2. This is the only difference in the proofs. ⊠

**Remark 3.4** *The famous for its simplicity and clarity Newton–Kantorovich hypothesis corresponding to (3.50) is given in (cf. [139], [163], [469]) is given by*

$$2\,h = 2\,\ell\,\eta \leq 1. \tag{3.61}$$

*It follows from (3.50) and (3.61) that*

$$h \leq \frac{1}{2} \Longrightarrow h_0 \leq \frac{1}{2},$$

*but not necessarily vice versa, unless if* $\ell = \ell_0$. *Comments similar to the ones in Remark 3.2 can follow for NM.*

We complete this section with numerical examples.

**Example 3.1** *Define the scalar function* $F$ *by* $F(x) = c_0\,x + c_1 + c_2\,\sin e^{c_3\,x}$, $x_0 = 0$, *where* $c_i$, $i = 0, 1, 2, 3$, *are given parameters. Define a linear operator* $\delta F(x,y)$ *by*

$$\delta F(x,y) = \int_0^1 F'(y + t\,(x-y))\,dt = c_0 + c_2\,\frac{\sin e^{c_3\,x} - \sin e^{c_3\,y}}{x-y}.$$

*Then, it can easily be seen that, for* $c_3$ *large and* $c_2$ *sufficiently small,* $\dfrac{\ell}{\ell_0}$ *can be arbitrarily large. That is, (3.9) may be satisfied, but not (3.2).*

**Example 3.2** *(a)* **NM case.** *See example in chapter 2 for the function* $F$ *defined by* $F(x) = (\xi_1^3 - \gamma, \xi_2^3 - \gamma)^T$, *where* $x = (\xi_1, \xi_2)^T$ *and* $\gamma$ *is a fixed positive real.*

*(b)* **Secant method case.** *Let us define divided difference operator of order one by*

$$\delta F(x,y) = \int_0^1 F'(y + t\,(x-y))\,dt. \tag{3.62}$$

*Choose* $x_{-1} = (1.01, 1.01)^T$ *and* $\gamma = .7$. *Then, we get for* $\eta = \parallel A^{-1} \parallel (1-\gamma)$, $\ell_0 = \dfrac{3}{2} \parallel A^{-1} \parallel (3-\gamma)$, $\ell = 3 \parallel A^{-1} \parallel (2-\gamma)$

$$\eta = .1006711407, \quad \ell = 1.308724829, \quad \ell_0 = 1.157718118.$$

*The function* $f$ *in (3.4) becomes*

$$f(t) = 1.157718118\,t^3 + 2.466442947\,t^2 - 1.308724829,$$

*and the root* $\delta$ *was found to be* $\delta = .6389028195$.

*It can easily be seen that (3.2), (3.4) and (3.50), (3.61) hold, since*

$$\ell\,c + 2\,\sqrt{\ell\,\eta} = .7390371677 \leq 1,$$

$$\delta_0 = .1661227048 \leq \delta = .6389028195 \leq \delta_1 = 1.227672515.$$

*Note that all assumptions of Theorem 3.1 are satisfied and Theorem 3.1 guarantee the convegence of (3.1) to* $x^\star = (.8366600265, .8366600265)^T$ *starting at* $x_0$.

*In the following table, we validate our Remarks 3.2 and 3.3.*

**Comparison table**

| | (3.10) | (3.39) | (3.43) |
|---|---|---|---|
| $n$ | $t_{n+1} - t_n$ | $w_{n+1} - w_n$ | $v_{n+1} - v_n$ |
| 0 | .1006711407 | .1006711407 | .1006711407 |
| 1 | .0167237622 | .0170505879 | .0147940973 |
| 2 | .0034912030 | .0037468515 | .0031752999 |
| 3 | .0001296208 | .0001513363 | .0001097280 |
| 4 | $8.671 \times 10^{-7}$ | .0000011544 | $6.976 \times 10^{-7}$ |
| 5 | $2 \times 10^{-10}$ | $3 \times 10^{-10}$ | $1 \times 10^{-10}$ |
| 6 | 0 | 0 | 0 |

*The table shows that our error bounds* $v_{n+1} - v_n$ *and* $t_{n+1} - t_n$ *are finer than* $w_{n+1} - w_n$ *given in (cf. [313], [427], [496], [596], [600], [668], [770]).*

## 3.2 Least Squares Problems

We propose in this section Secant type iterative method for approximating a solution of nonlinear least squares problems. The radius of convergence is determined as well as usable error estimates.

In this section, we are concerned with the problem of approximating a solution $x^\star$ of the nonlinear least square problem

$$\min_{x \in \mathbb{R}^i} \frac{1}{2} F(x)^T F(x), \tag{3.63}$$

where $F : D \subseteq \mathbb{R}^i \to \mathbb{R}^j$ is a twice continuously Fréchet-differentiable operator $((j \geq i)$ are natural numbers) and $D$ is an open convex subset of $\mathbb{R}^i$.

Nonlinear least square problem can be used to formulate many problems arising in applied sciences (cf. [569], [671], [793]). This problem can be solved by Gauss–Newton–type methods (cf. [163]). However, these methods use the derivative $F'(x)$ of operator $F$. The computation of the derivative $F'(x)$ is in general very expensive. That is why we propose the alternative iterative method

$$x_{n+1} = x_n - [A_n^T A_n]^{-1} A_n^T F(x_n) \quad (n \geq 0), \tag{3.64}$$

where $A_n$ is a linear operator, which is a divided difference or a linear combination of divided differences. We shall only consider the special case, when

$$A_n = [x_n, x_{n-1}; F] \quad (n \geq 0), \tag{3.65}$$

where $[.,.,F]$ is a divided difference of order one (cf. [157], [569]). Note that case (3.65) as well as cases

$$A_n = [2x_n - x_{n-1}, x_{n-1}; F] \quad (n \geq 0), \tag{3.66}$$

$$A_n = [x_n, x_{n-1}; F] + [x_{n-2}, x_n; F] - [x_{n-2}, x_{n-1}; F] \quad (n \geq 0), \tag{3.67}$$

were studied in (cf. [674]). However, the proof in (cf. [674, Theorem 1, p. 256]) using (3.65) is not completed (see Remark 3.6).

Here, we provide a new local convergence analysis for iterative method (3.64) under the special choice of $A_n$ given in (3.65). Numerical example further validating the theoretical results is also provided in this section. Note also that a local as well as a semilocal convergence analysis of more general method

$$y_{n+1} = y_n - A_n^{-1} F(y_n) \quad (n \geq 0), \tag{3.68}$$

where $F : D \subseteq B_1 \to B_2$ and $B_1$, $B_2$ are Banach spaces and $A_n$ is given by (3.65) can be found in (cf. [139], [95], [569], [632], [633], [793]).

We can show the following local convergence result for STM.

**Theorem 3.3** *Let $F : \mathbb{R}^i \to \mathbb{R}^j$ be a twice continuously differentiable mapping defined on an open convex subset $D$ of $\mathbb{R}^i$ with values in $\mathbb{R}^j$.*

*Assume that*

*(i) problem (3.63) has a solution $x^\star \in D$ and the inverse operator $L = (A_\star^T A_\star)^{-1} = [F'(x^\star)^T F'(x^\star)]^{-1}$ exists and*

$$\|L\| \leq B; \tag{3.69}$$

*(ii) $F$ has divided difference of order one and, for all $x, y \in D$*

$$\|[x, y; F] - [x, x^\star; F]\| \leq M \|y - x^\star\| \tag{3.70}$$

*and*

$$\|[x, y; F] - [x^\star, x^\star; F]\| \leq M_0 (\|x - x^\star\| + \|y - x^\star\|); \tag{3.71}$$

*(iii) the following hold*

$$\|F(x^\star)\| \leq \eta, \tag{3.72}$$

$$\|F'(x^\star)\| \leq \alpha, \tag{3.73}$$

$$2BM_0\eta < 1 \tag{3.74}$$

*and*

$$U(x^\star, r_\star) \subseteq D, \tag{3.75}$$

*where $r_\star$ is the unique positive root of polynomial $q$ given by*

$$q(r) = B[(\alpha + 2M_0 r)Mr + 2M_0\eta] + 4M_0 B(\alpha + M_0 r)r - 1. \tag{3.76}$$

*Then, for $x_{-1}, x_0 \in U(x^\star, r_\star)$, iterative process $\{x_n\}$ generated by STM for $A_n$ given by (3.65) is well defined, remains in $U(x^\star, r_\star)$ for all $n \geq 0$ and converges to $x^\star$. Moreover, the following error estimates hold for all $n \geq 0$*

$$\begin{aligned}\|x_{n+1} - x^\star\| &\leq C_0\|x_{n-1} - x^\star\| + C_1\|x_n - x^\star\| + C_2\|x_{n-1} - x^\star\|\|x_n - x^\star\| \\ &\quad + C_3\|x_{n-1} - x^\star\|\|x_n - x^\star\|^2 + C_4\|x_{n-1} - x^\star\|^2\|x_n - x^\star\|, \qquad (3.77)\end{aligned}$$

*where, for*

$$\begin{aligned}g^{-1}(r) &= B^{-1}(1 - 4M_0B(\alpha + M_0 r)r), \\ C_0 &= g(r_\star)M_0\eta, \quad C_1 = g(r_\star)M_0\eta, \quad C_2 = g(r_\star)\alpha M, \\ C_3 &= g(r_\star)M_0M, \quad C_4 = g(r_\star)M_0M. \qquad (3.78)\end{aligned}$$

**Proof** Polynomial $q$ has a positive root denoted by $r_\star$, by the intermediate value theorem applied on $[0, r]$, for sufficiently large $r > 0$ and (3.74). Moreover, $q'(r) \geq 0$, $r \geq 0$. That is, the graph of function $q$ crosses the positive $x - axis$ only once. By hypothesis, $x_{-1}, x_0 \in U(x^\star, r_\star)$. Using (3.69), (3.71), (3.73) and the definition of $r_\star$, we obtain in turn

$$\begin{aligned}\|I - LA_0^TA_0\| &= \|L[A_0^T(A_0 - A_\star) + (A_0 - A_\star)^TA_\star]\| \\ &\leq B(\|A_0^T\|\|A_0 - A_\star\| + \|(A_0 - A_\star)^T\|\|A_\star\|) \\ &= B(\|A_0\| + \|A_\star\|)\|A_0 - A_\star\| \\ &\leq B(\|A_\star\| + \|A_0 - A_\star\| + \|A_\star\|)\|A_0 - A_\star\| \\ &\leq B(2\alpha + M_0(\|x_0 - x^\star\| + \|x_{-1} - x^\star\|))M_0(\|x_0 - x^\star\| + \|x_{-1} - x^\star\|) \\ &\leq 4M_0B(\alpha + M_0r_\star)r_\star \\ &= 1 - B[(\alpha + 2M_0r_\star)Mr_\star + 2M_0\eta] < 1. \qquad (3.79)\end{aligned}$$

It follows from (3.79) and the Banach lemma on invertible operators that $(A_0^TA_0)^{-1}$ exists and

$$\|(A_0^TA_0)^{-1}\| \leq g_0 \leq g(r_\star) = \frac{B}{1 - 4M_0B(\alpha + M_0r_\star)r_\star}, \qquad (3.80)$$

where

$$g_0 = \frac{B}{1 - B(2\alpha + M_0(\|x_0 - x^\star\| + \|x_{-1} - x^\star\|))M_0(\|x_0 - x^\star\| + \|x_{-1} - x^\star\|)}.$$

Consequently, the iterate $x_1$ is well defined.

Next, we will show $x_1 \in U(x^\star, r_\star)$. Using the fact $A_\star^TF(x^\star) = 0$, STM,

(3.70), (3.71), (3.72), (3.73), (3.80) and the definition of $r_\star$, we get

$$\begin{aligned}
\|x_1 - x^\star\| &= \| - (A_0^T A_0)^{-1}(A_0^T F(x_0) - A_0^T A_0(x_0 - x^\star))\| \\
&= \|(A_0^T A_0)^{-1} A_0^T (F(x_0) - F(x^\star) + F(x^\star) - A_0(x_0 - x^\star))\| \\
&\le \|(A_0^T A_0)^{-1}\| \|(A_0^T - A_\star^T + A_\star^T)(([x_0, x^\star; F] - A_0)(x_0 - x^\star) + F(x^\star))\| \\
&\le g_0 \|(A_0 - A_\star)^T (([x_0, x^\star; F] - A_0)(x_0 - x^\star) + F(x^\star)) \\
&\quad + A_\star^T (([x_0, x^\star; F] - A_0)(x_0 - x^\star))\| \\
&\le g_0 ((\|A_0 - A_\star\| + \|A_\star\|)\|[x_0, x^\star; F] - A_0\| \|x_0 - x^\star\| + \|A_0 - A_\star\| \|F(x^\star)\|) \\
&\le g_0 [(\alpha + M_0(\|x_0 - x^\star\| + \|x_{-1} - x^\star\|)) M \|x_{-1} - x^\star\| \|x_0 - x^\star\| \\
&\quad + M_0(\|x_0 - x^\star\| + \|x_{-1} - x^\star\|)\eta] \\
&\le (g_\star)[(\alpha + M_0(\|x_0 - x^\star\| + \|x_{-1} - x^\star\|)) M \|x_{-1} - x^\star\| \|x_0 - x^\star\| \\
&\quad + M_0(\|x_0 - x^\star\| + \|x_{-1} - x^\star\|)\eta] \\
&< g(r_\star)[(\alpha + 2M_0 r_\star) M r_\star + 2M_0\eta] r_\star = p(r_\star) r_\star = r_\star,
\end{aligned} \tag{3.81}$$

which shows $x_1 \in U(x^\star, r_\star)$ and estimate (3.77) holds for $n = 0$, where

$$p(r) = g(r)[(\alpha + 2M_0 r) M r + 2M_0\eta]. \tag{3.82}$$

Let us assume $x_n \in U(x^\star, r_\star)$ for $n = 0, 1, \ldots, k$ and estimate (3.77) holds for $n = 0, 1, \ldots, k-1$ and $k \ge 1$ is an integer. We shall show: $x_{k+1} \in U(x^\star, r_\star)$ and estimate (3.77) holds for $n = k$. By simply replacing above $x_{-1}$, $x_0$ by $x_{k-1}$, $x_k$, respectively, we get

$$\begin{aligned}
\|I - L A_k^T A_k\| &= \|L[A_k^T (A_k - A_\star) + (A_k - A_\star)^T A_\star]\| \\
&\le B(\|A_k^T\| \|A_k - A_\star\| + \|(A_k - A_\star)^T\| \|A_\star\|) \\
&= B(\|A_k\| + \|A_\star\|)\|A_k - A_\star\| \\
&\le B(2\|A_\star\| + \|A_k - A_\star\|)\|A_k - A_\star\| \\
&\le B\big(2\alpha + M_0(\|x_k - x^\star\| + \|x_{k-1} - x^\star\|)\big) M_0(\|x_k - x^\star\| \\
&\quad + \|x_{k-1} - x^\star\|) \\
&\le 4M_0 B(\alpha + M_0 r_\star) r_\star < 1.
\end{aligned} \tag{3.83}$$

Hence, again $(A_k^T A_k)^{-1}$ exists and

$$\|(A_k^T A_k)^{-1}\| \le g_k \le g(r_\star), \tag{3.84}$$

where

$$g_k = \frac{B}{1 - B\big(2\alpha + M_0(\|x_k - x^\star\| + \|x_{k-1} - x^\star\|)\big) M_0(\|x_k - x^\star\| + \|x_{k-1} - x^\star\|)}.$$

That is, iterate $x_{k+1}$ is well defined. Moreover, as in (3.81) we obtain in turn that

$$\begin{aligned}
\|x_{k+1} - x^\star\| &= \|(A_k^T A_k)^{-1} A_k^T (F(x_k) - F(x^\star) + F(x^\star) - A_k(x_k - x^\star))\| \\
&\leq g_k \|(A_k^T - A_\star^T + A_\star^T)(([x_k, x^\star; F] - A_k)(x_k - x^\star) + F(x^\star))\| \\
&\leq g_k ((\|A_k - A_\star\| + \|A_\star\|)\|[x_k, x^\star; F] - A_k\|\|x_k - x^\star\| \\
&\quad + \|A_k - A_\star\|\|F(x^\star)\|) \\
&\leq g_k [(\alpha + M_0(\|x_k - x^\star\| + \|x_{k-1} - x^\star\|)M\|x_{k-1} - x^\star\|\|x_k - x^\star\|) \\
&\quad + M_0(\|x_k - x^\star\| + \|x_{k-1} - x^\star\|)\eta] \\
&\leq g(r_\star)[(\alpha + M_0(\|x_k - x^\star\| + \|x_{k-1} - x^\star\|)M\|x_{k-1} - x^\star\|\|x_k - x^\star\|) \\
&\quad + M_0(\|x_k - x^\star\| + \|x_{k-1} - x^\star\|)\eta] \\
&< p(r_\star) r_\star = r_\star,
\end{aligned} \tag{3.85}$$

which shows $x_{k+1} \in U(x^\star, r_\star)$ and estimate (3.77) holds for $n = k$. By induction, iterative process $\{x_n\}$ generated by STM for $A_n$ given by (3.65) is well defined, remains in $U(x^\star, r_\star)$ for all $n \geq 0$ and estimate (3.77) holds for all $n \geq 0$.

Next, we shall show $x_n \to x^\star$ as $n \to \infty$. Define functions $a$ and $b$ on $[0, r_\star]$ by

$$a(r) = g(r) M_0 \eta, \tag{3.86}$$

$$b(r) = g(r)(2M_0 M r^2 + \alpha M r + M_0 \eta). \tag{3.87}$$

By the choice of $r_\star$, we get

$$a(r_\star) \geq 0, \quad b(r_\star) \geq 0, \quad a(r_\star) + b(r_\star) = 1. \tag{3.88}$$

Using estimate (3.77), the definition of the $C$' constants and functions $a$, $b$, we obtain for $n \geq 0$

$$\begin{aligned}
\|x_{n+1} - x^\star\| &\leq C_0 \|x_{n-1} - x^\star\| + C_1 \|x_n - x^\star\| + C_2 r_\star \|x_{n-1} - x^\star\| \\
&\quad + C_3 r_\star^2 \|x_{n-1} - x^\star\| + C_4 r_\star^2 \|x_{n-1} - x^\star\| \\
&= a(r_\star)\|x_n - x^\star\| + b(r_\star)\|x_{n-1} - x^\star\|.
\end{aligned} \tag{3.89}$$

On the other hand, for the real number $r_\star > 0$ and initial points $x_0, x_{-1}$ belong to $U(x^\star, r_\star)$, there must exist a real number $r'$ such that $0 < r' < r_\star$, $x_0, x_{-1} \in U(x^\star, r')$ and all estimates about $\{x_n\}$ above will be true if $r_\star$ is replaced by $r'$. Specially, we get from (3.89) for $n \geq 0$ that

$$\|x_{n+1} - x^\star\| \leq a\|x_n - x^\star\| + b\|x_{n-1} - x^\star\|, \tag{3.90}$$

where

$$a = a(r'), \quad b = b(r'). \tag{3.91}$$

Clearly, we also have that

$$a \geq 0, \quad b \geq 0, \quad a + b < a(r_\star) + b(r_\star) = 1. \tag{3.92}$$

Define sequences $\{\theta_n\}$ and $\{\rho_n\}$ by

$$\theta_n = \frac{\|x_n - x^\star\|}{r'} \quad (n = -1, 0, 1, \ldots), \tag{3.93}$$

$$\rho_{-1} = \theta_{-1}, \quad \rho_0 = \theta_0, \quad \rho_{n+1} = a\rho_n + b\rho_{n-1} \quad (n = 0, 1, 2, \ldots). \tag{3.94}$$

Dividing two sides of inequality (3.90) by $r'$ gives

$$\theta_{n+1} \leq a\theta_n + b\theta_{n-1} \quad (n = 0, 1, 2, \ldots) \tag{3.95}$$

By the definition of sequence $\{\rho_n\}$, we get

$$0 \leq \theta_n \leq \rho_n \quad (n = -1, 0, 1, \ldots). \tag{3.96}$$

Moreover, it is known that an explicit formula of sequence $\{\rho_n\}$ can be given by

$$\rho_n = \omega_1 \lambda_1^n + \omega_2 \lambda_2^n \quad (n = -1, 0, 1, \ldots), \tag{3.97}$$

where

$$\lambda_1 = \frac{a - \sqrt{a^2 + 4b}}{2}, \quad \lambda_2 = \frac{a + \sqrt{a^2 + 4b}}{2}, \tag{3.98}$$

and

$$\omega_1 = \frac{\lambda_2^{-1}\rho_0 - \rho_{-1}}{\lambda_2^{-1} - \lambda_1^{-1}}, \quad \omega_2 = \frac{\rho_{-1} - \lambda_1^{-1}\rho_0}{\lambda_2^{-1} - \lambda_1^{-1}}. \tag{3.99}$$

Note that

$$0 \leq |\lambda_1| \leq |\lambda_2| < \frac{a + \sqrt{a^2 + 4(1-a)}}{2} = \frac{a + 2 - a}{2} = 1. \tag{3.100}$$

Using (3.96) and (3.100), we deduce that $\{\theta_n\} \to 0, (n \to \infty)$, that is, $x_n \to x^\star (n \to \infty)$. This completes the proof of the Theorem. ⊠

In the case of zero residual ($\eta = 0$), we have the special case of Theorem 3.3.

**Theorem 3.4** *Let $F : \mathbb{R}^i \to \mathbb{R}^j$ be a twice continuously differentiable mapping defined on an open convex subset $D$ of $\mathbb{R}^i$ with values in $\mathbb{R}^j$.*

*Assume that*

*(i) the problem (3.63) has a solution $x^\star \in D$ and the inverse operator $L$ exists and*

$$\|L\| \leq B, \tag{3.101}$$

*$F$ has divided differences of order one and for all $x, y \in D$*

$$\|[x, y; F] - [x, x^\star; F]\| \leq M\|y - x^\star\|, \tag{3.102}$$

$$\|[x, y; F] - [x^\star, x^\star; F]\| \leq M_0(\|x - x^\star\| + \|y - x^\star\|), \tag{3.103}$$

*(ii) the following hold*

$$\|F'(x^\star)\| \leq \alpha \tag{3.104}$$

*and*

$$U(x^\star, r_\star) \subseteq D, \tag{3.105}$$

*where $r_\star$ is the unique positive root of polynomial $q$ given by*

$$q(r) = BMr(\alpha + 2M_0 r) + 4M_0 B(\alpha + M_0 r)r - 1. \tag{3.106}$$

*Then, for $x_{-1}, x_0 \in U(x^\star, r_\star)$, iterative process $\{x_n\}$ generated by STM for $A_n$ given by (3.65) is well defined, remains in $U(x^\star, r_\star)$ for all $n \geq 0$ and converges to $x^\star$. Moreover, the following error estimates hold for all $n \geq 0$*

$$\begin{aligned}\|x_{n+1} - x^\star\| \leq{}& C_2\|x_{n-1} - x^\star\|\|x_n - x^\star\| + C_3\|x_{n-1} - x^\star\|\|x_n - x^\star\|^2 \\ &+C_4\|x_{n-1} - x^\star\|^2\|x_n - x^\star\|, \end{aligned}\tag{3.107}$$

$$\|x_{n+1} - x^\star\| \leq c(r_\star)\|x_{n-1} - x^\star\|\|x_n - x^\star\|, \tag{3.108}$$

*where*

$$c(r) = g(r)(2M_0 r + \alpha)M. \tag{3.109}$$

*Furthermore, the order of convergence of STM is $\dfrac{1+\sqrt{5}}{2}$.*

**Proof** All but the last estimate have been established in Theorem 3.3. In view of (3.77) and the definition of function $c$, we get

$$\begin{aligned}\|x_{n+1} - x^\star\| &\leq (C_2 + C_3 r_\star + C_4 r_\star)\|x_{n-1} - x^\star\|\|x_n - x^\star\| \\ &= c(r_\star)\|x_{n-1} - x^\star\|\|x_n - x^\star\|, \end{aligned}\tag{3.110}$$

which shows (3.108) for all $n \geq 0$. It follows by (3.110) and the study of the secant method (cf. [569]) that STM is of order $\dfrac{1+\sqrt{5}}{2}$. This completes the proof of Theorem 3.4. ⊠

**Remark 3.5** *(a) Hypotheses (3.70) and (3.71) can be replaced by the stronger but popular condition (yet not needed in the proof of Theorems 3.3 and 3.4)*

$$\|[x, y; F] - [u, v; F]\| \leq N(\|x - u\| + \|y - v\|).$$

*Note, however, that*

$$M \leq N,$$

$$M_0 \leq N$$

*hold in general and $\dfrac{N}{M}$ and $\dfrac{N}{M_0}$ can be arbitrarily large.*

*(b) A popular choice for the divided difference is given by*

$$[x, y; F] = \int_0^1 F'(y + t(x - y))dt.$$

**Remark 3.6** *A similar result was given in (cf. [674, Theorem 1, p. 256]). However, the proof is not complete. Indeed, they arrive at the estimate*

$$\|x_{n+1} - x^\star\| < \|x_{n-1} - x^\star\| < R_\star,$$

*which obviously shows $x_{n+1} \in U(x^\star, R_\star)$. However, it is not shown that $x_n \to x^\star$ as $n \to \infty$.*

We give now an example to show the application of our results.

**Example 3.3** *Consider function $F : D = \mathbb{R} \subseteq \mathbb{R} \to \mathbb{R}^2$ given by*

$$F(x) = \begin{pmatrix} x + 1 \\ \lambda x^2 + x - 1 \end{pmatrix}, \tag{3.111}$$

*where $\lambda \in \mathbb{R}$ is a parameter. It follows that*

$$F'(x) = \begin{pmatrix} 1 \\ 2\lambda x + 1 \end{pmatrix}, \tag{3.112}$$

*and $x^\star = 0$ is a solution of function*

$$F'(x)^T F(x) = \begin{pmatrix} 1 & 2\lambda x + 1 \end{pmatrix} \begin{pmatrix} x + 1 \\ \lambda x^2 + x - 1 \end{pmatrix}$$

$$= x + 1 + (2\lambda x + 1)(\lambda x^2 + x - 1). \tag{3.113}$$

*That is, $x^\star = 0$ is a solution of problem (3.63). In view of*

$$F(x^\star) = \begin{pmatrix} 1 \\ -1 \end{pmatrix}, \tag{3.114}$$

$$F'(x^\star) = \begin{pmatrix} 1 \\ 1 \end{pmatrix} \tag{3.115}$$

*and*

$$F'(x^\star)^T F'(x^\star) = \begin{pmatrix} 1 & 1 \end{pmatrix} \begin{pmatrix} 1 \\ 1 \end{pmatrix} = 2, \tag{3.116}$$

*we can set constants $B$, $\eta$ and $\alpha$ in Theorem 3.3 as follows*

$$B = \frac{1}{2}, \quad \eta = \sqrt{2}, \quad \alpha = \sqrt{2}. \tag{3.117}$$

*Moreover, using (3.112), we can get for all $x, y \in D \subseteq \mathbb{R}$*

$$\|F'(x) - F'(y)\| = \left\| \begin{pmatrix} 0 \\ 2\lambda(x - y) \end{pmatrix} \right\| = 2|\lambda||x - y|, \tag{3.118}$$

*Using divided difference of order $[x, y; F]$ as given in Remark 3.5 (b) (see also (cf. [632])), we can get for all $x, y, u, v \in D \subseteq \mathbb{R}$*

$$\begin{aligned}\|[x,y;F]-[u,v;F]\| &= \left\| \int_0^1 \left(F'(tx+(1-t)y)dt - F'(tu+(1-t)v)\right)dt \right\| \\ &\leq \int_0^1 2\,|\lambda||t(x-u)+(1-t)(y-v)|dt \\ &\leq |\lambda|\left(|x-u|+|y-v|\right). \end{aligned} \tag{3.119}$$

*That is, we can set constants $M = M_0 = |\lambda|$ in Theorem 3.3. Hence, the function $q(r)$ can be given by*

$$\begin{aligned} q(r) &= B[(\alpha+2M_0r)Mr+2M_0\eta]+4M_0B(\alpha+M_0r)r-1 \\ &= \frac{1}{2}(\sqrt{2}+2|\lambda|r)(|\lambda|r)+\frac{1}{2}(2|\lambda|)\sqrt{2}+\frac{1}{2}(\sqrt{2}+|\lambda|r)(4|\lambda|r)-1 \\ &= 3(|\lambda|r)^2+\frac{5\sqrt{2}}{2}|\lambda|r+\sqrt{2}|\lambda|-1, \end{aligned} \tag{3.120}$$

*which has a unique positive root $r_\star$ given by*

$$r_\star = \frac{\sqrt{98-48\sqrt{2}|\lambda|}-5\sqrt{2}}{12|\lambda|}. \tag{3.121}$$

*If we choose parameter $\lambda$ such that*

$$0 < |\lambda| < \frac{\sqrt{2}}{2}, \tag{3.122}$$

*then, $2BM\eta = \sqrt{2}|\lambda| < 1$ and $U(x^\star, r_\star) \subseteq D$. Therefore, all the conditions in Theorem 3.3 are satisfied and Theorem 3.3 applies.*

## 3.3 Nondiscrete Induction and Secant Method

We use in this section nondiscrete mathematical induction to extend the applicability of the Secant methods for solving equations in a Banach space setting. Our approach has the following advantages over earlier works under the same information: weaker sufficient convergence conditions; tighter error bounds on the distances involved and a more information on the location of the solution.

We are concerned with the problem of approximating a locally unique solution $x^\star$ of (2.1) using Secant method SM given by (3.1). SM is also known under the name of Regula Falsi or the method of chords. In the case of scalar equations, SM is better than NM from the efficiency index point of view (cf. [157], [605]).

Under certain assumptions and using the information provided by the triplet $(F, x_{-1}, x_0)$, sequence $\{x_n\}$ converges to $x^\star$ with $R$–order of convergence $\dfrac{1+\sqrt{5}}{2}$. SM requires the inversion of a linear operator $\delta F(x_{n-1}, x_n)$ at each step. This inversion is very expensive in general. That is way we can use the Modified Secant Method MSM

$$x_{n+1} = x_n - \delta F(x_{-1}, x_0)^{-1} \, F(x_n) \quad (n \geq 0, \; x_{-1}, x_0 \in \mathcal{D}) \tag{3.123}$$

whose convergence is linear but requires only one inversion.

The conditions usually associated with the semilocal convergence of SM using the triplet $(F, x_{-1}, x_0)$ (cf. [146], [202], [597], [608]) are given by

($\mathcal{H}_1$) $F$ is a nonlinear operator defined on a convex subset $\mathcal{D}$ of a Banach space $\mathcal{X}$ with values in a Banach space $\mathcal{Y}$;

($\mathcal{H}_2$) $x_{-1}$ and $x_0$ are two points of $\mathcal{D}$ ($x_{-1} \neq x_0$) such that

$$\| \, x_0 - x_{-1} \, \| \leq q_0 < b; \tag{3.124}$$

($\mathcal{H}_3$) $F$ is Fréchet–differentiable in the open ball $U = U(x_0, b)$ and continuous on its closure $\overline{U}$;

($\mathcal{H}_4$) There exists an operator $\delta F : U \times U \to \mathcal{L}(\mathcal{X}, \mathcal{Y})$ such that the linear operator $S_0 = \delta F(x_{-1}, x_0)$ is invertible, its inverse $T_0 = S_0^{-1} \in \mathcal{L}(\mathcal{Y}, \mathcal{X})$ and

$$\| \, T_0 \, (\delta F(x, y) - F'(z)) \, \| \leq L \, (\| \, x - z \, \| + \| \, y - z \, \|) \tag{3.125}$$

for all $x, y, z \in U$ and some $L > 0$;

($\mathcal{H}_5$) The following inequalities are satisfied

$$\| \, T_0 \, F(x_0) \, \| \leq r_0, \tag{3.126}$$

$$r_0 \leq r_{PP} = \frac{(1 - L \, q_0)^2}{4 \, L} \tag{3.127}$$

and

$$b_0 = \frac{1}{2 \, L} \left(1 - L \, q_0 - ((1 - L \, q_0)^2 - 4 \, L \, r_0)^{1/2}\right) \leq b. \tag{3.128}$$

If ($\mathcal{H}_1$)–($\mathcal{H}_5$) are satisfied, we say that $I$ belongs in the class $\mathcal{C}(L, q_0, r_0)$. Potra and Pták using the celebrated method of Nondiscrete Mathematical Induction NMI (cf. [605], [608]) have improved the error bounds of the discrete case (cf. [469]). We refer the reader to the excellent monograph by Potra and Pták (cf. [608]) for the motivation and the general principles of NMI. Potra and Pták developed in a series of works (cf. [597], [608]) the NMI concept and provided a posteriori estimates which are in general better than those given by Gragg and Tapia (cf. [395]). Other works on iterative methods and nondiscrete induction can be found in (cf. [608]). In the case of SM for $r \in \mathcal{T} = (0, r_0]$, $a \geq 0$, the rate of convergence

$$\omega(q, r) = \frac{r\,(q + r)}{r + 2\,(r\,(r + q) + a^2)^{1/2}} \tag{3.129}$$

with corresponding estimate function

$$s(q, r) = r - a + (r\,(q + r) + a^2)^{1/2} \tag{3.130}$$

have been used to provide error estimates on the distances $\| x_{n+1} - x_n \|$, $\| x_n - x^\star \|$ $(n \geq 0)$ (cf. [605], [608]). The corresponding functions for MSM are

$$\omega(r) = \left(1 + L\,r - 2\,L\left(a^2 + \frac{r}{L}\right)^{1/2}\right) r \tag{3.131}$$

and

$$s(r) = (a^2 + \frac{r}{L})^{1/2} - a. \tag{3.132}$$

The parameter "$a$" is chosen to be

$$a = a_{PP} = \frac{1}{2\,L}\left((1 - L\,q_0)^2 - 4\,L\,r_0\right)^{1/2}. \tag{3.133}$$

We remark that there are cases where (3.127) is viloated. Note also the larger parameter $a$ is the tighter the error bound. Moreover, in this case information on the location solution $x^\star$ is more precise.

We are motivated by the works of Potra and Pták (cf. [605], [608]), optimization considerations and the above remarks. We shall show that under the same information $(F, x_{-1}, x_0)$ used by Potra and Pták (cf. [605], [608]), we have certain advantages such as: hypothesis (3.127) is weakened; parameter "$a$" is enlarged; error bounds on the distances $\| x_{n+1} - x_n \|$, $\| x_n - x^\star \|$ $(n \geq 0)$ are tighter and the information on the location of the solution more precise. It follows from (3.125) that there exists $L_0 \geq 0$ such that

$$\| T_0\,(\delta F(x, y) - F'(x_0)) \| \leq L_0\,(\| x - x_0 \| + \| y - x_0 \|) \quad (x, y \in U). \tag{3.134}$$

Note that

$$L_0 \leq L \tag{3.135}$$

holds in general and $\dfrac{L}{L_0}$ can be arbitrarily large (cf. [139], [202]). Estimate (3.134) is not an additional hypothesis to (3.125), since in practice the computational of $L$ requires that of $L_0$. Using (3.134) in the case when $L_0 < L$, the upper bounds on the distances $\| \delta F(x_{n-1}, x_n)^{-1}\,S_0 \|$ $(n \geq 0)$ are more precise than if (3.125) is used. This observation has already lead to advantages such as the ones mentioned above, in the discrete case but not always (cf. [157], [202]).

We shall study the semilocal convergence of MSM (3.123) for the class $\mathcal{C}_M(L_0, q_0, r_0)$ defined by the following.

**Definition 3.4** *Let $L_0 \geq 0$, $q_0 \geq 0$ and $r_0 \geq 0$ be given. We say triplet $(F, x_{-1}, x_0)$ belongs to the class $\mathcal{C}_M(L_0, q_0, r_0)$ if*

$(\mathcal{H}_1)$ *$F$ is a nonlinear operator defined on a convex subset $\mathcal{D}$ of a Banach space $\mathcal{X}$ with values in a Banach space $\mathcal{Y}$;*

$(\mathcal{H}_2)$ *$x_{-1}$ and $x_0$ are two points of $\mathcal{D}$ $(x_{-1} \neq x_0)$ such that*

$$\| x_0 - x_{-1} \| \leq q_0 < \overline{b};$$

$(\mathcal{H}_3)$ *$F$ is Fréchet–differentiable in the open ball $U_0 = U(x_0, \overline{b})$ and continuous on its closure $\overline{U_0}$;*

$(\mathcal{H}_6)$ *There exists an operator $\delta F : U_0 \times U_0 \to \mathcal{L}(\mathcal{X}, \mathcal{Y})$ such that the linear operator $S_0 = \delta F(x_{-1}, x_0)$ is invertible, its inverse $T_0 = S_0^{-1} \in \mathcal{L}(\mathcal{Y}, \mathcal{X})$ and*

$$\| T_0 \, (\delta F(x, y) - F'(x_0)) \| \leq L_0 \, (\| x - x_0 \| + \| y - x_0 \|) \quad (x, y \in U_0);$$

$(\mathcal{H}_7)$ *The following inequalities are satisfied*

$$\| T_0 \, F(x_0) \| \leq r_0,$$

$$r_0 \leq \overline{r}_{AH} = \frac{(1 - L_0 \, q_0)^2}{4 \, L_0}$$

*and*

$$\overline{b_0} = \frac{1}{2L_0} \left(1 - L_0 \, q_0 - ((1 - L_0 \, q_0)^2 - 4L_0 \, r_0)^{1/2}\right) \leq \overline{b}.$$

Usually, hypothesis (3.125) is used together with the Banach lemma on invertible operator to provide upper bounds on the norms $\| \delta F(x_n, x_{n-1})^{-1} \, S_0 \|$ $(n \geq 0)$ (cf. [597], [608]). However, only the more precise (since $L_0 \leq L$) than (3.125), hypothesis $(\mathcal{H}_6)$ is needed to derive such bounds (see also (3.153) that follows). That is the only difference between our proofs and the corresponding ones in the literature (cf. [597], [608]). Hence, we omit the proofs.

**Theorem 3.5** *Assume that $(F, x_{-1}, x_0) \in \mathcal{C}_M(L_0, q_0, r_0)$. Then, sequence $\{x_n\}$ $(n \geq 0)$ generated by MSM is well defined, remains in $U(x_0, \overline{b_0})$ for all $n \geq 0$ and converges to a solution $x^\star$ of equation (2.1). Moreover, the following error estimates hold for all $n \geq 1$*

$$\| x_n - x_{n-1} \| \leq \omega^{(n-1)}(r_0),$$

$$\| x_n - x^\star \| \leq s(\omega^{(n)}(r_0)),$$

$$\| x_n - x^\star \| \leq s(\| x_n - x_{n-1} \|) - \| x_n - x_{n-1} \|,$$

$$\frac{2 \, r_0}{1 - q_0 - 2 \, L_0 \, \overline{a}} \, (L_0 \, (q_0 + r_0))^n \leq s(\omega^{(n)}(r_0)) \leq \frac{r_0}{2 \, L_0 \, \overline{a}} \, (1 - 2 \, L_0 \, \overline{a})^n \quad (a > 0)$$

*and*

$$\frac{1}{n+1} \left( \frac{r_0}{L_0} \right)^{1/2} \leq s(\omega^{(n)}(r_0)) \leq \frac{1}{L_0 \, (n+2)} \quad (a = 0),$$

*where $\omega$, $s_1$ are given by (3.131) and (3.132), respectively, with*

$$a = \overline{a} = \frac{1}{2\,L_0}\,((1 - L_0\,q_0)^2 - 4\,L_0\,r_0)^{1/2}.$$

*Furthermore, if $\overline{a} > 0$, then $x^\star$ is the unique solution of equation $F(x) = 0$ in $\overline{U_0} \cap U(x_0, \overline{b_0} + 2\,\overline{a})$. If $a = 0$, then, $x^\star$ is the unique solution of equation $F(x) = 0$ in $\overline{U}(x_0, \overline{b_0})$.*

**Remark 3.7** *If $L_0 = L$, the results coincide with the corresponding ones in (cf. [597], [608]). Otherwise, they contitute an improvement. Notice that*

$$\overline{b_0} < b_0, \quad r_{PP} < \overline{r} \quad \text{and} \quad a_{PP} < \overline{a}.$$

We shall study the semilocal convergence of SM (3.1) for the class $\mathcal{C}^\star(L, q_0, r_0)$ defined as follows.

**Definition 3.5** *Let $L \geq L_0 \geq 0$, $q_0 \geq 0$ and $r_0 \geq 0$ be given. We say triplet $(F, x_{-1}, x_0)$ belongs to the class $\mathcal{C}^\star(L_0, q_0, r_0)$ if:*

($\mathcal{H}_1$) *$F$ is a nonlinear operator defined on a convex subset $\mathcal{D}$ of a Banach space $\mathcal{X}$ with values in a Banach space $\mathcal{Y}$;*

($\mathcal{H}_2$) *$x_{-1}$ and $x_0$ are two points of $\mathcal{D}$ ($x_{-1} \neq x_0$) such that*

$$\| x_0 - x_{-1} \| \leq q_0 < \overline{\overline{b}};$$

($\mathcal{H}_3$) *$F$ is Fréchet–differentiable in the open ball $U_1 = U(x_0, \overline{\overline{b}})$ and continuous on its closure $\overline{U_1}$;*

($\mathcal{H}_4$) *There exists an operator $\delta F : U_1 \times U_1 \to \mathcal{L}(\mathcal{X}, \mathcal{Y})$ such that the linear operator $S_0 = \delta F(x_{-1}, x_0)$ is invertible, its inverse $T_0 = S_0^{-1} \in \mathcal{L}(\mathcal{Y}, \mathcal{X})$ and*

$$\| T_0\,(\delta F(x, y) - F'(z)) \| \leq L\,(\| x - z \| + \| y - z \|) \quad (x, y, z \in U_1);$$

($\mathcal{H}_5$)

$$\| T_0\,(\delta F(x, y) - F'(x_0)) \| \leq L_0\,(\| x - x_0 \| + \| y - x_0 \|) \quad (x, y \in U_1);$$

($\mathcal{H}_6$) *The following inequalities are satisfied*

$$\| T_0\,F(x_0) \| \leq r_0,$$

$$L_0\,q_0 + (L_0 + L)\,r_0 + 2\,L\,(r_0\,(r_0 + q_0) + a^2)^{1/2} \leq 1 \tag{3.136}$$

*for some $a \geq 0$ and*

$$\overline{\overline{b_0}} = r_0 - a + (r_0\,(r_0 + q_0) + a^2)^{1/2} < \overline{\overline{b}}.$$

Next, we provide cases when (3.136) is satisfied.

**Proposition 3.2** *Let $q_0 \geq 0$, $r_0 \geq 0$, $L_0 \geq 0$, $L > 0$ and $L \geq L_0$.*
*Set*

$$a_{PP} = \frac{1}{2L}\left((1 - L\,q_0)^2 - 4L\,r_0\right)^{1/2}, \tag{3.137}$$

$$r_{PP} = \frac{(1 - L\,q_0)^2}{4L}, \tag{3.138}$$

$$\alpha = a_{AH} = \frac{1}{2L}\left((1 - L_0\,q_0 - (L_0 + L)\,r_0)^2 - 4L^2\,r_0\,(r_0 + q_0)\right)^{1/2}, \tag{3.139}$$

$$r(L_0, L) = \begin{cases} \dfrac{1 - L_0\,q_0}{\Xi} & \text{if} \quad L_0 \neq L, \\ r_{PP} & \text{if} \quad L_0 = L, \end{cases} \tag{3.140}$$

*and*

$$r_{AH}(L_0, L) = \begin{cases} r(L_0, L) & \text{if} \quad L_0 \neq L, \\ r_{PP} & \text{if} \quad L_0 = L, \end{cases} \tag{3.141}$$

*where*

$$\Xi = (1 - L_0\,q_0)\,(L_0 + L) + 2\,q_0\,L^2 + \left(((1 - L_0\,q_0)\,(L_0 + L) + 2\,q_0\,L^2)^2 + (1 - L_0\,q_0)\,(4\,L^2 - (L_0 + L)^2)\right)^{1/2}.$$

(i) *If $L = L_0$ and $r_0 \leq r_{PP}$, then, parameters $a_{PP}$, $r_{PP}$, $a_{AH}(L_0, L)$, $r(L, L)$ are non–negative and*

$$r_{PP} = r(L, L). \tag{3.142}$$

(ii) *If $r_0 \leq r_{AH}(L_0, L)$ for $L_0 < L$, then, parameters $a_{AH}(L_0, L)$ and $r(L_0, L)$ are non–negative.*

(iii) *If $r_0 \leq r_{PP}$, then, $r_{PP} \leq r(L_0, L)$.*

*Moreover, if (i) and (iii) hold, then $a_{PP} \leq a_{AH}(L_0, L)$. Furthemore, if $r_0 < q_0$, then $a_{PP} < a_{AH}(L_0, L)$.*

**Proof**

(i) The quantity $\Xi$ in (3.139) is non–negative if and only if, for $r \in [0, r_0]$

$$\begin{aligned} P_0(r) &= (4\,L^2 - (L_0 + L)^2)\,r^2 + 2\,((1 - L_0\,q_0)\,(L_0 + L) + 2\,q_0\,L^2)\,r \\ &\quad - (1 - L_0\,q_0)^2 \\ &\leq 0 \end{aligned} \tag{3.143}$$

at $r = r_0$.

In the case $L_0 = L$, (3.143) reduces to $r_0 \leq r_{PP}$. Hence we deduce $a_{AH}(L, L) \geq 0$. Clearly, $r_{PP} \geq 0$ and, by the definition of $r(L, L)$, we have $r_{PP} = r(L, L)$. It also follows from $r_0 \leq r_{PP}$ that $a_{PP} \geq 0$. Define a function $P : [0, L] \longrightarrow (-\infty, +\infty)$ by

$$P(\lambda) = (1 - (q_0 + r_0)\,\lambda - L\,r_0)^2 - 4\,L^2\,r_0\,(r_0 + q_0).$$

We have that

$$P'(\lambda) = -2\,(q_0 + r_0)\,(1 - (q_0 + r_0)\,\lambda - L\,r_0)\,\lambda < 0,$$

so function $P$ is decreasing on interval $[0, L]$ and $P(0)$ is the smallest value of $P$. Note that

$$a_{AH}^2(0, L) = \frac{P(0)}{4\,L^2}$$

and (3.143) holds.

(ii) It follows from $L_0 < L$ and (3.143) that quadratic polynomial $P_0$ has a unique positive root $r_{AH}(L_0, L)$. Then, the quantity $\Xi$ is non–negative if $r_0 \le r_{AH}(L_0, L)$ holds.

(iii) This part follows from $r_0 \le r_{PP}$ using the definition of $r_{PP}$, $r(L_0, L)$, $a_{PP}$ and $a_{AH}(L_0, L)$.

This completes the proof of Proposition 3.2. ⊠

The proof of the following semilocal result for SM (3.1) is similar to the one in (cf. [608, p. 69] or [605, p. 438]), but there is a crutial difference where (3.136) is used instead (3.127).

**Theorem 3.6** *Assume that* $(F, x_{-1}, x_0) \in C^\star(L, q_0, r_0)$. *Then, sequence* $\{x_n\}$ $(n \ge 0)$ *generated by SM is well defined, remains in* $U(x_0, \overline{\overline{b_0}})$ *for all* $n \ge 0$ *and converges to a solution* $x^\star$ *of equation (2.1). Moreover, the following error estimates hold for all* $n \ge 1$

$$\| \, x_{n+1} - x_n \, \| \le \omega^{(n)}(q_0, r_0), \tag{3.144}$$

$$\| \, x_n - x^\star \, \| \le s(\tilde{\omega}^{(n)}(q_0, r_0)) \tag{3.145}$$

*and*

$$\| \, x_n - x^\star \, \| \le \left(\alpha^2 + \| \, x_n - x_{n-1} \, \| \, \left(\| \, x_{n-1} - x_{n-2} \, \| + \| \, x_n - x_{n-1} \, \|\right)\right)^{1/2} - \alpha, \tag{3.146}$$

*where functions* $\omega$, $s$ *and parameter* $\alpha$ *are given by (3.129), (3.130), (3.139), respectively and* $\tilde{\omega}(q, r) = \omega(r, \omega(q, r))$.

**Proof** We shall use (cf. [608, Proposition 3.4, p. 53]) for $p = 2$ and $m = 1$.
Define set $\mathcal{D}$ and operator $G$ by

$$\mathcal{D} = \{u = (y, x) \in U_1^2 \, : \, \delta F(x, y) \ \text{ is invertible}\}, \tag{3.147}$$

$$G\,u = x - \delta F(x, y)^{-1}\,F(x), \quad u = (y, x) \in U_1^2. \tag{3.148}$$

Set, for $q, r \in \mathcal{T}$, $t = (q, r)$. Define approximate set

$$\mathbb{Z}(t) = \{u = (y, x) \in \mathcal{X}^2 \, : \, y \in U_1, \ \| \, x - y \, \| \le q, \ \| \, x - x_0 \, \| \le \overline{\overline{b}} - s(t),$$

$$S = \delta F(x, y) \ \text{ is invertible } \ \textit{and} \ \| \, S^{-1}\,F(x) \, \| \le r\}. \tag{3.149}$$

Using (3.130), we see function $s$ is increasing, $\bar{\bar{b}} = s(t_0)$, $t_0 = (q_0, r_0)$ and

$$0 < s(t) < \bar{\bar{b}} \quad (0 < t < t_0).$$

Clearly, $\mathbb{Z}(t) \subseteq \mathcal{D}$ and $u_0 = (x_{-1}, x_0) \in \mathbb{Z}(t_0)$. We must show that, if $u = (y, x) \in \mathbb{Z}(t)$ then,

$$(x, G\,u) \in \mathbb{Z}(\tilde{\omega}(t)). \tag{3.150}$$

Set $z = G\,u$. We can show instead of implication (3.150) that

$$x \in U_1, \quad \| \, z - x \, \| \leq r, \tag{3.151}$$

$$\| \, z - x_0 \, \| \leq \bar{\bar{b}} - s(\tilde{\omega}(t)). \tag{3.152}$$

Operator $S_1 = \delta F(x, y)$ is invertible and $\| \, S_1^{-1} \, F(z) \, \| \leq \omega(t)$. By $z - x = -S^{-1}\,F(x)$ and $u \in \mathbb{Z}(t)$, we obtain (3.151). We also have that

$$\| \, z - x_0 \, \| \leq \| \, z - x \, \| + \| \, x - x_0 \, \| \leq r + \bar{\bar{b}} - s(t) = \bar{\bar{b}} - s(\tilde{\omega}(t))$$

(since $s(\tilde{\omega}(t)) = s(t) - r$), which shows (3.152).

Using (3.134) (not (3.125) used in (cf. [608])) and (3.136) (not (3.127) used in (cf. [608])), we get in turn that

$$\begin{aligned}
\| \, T_0\,(S_0 - S_1) \, \| &\leq \| \, T_0\,(S_0 - F'(x_0)) \, \| + \| \, T_0\,(F'(x_0) - S_1) \, \| \\
&\leq L_0\,(\| \, x_0 - x_{-1} \, \| + \| \, x - x_0 \, \| + \| \, z - x_0 \, \|) \\
&\leq L_0\,(q_0 + \| \, x - z \, \| + 2 \, \| \, z - x_0 \, \|) \\
&\leq L_0\,(q_0 + r + 2\,(\bar{\bar{b}} - s(t))) \\
&\leq L_0\,(q_0 + r + 2\,(r_0 + (r_0\,(q_0 + r_0) + a^2)^{1/2} - r \\
&\quad - (r(q + r) + a^2)^{1/2}) \\
&\leq 1 - L\,(r + 2\,(r(q + r) + a^2)^{1/2}) \quad (r \in [0, r_0]).
\end{aligned} \tag{3.153}$$

It follows from (3.153) and the Banach lemma on invertible operators that

$$\| \, (T_0\,S_1)^{-1} \, \| \leq \frac{1}{L\,(r + 2\,(r\,(q + r) + \alpha^2)^{1/2})}. \tag{3.154}$$

We also have by (3.125) and identity

$$F(z) = F(z) - F(x) - \delta F(y, x)\,(z - x)$$

that

$$\begin{aligned}
&\| \, T_0\,(F(x) - F(y) - \delta F(u, v)\,(x - y)) \, \| \\
&\leq L\,(\| \, x - v \, \| + \| \, y - v \, \| + \| \, u - v \, \|) \, \| \, x - y \, \| \, .
\end{aligned}$$

Consequently, we get that

$$\begin{aligned} \| T_0\, F(z) \| &\leq L\, (\| z - x \| + \| x - y \|)\; \| z - x \| \\ &\leq L\, (q + r)\, r. \end{aligned} \tag{3.155}$$

In view of (3.154) and (3.155), we obtain that

$$\begin{aligned} \| S^{-1}\, F(z) \| &\leq \| (T_0\, S_1)^{-1} \| \, \| T_0\, F(z) \| \\ &\leq \frac{L\, (q + r)\, r}{L\, (r + 2\, (r(q + r) + \alpha^2)^{1/2})} \\ &= \omega(t). \end{aligned} \tag{3.156}$$

The rest follows from (cf. [608, Proposition 3.4 and Theorem 4.11]). This completes the proof of Theorem 3.6. ⊠

**Remark 3.8** *In view of Proposition 3.2 and Theorem 3.6, we have the already mentioned advantages over the corresponding (cf. [608, Theorem 4.11]) or [605, Theorem 3]). Note also*

$$\overline{\overline{b_0}} < b_0.$$

We provide example where our results apply, but earlier ones do not. In case all results apply, we show that ours provide tighter error results bounds and better information on the location of the solution.

**Example 3.4** *Let $\mathcal{X} = \mathcal{Y} = \mathbb{R}$ be equipped with the max–norm, $(x_{-1}, x_0) = (.99, 1)$, $\mathcal{D} = [\varrho, 2 - \varrho]$, $\varrho \in [0, 1)$ and define a function $F$ on $\mathcal{D}$ by*

$$F(x) = x^3 - \varrho. \tag{3.157}$$

*Define a linear operator $\delta F(x, y)$ by*

$$\delta F(x, y) = \int_0^1 F'(y + t\,(x - y))\, dt = x^2 + y^2 + x\, y. \tag{3.158}$$

*Using (3.157) and (3.158), we can write for all $(x, y, z) \in \mathcal{D}^3$ that*

$$\delta F(x, y) - F'(z) = (x - y + 3z)\,(x - z) + (y + 2\,x)\,(y - z).$$

*Then, using Definition 3.4 and Maple 13, we get that*

$$r_0 = \frac{1}{2.9701}\,(1 - \varrho), \quad L_0 = \frac{1}{2.9701}\,(6 - 3\,\varrho) \quad \text{and} \quad L = \frac{1}{2.9701}\,(8 - 5\,\varrho).$$

*The second inequality in $(\mathcal{H}_7)$ of Definition 3.4 holds if*

$$\varrho \in (.5185273827, 1).$$

**The MSM case for $\varrho = .52$.** *By Definition 3.4 and using Maple 13, we get that*

$$r_0 = .1616107202, \quad L_0 = 1.494899162, \quad L = 1.818120602$$

*and*

$$r_{PP} = .1325500827, \quad \overline{r}_{AH} = .1622727328.$$

*Condition (3.127) is violated since*

$$r_0 = .1616107202 > r_{PP} = .1325500827.$$

*Hence, there is no guarantee that MSM converges to $x^\star = .8041451517$, starting at $(x_{-1}, x_0) = (.99, 1)$. Moreover, all the condition of Theorem 3.5 hold and we can apply the MSM for solving equation (3.157).*

**The SM case for $\varrho = .57$.** *Using Definition 3.5 and Maple 13, we obtain*

$$r_0 = .1447762702, \quad L_0 = 1.444395812, \quad L = 1.733948352,$$

$$r_{PP} = .1392229604$$

*and condition (3.136) holds for $a = .01$ since*

$$L_0\, q_0 + (L_0 + L)\, r_0 + 2\, L\, (r_0\, (r_0 + q_0) + .01^2)^{1/2} = .9948689500 < 1.$$

*However, condition (3.127) is violated, since*

$$r_0 = .1447762702 > r_{PP} = .1392229604.$$

*Hence, there is no guarantee that SM converges to $x^\star = .8291344342$, starting at $(x_{-1}, x_0) = (.99, 1)$. Moreover, all the condition of Theorem 3.6 hold and we can apply the SM for solving equation (3.157).*

Other applications and examples including the solution of nonlinear Chandrasekhar–type integral equations appearing in radiative transfer can be found in Chapter 2.

## 3.4 Nondiscrete Induction and a Double Step Secant Method

We provide in this section a semilocal convergence analysis for a double step Secant method using nondiscrete induction. Special cases and numerical examples where our results apply but earlier ones fail to solve equation as well as tighter error bounds are also provided.

Pták in his celebrated Gatlinburg lecture (cf. [611]) initiated the method of Nondiscrete Mathematical Induction NMI as a tool for improving error bounds of discrete case for iterative procedures but not necessarily the sufficient convergence conditions for such methods. We find in the excellent monograph by Potra and Pták (cf. [608]) more details about the motivation, general principles for NMI. Potra and Pták in a plethora of papers provided semilocal convergence results for Newton's method (cf. [605]), modified Newton method (cf. [604]), Secant method (cf. [597]), Secant–type methods (cf. [608]) and Traub–type method (cf. [606], [608]).

In particular, they used the double step Secant method DSSM

$$y_{n+1} = y_n - [y_n, x_n; f]^{-1} f(y_n),$$

$$x_{n+1} = y_{n+1} - [y_n, x_n, f]^{-1} f(y_{n+1}) \quad (n \geq 0, \quad x_0, y_0 \in \mathcal{D}) \tag{3.159}$$

to generate two sequences $\{x_n\}$, $\{y_n\}$ $(n \geq 0)$ approximating a locally unique solution $x^\star$ of (2.1). Here, for $x, y \in \mathcal{D}$, $[x, y; f] \in \mathcal{L}(\mathcal{X}, \mathcal{Y})$ the space of bounded linear operators from $\mathcal{X}$ into $\mathcal{Y}$ denotes a divided difference of order one for operator $f$ at the points $x$ and $y$. DSSM uses one inversion of a linear operator at each step and it is of convergence order $1+\sqrt{2}$. The discrete case has been studied by Schmidt and Schwetlick (cf. [670]), Laasonen (cf. [496]) (see also (cf. [157], [605], [608])). Potra and Ptàk (cf. [603], [608]) improved the error bounds using NMI with the pair of functions $\omega = (\omega_1, \omega_2)$ defined as the following:

$$\omega_1(q, r) = \frac{r\,(q+r)}{q + 2\,r + 2\,(r\,(r+q) + a^2)^{1/2}}, \tag{3.160}$$

$$\omega_2(q, r) = \omega_1(q, r)\, \frac{q + 2\,r + \omega_1(q, r)}{2\,(r\,(r+q) + a^2)^{1/2} - \omega_1(q, r)} \tag{3.161}$$

as a rate of convergence on $\mathcal{T} = (0, q_0) \times (0, r_0)$ with corresponding estimate function $\sigma = (\sigma_1, \sigma_2)$ given for

$$g(q, r) = (r\,(r+q) + a^2)^{1/2} - a \tag{3.162}$$

by

$$\sigma_1(q, r) = g(q, r) + q + r \tag{3.163}$$

and

$$\sigma_2(q, r) = g(q, r) + r, \tag{3.164}$$

under the sufficient convergence condition

$$(r_0^{1/2} + (q_0 + r_0)^{1/2})^2 \leq \frac{d_0}{L}, \tag{3.165}$$

where

$$a = a_{PP} = \frac{1}{2\,L}\left(d_0^2 + L^2\,q_0^2 - 2\,d_0\,L\,(q_0 + 2\,r_0)\right)^{1/2}. \tag{3.166}$$

Here, $r_0 > 0$, $q_0 > 0$, $L > 0$ and $d_0 > 0$ depend on the initial data (to be also precised later). Inequality (3.165) is the sufficient convergence condition for the discrete case (cf. [597], [608]). Parameter $a_{PP}$ is chosen to be the largest possible. If $q_0 = 0$, (3.165) and (3.166) become respectively

$$4\,r_0 \le \frac{d_0}{L} \tag{3.167}$$

and

$$a = a_{PP} = \frac{1}{2L}\left(d_0^2 - 4\,d_0\,L\,r_0\right)^{1/2}. \tag{3.168}$$

That is, DSSM specializes to the corresponding double step Newton's method DSNM given by

$$\begin{aligned} y_{n+1} &= y_n - f'(y_n)^{-1}\,f(y_n), \\ x_{n+1} &= y_{n+1} - f'(y_n)^{-1}\;f(y_{n+1}) \quad (n \ge 0,\; x_0, y_0 \in \mathcal{D}). \end{aligned} \tag{3.169}$$

We are motivated by optimization considerations, the works of Potra, Ptàk (cf. [597], [608]) and Amat et. al. (cf. [11]). We would like to know if it is possible to expand the applicability of DSSM by weakening (3.165) and enlarging $a_{PP}$ using the same information $(f, y_0, x_0, q_0, r_0, L)$. It turns out that this can be done. Note that the larger the "$a$" is the tighter of error bounds. The main idea is to use center–Lipschitz conditions (3.178) or (3.179) instead of the less precise Lipschitz condition (3.176) (used so far (cf. [603])) for the computation of the inverses of linear operators. This technique has been used by us in the discrete case on Newton–type methods to produce weaker sufficient convergence conditions and tighter error bounds under the same information as before (cf. [157], [209]).

In order for us to make the section as self contained as possible, we need to re–introduce some concepts (cf. [157], [209], [605], [603], [608]). Let $(\mathcal{X}, d)$ be a metric space, where

$$d(A) = \inf_{\|x\| \ge 1} \|\, A\,x \,\|\,. \tag{3.170}$$

If $\mathcal{X}_0$ is a normed vector space, then $\|\, y \,\|$ denotes the norm of $y \in \mathcal{X}_0$. Let $A$ be a linear operator on a Banach space $\mathcal{X}$. If $A$ is invertible, then, $d(A) = \|\, A^{-1} \,\|^{-1}$. For any perturbation by a bounded linear operator $B$, we have $d(A+B) \ge d(A) - \|\, B \,\|$. Moreover, by Banach lemma on invertible operator (cf. [157], [469]), if $A$ is invertible and $d(A) - \|\, B \,\| > 0$, then, so is $A + B$. Let $i = 1, 2$. Let $\mathcal{T}_i$ be the set of all positive numbers or an open interval of the form $(0, t_i)$, $t_i > 0$.

Set $\mathcal{T} = \mathcal{T}_1 \times \mathcal{T}_2$. We need the definition of rate of convergence $\omega$ and corresponding estimate function $\sigma$.

**Definition 3.6** *A function $\omega : \mathcal{T} \longrightarrow \mathcal{T}$ is called a rate of convergence on $\mathcal{T}$ if the series $\sum_{n=0}^{\infty} \omega^{(n)}(r)$ is convergent for each $r \in \mathcal{T}$. We have*

$$\omega^{(0)}(r) = r, \quad \omega^{(n+1)}(r) = \omega(\omega^{(n)}(r))$$

*and*

$$\omega^{(n)}(r) = (\omega_1^{(n)}(r), \omega_2^{(n)}(r)). \tag{3.171}$$

*Define vector function p for $r \in \mathcal{T}$ by*

$$p(r) = (r_1 + r_2, r_2 + \omega_1(r)). \tag{3.172}$$

*Then, we can attach to $\omega$ the estimate function*

$$\sigma(r) = \sum_{n=0}^{\infty} p(\omega^{(n)}(r)). \tag{3.173}$$

*Clearly, we have*

$$\sigma(r) = \sigma(\omega(r)) + p(r) \quad (r \in \mathcal{T}). \tag{3.174}$$

It was shown in (cf. [603, p. 242]) that $\omega$, $\sigma$ defined in the Introduction satisfy (3.174). We need the definition of a divided difference of order one (cf. [157], [209], [469]).

**Definition 3.7** *A mapping $(x, y) \longrightarrow [x, y; f]$ is a divided difference of $f$ on $\mathcal{D}$ if the following conditions are satisfied:*

*(1) $[x, y; f] \in \mathcal{L}(\mathcal{X}, \mathcal{Y})$ such that*

$$[x, y; f]\,(x - y) = f(x) - f(y) \quad (x \neq y). \tag{3.175}$$

*(2) For all $x, y, \overline{x}, \overline{y} \in \mathcal{D}$, we have*

$$\|\, [x, y; f] - [\overline{x}, \overline{y}; f] \,\| \leq L\,(\| x - \overline{x} \| + \| y - \overline{y} \|). \tag{3.176}$$

It follows from (3.176) that, for each $x \in \mathcal{D}$, the Fréchet–derivative of $f$ exists and we can set

$$[x, x; f] = f'(x). \tag{3.177}$$

In view of (3.176), there exist $L_0 \geq 0$, $L_1 > 0$ such that

$$\|\, [y, x; f] - f'(x_0) \,\| \leq L_0\,(\| y - x_0 \| + \| x - x_0 \|) \quad (x, y \in \mathcal{D}) \tag{3.178}$$

and

$$\|\, [y, x; f] - [y_0, x_0; f] \,\| \leq L_1\,(\| y - y_0 \| + \| x - x_0 \|) \quad (x, y \in \mathcal{D}). \tag{3.179}$$

Note that

$$L_0 \leq L_1 \leq L \tag{3.180}$$

holds in general and $\dfrac{L}{L_0}, \dfrac{L}{L_1}, \dfrac{L_1}{L_0}$ can be arbitrarily large. Center–Lipschitz conditions (3.178) or (3.179) are not an additional hypotheses to (3.176), since in practice the computation of Lipschitz constant $L$ requires that of $L_0$ and $L_1$.

We provide now a semilocal convergence analysis of DSSM using NMI. We assume familiarily with the generalization of the induction theorem (see (cf. [603, p. 238])). We shall follow the proof of (cf. [603, Theorem 3.2, p. 243]), we use (3.178) or (3.179) instead of (3.176) for the computation of the inverses involved.

First, we need three auxilliary results for the functional inequalities and the "$a$" parameters used in our main Theorem 3.7. We are interested in finding solutions of a certain type for functional inequations.

**Lemma 3.3** *Let $L_0 > 0$, $L > 0$, $q_0 \geq 0$, $r_0 \geq 0$ and $a \geq 0$. Let also functions $\omega_1$, $\omega_2$, $\omega$, $\sigma_1$ and $\sigma_2$ be as in the Introduction. For all $r \in [0, r_0]$ and $q \in [0, q_0]$, define functional inequality on $[0, q_0) \times [0, r_0)$ by*

$$\begin{aligned} L\,M - L_0\,\big(q + 2\,r + \omega_1(q,r) + 2\,((\sigma_2(q,r) - \sigma_1(\omega(q,r)))\\ +(\sigma_2(q,r) - \sigma_2(\omega(q,r))))\big) \geq L\,M \circ \omega, \end{aligned} \tag{3.181}$$

*where $M = M(q,r)$ and $M \circ \omega$ denote the composition between $M$ and $\omega$ given by*

$$M \circ \omega = (M\,\omega_1, M\,\omega_2). \tag{3.182}$$

*We are interested in non–negative solutions of (3.181) in the form*

$$M(q,r) = \alpha\,q + \beta\,r + \gamma\,(r\,(r+q) + a^2)^{1/2} - \delta, \tag{3.183}$$

*where $\alpha$, $\beta$ and $\gamma$ are non–negative numbers. Then, the following assertions*

*(1) Inequation (3.181) has a non–negative solution $K_0(q,r)$ given by*

$$K_0(q,r) = \frac{3\,L_0}{L}\,(q + 2\,r + 2\,(r\,(r+q) + a^2)^{1/2} - \delta_0) \tag{3.184}$$

*for $0 \leq \delta_0 \leq 2\,a$. Function $K_0$ solves (3.181) as equation.*

*(2) Inequation (3.181) has a non–negative solution $K_1(q,r)$ given by*

$$K_1(q,r) = \frac{L_0}{L}\,(q + 6\,r + 6\,(r\,(r+q) + a^2)^{1/2} - \delta_1) \tag{3.185}$$

*for $0 \leq \delta_1 \leq 6\,a$. Function $K_1$ solves (3.181) as inequality.*

**Proof** We need certain estimates. For simplicity, we denote $s := (r\,(r+q) + a^2)^{1/2}$. By the definition of functions $\omega_1$ and $\omega_2$, we have

$$\begin{aligned} & r(q+r) + \omega_1^2 - 2\,\omega_1\,s = \omega_2\,(\omega_1 - 2\,s)\\ & \Longrightarrow s^2 + (\omega_1+\omega_2)^2 - 2\,(\omega_1+\omega_2)\,s = \omega_2\,(\omega_1+\omega_2) + a^2\\ & \Longrightarrow s = \omega_1 + \omega_2 + (\omega_2\,(\omega_1+\omega_2) + a^2)^{1/2}. \end{aligned} \tag{3.186}$$

Using the definition of $\sigma$, $\omega_i$ $(i = 1,2)$, $\omega$ functions, (3.182) and (3.186), we get

$$M \circ \omega = \alpha\,\omega_1 + \beta\,\omega_2 + \gamma\,(\omega_2\,(\omega_1+\omega_2) + a^2)^{1/2} - \delta = (\alpha - \gamma)\,\omega_2 + \gamma\,s - \delta,$$

$$\sigma_1(\omega(q,r)) = \sigma_1(\omega_1,\omega_2) = (\omega_2\,(\omega_1+\omega_2)+a^2)^{1/2}+\omega_1+\omega_2-a$$
$$= s-a, \tag{3.187}$$
$$\sigma_2(\omega(q,r)) = \sigma_2(\omega_1,\omega_2) = (\omega_2\,(\omega_1+\omega_2)+a^2)^{1/2}+\omega_2-a$$
$$= s-\omega_1-a, \tag{3.188}$$
$$\sigma_2(q,r)-\sigma_1(\omega(q,r)) = s+r-a-s+a = r \tag{3.189}$$

and

$$\sigma_2(q,r)-\sigma_2(\omega(q,r)) = s+r-a-s+\omega_1+a = r+\omega_1. \tag{3.190}$$

So, we have

$$q+2\,r+\omega_1+2\,((\sigma_2(q,r)-\sigma_1(\omega(q,r)))+(\sigma_2(q,r)-\sigma_2(\omega(q,r))))$$
$$= q+2\,r+\omega_1+2\,(2\,r+\omega_1) = q+6\,r+3\,\omega_1. \tag{3.191}$$

(1) Inequation (3.181) holds as equation if there exists $b \geq 0$ such that

$$M(q,r)+\delta = \frac{1}{L}\,\big(b+L\,(\alpha-\gamma)\,\omega_1+L\,(\beta-\gamma)\,\omega_2$$
$$+\,L\,\gamma\,s+L_0\,(q+6\,r+3\,\omega_1)\big) \tag{3.192}$$

or

$$\alpha\,q+\beta\,r+\gamma\,s = \frac{1}{L}\,b+(\alpha-\gamma)\,\omega_1+(\beta-\gamma)\,\omega_2+$$
$$\gamma\,s+\frac{L_0}{L}\,q+\frac{6\,L_0}{L}\,r+\frac{3\,L_0}{L}\,\omega_1$$

or

$$(\beta-\gamma)\,\omega_2+\left(\alpha-\gamma+\frac{3\,L_0}{L}\right)\omega_1+\left(\frac{6\,L_0}{L}-\beta\right)r+\left(\frac{L_0}{L}-\alpha\right)q+\frac{b}{L}=0, \tag{3.193}$$

which leads to the choice

$$\alpha = \frac{3\,L_0}{L}, \quad \beta=\gamma=\frac{6\,L_0}{L} \quad \textit{and} \quad b = 2\,L_0\,q. \tag{3.194}$$

That is, we obtain function $K_0$ for $\delta = \alpha\,\delta_0$. Note that condition $\delta_0 \leq 2\,a$ guarantees the non–negativity of function $K_0$ on $[0,q_0]\times[0,r_0]$.

(b) This time if we choose

$$b=0, \quad \alpha = \frac{L_0}{L}, \quad \beta=\gamma=\frac{6\,L_0}{L}, \tag{3.195}$$

we have by (3.193)

$$-\frac{2\,L_0}{L}\,\omega_1 \le 0 \quad ((q,r)\in[0,q_0]\times[0,r_0]). \tag{3.196}$$

These choices lead to function $K_1$ if we consider $\delta = \alpha\,\delta_1$, which is non–negative, since $\delta_1 \le 6\,a$.

This completes the proof of Lemma 3.3. ☒

Next, we compare "$a$" parameters as well as various sufficient convergence conditions of the form (3.165).

**Lemma 3.4** *Let $L_0>0$, $L>0$, $d_0>0$, $r_0\ge 0$ and $q_0\ge 0$ be given parameters. Let also functions $K_0$ and $K_1$ be given by (3.184) and (3.185), respectively. Then, the following assertions hold*

*(1) If*

$$(r_0^{1/2}+(q_0+r_0)^{1/2})^2 \le \frac{1}{L}\left(\frac{d_0}{3\,L_0}+\delta_0\right),\quad 0\le\delta_0\le 2\,a_{AH}^0 \tag{3.197}$$

*and*

$$a=a_{AH}^0=\frac{1}{2\,L}\left(\left(\frac{d_0}{3\,L_0}+\delta_0\right)^2+L^2\,q_0^2-2\left(\frac{d_0}{3\,L_0}+\delta_0\right)L\,(q_0+2\,r_0)\right)^{1/2}, \tag{3.198}$$

*then,*

$$L\,K_0(q_0,r_0)\le d_0. \tag{3.199}$$

*(2) If*

$$12\,r_0\left(\frac{d_0}{L_0}+\delta_1+2\,q_0\right)\le\left(\frac{d_0}{L_0}+\delta_1-q_0\right)^2,\quad 0\le\delta_1\le 6\,a_{AH}^1 \tag{3.200}$$

*and*

$$a=a_{AH}^1=\frac{1}{6}\left(\left(\frac{d_0}{L_0}+\delta_1-q_0\right)^2-12\,r_0\left(\frac{d_0}{L_0}+\delta_1+2\,q_0\right)\right)^{1/2}, \tag{3.201}$$

*then,*

$$L\,K_1(q_0,r_0)\le d_0. \tag{3.202}$$

*(3) If (3.165), (3.197) and*

$$d_0<\frac{d_0}{3\,L_0}+\delta_0 \tag{3.203}$$

*hold, then,*

$$a_{PP}<a_{AH}^0 \tag{3.204}$$

*and*

$$(3.165)\Longrightarrow(3.197), \tag{3.205}$$

*but not necessarily vice versa.*

The results of Lemma 3.4 are obtained by simply solving inequalities (3.199) and (3.202). Note that we can always choose the worst values for $\delta_0$ and $\delta_1$, which are $\delta_0 = \delta_1 = 0$. Estimate (3.203) always holds provided that

$$3\,L_0 < 1. \tag{3.206}$$

**Lemma 3.5** *Let $L_1 > 0$, $L > 0$, $d_0 > 0$, $r_0 \geq 0$, $q_0 \geq 0$ and $a \geq 0$. Let also functions $\omega_1$, $\omega_2$, $\omega$, $\sigma_1$, $\sigma_2$ be as in the Introduction for all $r \in [0, r_0]$, $q \in [0, q_0]$. Define a functional inequation on $[0, q_0] \times [0, r_0]$ by*

$$d_0 - L_1\,(q_0 + q + 4\,r + 2\,\omega_1) \geq L\,K \circ \omega. \tag{3.207}$$

*Then, the following hold*

*(1) If*

$$(r_0\,(q_0 + r_0))^{1/2} + 2\,r_0 + q_0 \leq \delta_2 + \frac{d_0}{2L_1} \tag{3.208}$$

*and*

$$a = a^\star_{AH} = \frac{1}{2L_1}\left((d_0 + 2\,L_1\,\delta_2 - 2\,L_1\,(q_0 + 2\,r_0))^2 - 4\,L_1^2\,r_0\,(r_0 + q_0)\right)^{1/2}, \tag{3.209}$$

*then, inequality (3.207) has a non–negative solution $K_2(q,r)$ given by*

$$K_2(q,r) = \frac{2\,L_1}{L}\,(r + (r(q+r) + a^2)^{1/2} - \delta_2) \tag{3.210}$$

*for $0 \leq \delta_2 \leq a$.*

*(2)*

$$L\,K_2(q_0, r_0) \leq d_0. \tag{3.211}$$

**Proof**

(1) Using (3.183), we see that (3.207) is satisfied if

$$\begin{aligned} d_0 - 2\,L_1\,q_0 + L\,\delta - (2\,L_1 + L\,(\alpha - \gamma))\,\omega_1 - L\,(\beta - \gamma)\,\omega_2 \\ \geq 4\,L_1\,r_0 + L\,\gamma\,(r_0\,(q_0 + r_0) + a^2)^{1/2}. \end{aligned} \tag{3.212}$$

Choose

$$\beta = \gamma, \quad 2\,L_1 + L\,(\alpha - \gamma) = 0 \quad \textit{and} \quad \delta = \frac{2\,L_1}{L}\,\delta_2.$$

Then, we get $\alpha = \gamma - \dfrac{2\,L_1}{L}$. We must have $\alpha \geq 0$, so $\gamma = \dfrac{2\,L_1}{L}$ to obtain $\beta = \dfrac{2\,L_1}{L}$ and set $\alpha = 0$. That is (3.212) is satisfied by (3.208) and the choice of $K_2$ and $a^\star_{AH}$.

(2) This part follows from the choices of $K_2$, $a^\star_{AH}$ and (3.208).

The proof of Lemma 3.5 is complete. ⊠

We can certainly choose $\delta_2 = 0$ in the Lemma.

We can show the following semilocal convergence result for DSSM.

**Theorem 3.7** *Let $f : \mathcal{D} \subseteq \mathcal{X} \longrightarrow \mathcal{Y}$. Assume*

($\mathcal{H}_1$) *There exists a divided difference $[x, y; f]$ $(x, y \in \mathcal{D})$ such that (3.175) and (3.176) are satisfied for all $x$, $y \in \mathcal{D}$;*

($\mathcal{H}_2$) *For $x_0$, $y_0 \in \mathcal{D}$, $A_0 = [y_0, x_0; f]^{-1} \in \mathcal{L}(\mathcal{Y}, \mathcal{X})$ and $d(A_0^{-1}) \geq d_0$;*

($\mathcal{H}_3$)
$$\| y_0 - x_0 \| \leq q_0;$$

($\mathcal{H}_4$)
$$\| A_0 f(x_0) \| \leq r_0;$$

($\mathcal{H}_5$) *Hypotheses (3.197) or (3.200) or (3.208) are satisfied;*

($\mathcal{H}_6$)
$$R_0 \geq \sigma_2(q_0, r_0)$$

*and*
$$\overline{U}(x_0, R_0) \subseteq D.$$

*Then, sequences $\{y_n\}$, $\{x_n\}$ generated by DSSM are well defined, remain in $\overline{U}(x_0, R_0)$ for all $n \geq 0$ and converge to a solution $x^\star$ of equation (2.1). Moreover, the following error estimates hold for all $n \geq 1$*

$$\| x^\star - x_0 \| \leq \sigma_2(q_0, r_0),$$

$$\| y_n - x^\star \| \leq \sigma_1(\omega^{(n)}(q_0, r_0)),$$

$$\| x_n - x^\star \| \leq \sigma_2(\omega^{(n)}(q_0, r_0)),$$

$$\| y_n - x^\star \| \leq g(\| y_{n-1} - x_{n-1} \|, \| x_{n-1} - y_n \|)$$

*and*

$$\| x_n - x^\star \| \leq g(\| y_{n-1} - x_{n-1} \|, \| x_{n-1} - y_n \|) - \omega_1(\| y_{n-1} - x_{n-1} \|, \| x_{n-1} - y_n \|),$$

*where $\omega_1$, $\omega_2$, $g$, $\sigma_1$, $\sigma_2$ are given in the Introduction with "a" defined in (3.198) or (3.201) or (3.209) depending on which hypotheses (3.197) or (3.200) or (3.208) are assumed.*

**Proof** Let $q > 0$, $r > 0$. Define the family of sets

$$\mathbb{Z}(q, r) = \{u = (y, x) \in \mathcal{X}^2 \;\; \| x - y \| \leq q, \;\; S = [y, x; f]^{-1} \in \mathcal{L}(\mathcal{Y}, \mathcal{X}),$$
$$d(S^{-1}) \geq h(q, r) \text{ and } \| S\, f(x) \| \leq r\} \tag{3.213}$$

for some positive function $h$ to be determined later. According to the generalization of the induction theorem (see (cf. [608] or [603, p. 238])), we must show

$$(y_0, x_0) \in \mathbb{Z}(q_0, r_0), \tag{3.214}$$

$$(y_+, x_+) \in \mathbb{Z}(\omega(q,r)) \cap U((y,x), g(q,r)) \quad ((y,x) \in \mathbb{Z}(q,r)) \tag{3.215}$$

and

$$y_+ = y - [y,x;f]^{-1} f(y), \tag{3.216}$$

$$x_+ = y_+ - [y,x;f]^{-1} f(y_+). \tag{3.217}$$

But inclusion (3.215) is equivalent to the set of conditions

$$\| y_+ - x_+ \| \leq \omega_1(q,r), \tag{3.218}$$

$$d([y_+, x_+; f]) \geq h(\omega(q,r)), \tag{3.219}$$

$$\| [y_+, x_+; f]^{-1} f(x_+) \| \leq \omega_2(q,r), \tag{3.220}$$

$$\| y_+ - y \| \leq q + r \tag{3.221}$$

and

$$\| x_+ - x \| \leq r + \omega_1(q,r). \tag{3.222}$$

Our proof differs from the one given in (cf. [603, p. 245]) only in the verification of (3.219). Therefore, we shall only show (3.219).

**Case 1.** We have using (3.178)

$$\begin{aligned} d([y_+, x_+; f]) &\geq d([y,x;f]) - \| [y_+, x_+; f] - [y,x;f] \| \\ &\geq d([y,x;f]) - (\| [y_+, x_+; f] - [x_0, x_0; f] \| \\ &+ \| [x_0, x_0; f] - [y,x;f] \|) \\ &\geq d([y,x;f]) - L_0 (\| y_+ - x_0 \| + \| x_+ - x_0 \| \\ &+ \| y - x_0 \| + \| x - x_0 \|) \\ &\geq d([y,x;f]) - L_0 (\| y_+ - y \| + \| x_+ - x \| \\ &+ 2 (\| y - x_0 \| + \| x - x_0 \|)) \\ &\geq h(q,r) - L_0 \left(q + 2r + \omega_1(q,r) + 2 ((\sigma_2(q,r) - \sigma_1(\omega(q,r))) \right. \\ &\left. + (\sigma_2(q,r) - \sigma_2(\omega(q,r))))\right). \end{aligned} \tag{3.223}$$

Set $h = L\,K$. We must show

$$\begin{aligned} &h(q,r) - L_0 \left(q + 2r + \omega_1(q,r) + 2 ((\sigma_2(q,r) - \sigma_1(\omega(q,r))) \right. \\ &\left. +(\sigma_2(q,r) - \sigma_2(\omega(q,r))))\right) \geq h \circ \omega. \end{aligned} \tag{3.224}$$

Estimate (3.224) is true for $h_0 = L\,K_0$ or $h_1 = L\,K_1$. We also have $h_0 \leq d_0$ and $h_1 \leq d_0$ (see also Lemma 3.3 or 3.4).

**Case 2.** Using (3.179), we get in turn

$$\begin{aligned}
d([y_+, x_+; f]) &\geq d([y_0, x_0; f]) - \| [y_+, x_+; f] - [y_0, x_0; f] \| \\
&\geq d_0 - L_1 \, (\| y_+ - x_0 \| + \| x_+ - x_0 \|) \\
&\geq d_0 - L_1 \, (\| y_0 - x_0 \| + \| y_+ - x_0 \| + \| x_+ - x_0 \|) \\
&\geq d_0 - L_1 \, (q_0 + \| y_+ - y \| + \| y - x_0 \| + \\
&\| x_+ - x \| + \| x - x_0 \|) \\
&\geq d_0 - L_1 \left( q_0 + q + 2\, r + \omega_1(q, r) + (\sigma_2(q, r) - \sigma_1(\omega(q, r))) \right. \\
&\left. + \, (\sigma_2(q, r) - \sigma_2(\omega(q, r))) \right) \\
&\geq d_0 - L_1 \, (q_0 + q + 4\, r + 2\, \omega_1) \geq h \circ \omega \qquad (3.225)
\end{aligned}$$

if we choose $h_2 = L\, K_2$. We also have $h_2 \leq d_0$ (see also Lemma 3.5).

This completes the proof of Theorem 3.7. ⊠

Concerning the uniqueness of the solution $x^\star$, we have the following result.

**Proposition 3.3** *Under the hypotheses of Theorem 3.7, further assume there exists $R \geq R_0$ such that*

$$\frac{L_0}{d_0} \, (q_0 + \sigma_2(q_0, r_0) + R) < 1 \qquad (3.226)$$

*and*

$$\overline{U}(x_0, R) \subseteq \mathcal{D}. \qquad (3.227)$$

*Then, limit point $x^\star$ is the only solution of equation $f(x) = 0$ in $\overline{U}(x_0, R)$.*

**Proof** Let $y^\star \in \overline{U}(x_0, R)$ be such that $f(y^\star) = 0$. For simplicity, we denote $A_0 = [y_0, x_0; f]$ and $\mathcal{Q} = [y^\star, x^\star; f]$. Using (3.178), (3.226) and (3.227), we have in turn that

$$\begin{aligned}
\| A_0^{-1} \, (A_0 - \mathcal{Q}) \| &\leq \| A_0^{-1} \, (A_0 - [x_0, x_0; f]) + ([x_0, x_0; f] - \mathcal{Q}) \| \\
&\leq \frac{L_0}{d_0} \, (\| y_0 - x_0 \| + \| x_0 - y^\star \| + \| x_0 - x^\star \|) \\
&\leq \frac{L_0}{d_0} \, (q_0 + \sigma_2(q_0, r_0) + R) < 1. \qquad (3.228)
\end{aligned}$$

It follows from (3.228) and the Banach lemma on invertible operators that $\mathcal{Q}^{-1}$ exists. Hence, from the identity $y^\star - x^\star = \mathcal{Q}^{-1} \, (f(y^\star) - f(x^\star))$, we deduce $x^\star = y^\star$.

This completes the proof of Proposition 3.3. ⊠

**Remark 3.9** *Define*

$$a_0 = \frac{1}{4}\left(\frac{d_0}{L_0} - q_0\right) + \frac{\left(\dfrac{d_0}{L_0} - 2\,q_0\right) r_0}{\dfrac{d_0}{L_0} - q_0}. \tag{3.229}$$

*Set*

$$a_1 < \min\{a^0_{AH}, a_0\} \tag{3.230}$$

*and*

$$a_2 < \min\{a^1_{AH}, a_0\}. \tag{3.231}$$

*If* $2\,q_0\,L_0 \leq d_0$, *then,* $a_0 > 0$. *We also have*

$$\sigma_2(q_0, r_0) < \frac{d_0}{L_0} - (q_0 + \sigma_2(q_0, r_0)). \tag{3.232}$$

*Then, we can choose R such that*

$$\sigma_2(q_0, r_0) \leq R_0 \leq R < \frac{d_0}{L_0} - (q_0 + \sigma_2(q_0, r_0)), \tag{3.233}$$

*which implies (3.226).*

Let

$$\mathcal{V} = \{x \in \mathcal{X} : \| x - x_0 \| < R_1 = \frac{1}{L_0}\,((2\,L - L_0)\,\sigma_2(q_0, r_0) + (L - L_0)\,q_0 + 2\,L\,a_{PP})\}, \tag{3.234}$$

where $a_{PP}$ is given by (3.166). We can improve the uniqueness result (cf. [603, p. 247]) as follows.

**Proposition 3.4** *If strict inequality holds in (3.165), then,* $x^\star$ *is the only solution of equation* $f(x) = 0$ *in* $U(x_0, \sigma_2(q_0, r_0)) \cap \mathcal{V}$.

**Proof** Using the same notations as Proposition 3.3, (3.228) and (3.234), we get

$$\begin{aligned}
\| A_0^{-1}\,(A_0 - \mathcal{Q}) \| &\leq \frac{L_0}{d_0}\,(\| y_0 - x_0 \| + \| x_0 - y^\star \| + \| x_0 - x^\star \|) \\
&\leq \frac{L_0}{d_0}\,(q_0 + \sigma_2(q_0, r_0) + R_1) \\
&\leq \frac{L_0}{d_0}\,(q_0 + 2\,\sigma_2(q_0, r_0) + 2\,a_{PP}) \qquad (3.235) \\
&= \frac{L_0}{d_0}\,(q_0 + 2\,r_0 + a_{PP} + 2\,(r_0\,(r_0 + q_0) + a_{PP})^{1/2}) = 1.
\end{aligned}$$

This completes the proof of Proposition 3.4. ⊠

**Remark 3.10** *(a) If $L = L_0$, Proposition 3.4 reduces to (cf. [603, Theorem 3.3, p. 247]). Otherwise, it constitutes an improvement, since*

$$R_1 > \sigma_2(q_0, r_0) + 2\, a_{PP}. \tag{3.236}$$

*(b) A similar approach as this section can extend the results obtained in (cf. [11]) for Steffensen–type method.*

We provide examples where our results apply but earlier ones do not. In case all results apply we show that ours provide tighter error results bounds and better information on the location of the solution.

**Example 3.5** *Let $\mathcal{X} = \mathcal{Y} = \mathbb{R}$ be equipped with the max–norm, $y_0 = .99$, $x_0 = 1$, $\mathcal{D} = [\varrho, 2 - \varrho]$, $\varrho \in [0, 1)$, and define a function $f$ on $\mathcal{D}$ by*

$$f(x) = x^3 - \varrho. \tag{3.237}$$

*Define a linear operator $[x, y; f]$ by*

$$[x, y; f] = \int_0^1 f'(y + t\,(x - y))\, dt = x^2 + y^2 + x\,y. \tag{3.238}$$

*Using (3.237) and (3.238), we can write for all $(x, y, z) \in \mathcal{D}^3$*

$$[y, x; f] - [\overline{y}, \overline{x}; f] = (x + \overline{x} + y)\,(x - \overline{x}) + (y + \overline{y} + \overline{x})\,(y - \overline{y})$$

*and*

$$[y, x; f] - f'(x_0) = (y - x + 3)\,(x - 1) + (y + 2\,x)\,(y - 1).$$

*Then, using Definition 3.7 and $(\mathcal{H}_2)$–$(\mathcal{H}_4)$, we get that*

$$r_0 = \frac{1}{2.9701}\,(1 - \varrho), \quad L_0 = L = 6 - 3\,\varrho, \quad L_1 = 5 - 2\,\varrho,$$

$$d_0 = 2.9701 \quad \text{and} \quad q_0 = .01.$$

*Consider $\delta_1 = \delta_2 = \delta_3 = 0$. Then, inequalities (3.165), (3.197), (3.200) hold for*

$$\varrho \in (.5185273826, 1),$$

$$\varrho \in (.2796509993, 1)$$

*and*

$$\varrho \in (.7921720194, 1),$$

*respectively.*

**The case $\varrho = .29$.** *Using Maple 13, we get that*

$$r_0 = .2390491903, \quad L = L_0 = 5.13 \quad \text{and} \quad L_1 = 4.42.$$

*Condition (3.165) is violated since*

$$(r_0^{1/2} + (q_0 + r_0)^{1/2})^2 = .9760943119 > \frac{d_0}{L} = .5789668616.$$

*Hence, (cf. [603, Theorem 3.2, p. 243]) does not guarantees that DSSM converges to* $x^\star = .6619105948$*, starting at* $y_0 = .99$ *and* $x_0 = 1$*. Moreover, all condition of Theorem 3.7 hold and we can apply the DSSM for solving equation (3.237).*

**The case** $\varrho = .95$**.** *Using Maple 13, we get that*

$$r_0 = .01683445002, \quad L = L_0 = 3.15, \quad L_1 = 3.10,$$

$$a_{PP} = .4491073412, \quad a_{AH}^0 = .01831052558 \quad \textit{and} \quad a_{AH}^1 = .1370082338.$$

*All the conditions (3.165), (3.197) and (3.200) hold. Hence, (cf. [603, Theorem 3.2, p. 243]) and our Theorem 3.7 guarantee that DSSM converges to* $x^\star = .9830475725$*, starting at* $y_0 = .99$ *and* $x_0 = 1$*.*

Applications and examples where the inequality is strict between the Lipschitz and the center–Lipschitz constants (i.e., $L_0 < L$), including the solution of nonlinear Chandrasekhar–type integral equations appearing in radiative transfer can be found in section 2.1.

## 3.5 Directional Secant Methods

A semilocal convergence analysis for directional Secant–type methods in $n$–variables is provided in this section. Using weaker hypotheses than in the related work by An and Bai (cf. [10]), and motivated by optimization considerations, we provide under the same computational cost a semilocal convergence analysis with the following advantages: weaker convergence conditions; larger convergence domain; finer error estimates on the distances involved and an at least as precise information on the location of the solution. A numerical example where our results apply to solve an equation but not the ones in (cf. [10]) is also provided. In a second example, we show how to implement the method.

We are concerned with the problem of approximating a solution $x^\star$ of equation (2.1), where $F$ is a differentiable mapping defined on a convex subset $\mathcal{D}$ of $\mathbb{R}^n$ ($n$ a natural number) with values in $\mathbb{R}$. In computer graphics, we usually compute the intersection $\mathcal{C} = \mathcal{A} \cap \mathcal{B}$ of two surfaces $\mathcal{A}$ and $\mathcal{B}$ in $\mathbb{R}^3$ (cf. [228], [510]). If the two surfaces are explicitly given by

$$\mathcal{A} = \{(u, v, w)^T \,:\, w = F_1(u, v)\}$$

and

$$\mathcal{B} = \{(u, v, w)^T \,:\, w = F_2(u, v)\},$$

then, the solution $x^\star = (u^\star, v^\star, w^\star)^T \in \mathcal{C}$ must satisfy the nonlinear equation

$$F_1(u^\star, v^\star) = F_2(u^\star, v^\star)$$

and

$$w^\star = F_1(u^\star, v^\star).$$

Hence, we must solve a nonlinear equation in two variables $x = (u, v)^T$ of the form

$$F(x) = F_1(x) - F_2(x) = 0,$$

which is a special case of equation (2.1). In mathematical programming (cf. [592]), for an equality–constraint optimization problem, e.g.,

$$\begin{array}{ll} \min & \psi(x) \\ s.t. & F(x) = 0, \end{array}$$

where $\psi$, $F : \mathcal{D} \subseteq \mathbb{R}^n \longrightarrow \mathbb{R}$ are nonlinear mappings, we usually seek a feasible point to start a numerical algorithm, which again requires the determination of $x^\star$. An and Bai (cf. [10]) used the directional Secant method DSM

$$\begin{aligned} & x_{k+1} = x_k + h_k, \\ & h_k = -A_k\, F(x_k), \quad A_k = \frac{\theta_k}{F(v_k) - F(x_k)} \cdot d_k, \quad v_k = x_k + \theta_k\, d_k \\ & (k \geq 0, \quad x_0 \in \mathbb{R}^n, \quad d_k \in \mathbb{R}^n, \quad \| d_k \| = 1, \quad \theta_k \geq 0) \end{aligned}$$

to generate a sequence $\{x_k\}$ converging to $x^\star$. DSM is a usefull alternative to DNM (cf. [170], [228], [499])

$$x_{k+1} = x_k - \frac{F(x_k)}{\nabla F(x_k) \cdot d_k}\, d_k \quad (k \geq 0),$$

where

$$\nabla F(x_k) = \left( \frac{\partial F(x_k)}{\partial x_1}, \frac{\partial F(x_k)}{\partial x_2}, \ldots, \frac{\partial F(x_k)}{\partial x_n} \right)$$

is the gradient of $F$ and $d_k$ is a direction at $x_k$. DNM converges quadratically to $x^\star$ if $x_0$ is close enough to $x^\star$ (cf. [499]). However, as already noted in (cf. [499]), the computation of the gradient $\nabla F(x_k)$ may be too expensive as it is the case when the number $n$ of unknowns is large. In some applications, the mapping $F$ may not be differentiable, or the gradient is impossible to compute. The DSM avoids these obstacles. Note that, if $n = 1$, DSM reduces to the classical Secant method (cf. [139], [597], [600]).

We consider two choices of the directions $d_k$ for locating $x^\star$. The first is called the directional near–gradient Secant method DNGSM and the second is called the directional maximal–component–modulus Secant method DMCMSM.

The quadratic convergence of DSM was established for directions $d_k$ sufficiently close to the gradients $\nabla F(x_k)$ and under standard Newton–Kantorovich–type hypotheses (cf. [10]).

We are motivated by the papers ([10], [170], [499]), and optimization considerations. By introducing the center–Lipschitz condition and using it, in combination with the Lipschitz condition (along the lines of our works in (cf. [136], [194])), we provide a semilocal convergence analysis with the following advantages over the work in (cf. [10])

1. Weaker hypotheses;
2. Larger convergence domain for DSM;
3. Finer error bounds on the distances $\| x_{k+1} - x_k \|$, $\| x_k - x^\star \|$ $(k \geq 0)$;
4. An at least as precise information on the location of the solution $x^\star$.

We use the Euclidean norms for both vector and matrix. The unit direction $d_k$ is chosen such that

$$d_k \approx \frac{\nabla F(x_k)}{\| \nabla F(x_k) \|}.$$

As in (cf. [10]), we need two lemmas to establish semilocal convergence of DNGSM.

**Lemma 3.6** *(cf. [569, 3.2.2]) Let $\mathcal{C} \subseteq \mathbb{R}^n$ be a convex set and $F : \mathcal{C} \longrightarrow \mathbb{R}$ be a differentiable mapping. Then, for any $x, y \in \mathcal{C}$, there exists a vector $\mu \in (x, y)$ such that*

$$F(y) - F(x) = \nabla F(\mu)\ (y - x), \tag{3.239}$$

*where*

$$(x, y) = \{z\ :\ z = x + \theta\, y, \quad 0 < \theta < 1\}, \tag{3.240}$$

*represents the open straight line between the points $x$ and $y$.*

**Lemma 3.7** *(cf. [569, 3.3.10]) Assume that $F : \mathcal{C} \longrightarrow \mathbb{R}$ is a twice differentiable mapping. Then, for any $x, y \in \mathcal{C}$, there exists a vector $\lambda \in (x, y)$ such that*

$$F(y) - F(x) - \nabla F(x)\,(y - x) = \frac{1}{2}\ (y - x)^T\ F''(\lambda)\ (y - x). \tag{3.241}$$

We need also the definition of Lipschitz and center–Lipschitz continuity.

**Definition 3.8** *Let $F : U_0 \subseteq \mathbb{R}^n \longrightarrow \mathbb{R}$ be a differentiable mapping. If there exists a constant $\overline{M} \geq 0$ such that*

$$\| \nabla F(x) - \nabla F(y) \| \leq \overline{M}\ \ \| x - y \| \quad (x, y \in U_0), \tag{3.242}$$

*then we say $\nabla F$ is Lipschitz continuous on $U_0$ with constant $\overline{M}$. Note that, in view of (3.242), there exists $M_0 \geq 0$ such that*

$$\| \nabla F(x) - \nabla F(x_0) \| \leq M_0\ \ \| x - x_0 \| \quad (x \in U_0). \tag{3.243}$$

*Then, we say $\nabla F$ is center–Lipschitz continuous on $U_0$ with constant $M_0$.*

Clearly,

$$M_0 \leq \overline{M} \tag{3.244}$$

holds in general and $\dfrac{\overline{M}}{M_0}$ can be arbitrarly large (cf. [136], [163]). We can state the main semilocal convergence result for DNGSM.

**Theorem 3.8** *Let $\mathcal{D}_0 \subseteq \mathbb{R}^n$ be a convex set, $F : \mathcal{D}_0 \subseteq \mathbb{R}^n \longrightarrow \mathbb{R}$ be a differentiable mapping. Let $x_0 \in \mathcal{D}_0$ satisfy*

$$F(x_0) \neq 0 \quad \text{and} \quad \nabla F(x_0) \neq 0, \tag{3.245}$$

*$d_0 \in \mathbb{R}^n$ be a given unit vector, $\theta_0$ be a given positive parameter, $h_0$ and $x_1$ be defined as*

$$h_0 = \frac{\theta_0 \ F(x_0)}{F(x_0 + \theta_0 \ d_0) - F(x_0)} \ d_0, \tag{3.246}$$

$$x_1 = x_0 + h_0. \tag{3.247}$$

*We suppose that, for $U_0 \subseteq \mathcal{D}_0$, $\nabla F$ is center–Lipschitz on $U_0$ with constant $M_0$ and $F \in \mathcal{C}^2[U_0]$. Let*

$$M = \sup_{x \in U_0} \ \| F''(x) \| \,. \tag{3.248}$$

*Moreover, let the sequence $\{x_k\}$ be generated by DNGSM. Further, assume that the unit direction $d_k$ and positive parameters $\alpha$, $\beta$, $\theta_k$ satisfy for all $k \geq 0$*

$$\frac{|\nabla F(x_k) \cdot d_k|}{\| \nabla F(x_k) \|} \geq \frac{|\nabla F(x_0) \cdot d_0|}{\| \nabla F(x_0) \|}, \tag{3.249}$$

$$\theta_k \leq \alpha \ \| h_k \|, \tag{3.250}$$

$$|\nabla F(x_k) \cdot d_k| \leq \beta \ |\nabla F(x) \cdot d_k| \qquad (x \in (x_k, x_k + \theta_k \, d_k)) \tag{3.251}$$

*and*

$$p = \overline{L} \ \| h_0 \| \leq \frac{1}{2}, \tag{3.252}$$

*where*

$$L_0 = M_0 \ \| \nabla F(x_0) \|^{-1}, \tag{3.253}$$

$$L = \beta \ (1 + \alpha) \ M \ |\nabla F(x_0) \cdot d_0|^{-1}, \tag{3.254}$$

$$U_0 = \{x \in \mathbb{R}^n : \| x - x_1 \| \leq t^\star\} \subseteq \mathcal{D}_0, \tag{3.255}$$

$$t_1 = \| h_0 \| \tag{3.256}$$

*and $\delta$, $\overline{L}$ are given in Lemma 2.11. Then, the sequence $\{x_k\}$ generated by DNGSM is well defined, remains in $U_0$ for all $k \geq 0$ and converges to a solution $x^\star \in U_0$ of equation $F(x) = 0$. Moreover, the following hold*

$$\begin{aligned} &\nabla F(x) \neq 0 \quad (x \in U_0), \\ &\nabla F(x^\star) \neq 0 \quad \text{unless} \quad \| x^\star - x_0 \| = t^\star. \end{aligned} \tag{3.257}$$

*Furthemore, the following estimates hold for all* $k \geq 0$

$$\| x_{k+1} - x_k \| \leq t_{k+1} - t_k \leq \left(\frac{\delta}{2}\right)^k (2\,p)^{2^k - 1} \ \| h_0 \|, \tag{3.258}$$

$$\| x_k - x^\star \| \leq t^\star - t_k \leq \left(\frac{\delta}{2}\right)^k \frac{(2\,p)^{2^k - 1} \ \| h_0 \|}{1 - (2\,p)^{2^k}} \quad (2\,p < 1), \tag{3.259}$$

*where* $t^\star$, $\{t_k\}$ *are given in Lemma 2.11,*

$$\eta = \| h_0 \| \quad \text{and} \quad q = p. \tag{3.260}$$

**Proof** We shall show using mathematical induction on $k \geq 0$

$$\| x_{k+1} - x_k \| \leq t_{k+1} - t_k \tag{3.261}$$

and

$$\overline{U}(x_{k+1}, t^\star - t_{k+1}) \subseteq \overline{U}(x_k, t^\star - t_k). \tag{3.262}$$

For every $z \in \overline{U}(x_1, t^\star - t_1)$,

$$\| z - x_0 \| \leq \| z - x_1 \| + \| x_1 - x_0 \| \leq t^\star - t_1 + t_1 - t_0 = t^\star - t_0,$$

which shows $z \in \overline{U}(x_0, t^\star - t_0)$. Since also

$$\| x_1 - x_0 \| = \| h_0 \| \leq \eta = t_1 - t_0,$$

estimates (3.261) and (3.262) hold for $k = 0$. Assume (3.261) and (3.262) hold for all $i \leq k$. Then, we have that

$$\begin{aligned} \| x_{i+1} - x_0 \| &\leq \| x_{i+1} - x_i \| + \| x_i - x_{i-1} \| + \cdots + \| x_1 - x_0 \| \\ &\leq (t_{i+1} - t_i) + (t_i - t_{i-1}) + \cdots + (t_1 - t_0) = t_{i+1} \end{aligned}$$

and

$$\| x_i + t\,(x_{i+1} - x_i) - x_0 \| \leq t_i + t\ (t_{i+1} - t_i) \leq t^\star \quad (t \in [0,1]).$$

Using condition (3.243) and (3.253) for $x = x_i$, we get in turn that

$$\begin{aligned} \| \nabla F(x_i) \| &\geq \| \nabla F(x_0) \| - \| \nabla F(x_i) - \nabla F(x_0) \| \\ &\geq \| \nabla F(x_0) \| - M_0 \ \| x_i - x_0 \| \\ &\geq \| \nabla F(x_0) \| - M_0 \ (t_i - t_0) \\ &\geq \| \nabla F(x_0) \| \ (1 - L_0\, t_i) > 0 \quad \text{(by Lemma 2.11 and (3.245))}. \end{aligned} \tag{3.263}$$

We need the approximation

$$F(x_i) = \int_{x_{i-1}}^{x_i} (x_i - x)^T \ F''(x)\ dx + r_{i-1}, \tag{3.264}$$

where

$$r_i = F(x_i) + \nabla F(x_i)\ h_i \qquad (i \geq 0). \tag{3.265}$$

Using (3.248), (3.250) and the induction hypotheses, we have that

$$\begin{aligned}\left|\int_{x_{i-1}}^{x_i} (x_i - x)^T\ F''(x)\ dx\right| &= \left|\int_0^1 (1-t)\ (h_{i-1})^T\ F''(x_{i-1} + t\ h_{i-1})\ h_{i-1}\ dt\right| \\ &\leq \frac{1}{2}\ M\ \ \| h_{i-1} \|^2 \end{aligned} \tag{3.266}$$

and

$$\begin{aligned}|r_{i-1}| &= \left|F(x_{i-1}) - \frac{\theta_{i-1}\ F(x_{i-1})\ \nabla F(x_{i-1}) \cdot d_{i-1}}{F(x_{i-1} + \theta_{i-1}\, d_{i-1}) - F(x_{i-1})}\right| \\ &= \left|\frac{F(x_{i-1})}{2}\ \frac{(\theta_{i-1}\ d_{i-1})^T\ F''(\lambda)\ (\theta_{i-1}\ d_{i-1})}{F(x_{i-1} + \theta_{i-1}\, d_{i-1}) - F(x_{i-1})}\right| \quad (\lambda \in (x_{i-1}, x_{i-1} + \theta_{i-1}\, d_{i-1})) \\ &\leq \frac{1}{2}\ M\ \theta_{i-1} \left|\frac{\theta_{i-1}\ F(x_{i-1})}{F(x_{i-1} + \theta_{i-1}\, d_{i-1}) - F(x_{i-1})}\right| \\ &\leq \frac{1}{2}\ M\ \alpha\ \ \| h_{i-1} \|^2 . \end{aligned} \tag{3.267}$$

In view of (3.251) and (3.263)–(3.267), we get

$$|F(x_i)| \leq \frac{1}{2}\ M\ \ \| h_{i-1} \|^2 + \frac{1}{2}\ \alpha\ M\ \ \| h_{i-1} \|^2 = \frac{1}{2}\ (1+\alpha)\ M\ \ \| h_{i-1} \|^2 . \tag{3.268}$$

Hence, we have that

$$\begin{aligned}\| h_i \| &\leq \left|\frac{\theta_i\ F(x_i)}{F(x_i + \theta_i\ d_i) - F(x_i)}\right| \\ &\leq \frac{1}{2}\ (1+\alpha)\ M\ \ \| h_{i-1} \|^2\ \left|\frac{\theta_i}{F(x_i + \theta_i\ d_i) - F(x_i)}\right| \\ &\leq \frac{1}{2}\ (1+\alpha)\ M\ \ \| h_{i-1} \|^2\ \frac{1}{|\nabla F(\mu_i) \cdot d_i|} \quad (\mu_i \in (x_i, x_i + \theta_i\ d_i)) \\ &\leq \frac{\beta}{2}\ \frac{(1+\alpha)\ M\ \ \| h_{i-1} \|^2}{|\nabla F(x_i) \cdot d_i|} \\ &\leq \frac{\beta}{2}\ \frac{(1+\alpha)\ M\ \ \| h_{i-1} \|^2}{|\nabla F(x_0) \cdot d_0|}\ \frac{\| \nabla F(x_0) \|}{\| \nabla F(x_i) \|} \\ &\leq \frac{\beta\ (1+\alpha)\ M\ \ \| h_{i-1} \|^2}{2\ |\nabla F(x_0) \cdot d_0|}\ \frac{\| \nabla F(x_0) \|}{\| \nabla F(x_0) \|\ \ (1 - L_0\ t_i)} \\ &\leq \frac{L}{2}\ \frac{\| h_{i-1} \|^2}{1 - L_0\ t_i} \leq \frac{L}{2}\ \frac{(t_i - t_{i-1})^2}{1 - L_0\ t_i} = t_{i+1} - t_i, \end{aligned} \tag{3.269}$$

which shows (3.261) for all $i$. Moreover, for every $w \in \overline{U}(x_{i+2}, t^\star - t_{i+2})$, we obtain that

$$\begin{aligned} \| w - x_{i+1} \| &\leq \| w - x_{i+2} \| + \| x_{i+2} - x_{i+1} \| \\ &\leq t^\star - t_{i+2} + t_{i+2} - t_{i+1} = t^\star - t_{i+1}, \end{aligned}$$

showing (3.262) for all $k \geq 0$. Lemma 2.11 implies that $\{t_n\}$ is a Cauchy sequence. It follows from (3.261), (3.262), that $\{x_n\}$ is a Cauchy sequence too and as such it converges to some $x^\star \in U_0$ (since $U_0$ is a closed set). The vector $x^\star$ is a zero of $F$, since by (3.268), we get that

$$0 \leq |F(x_i)| \leq \frac{1}{2} (1+\alpha) M (t_i - t_{i-1})^2 \longrightarrow 0 \quad \text{as} \quad i \to \infty. \tag{3.270}$$

Furthermore, we shall show (3.257). Using (3.243) for $x \in U_0$, (2.447) and the definition of $L_0$, we get

$$\| \nabla F(x) - \nabla F(x_0) \| \leq M_0 \| x - x_0 \| \leq M_0 t^\star \leq \| \nabla F(x_0) \| .$$

If $\| x - x_0 \| < t^\star$, then, we have that

$$\| \nabla F(x) - \nabla F(x_0) \| \leq M_0 \| x - x_0 \| < M_0 t^\star \leq \| \nabla F(x_0) \|$$

or

$$\| \nabla F(x_0) \| > \| \nabla F(x) - \nabla F(x_0) \|,$$

which shows $\nabla F(x) \neq 0$.

Finally, the left hand side inequality in (3.259) follows from (3.258) by using standard majorization techniques (cf. [163], [569], [746]). This completes the proof of Theorem 3.8. ⊠

We state Theorems 3.1 and 3.2 in a condensed form (cf. [10, p. 294]), so we can compare them with our Theorem 3.8.

**Theorem 3.9** *Let $\mathcal{D}_0 \subseteq \mathbb{R}^n$ be a convex set and $F : \mathcal{D}_0 \subseteq \mathbb{R}^n \longrightarrow \mathbb{R}$ be a differentiable mapping. Let $x_0 \in \mathcal{D}_0$ satisfy*

$$F(x_0) \neq 0 \quad \text{and} \quad \nabla F(x_0) \neq 0,$$

*$d_0 \in \mathbb{R}^n$ be a given unit vector, $\theta_0$ be a given positive parameter, $h_0$ and $x_1$ be defined by (3.246) and (3.247), respectively. Let $U_0 \subseteq \mathcal{D}_0$, $F \in \mathcal{C}^2[U_0]$ and $M = \sup\limits_{x \in U_0} \| F''(x) \|$ be such that*

$$\varsigma_0 = \ell_0 \| h_0 \| \leq \frac{1}{2}, \tag{3.271}$$

*where*

$$\ell_0 = 2 M |\nabla F(x_0) \cdot d_0|^{-1} \tag{3.272}$$

*and*

$$U_0 = \{x \in \mathbb{R}^n : \| x - x_1 \| \leq \| h_0 \|\}.$$

*In addition, let the sequence $\{x_k\}$ be generated by DNGSM and assume that the unit direction $d_k$ and positive parameter $\theta_k$ satisfy for all $k \geq 0$*

$$\frac{|\nabla F(x_{k+1}) \cdot d_{k+1}|}{\| \nabla F(x_{k+1}) \|} \geq \frac{|\nabla F(x_k) \cdot d_k|}{\| \nabla F(x_k) \|}, \tag{3.273}$$

$$\theta_k \leq \frac{1}{2} \; \| h_k \|$$

*and*

$$|\nabla F(x_k) \cdot d_k| \leq 2 \; |\nabla F(x) \cdot d_k| \qquad (x \in (x_k, x_k + \theta_k \, d_k)).$$

*Then, the following hold*

(a) *$\{x_k\} \subseteq U_0$ and there exists $x^\star \in U_0$, such that $x_k \longrightarrow x^\star$ and $F(x^\star) = 0$;*

(b)

$$\nabla F(x) \neq 0 \qquad (x \in U_0);$$

(c) *For $k \geq 0$,*

$$\nabla F(x_k) \cdot d_k \neq 0, \quad \| h_k \| \leq \frac{1}{2} \; \| h_{k-1} \|, \tag{3.274}$$

$$\begin{aligned} \| x_{k+1} - x_k \| &\leq \frac{3M}{2 \; |\nabla F(x_k) \cdot d_k|} \; \| x_k - x_{k-1} \|^2 \\ &\leq \frac{3M \; \| \nabla F(x_0) \|}{2 \, a \, |\nabla F(x_0) \cdot d_0|} \; \| x_k - x_{k-1} \|^2 \end{aligned} \tag{3.275}$$

*and*

$$\begin{aligned} \| x_{k+1} - x^\star \| &\leq \frac{3M}{2 \; |\nabla F(x_k) \cdot d_k|} \; \| x_k - x_{k-1} \|^2 \\ &\leq \frac{3 \; M \; \| \nabla F(x_0) \|}{2 \, a \, |\nabla F(x_0) \cdot d_0|} \; \| x_k - x_{k-1} \|^2, \end{aligned} \tag{3.276}$$

*where*

$$a = \min_{x \in U_0} \| \nabla F(x) \| > 0; \tag{3.277}$$

(d) *If for*

$$\gamma = \frac{\| \nabla F(x_0) \| \;\; |\nabla F(x_0) \cdot d_0|^{-1}}{a}, \tag{3.278}$$

$$\ell = \frac{3 \, \gamma \, M}{4}, \tag{3.279}$$

$$r = \ell \;\; \| h_0 \| < \frac{1}{2}, \tag{3.280}$$

*then*

$$\| x_{k+1} - x_k \| \leq \frac{(2\,q)^{2^k-1}}{1-(2\,q)^{2^k-1}} \;\; \| h_0 \| \tag{3.281}$$

*and*

$$\| x_k - x^\star \| \leq \frac{(2\,q)^{2^k-1}}{1-(2\,q)^{2^k-1}} \;\; \| h_0 \| \,. \tag{3.282}$$

Note that, if $F \in C^2[U_0]$ and (3.242) holds, then $\overline{M}$ can replace $M$ in hypothesis (3.248). Moreover, we have

$$M_0 \leq M \tag{3.283}$$

holds in general and $\dfrac{M}{M_0}$ can also be arbitrarily large.

Next, we provide a comparison between Theorems 3.8 and 3.9.

**Remark 3.11** *(a) We compare sufficient convergence condition (3.252) to (3.271) for* $\alpha = \dfrac{1}{2}$ *and* $\beta = 2$. *We have*

$$\varsigma_0 \leq \frac{1}{2} \Longrightarrow p \leq \frac{1}{2} \tag{3.284}$$

*but not necessarily vice verca, unless if* $M_0 = M$. *Hence, condition (3.271) can always be replaced by (3.252).*

*That is, the applicability of DNGSM has been extended. Note also that (3.243) is not an additional hypothesis, since in practice the computation of* $M$ *requires that of constant* $M_0$. *According to (3.274), only linear convergence is shown provided that hypothesis (3.271) holds. However, in Theorem 3.8, we showed quadratic convergence. Theorem 3.8 avoids the computation of a given (3.277), which is expensive in general and also uses (3.249) instead of the more difficult to verify (3.273).*

*Finally, note that, if* $p \in [0, \dfrac{1}{2})$ *and* $\beta \in [0, 2)$, *then, condition (3.252) is even weaker than (3.271).*

*(b) Condition (3.249) is equivalent to* $c_{k+1} \leq c_0$, *since* $\| d_k \| = 1$, *where*

$$c_k = \angle(\nabla F(x_k), d_k)$$

*with* $\angle(.,.)$ *being the angle between two vectors* $u$, $v$ *given by*

$$\angle(u,v) = \arccos \frac{u \cdot v}{\| u \| \cdot \| v \|} \qquad (u \neq 0,\;\; v \neq 0).$$

*(c) Condition (3.250) is equivalent to*

$$|F(x_k + \theta_k\, d_k) - F(x_k)| \leq \alpha\, |F(x_k)|.$$

(d) *If $M\,\theta_k \leq b|\nabla F(x_k)\cdot d_k|$, $b \in [0,1)$, then condition (3.251) holds for $\beta = 1-b$. Indeed, we have*

$$|\nabla F(x)\cdot d_k - \nabla F(x_k)\cdot d_k| \leq M\ \theta_k,$$

*so*

$$\begin{aligned}|\nabla F(x)\cdot d_k| &\geq |\nabla F(x_k)\cdot d_k| - |\nabla F(x)\cdot d_k - \nabla F(x_k)\cdot d_k|\\ &\geq |\nabla F(x_k)\cdot d_k| - b\ |\nabla F(x_k)\cdot d_k| = \beta\ |\nabla F(x_k)\cdot d_k|.\end{aligned}$$

We use now $\infty$–norm for both vector and matrix. By simply replacing the Euclidean norm by $\infty$–norm and $|\nabla F(x_k)\cdot d_k|$ by $\|\ \nabla F(x_k)\ \|_\infty$ in the proof of Theorem 3.8, we arrive at the following corresponding semilocal convergence result for DMCMCM

**Theorem 3.10** *Let $\mathcal{D}_0 \subseteq \mathbb{R}^n$ be a convex set and $F : \mathcal{D}_0 \subseteq \mathbb{R}^n \longrightarrow \mathbb{R}$ be a differentiable mapping. Let $x_0 \in \mathcal{D}_0$ satisfy*

$$F(x_0) \neq 0 \quad \text{and} \quad \nabla F(x_0) \neq 0, \tag{3.285}$$

*$\theta_0$ be a given positive parameter, $h_0$ and $x_1$ be defined as*

$$h_0 = \frac{\theta_0\ F(x_0)}{F(x_0+\theta_0\ d_0) - F(x_0)}\ d_0,$$

$$x_1 = x_0 + h_0,$$

*where*

$$d_0 = e_{m(0)}$$

*is the $m(0)$th unit vector in $\mathbb{R}^n$. We suppose that, for $U_0 \subseteq \mathcal{D}_0$ and $F \in \mathcal{C}^2[U_0]$, there exist constants $M_0$ and $M$ such that*

$$\|\ \nabla F(x) - \nabla F(x_0)\ \|_\infty \leq M_0\ \|\ x - x_0\ \|_\infty \qquad (x \in U_0)$$

*and*

$$M = \sup_{x\in U_0}\ \|\ F''(x)\ \|_\infty\ .$$

*Moreover, let the sequence $\{x_k\}$ be generated by DMCMSM and further assume direction $d_k$ and positive parameters $\alpha$, $\beta$, $\theta_k$ satisfy for all $k \geq 0$*

$$\|\ \nabla F(x_k)\ \|_\infty \leq \beta\ \|\ \nabla F(x)\cdot e_{m(k)}\ \|_\infty \quad (x \in (x_k, x_k+\theta_k\, d_k)),$$

$$\theta_k \leq \alpha\ \|\ h_k\ \|_\infty$$

*and*

$$p = \overline{L}\ \|\ h_0\ \|_\infty \leq \frac{1}{2},$$

*where,*

$$L_0 = M_0 \ \| \nabla F(x_0) \|_\infty^{-1},$$

$$L = \beta \ (1+\alpha) \ M \ \ \| \nabla F(x_0) \|_\infty^{-1},$$

$$U_0 \subseteq \mathcal{D}_0$$

*with $\delta$ and $\overline{L}$ are given in Lemma 2.11. Then, the sequence $\{x_k\}$ generated by DMCMSM is well defined, remains in $U_0$ for all $k \geq 0$ and converges to a solution $x^\star \in U_0$ of equation $F(x) = 0$. Moreover, the following hold*

$$\nabla F(x) \neq 0 \quad (x \in U_0),$$

$$\nabla F(x^\star) \neq 0 \quad \text{unless} \quad \| x^\star - x_0 \|_\infty = t^\star.$$

*Furthermore, the following estimates hold for all $k \geq 0$*

$$\| x_{k+1} - x_k \|_\infty \leq t_{k+1} - t_k \leq \left(\frac{\delta}{2}\right)^k \ (2\,p)^{2^k-1} \ \ \| h_0 \|_\infty$$

*and*

$$\| x_k - x^\star \|_\infty \leq t^\star - t_k \leq \left(\frac{\delta}{2}\right)^k \ \frac{(2\,p)^{2^k-1} \ \ \| h_0 \|_\infty}{1-(2\,p)^{2^k}} \qquad (2\,p < 1),$$

*where $t^\star$, $\{t_k\}$ are given in Lemma 2.11 and*

$$\eta = \| h_0 \|_\infty, \quad q = p.$$

As in Remark 3.11, Theorem 3.10 compares favorably to Theorem 3.3 in (cf. [10, p. 300]).

So far we showed that the results in (cf. [10]) can be considerably improved and extended. However, although our approach can be implemented (see Example 3.7) there is a difficulty in general verifying condition (3.252) since to calculate $h_k$, we must first determine $\theta_k$. In the following section, we show how to overcome this obstacle in case the verification of (3.252) is very difficult or impossible.

As before in this section, we assume now $F : \mathcal{D} \subseteq \mathbb{R}^n \longrightarrow \mathbb{R}$ be a nonlinear differentiable mapping, $x_0 \in \mathcal{D}$, $d_k \in \mathbb{R}^n$ with $\| d_k \| = 1$ and $\theta_0 \in \mathbb{R}$.

We shall show the semilocal convergence of DNGSM under the following set $(\mathcal{C})$ conditions:

$(\mathcal{C}_1)$

$$|F(x_0)| \leq \eta, \quad \| \nabla F(x_0) \| \geq \gamma > 0$$

and

$$|\nabla F(x_0) \cdot d_0| \geq \delta \ \| \nabla F(x_0) \| \quad (\delta \in [0,1]);$$

$(\mathcal{C}_2)$ $d_k$ $(k \geq 0)$ satisfies

($\mathcal{C}_{21}$)

$$\varrho\,|\nabla F(x_0)\cdot d_0| \le |\nabla F(x)\cdot d_0| \quad (x\in(x_0,\,x_0+\theta_0\,d_0),\ \varrho\in(0,1));$$

($\mathcal{C}_{22}$)

$$\angle(d_{k+1},\nabla F(x_{k+1})) \le \angle(d_k,\nabla F(x_k));$$

($\mathcal{C}_3$) $\theta_k$ satisfies

$$\theta_{k+1}\le \xi\,\theta_k\ \parallel x_{k+1}-x_k\parallel \quad (\xi\in(0,1));$$

($\mathcal{C}_4$) Mapping $\nabla F$ is Lipschitz with constant $M$ on $\mathcal{D}$;

($\mathcal{C}_5$) Mapping $F$ has a divided difference of order one $[x,y;F]$ at the points $(y,y)\in\mathcal{D}^2$ satisfying

($\mathcal{C}_{51}$)

$$[x,y;F]\,(x-y)=F(y)-F(x)$$

and

($\mathcal{C}_{52}$)

$$\parallel [x,z;F]-[y,z;F]\parallel \le N\parallel x-y\parallel \quad ((x,y,z)\in\mathcal{D}^3).$$

**Remark 3.12** *Note that* ($\mathcal{C}_{22}$) *implies condition (3.273).*

We need two auxiliary lemmas before we show the semilocal convergence of DNGSM.

**Lemma 3.8** *Assume* $F:\mathcal{D}\subseteq\mathbb{R}^n\longrightarrow\mathbb{R}$ *is a nonlinear differentiable mapping and sequence* $\{x_n\}$ *generated by DNGSM is well defined and remains in* $\mathcal{D}$. *Then, the following hold*

$$|\nabla F(x_k)\cdot d_k|\ge\delta\parallel\nabla F(x_k)\parallel \quad (k\ge 0) \tag{3.286}$$

*and*

$$F(x_{k+1})=([x_{k+1},x_k;F]-[v_k,x_k;F])\,(x_{k+1}-x_k). \tag{3.287}$$

**Proof** We shall use induction on $k$. Estimate (3.286) holds for $k=0$ by ($\mathcal{C}_1$).

Assume (3.286) holds for all integers $j$ with $j\le k$. Then, by ($\mathcal{C}_{22}$) and Remark 3.12, we have that

$$\frac{\delta\parallel\nabla F(x_{k+1})\parallel}{|\nabla F(x_{k+1})\cdot d_{k+1}|}\le\frac{\delta\parallel\nabla F(x_k)\parallel}{|\nabla F(x_k)\cdot d_k|}\le 1\Longrightarrow (3.286).$$

Using DNGSM and ($\mathcal{C}_{51}$), we have

$$\begin{aligned}F(x_{k+1})&=F(x_{k+1})-F(x_k)-[v_k,x_k;F]\,(x_{k+1}-x_k)\\&=([x_{k+1},x_k;F]-[v_k,x_k;F])\,(x_{k+1}-x_k).\end{aligned}$$

This completes the proof of Lemma 3.8. ⊠

It is convenient for us to introduce the initial conditions:

$$a_0 = \frac{1}{\rho\,\gamma\,\delta}, \quad c_0 = a_0\,\eta, \quad b_0 = \frac{M\,\eta}{\rho\,\gamma^2\,\delta},$$

$$N\,a_0\,\eta = r_0, \quad N\,\theta_0 = t_0 \quad \text{and} \quad s_0 = (r_0 + t_0)\,a_0.$$

Then, we can have in turn that

$$\parallel A_0 \parallel \leq \left| \frac{\theta_0}{F(v_0) - F(x_0)} \right| \leq \frac{1}{\rho\,\gamma\,\delta} = a_0 < R$$

for some $R > 0$ to be determined later,

$$\parallel x_1 - x_0 \parallel \leq \parallel A_0 \parallel \; |F(x_0)| \leq a_0\,\eta,$$

$$N \, \parallel x_1 - x_0 \parallel \, \leq N\,a_0\,\eta = r_0,$$

$$N \, \parallel x_0 - v_0 \parallel \, \leq N\,\theta_0 = t_0,$$

$$N \, \parallel A_0 \parallel \parallel x_0 - v_0 \parallel \, \leq (r_0 + t_0)\,a_0 = s_0,$$

$$M \, \parallel \nabla F(x_0) \parallel^{-1} \parallel x_1 - x_0 \parallel \, \leq \frac{M\,\eta}{\gamma^2\,\rho\,\delta} = b_0,$$

so, we get

$$N \, \parallel x_1 - v_0 \parallel \leq N \, \parallel x_1 - x_0 \parallel + N\,\theta_0 \leq r_0 + t_0$$

and

$$\parallel F(x_1) \parallel \, \leq N \, \parallel x_1 - v_0 \parallel \parallel x_1 - x_0 \parallel \, \leq (r_0 + t_0) \, \parallel x_1 - x_0 \parallel .$$

We consider the auxiliary function

$$f(x) = \frac{1}{1-x}$$

and sequences ($k \geq 1$)

$$r_k = r_{k-1}\,f(b_{k-1})\,s_{k-1}, \quad c_k = c_{k-1}\,f(b_{k-1})\,s_{k-1}, \quad t_k = t_{k-1}\,c_{k-1}\,\xi,$$

$$a_k = a_{k-1}\,f(b_{k-1}), \quad s_k = (r_k + t_k)\,a_k, \quad b_k = b_{k-1}\,f^2(b_{k-1})\,s_{k-1}.$$

Then, using induction, we show the following recurrence relations.

**Lemma 3.9** *Let us suppose that $x_0$, $v_0$, $x_k$, $v_k \in \mathcal{D}$ ($k \in \mathbb{N}$). If*

$$f^2(b_0)\,s_0 < 1$$

*and*

$$\xi\,f(b_0)\,c_0 < 1,$$

*then, the following relations are satisfied for ($k \geq 0$)*

$(I_k)$

$$\frac{1}{\rho\,|\nabla F(x_k)\cdot d_k|}\leq a_k \quad and \quad \| A_k \| \leq a_k,$$

$(II_k)$

$$\| x_{k+1}-x_k \| \leq f(b_{k-1})\, s_{k-1} \;\| x_k - x_{k-1} \| \leq c_k,$$

$(III_k)$

$$N \;\| A_k \| \| x_k - v_k \| \leq s_k,$$

$(IV_k)$

$$M \;\| \nabla F(x_k) \|^{-1} \| x_{k+1}-x_k \| \leq b_k,$$

$(V_k)$

$$\| x_{k+1}-x_0 \| \leq \xi\,\theta_0\,\frac{1-(f(b_0)\,s_0)^{k+1}}{1-f(b_0)\,s_0}\;\| x_1 - x_0 \|\,.$$

**Proof** We first show conditions $(I_k)$–$(V_k)$ are satisfied for $k=1$. We have in turn that

$$\| v_1 - x_0 \| \leq \| x_1 - x_0 \| + \theta_1 \leq (1+\xi\,\theta_0)\;\| x_1 - x_0 \|,$$

$$\begin{aligned}\| \nabla F(x_1) \| &\geq \| \nabla F(x_0) \| - \| \nabla F(x_1) - \nabla F(x_0) \| \\ &\geq \| \nabla F(x_0) \| - M \;\| x_1 - x_0 \| \\ &= \| \nabla F(x_0) \| \;(1 - M \;\| \nabla F(x_0) \|^{-1} \| x_1 - x_0 \|) \\ &\geq \| \nabla F(x_0) \| \;(1-b_0),\end{aligned}$$

so

$$\| \nabla F(x_1) \|^{-1} \leq \| \nabla F(x_0) \|^{-1}\; f(b_0),$$

$(I_1)$

$$\begin{aligned}\| A_1 \| &\leq \left|\frac{\theta_1}{F(v_1)-F(x_1)}\right| \leq \left|\frac{\theta_1}{\nabla F(\lambda_1)\theta_1\cdot d_1}\right| \\ &\leq \frac{1}{\rho\,|\nabla F(x_1)\cdot d_1|} \leq \frac{\| \nabla F(x_0) \|}{\rho\,|\nabla F(x_0)\cdot d_0|\;\| \nabla F(x_1) \|} \\ &\leq \frac{f(b_0)}{\rho\,|\nabla F(x_0)\cdot d_0|} \leq a_0\, f(b_0) = a_1,\end{aligned}$$

$(II_2)$

$$\begin{aligned}\| x_2 - x_1 \| &\leq \| A_1 \| \;|F(x_1)| \leq a_0\, f(b_0)\,(r_0+t_0)\;\| x_1 - x_0 \| \\ &= f(b_0)\, s_0 \;\| x_1 - x_0 \|, \\ \| x_2 - x_0 \| &\leq \| x_2 - x_1 \| + \| x_1 - x_0 \| \\ &\leq f(b_0)\, s_0 \;\| x_1 - x_0 \| + \| x_1 - x_0 \| \\ &\leq (f(b_0)\, s_0 + 1)\;\| x_1 - x_0 \|\end{aligned}$$

$$N \parallel x_2 - x_1 \parallel \leq r_0\, f(b_0)\, s_0 = r_1,$$

$$N\,\theta_1 \leq N\,\xi\,\theta_0 \parallel x_1 - x_0 \parallel \leq N\,\xi\,\theta_0 \parallel x_1 - x_0 \parallel \leq t_0\,\xi\,a_0\,\eta = t_1,$$

$(III_1)$

$$N \parallel A_1 \parallel \parallel x_1 - v_1 \parallel \leq a_1\,(N \parallel x_2 - x_1 \parallel + N\,\theta_1) \leq a_1\,(r_1 + t_1) = s_1,$$

$(IV_1)$

$$M \parallel \nabla F(x_1) \parallel^{-1} \parallel x_2 - x_1 \parallel \leq M \parallel \nabla F(x_0) \parallel^{-1} f^2(b_0)\, s_0 \parallel x_1 - x_0 \parallel$$
$$\leq b_0\, f^2(b_0)\, s_0 = b_1,$$

$(V_1)$

$$\begin{aligned} \parallel x_2 - x_0 \parallel &\leq \parallel x_2 - x_1 \parallel + \parallel x_1 - x_0 \parallel \\ &\leq (f(b_0)\, s_0 + 1) \parallel x_1 - x_0 \parallel \\ &\leq \xi\,\theta_0\, \frac{1 - (f(b_0)\, s_0)^2}{1 - f(b_0)\, s_0} \parallel x_1 - x_0 \parallel . \end{aligned}$$

The induction process now proves the result. This completes the proof of Lemma 3.9. ⊠

We can show the following semilocal convergence result for DNGSM.

**Theorem 3.11** *Let $F : \mathcal{D} \subseteq \mathbb{R}^n \longrightarrow \mathbb{R}$ be a nonlinear differentiable mapping. Suppose that $(\mathcal{C})$ conditions,*

$$f^2(b_0)\, s_0 < 1,$$

$$\xi\, f(b_0)\, c_0 < 1$$

*and*

$$\overline{U}(x_0, R) \subseteq \mathcal{D},$$

*where*

$$R = \frac{(1 + \xi\,\theta_0)\, a_0\,\lambda}{1 - f(b_0)\, s_0}.$$

*Then, the sequence $\{x_k\}$ generated by DNGSM is well defined, remains in $\overline{U}(x_0, R)$ for all $k \geq 0$ and converges to a solution $x^\star \in \overline{U}(x_0, R)$ of equation $F(x) = 0$.*

**Proof** We have in turn that

$$\parallel v_0 - x_0 \parallel \leq \theta_0 < R \Longrightarrow v_0 \in \mathcal{D},$$

$$\parallel x_1 - x_0 \parallel < R \Longrightarrow x_1 \in \mathcal{D}.$$

Similarly, $v_1$, $x_2 \in \mathcal{D}$. Assume $v_k$, $x_{k+1} \in \mathcal{D}$, $k = 1, \cdots, n$. We shall show using induction that $v_{k+1}$, $x_{k+2} \in \mathcal{D}$.

Using the recurrence relations, we get that

$$\| v_{k+1} - x_k \| \leq (1 + \xi\,\theta_k) \ \| x_{k+1} - x_k \| \leq (f(b_0)\,s_0)^k\,(1 + \xi\,\theta_0) \ \| x_1 - x_0 \|,$$

$$\begin{aligned} \| x_{k+1} - x_0 \| &\leq (f(\xi_0)\,s_0)^k \ (1 + \xi\,\theta_0) \ \| x_1 - x_0 \| \\ &\quad + \frac{1 - (f(b_0)\,s_0)^k}{1 - f(b_0)\,s_0}\,(1 + \xi\,\theta_0) \ \| x_1 - x_0 \| \\ &\leq (1 + \xi\,\theta_0)\,\frac{1 - (f(b_0)\,s_0)^{k+1}}{1 - f(b_0)\,s_0} \ \| x_1 - x_0 \| < R, \end{aligned}$$

$$\begin{aligned} &\| x_{k+2} - x_0 \| \\ &\leq \| x_{k+2} - x_{k+1} \| + \| x_{k+1} - x_0 \| \\ &\leq f(b_k)\,s_k \ \| y_k - x_k \| + (1 + \xi\,\theta_0)\,\frac{1 - (f(b_0)\,s_0)^{k+1}}{1 - f(b_0)\,s_0} \ \| x_1 - x_0 \| \\ &\leq (f(b_0)\,s_0)^{k+1} \ \| x_1 - x_0 \| + (1 + \xi\,\theta_0)\,\frac{1 - (f(b_0)\,s_0)^{k+1}}{1 - f(b_0)\,s_0} \ \| x_1 - x_0 \| \\ &\leq (1 + \xi\,\theta_0)\,\frac{1 - (f(b_0)\,s_0)^{k+2}}{1 - f(b_0)\,s_0} \ \| x_1 - x_0 \| < R. \end{aligned}$$

Hence, we deduce $v_{k+1}$, $v_{k+2} \in \mathcal{D}$.

Next, we shall show the convergence of sequence $\{x_k\}$. We have

$$\| x_{k+1} - x_k \| \leq (f(b_0)\,s_0)^k \ \| x_1 - x_0 \|,$$

$$\begin{aligned} \| x_{k+j} - x_k \| &\leq \sum_{i=k}^{i=k+j-1} \ \| x_{i+1} - x_i \| \\ &\leq \sum_{i=k}^{i=k+j-1} \ (f(b_0)\,s_0)^i \ \| x_1 - x_0 \| \\ &\leq (f(b_0)\,s_0)^k\,\frac{1 - (f(b_0)\,s_0)^k}{1 - f(b_0)\,s_0} \ \| x_1 - x_0 \| \,. \end{aligned} \tag{3.288}$$

By hypotheses $f(b_0)^2\,s_0 < 1$, we get $f(b_0)\,s_0 < 1$, which together with (3.288) imply that sequence $\{x_k\}$ is Cauchy in a complete space and as such it converges to some $x^\star \in \overline{U}(x_0, R)$ (since $\overline{U}(x_0, R)$ is a closed set).

We shall show that $x^\star$ is a solution of equation $F(x) = 0$. We have

$$|F(x_{k+1})| \leq (r_k + t_k) \ \| x_{k+1} - x_k \| \leq (r_0 + t_0)\,(f(b_0)\,s_0)^k \ \| x_1 - x_0 \| \longrightarrow 0$$

as $k \longrightarrow \infty$, since $(r_0 + t_0) \ \| x_1 - x_0 \|$ is bounded and $f(b_0)\,s_0 < 1$. Hence, we deduce $|F(x^\star)| = 0$. This completes the proof of Theorem 3.11. ⊠

We also can provide a second semilocal convergence result. The proofs are omitted as almost identical to the corresponding previous ones. First, we need an auxiliary lemma.

**Lemma 3.10** *Let* $\xi \in (0,1)$ *and* $k \geq 0$*. If* $\theta_k \in [0, \tau_k]$*, where,*

$$\tau_k = \frac{\sqrt{\| \nabla F(x_k) \|^2 + 2\, M\, \xi\, |F(x_k)|} - \| \nabla F(x_k) \|}{M},$$

*then, the following holds*

$$\theta_k \leq \xi \; \| x_{k+1} - x_k \| .$$

**Proof** We have in turn

$$\begin{aligned}
&|F(v_k) - F(x_k)| \\
&= \left| \int_{x_k}^{v_k} \nabla F(y)\, dy \right| \\
&= \left| \int_0^1 (\nabla F(x_k + \theta\, \theta_k\, d_k)\, \theta_k\, d_k - \nabla F(x_k)\, \theta_k\, d_k + \nabla F(x_k)\, \theta_k\, d_k)\, d\theta \right| \\
&\leq \theta_k\, (M \int_0^1 \theta\, \theta_k\, d\theta + \| \nabla F(x_k) \|) \\
&= \theta_k \left( \frac{M}{2}\, \theta_k + \| \nabla F(x_k) \| \right) = \frac{M}{2}\, \theta_k^2 + \| \nabla F(x_k) \|\; \theta_k.
\end{aligned}$$

Hence, we have

$$\frac{M}{2}\, \theta_k^2 + \| \nabla F(x_k) \|\; \theta_k \leq \xi\, |F(x_k)|$$

provided that $\theta_k \in [0, \tau_k]$. Therefore, we get in turn

$$\| x_{k+1} - x_k \| = \frac{\theta_k\, |F(x_k)|}{|F(x_k + \theta_k\, d_k) - F(x_k)|} \geq \frac{\theta_k\, |F(x_k)|}{\xi\, |F(x_k)|} = \frac{\theta_k}{\xi}.$$

This completes the proof of Lemma 3.10. ⊠

We shall assume that the $(\mathcal{C}^\star)$ conditions which are: $(\mathcal{C}_1)$, $(\mathcal{C}_2)$, $(\mathcal{C}_4)$, $(\mathcal{C}_5)$ and

$(\mathcal{C}_3^\star)$ $\theta_0 \in [0, \tau_0]$ and $0 < \theta_k \leq \min\{\tau_k,\, \xi\, \| x_k - x_{k-1} \|\}$ $(k \geq 1)$.

Define scalar sequences

$$s_k = f^2(t_{k-1})\,(1+\xi)\, s_{k-1}^2, \quad s_0 = \frac{N\, \mu^2\, \eta}{\gamma^2}, \quad \mu = \frac{1}{\rho\, \delta},$$

$$t_k = f^2(t_{k-1})\,(1+\xi)\, s_{k-1}\, t_{k-1}, \quad t_0 = \frac{M\, \mu\, \eta}{\gamma^2}.$$

The corresponding recurrence relations are

$$\| A_k \| \leq \mu \| \nabla F(x_k) \|^{-1} \leq \gamma \| \nabla F(x_{k-1}) \|^{-1} f(t_{k-1}),$$

$$\| x_{k+1} - x_k \| \leq f(t_{k-1})\,(1+\xi)\, s_{k-1} \| x_k - x_{k-1} \|,$$

$$N \| A_k \| \| x_{k+1} - x_k \| \leq s_k,$$

$$M \| \nabla F(x_k) \|^{-1} \| x_{k+1} - x_k \| \leq t_k,$$

$$\| x_{k+1} - x_0 \| \leq (1+\xi)\, \frac{1 - (f(t_0)\,(1+\xi)\, s_0)^k}{1 - f(t_0)\,(1+\xi)\, s_0} \| x_1 - x_0 \|$$

and

$$|F(x_k)| \leq (1+\xi)\, N \| x_k - x_{k-1} \|^2 .$$

Then, we have the following semilocal convergence result for DNGSM.

**Theorem 3.12** *Let $F : \mathcal{D} \subseteq \mathbb{R}^n \longrightarrow \mathbb{R}$ be a nonlinear differentiable mapping. Suppose that $(\mathcal{C}^\star)$ conditions,*

$$f^2(t_0)\,(1+\xi)\, s_0 < 1$$

*and*

$$\overline{U}(x_0, R) \subseteq \mathcal{D},$$

*where*

$$R = \frac{1+\xi}{1 - f(t_0)\,(1+\xi)\, s_0} \| x_1 - x_0 \| .$$

*Then, the sequence $\{x_k\}$ generated by DNGSM is well defined, remains in $\overline{U}(x_0, R)$ for all $k \geq 0$ and converges to a solution $x^\star \in \overline{U}(x_0, R)$ of equation $F(x) = 0$.*

The results for DMCMSM can now follow. We provide now two numerical examples. In the first one, we show that hypotheses of Theorem Theorem 3.8 are satisfied, but not corresponding hypotheses of Theorem 3.9.

**Example 3.6** *Let $n = 2$. Here, we use the Euclidean inner product and the corresponding norm for both vector and matrix. Choose*

$$x_0 = (1,1)^T, \quad \mathcal{D}_0 = \{x \in \mathbb{R}^2 : \| x - x_0 \| \leq 1 - b\} \quad (b \in [0,1))$$

*and define a mapping $F$ on $\mathcal{D}_0$ by*

$$F(x) = \frac{\lambda_1^3 + \lambda_2^3}{2} - 2\, b, \quad x = (\lambda_1, \lambda_2)^T. \tag{3.289}$$

*Then, the gradient $\nabla F$ of mapping $F$ is given by*

$$\nabla F(x) = \frac{3}{2}\, (\lambda_1^2, \lambda_2^2)^T. \tag{3.290}$$

*Using (3.246), (3.243), (3.248), (3.253), (3.254), we obtain for* $\alpha = \frac{1}{2}$, $\beta = 2$

$$M_0 = \frac{3\sqrt{2}}{2}\,(3-b), \quad M = 3\sqrt{2}\,(2-b),$$

$$L_0 = 3-b, \quad L = 3\,(2-b), \quad \|\,\nabla F(x_0)\,\| = \frac{3\sqrt{2}}{2},$$

$$F(x_0) = 1-\sqrt{2}\,b \quad \text{and} \quad \eta = \frac{\sqrt{2}}{3}\,(1-2\,b).$$

*We can choose the directions* $d_k$ *by*

$$d_k = \frac{\nabla F(x_k)}{\|\,\nabla F(x_k)\,\|},$$

*so that conditions (3.249) and (3.273) are satisfied as equalities. Let* $b = .4162$*. Then, condition (3.271) is violated, since we have* $\eta = .079007397$, $M = 6.71949432$ *and*

$$\frac{4\,\sqrt{2}}{3}\,(2-b)\,(1-2\,b) = .500527666 > .5.$$

*Hence, there is no guarantee that DNGSM starting at* $x_0$ *converges to the solution* $x^\star$ *of equation* $F(x) = 0$*. Moreover, we obtain*

$$L_0 = 2.5838, \quad L = 4.7514, \quad \overline{L} = 3.259626413,$$

$$\delta = 1.207332475, \quad t^{\star\star} = .199345613 < 1-b = .5838, \quad \text{and} \quad q = .257534598 < .5.$$

*That is, condition (3.252) is satisfied and* $U_0 \subseteq \mathcal{D}$*. The conclusions of Theorem 3.8 apply for solving equation* $F(x) = 0$.

*We found* $x^\star = (.940684577, .940684577)$.

**Example 3.7** *Let* $n = 20$ *and the starting point* $x_0 = (0.1, 0.1, \ldots, 0.1)$*. Here, we use the Euclidean inner product and the corresponding norm for both vector and matrix. We consider the following three nonlinear problems considered in (cf. [10])*

$$F_1(x) = \sum_{i=1}^{20} x_i^2\, e^{1-x_i^2}, \tag{3.291}$$

$$F_2(x) = \sum_{i=1}^{20} (\sin x_i)^2 \tag{3.292}$$

*and*

$$F_3(x) = \sum_{i=1}^{5} (\sin x_i)^2 + \sum_{i=6}^{20} (\tan x_i)^2. \tag{3.293}$$

*Note that* $x^\star = 0$ *is a locally unique solution for each of* $F_i(x) = 0$ $(i = 1, 2, 3)$*. We apply in this example the convergence results previously obtained* *for DSM. We also compare DSM with the directional Newton method DNM . The direction* $d_k$ $(k \geq 0)$ *is chosen such that is sufficiently close to the gradient* $\nabla F(x_k)$ *of* $F$ *in each iteration* $x_k$. *In the implementation using Maple 13, we consider two cases for direction* $d_k$. *In the first case:*

$$d_k := \nabla F(x_k)/\|\nabla F(x_k)\| \tag{3.294}$$

*and, in the second case:*

$$d_k := \frac{p_k}{\| p_k \|}, \tag{3.295}$$

*where*

$$p_k := \left( \frac{F(x_k + F(x_k)e_1)}{F(x_k)} - 1, \frac{F(x_k + F(x_k)e_2)}{F(x_k)} - 1, \ldots, \frac{F(x_k + F(x_k)e_{20})}{F(x_k)} - 1 \right).$$

$e_k$ $(k = 1, \cdots, 20)$ *is the kth unit vector of* $\mathbb{R}^{20}$. *We give three comparison tables for error bounds* $\| x_k - x^\star \|$ $(k \geq 0)$, *where the stopping criterion* $|F_i(x_k)| < 10^{-12}$ $(i = 1, 2, 3)$ *is used.*

**Comparison Table for $F_1$**

| $k$ | DNM with (3.294) | DSM with (3.294) | DSM with (3.295) |
|---|---|---|---|
| 1 | .2213481425 | .2334571896 | .2330073138 |
| 2 | .1104022818 | .1150173101 | .1215974906 |
| 3 | .05516747919 | .05809352097 | .06486112837 |
| 4 | .02757954151 | .02908230293 | .03635275258 |
| 5 | .01378924624 | .01455220678 | .01915900153 |
| 6 | .006894557562 | .007272642285 | .009630928543 |
| 7 | .003447270599 | .003636811795 | .004812897796 |
| 8 | .001723634277 | .001818963011 | .002406148658 |
| 9 | .0008618170085 | .0009092235709 | .001203624956 |
| 10 | .0004309084889 | .0004545980359 | .0006016686952 |
| 11 | .0002154542425 | .0002273082528 | .0003008770080 |
| 12 | .0001077271212 | .0001136578186 | .0001504472661 |
| 13 | .00005386356060 | .00005683020130 | .00007522944185 |
| 14 | .00002693178026 | .00002841754599 | .00003761831663 |
| 15 | .00001346589015 | .00001421126889 | .00001881121506 |
| 16 | .000006732945077 | .000007108122620 | .000009406895828 |
| 17 | .000003366472543 | .000003556547452 | .000004704237278 |
| 18 | $\sim$ | $\sim$ | .000002352609769 |

**Comparison Table for $F_2$**

| $k$ | DNM with (3.294) | DSM with (3.294) | DSM with (3.295) |
|---|---|---|---|
| 1 | .2228584479 | .2185022023 | .2291197948 |
| 2 | .1113368951 | .1093252604 | .1182527578 |
| 3 | .05565694383 | .05469418320 | .06314434968 |
| 4 | .02782703509 | .02737622640 | .03280017538 |
| 5 | .01391333797 | .01368862309 | .01665214627 |
| 6 | .006956646548 | .006841851707 | .008367425850 |
| 7 | .003478320457 | .003419925369 | .004195078981 |
| 8 | .001739159878 | .001709963795 | .002099295662 |
| 9 | .0008695798963 | .0008549568094 | .001049898919 |
| 10 | .0004347899430 | .0004274834619 | .0005250947457 |
| 11 | .0002173949714 | .0002137417329 | .0002625837980 |
| 12 | .0001086974858 | .0001068733876 | .0001313046567 |
| 13 | .00005434874294 | .00005343921465 | .00006565673218 |
| 14 | .00002717437149 | .00002672207772 | .00003283067840 |
| 15 | .00001358718574 | .00001336351914 | .00001641606330 |
| 16 | .000006793592895 | .000006684239494 | .000008208345478 |
| 17 | .000003396796451 | .000003344597729 | .000004104300066 |

**Comparison Table for $F_3$**

| $k$ | DNM with (3.294) | DSM with (3.294) | DSM with (3.295) |
|---|---|---|---|
| 1 | .2245686780 | .2285067687 | .2308497146 |
| 2 | .1123990871 | .1139792812 | .1181539085 |
| 3 | .05621369955 | .05721706204 | .06282663095 |
| 4 | .02810861328 | .02861766738 | .03266887733 |
| 5 | .01405452689 | .01431805650 | .01661086509 |
| 6 | .007027290967 | .007160297337 | .008356807613 |
| 7 | .003513648921 | .003579879135 | .004185849468 |
| 8 | .001756824890 | .001790209649 | .002093848982 |
| 9 | .0008784124981 | .0008951552522 | .001047627095 |
| 10 | .0004392062563 | .0004476150645 | .0005239693784 |
| 11 | .0002196031293 | .0002238285767 | .0002619979495 |
| 12 | .0001098015645 | .0001119258299 | .0001310142103 |
| 13 | .00005490078249 | .00005597005369 | .00006551199164 |
| 14 | .00002745039116 | .00002798994481 | .00003275805190 |
| 15 | .00001372519562 | .00001399867952 | .00001637986395 |
| 16 | .000006862597819 | .000007002457144 | .000008190267678 |
| 17 | .000003431298914 | .000003504042798 | .000004095270726 |

**Remark 3.13** *The results extend in a Hilbert space setting. Indeed, let $F$ be a differentiable operator defined on a convex subset $\mathcal{D}$ of a Hilbert space $\mathcal{H}$ with values in $\mathbb{R}$. Here $x \cdot y$ denotes the inner product of elements $x$ and $y$ in $\mathcal{H}$. We also use $\| x \| = (x \cdot x)^{1/2}$. Moreover, instead of condition (3.249), assume*

$$\| d_k \| = 1$$

*and there exists $\kappa \in [0, 1]$ such that*

$$|\nabla F(x_k) \cdot d_k| \geq \kappa \ \| \nabla F(x_k) \| .$$

*Note that, in the case of (3.249), we can set*

$$\kappa = \frac{|\nabla F(x_k) \cdot d_k|}{\| \nabla F(x_0) \|} \leq 1.$$

*Define*

$$L_0 = \frac{M_0}{\kappa \ \| \nabla F(x_0) \|}, \quad L = \frac{\beta \ (1+\alpha) \ M}{\kappa \ \| \nabla F(x_0) \|}, \quad \eta = \| \ h_0 \ \|$$

*in the case of Theorem 3.8 and*

$$L_0 = \frac{M_0}{\kappa \ \| \nabla F(x_0) \|_\infty}, \quad L = \frac{\beta \ (1+\alpha) \ M}{\kappa \ \| \nabla F(x_0) \|_\infty}, \quad \eta = \| \ h_0 \ \|_\infty$$

*in the case of Theorem 3.10. Then, due to the proofs of Theorems 3.8 and 3.10, the results hold in this more general setting.*

## 3.6 Efficient Three Step Secant Methods

A local convergence analysis for an efficient Secant–type method STM for solving nonlinear equations is given using both the Lipschitz continuous and center–Lipshchitz continuous divided differences of order one. An estimate of the radius of the convergence ball of STM is provided, the error estimate matching its convergence order is established. Numerical examples validating the theoretical results are also provided in this section.

We are concerned with the Seccant method

$$x_{n+1} = x_n - [x_{n-1}, x_n; F]^{-1} F(x_n) \qquad (n \geq 0, \ x_{-1}, x_0 \in \Omega) \tag{3.296}$$

for approximating a locally unique solution $x^\star$ of (2.1). Here, $F$ is an operator defined on an open convex subset $\Omega$ of a Banach space $X$ with values in a Banach space $Y$, $[x, y; F] \in L(X, Y)$, the space of bounded linear operators from $X$ to $Y$, is a divided difference of order one for the operator $F$ at the points $x, y \in \Omega$. Another famous iterative process for solving nonlinear equations is Chebyshev's method (see (cf. [163]))

$$\begin{cases} y_n = x_n - F'(x_n)^{-1} \ F(x_n), \\ x_{n+1} = y_n - \dfrac{1}{2} \ F'(x_n)^{-1} F''(x_n)(y_n - x_n)^2 \quad (n \geq 0, \ x_0 \in \Omega), \end{cases} \tag{3.297}$$

which has cubic convergence. This one–point iterative method depends explicitly on the first and second derivatives of $F$. Ezquerro and Hernández proposed in (cf. [360]) some modifications of Chebyshev's method that avoid the computation of the second derivative of $F$. We recall this method as the Chebyshev–Newton–type method CNTM, which has the form:

$$\begin{cases} y_n = x_n - F'(x_n)^{-1} \ F(x_n), \\ z_n = x_n + a \ (y_n - x_n), \\ x_{n+1} = x_n - \dfrac{1}{a^2} \ F'(x_n)^{-1} \ ((a^2 + a - 1) \ F(x_n) + F(z_n)) \quad (n \geq 0, \ x_0 \in \Omega), \end{cases} \tag{3.298}$$

where $a \in (0, 1]$ is a parameter.

Recently, Argyros et. al in (cf. [177]) introduced a new family of iterative methods free from derivatives by replacing the first derivative $F'$ with a divided difference of order one for the operator $F$. We call this new family as the Chebyshev–Secant–type method CSTM:

$$\begin{cases} y_n = x_n - A_n^{-1}\, F(x_n), \quad A_n = [x_{n-1}, x_n; F], \\ z_n = x_n + a\, (y_n - x_n), \\ x_{n+1} = x_n - A_n^{-1}\, (b\, F(x_n) + c\, F(z_n)) \quad (n \geq 0,\ x_{-1}, x_0 \in \Omega), \end{cases} \tag{3.299}$$

where $a \in [0,1]$, $b \geq 0$ and $c \geq 0$ are parameters. A semilocal convergence theorem for CSTM is established in (cf. [177]) under relation

$$(1-a)\, c \;=\; 1-b \tag{3.300}$$

and some other mild conditions. Moreover, several numerical examples were also provided in (cf. [177]) to show the efficiency of CSTM. However, the order of convergence of $\{x_n\}$ to $x^\star$ generated by CSTM in the second example in (cf. [177]) has only 1.625, which is near 1.618 of the Secant method (3.296).

In this section, a more precise local convergence analysis for the CSTM is given. In fact, we show that in order to achieve higher order, apart from (3.300), the following relation

$$ac = 1 \tag{3.301}$$

is also needed. Under relations (3.300) and (3.301), we can rewrite (3.300) as follows

$$\begin{cases} y_n = x_n - A_n^{-1}\, F(x_n), \quad A_n = [x_{n-1}, x_n; F], \\ z_n = x_n + a\, (y_n - x_n), \\ x_{n+1} = x_n - A_n^{-1}\, ((2 - \dfrac{1}{a})\, F(x_n) + \dfrac{1}{a}\, F(z_n)) \quad (n \geq 0,\ x_{-1}, x_0 \in \Omega), \end{cases} \tag{3.302}$$

where $a \in (0,1]$ is a parameter. We shall call (3.302) Secant–type method STM. Note that CSTM and STM are useful alternatives of CNTM, especially in cases where the computation of $F''(x_n)$ or $F'(x_n)^{-1}$ are expensive or impossible to compute or their analytic representation is unavailable. Our results show that (3.302) is a family of methods with order two for any choice of $a \in (0,1]$. Moreover, a unified estimate of radius of convergence ball of STM as well as a unified error analysis matching its convergence order are given.

We shall show a lemma needed for our main theorem.

**Lemma 3.11** *Assume $a \in [0,1]$, $c \geq 0$ are parameters, $M$ and $M_0$ are positive constants with $M_0 \leq M$. Let $q$ be a function defined on $[0,+\infty)$ by*

$$\begin{aligned} q(r) &= a^2 c (Mr)^3 + 4ac(Mr)^2(1 - M_0 r) \\ &\quad + 4(|1-ac| + (1-a)ac) Mr (1-M_0 r)^2 - 8(1-M_0 r)^3. \end{aligned} \tag{3.303}$$

*Then, $q$ has a least positive zero in $(0, R_0]$ with $R_0$ given by*

$$R_0 = \frac{2}{M + 2M_0}. \tag{3.304}$$

**Proof** For the case of $ac = 0$, $q$ can be written as

$$q(r) = 4Mr(1 - M_0 r)^2 - 8(1 - M_0 r)^3 = 4(1 - M_0 r)^2(Mr - 2(1 - M_0 r)), \tag{3.305}$$

which has the least positive zero $R_0$ given by (3.304) since

$$R_0 < \frac{1}{M_0}. \tag{3.306}$$

Next, we assume $ac > 0$. It follows from the definition of $R_0$ that $1 - M_0 R_0 = \frac{1}{2} M R_0$ and thus

$$\begin{aligned} q(R_0) &= a^2c(MR_0)^3 + 2ac(MR_0)^3 + (|1 - ac| + (1 - a)ac)(MR_0)^3 - (MR_0)^3 \\ &= (3ac + |1 - ac| - 1)(MR_0)^3 \geq 2\min\{ac, 1\}(MR_0)^3 > 0. \end{aligned} \tag{3.307}$$

In view of $q(0) = -8 < 0$, by the intermediate value theorem, there is a zero of $q$ in $(0, R_0)$ and the least positive zero $R$ of $q$ must satisfy $0 < R < R_0$. This completes the proof of the lemma. ⊠

We can show the following local convergence theorem for CSTM.

**Theorem 3.13** *Assume that*

(a) *$F : \Omega \subseteq X \to Y$ is a Fréchet-differentiable operator and there exists divided difference $[x,y;F]$ satisfying*

$$[x, y; F](x - y) = F(x) - F(y) \quad (x, y \in \Omega); \tag{3.308}$$

(b) *$x^\star$ is a solution of equation (2.1), $F'(x^\star)^{-1}$ exists and there is a constant $M > 0$ such that*

$$\|F'(x^\star)^{-1}([x, y; F] - [u, v; F])\| \leq \frac{M}{2}(\|x - u\| + \|y - v\|) \quad (x, y, u, v \in \Omega); \tag{3.309}$$

(c) *There is exists constant $M_0 > 0$ such that*

$$\|F'(x^\star)^{-1}([x, y; F] - F'(x^\star))\| \leq \frac{M_0}{2}(\|x - x^\star\| + \|y - x^\star\|) \quad (x, y \in \Omega); \tag{3.310}$$

*In view of (3.309), condition (3.310) always holds. Moreover $M_0 \leq M$ holds in general and $\frac{M}{M_0}$ can be arbitrarily large (cf. [209], [177]). That is, (3.310) is not an additional to (3.309) hypothesis.*

(d) *The relation (3.300) is true;*

*(e) $U(x^\star, R) \subseteq \Omega$, where $R$ is the least positive zero of function $q$ defined by (3.303).*

*Then, sequence $\{x_n\}$ generated by CSTM from any initial points $x_{-1}$ and $x_0$ in $U(x^\star, R)$ is well defined, remains in $U(x^\star, R)$ and converges to $x^\star$ with order of 1.618... at least. Moreover, the following error estimates are satisfied*

$$\|e_{n+1}\| \leq \frac{1}{R}\|e_n\|\|e_{n-1}\| \quad (n \geq 1), \tag{3.311}$$

*where $e_n = x_n - x^\star$ $(n \geq -1)$. Furthermore, if relation (3.301) is also true, the least positive zero of function $q$ can be expressed explicitly by*

$$R = \frac{1}{\dfrac{\sqrt{4+a^2}+2-a}{4}M + M_0}, \tag{3.312}$$

*$\{x_n\}$ converges to $x^\star$ with order at least two and the following estimates are satisfied*

$$\|e_{n+1}\| \leq \frac{1}{R}\left(\frac{\gamma}{R}\|e_n\|\|e_{n-1}\|^2 + (1-\gamma)\|e_n\|^2\right) \quad (n \geq 1), \tag{3.313}$$

*where*

$$\gamma = \left(\frac{\sqrt{4+a^2}+a-2}{2a}\right)^2. \tag{3.314}$$

**Proof** At first, let us assume conditions (a)–(e) are satisfied. We will prove all the results before (3.312) of the theorem by induction. Denote by $u_n = y_n - x^\star$, $v_n = z_n - x^\star$ for $n \geq 0$. Using the basic identify (cf. [209])

$$F'(x^\star) = [x^\star, x^\star; F], \tag{3.315}$$

condition (3.310), $x_{-1}, x_0 \in U(x^\star, R)$ and Lemma 3.11, we have

$$\begin{aligned} \|I - F'(x^\star)^{-1}A_0\| &= \|F'(x^\star)^{-1}([x_{-1}, x_0; F] - [x^\star, x^\star; F])\| \\ &\leq \frac{M_0}{2}(\|e_{-1}\| + \|e_0\|) < M_0 R \leq M_0 R_0 < 1. \end{aligned} \tag{3.316}$$

By the Banach lemma, $A_0$ is invertible and

$$\|A_0^{-1}F'(x^\star)\| \leq \frac{1}{1 - \dfrac{M_0}{2}(\|e_{-1}\| + \|e_0\|)} < \frac{1}{1 - M_0 R}. \tag{3.317}$$

Thus $y_0$ is well defined. By the definition of divided difference, we have

$$\begin{aligned} F(x_0) &= F(x_0) - F(x^\star) = -(F(x^\star) - F(x_0)) \\ &= -[x^\star, x_0; F](x^\star - x_0) = [x^\star, x_0; F]e_0. \end{aligned} \tag{3.318}$$

Using (3.299) and (3.318), we have

$$\begin{aligned} u_0 &= y_0 - x^\star = x_0 - x^\star - A_0^{-1}F(x_0) \\ &= A_0^{-1}F'(x^\star)\, F'(x^\star)^{-1}([x_{-1}, x_0; F] - [x^\star, x_0; F])e_0. \end{aligned} \tag{3.319}$$

Combining (3.317), condition (3.309) and Lemma 3.11 yields

$$\begin{aligned} \|u_0\| &\le \frac{\frac{M}{2}\|e_{-1}\|\|e_0\|}{1 - \frac{M_0}{2}(\|e_{-1}\| + \|e_0\|)} \le \frac{M\|e_{-1}\|\|e_0\|}{2(1 - M_0 R)} \le \frac{MR\|e_0\|}{2(1 - M_0 R)} \\ &\le \frac{MR_0\|e_0\|}{2(1 - M_0 R_0)} = \|e_0\| < R. \end{aligned} \tag{3.320}$$

That is, $y_0 \in U(x^\star, R)$. Since $z_0$ is a convex combination of $x_0$ and $y_0$ and $U(x^\star, R) \subseteq \Omega$ is a convex set, we can deduce that $z_0 \in U(x^\star, R)$. In fact, we have

$$v_0 = au_0 + (1 - a)e_0 \tag{3.321}$$

and thus

$$\|v_0\| \le a\|u_0\| + (1 - a)\|e_0\| \le \|e_0\| < R. \tag{3.322}$$

Using the same technique as in (3.318), we have

$$F(z_0) = [x^\star, z_0; F]v_0. \tag{3.323}$$

Using (3.299), (3.300), (3.321) and (3.323), we get

$$\begin{aligned} e_1 &= e_0 - A_0^{-1}(bF(x_0) + cF(z_0)) \\ &= A_0^{-1}([x_{-1}, x_0; F]e_0 - (b[x^\star, x_0; F]e_0 + c[x^\star, z_0; F]v_0)) \\ &= A_0^{-1}([x_{-1}, x_0; F]e_0 - b[x^\star, x_0; F]e_0 - c[x^\star, z_0; F](au_0 + (1 - a)e_0)) \\ &= A_0^{-1}([x_{-1}, x_0; F]e_0 - b[x^\star, x_0; F]e_0 - (1 - b)[x^\star, z_0; F]e_0 - ac[x^\star, z_0; F]u_0) \\ &= A_0^{-1}(([x_{-1}, x_0; F] - [x^\star, x_0; F])e_0 + (1 - b)([x^\star, x_0; F] - [x^\star, z_0; F])e_0) \\ &+ acA_0^{-1}([x^\star, x_0; F] - [x^\star, z_0; F] + [x_{-1}, x_0; F] - [x^\star, x_0; F])u_0 - acu_0. \end{aligned} \tag{3.324}$$

Define

$$\begin{aligned} D_0 &= A_0^{-1}([x_{-1}, x_0; F] - [x^\star, x_0; F]), \\ E_0 &= A_0^{-1}([x^\star, x_0; F] - [x^\star, z_0; F]), \end{aligned} \tag{3.325}$$

then, we have from (3.319) that

$$u_0 = D_0 e_0. \tag{3.326}$$

Moreover, we can rewrite (3.324) as

$$\begin{aligned} e_1 &= D_0e_0 + (1-b)E_0e_0 + acE_0u_0 + acD_0u_0 - acu_0 \\ &= (1-ac)D_0e_0 + (1-b)E_0e_0 + acE_0D_0e_0 + acD_0^2e_0. \end{aligned} \tag{3.327}$$

On the other hand, by condition (3.309), we have from (3.317) and (3.325) that

$$\begin{aligned} \|D_0\| &= \|A_0^{-1}F'(x^\star)\,F'(x^\star)^{-1}([x_{-1},x_0;F]-[x^\star,x_0;F])\| \\ &\le \frac{\frac{M}{2}\|e_{-1}\|}{1-\frac{M_0}{2}(\|e_{-1}\|+\|e_0\|)}, \end{aligned} \tag{3.328}$$

$$\begin{aligned} \|E_0\| &= \|A_0^{-1}F'(x^\star)\,F'(x^\star)^{-1}([x^\star,x_0;F]-[x^\star,z_0;F])\| \\ &\le \frac{\frac{M}{2}\|z_0-x_0\|}{1-\frac{M_0}{2}(\|e_{-1}\|+\|e_0\|)} = \frac{\frac{aM}{2}\|y_0-x_0\|}{1-\frac{M_0}{2}(\|e_{-1}\|+\|e_0\|)} \\ &\le \frac{\frac{aM}{2}(\|u_0\|+\|e_0\|)}{1-\frac{M_0}{2}(\|e_{-1}\|+\|e_0\|)} = \frac{\frac{aM}{2}(\|D_0e_0\|+\|e_0\|)}{1-\frac{M_0}{2}(\|e_{-1}\|+\|e_0\|)} \\ &\le \frac{\frac{aM}{2}(\|D_0\|+1)\|e_0\|}{1-\frac{M_0}{2}(\|e_{-1}\|+\|e_0\|)} \le \frac{\frac{aM}{2}\left(\frac{\frac{M}{2}\|e_{-1}\|}{1-\frac{M_0}{2}(\|e_{-1}\|+\|e_0\|)}+1\right)\|e_0\|}{1-\frac{M_0}{2}(\|e_{-1}\|+\|e_0\|)} \\ &= \frac{aM^2\|e_{-1}\|\|e_0\|}{4\left(1-\frac{M_0}{2}(\|e_{-1}\|+\|e_0\|)\right)^2} + \frac{aM\|e_0\|}{2\left(1-\frac{M_0}{2}(\|e_{-1}\|+\|e_0\|)\right)}. \end{aligned} \tag{3.329}$$

Therefore, we have from (3.327)–(3.329) that

$$\begin{aligned} \|e_1\| &\le |1-ac|\|D_0\|\|e_0\| + (1-b)\|E_0\|\|e_0\| + ac\|E_0\|\|D_0\|\|e_0\| + ac\|D_0\|^2\|e_0\| \\ &\le \frac{|1-ac|M\|e_{-1}\|\|e_0\|}{2\left(1-\frac{M_0}{2}(\|e_{-1}\|+\|e_0\|)\right)} + \frac{a(1-b)M^2\|e_{-1}\|\|e_0\|^2}{4\left(1-\frac{M_0}{2}(\|e_{-1}\|+\|e_0\|)\right)^2} \end{aligned}$$

$$+\frac{a(1-b)M\|e_0\|^2}{2\left(1-\dfrac{M_0}{2}(\|e_{-1}\|+\|e_0\|)\right)}+\frac{a^2cM^3\|e_{-1}\|^2\|e_0\|^2}{8\left(1-\dfrac{M_0}{2}(\|e_{-1}\|+\|e_0\|)\right)^3}$$

$$+\frac{a^2cM^2\|e_{-1}\|\|e_0\|^2}{4\left(1-\dfrac{M_0}{2}(\|e_{-1}\|+\|e_0\|)\right)^2}+\frac{acM^2\|e_{-1}\|^2\|e_0\|}{4\left(1-\dfrac{M_0}{2}(\|e_{-1}\|+\|e_0\|)\right)^2}. \tag{3.330}$$

Define

$$h_0=\frac{M}{2\left(1-\dfrac{M_0}{2}(\|e_{-1}\|+\|e_0\|)\right)}, \tag{3.331}$$

$$H=\frac{M}{2(1-M_0R)}, \tag{3.332}$$

then

$$h_0\leq H \tag{3.333}$$

and we have from (3.330) that

$$\begin{aligned}\|e_1\| &\leq |1-ac|h_0\|e_{-1}\|\|e_0\|+a(1-b)h_0^2\|e_{-1}\|\|e_0\|^2+a(1-b)h_0\|e_0\|^2\\ &+a^2ch_0^3\|e_{-1}\|^2\|e_0\|^2+a^2ch_0^2\|e_{-1}\|\|e_0\|^2+ach_0^2\|e_{-1}\|^2\|e_0\|\\ &\leq |1-ac|H\|e_{-1}\|\|e_0\|+a(1-b)H^2\|e_{-1}\|\|e_0\|^2+a(1-b)H\|e_0\|^2\\ &+a^2cH^3\|e_{-1}\|^2\|e_0\|^2+a^2cH^2\|e_{-1}\|\|e_0\|^2+acH^2\|e_{-1}\|^2\|e_0\|\\ &\leq \left([|1-ac|+a(1-b)]HR+[a(1-b)+a^2c+ac](HR)^2+a^2c(HR)^3\right)\|e_0\|\\ &=\frac{q(R)+8(1-M_0R)^3}{8(1-M_0R)^3}\|e_0\|=\|e_0\|<R.\end{aligned} \tag{3.334}$$

This means $x_1\in U(x^\star,R)$.

Now, we suppose that $\{x_n\}$ $(0\leq n\leq k)$ is well defined and $x_n\in U(x^\star,R)(0\leq n\leq k)$, where $k$ is a fixed integer and $k\geq 1$. Using a similar argumentation about $x_{-1}$ and $x_0$, we can deduce the following results:

(i) $A_k$ is invertible and

$$\|A_k^{-1}F'(x^\star)\| \leq \frac{1}{1-\dfrac{M_0}{2}(\|e_{k-1}\|+\|e_k\|)}<\frac{1}{1-M_0R}; \tag{3.335}$$

(ii) $y_k$ is well defined, $y_k\in U(x^\star,R)$ and

$$\|u_k\| \leq \frac{\dfrac{M}{2}\|e_{k-1}\|\|e_k\|}{1-\dfrac{M_0}{2}(\|e_{k-1}\|+\|e_k\|)}\leq\frac{M\|e_{k-1}\|\|e_k\|}{2(1-M_0R)}\leq\frac{MR\|e_k\|}{2(1-M_0R)}\leq\|e_k\|<R; \tag{3.336}$$

(iii) $z_k$ is well defined, $z_k \in U(x^\star, R)$;

(iv) $x_{k+1}$ is well defined and

$$\begin{aligned}
&\|e_{k+1}\| \le |1-ac|H\|e_{k-1}\|\|e_k\| + a(1-b)H^2\|e_{k-1}\|\|e_k\|^2 + a(1-b)H\|e_k\|^2 \\
&+ a^2cH^3\|e_{k-1}\|^2\|e_k\|^2 + a^2cH^2\|e_{k-1}\|\|e_k\|^2 + acH^2\|e_{k-1}\|^2\|e_k\| \\
&\le \frac{([|1-ac| + a(1-b)]HR + [a(1-b) + a^2c + ac](HR)^2 + a^2c(HR)^3)\|e_k\|\|e_{k-1}\|}{R} \\
&= \frac{\dfrac{q(R) + 8(1-M_0R)^3}{8(1-M_0R)^3}\|e_k\|\|e_{k-1}\|}{R} = \frac{\|e_k\|\|e_{k-1}\|}{R} \le \|e_k\| < R. \qquad (3.337)
\end{aligned}$$

Therefore, (3.311) holds for $n = k$. By induction, the sequence $\{x_n\}$ generated by CSTM from any initial points $x_{-1}$ and $x_0$ in $U(x^\star, R)$ is well defined, remains in $U(x^\star, R)$ and the error estimates (3.311) are satisfied. Using the classical technique as the Secant method, we can deduce that sequence $\{x_n\}$ converges to $x^\star$ with order of 1.618... at least.

In what follows, we assume apart from conditions (a)–(e) are satisfied, relation (3.301) is also true. By (3.303), we can rewrite function $q$ as

$$\begin{aligned}
q(r) &= a(Mr)^3 + 4(Mr)^2(1-M_0r) + 4(1-a)Mr(1-M_0r)^2 - 8(1-M_0r)^3 \\
&= 8(1-M_0r)^3\left(a\left(\frac{Mr}{2(1-M_0r)}\right)^3 + 2\left(\frac{Mr}{2(1-M_0r)}\right)^2 + (1-a)\left(\frac{Mr}{2(1-M_0r)}\right) - 1\right) \\
&= 8(1-M_0r)^3\left(\frac{Mr}{2(1-M_0r)} + 1\right)\left(a\left(\frac{Mr}{2(1-M_0r)}\right)^2 + (2-a)\left(\frac{Mr}{2(1-M_0r)}\right) - 1\right) \\
&= 8a(1-M_0r)^3\left(\frac{Mr}{2(1-M_0r)} + 1\right)\left(\frac{Mr}{2(1-M_0r)} - \frac{\sqrt{4+a^2}+a-2}{2a}\right) \\
&\times \left(\frac{Mr}{2(1-M_0r)} + \frac{\sqrt{4+a^2}+a-2}{2a}\right). \qquad (3.338)
\end{aligned}$$

That is to say, the least positive zero $R$ can be solved by

$$\frac{MR}{2(1-M_0R)} - \frac{\sqrt{4+a^2}+a-2}{2a} = 0 \qquad (3.339)$$

and (3.312) is obtained at once. Returning to (3.337), we have from relations (3.300) and (3.301) that for $n \ge 1$

$$\begin{aligned}
\|e_{n+1}\| &\le (1-a)H^2\|e_{n-1}\|\|e_n\|^2 + (1-a)H\|e_n\|^2 \\
&+ aH^3\|e_{n-1}\|^2\|e_n\|^2 + aH^2\|e_{n-1}\|\|e_n\|^2 + H^2\|e_{n-1}\|^2\|e_n\| \\
&\le ((1-a)H^2R + (1-a)H + aH^3R^2 + aH^2R)\|e_n\|^2 + H^2\|e_n\|\|e_{n-1}\|^2 \\
&\le H^2\|e_n\|\|e_{n-1}\|^2 + (H^2R + (1-a)H + aH^3R^2)\|e_n\|^2. \qquad (3.340)
\end{aligned}$$

Using (3.339) and the definitions of $H$ and $\gamma$, we get

$$H^2 = \frac{\gamma}{R^2}. \qquad (3.341)$$

From the derivation of (3.338), we can deduce that

$$a(HR)^3 + 2(HR)^2 + (1-a)HR - 1 = 0. \tag{3.342}$$

In view of (3.341) and (3.342), we have

$$H^2 R + (1-a)H + aH^3 R^2 = \frac{1-\gamma}{R}. \tag{3.343}$$

This means (3.314) is true.

At last, we shall show $\{x_n\}$ converges to $x^\star$ with order at least two. In fact, to discover the order of convergence of CSTM, we postulate the following asymptotic relationship

$$\|e_{n+1}\| \sim C\|e_n\|^p \tag{3.344}$$

where $C$ is a positive constant. This means that the radio $\dfrac{|e_{n+1}|}{C|e_n|^p}$ tends to 1 as $n \to \infty$ and implies $p-order$ convergence. Hence, we get that

$$\|e_n\| \sim C\|e_{n-1}\|^p \quad and \quad \|e_{n-1}\| \sim (C^{-1}\|e_n\|)^{\frac{1}{p}}. \tag{3.345}$$

Clearly, we have $p \geq 1.618...$ as proved above. Suppose $p < 2$. We can rewrite (3.313) as

$$\frac{\|e_{n+1}\|}{\|e_n\|^p} \leq \frac{1}{R}\big(\frac{\gamma}{R}\|e_n\|^{1-p}\|e_{n-1}\|^2 + (1-\gamma)\|e_n\|^{2-p}\big) \quad (n \geq 1). \tag{3.346}$$

Then, we have

$$\frac{\|e_{n+1}\|}{\|e_n\|^p} \to C > 0 \quad as \quad n \to \infty \tag{3.347}$$

and

$$\begin{aligned} &\frac{1}{R}\big(\frac{\gamma}{R}\|e_n\|^{1-p}\|e_{n-1}\|^2 + (1-\gamma)\|e_n\|^{2-p}\big) \\ &\sim \frac{1}{R}\left(\frac{\gamma}{RC^2}\|e_n\|^{\frac{(2-p)(1+p)}{p}} + (1-\gamma)\|e_n\|^{2-p}\right) \\ &\to 0 \quad as \quad n \to \infty. \end{aligned} \tag{3.348}$$

In view of (3.347) and (3.348), we will deduce that $C \leq 0$ by letting $n \to \infty$ in both sides of (3.346), this is a contradiction since $C$ is a positive constant. Therefore, $p \geq 2$ is obtained. This completes the proof of the theorem. ⊠

**Remark 3.14** *(a) Estimate (3.304) is the radius of convergence for Newton's method obtained by Argyros in (cf. [139], [146]) that has enlarged the corresponding one given by Traub and Wozniakowski (cf. [710])*

$$R_1 = \frac{2}{3M} \tag{3.349}$$

*provided that* $M_0 < M$. *That is, we have* $R_1 < R_0$ *if* $M_0 < M$.

(b) *Note that CSTM is reduced to the Secant method SM if $a = 0$ and any choices of $b$ and $c$ satisfying (3.300). By Theorem 3.13 and the proof process of Lemma 3.11, we can see that estimate (3.304) is also the radius of convergence for the Secant method.*

(c) *Using Theorem 3.13, STM is a family of methods with quadratic convergence for any parameter $a \in (0,1]$. If we set $a = 1$ and $a = \frac{1}{2}$ in (3.302) in turn, we can obtain two useful formulas*

$$\begin{cases} y_n = x_n - A_n^{-1}\ F(x_n), \quad A_n = [x_{n-1}, x_n; F], \\ x_{n+1} = y_n - A_n^{-1}\ F(y_n)\ \ (n \geq 0,\ x_{-1}, x_0 \in \Omega) \end{cases} \tag{3.350}$$

*and*

$$\begin{cases} y_n = x_n - A_n^{-1}\ F(x_n), \quad A_n = [x_{n-1}, x_n; F], \\ x_{n+1} = x_n - 2A_n^{-1}\ F\left(\dfrac{x_n + y_n}{2}\right)\ \ (n \geq 0,\ x_{-1}, x_0 \in \Omega), \end{cases} \tag{3.351}$$

*respectively. Although the number of the evaluations of the function value increases by one at each step, the convergence order of (3.350) or (3.351) is raised from* 1.618... *to 2.*

(d) *Using estimate (3.312), we can deduced that among the members of family (3.302), method (3.350) has the biggest radius given by*

$$R = \frac{1}{\dfrac{\sqrt{5}+1}{4}M + M_0}, \tag{3.352}$$

*which shows method (3.350) allows a wider choice for the initial points than the other one in family (3.302).*

We provide now some examples to show the applications of our results.

**Example 3.8** *Let $X = Y = \mathbb{R}$, $\Omega = (-1,1)$ and define $F$ on $\Omega$ by*

$$F(x) = e^x - 1. \tag{3.353}$$

*Then, $x^\star = 0$ is a solution of Eq. (2.1) and $F'(x^\star) = 1$. Note that, for any $x, y, u, v \in \Omega$, we have*

$$|F'(x^\star)^{-1}([x,y;F] - [u,v;F])| = |\int_0^1 (F'(tx+(1-t)y) - F'(tu+(1-t)v))dt|$$

$$= \Big| \int_0^1 \int_0^1 (F''(\theta(tx+(1-t)y) + (1-\theta)$$

$$\times (tu+(1-t)v)\big)\big(tx+(1-t)y - (tu+(1-t)v)\big)d\theta dt\Big|$$

$$= \Big| \int_0^1 \int_0^1 (e^{\theta(tx+(1-t)y)+(1-\theta)(tu+(1-t)v)}\big(tx+(1-t)y - (tu+(1-t)v)\big)d\theta dt\Big|$$

$$\leq \int_0^1 e|t(x-u) + (1-t)(y-v)|dt \leq \frac{e}{2}(|x-u| + |y-v|) \tag{3.354}$$

*and*

$$|F'(x^\star)^{-1}([x,y;F]-[x^\star,x^\star;F])| = \left|\int_0^1 F'(tx+(1-t)y)dt - F'(x^\star)\right|$$

$$= \left|\int_0^1 (e^{tx+(1-t)y}-1)dt\right|$$

$$= \left|\int_0^1 (tx+(1-t)y)\left(1+\frac{tx+(1-t)y}{2!}+\frac{(tx+(1-t)y)^2}{3!}+\cdots\right)dt\right|$$

$$\leq \left|\int_0^1 (tx+(1-t)y)\left(1+\frac{1}{2!}+\frac{1}{3!}+\cdots\right)dt\right|$$

$$\leq \frac{e-1}{2}(|x-x^\star|+|y-x^\star|). \tag{3.355}$$

*That is to say, the Lipschitz condition (3.309) and the center-Lipschitz condition (3.310) are true for $M = e$ and $M_0 = e-1$, respectively. Using Theorem 3.13, we can deduce that the radius of convergence for method (3.350) is given by*

$$R = \frac{1}{\frac{\sqrt{5}+1}{4}M+M_0} = \frac{1}{\frac{\sqrt{5}+1}{4}e+e-1} = .255270179, \tag{3.356}$$

*which is bigger than the corresponding radius*

$$R' = \frac{1}{\frac{\sqrt{5}+1}{4}M+M} = \frac{1}{\frac{\sqrt{5}+1}{4}e+e} = .203358754 \tag{3.357}$$

*obtained by only using the Lipschitz condition (3.309).*

*Let us choose $x_{-1} = .199, x_0 = .2$. Suppose sequence $\{x_n\}$ is generated by method (3.350). Table 3.1 gives a comparison results of error estimates for Example 3.8, which shows that tighter error estimates can be obtained from (3.313) by using both the Lipschitz condition (3.309) and the center–Lipschitz condition (3.310) instead of by using only the Lipschitz condition (3.309).*

**Table 3.1: Comparison results of error estimates for Ex. 3.8**

| $n$ | the right-side of (3.313) by using both $M$ and $M_0$ | the right-side of (3.313) by using only $M$ |
|---|---|---|
| 0 | .142303821 | .193506148 |
| 1 | .097600776 | .122757871 |
| 2 | 2.52305E-05 | 3.16716E-05 |
| 3 | 2.25904E-09 | 2.83570E-09 |
| 4 | 1.55876E-20 | 1.95667E-20 |

**Example 3.9** *Let $X = Y = C[0,1]$, the space of continuous functions defined on $[0,1]$, equipped with the max–norm and $\Omega = \overline{U}(0,1)$. Define a function $F$ on $\Omega$ given by*

$$F(x)(s) = x(s) - 5\int_0^1 stx^3(t)dt \tag{3.358}$$

*and the divided difference of $F$ is defined by*

$$[x, y; F] = \int_0^1 F'(tx + (1-t)y)dt. \tag{3.359}$$

*Then, we have*

$$[F'(x)y](s) = y(s) - 15\int_0^1 stx^2(t)y(t)dt \ \ (y \in \Omega). \tag{3.360}$$

*We have $x^\star(s) = 0$ for all $s \in [0,1]$, $M_0 = 7.5$ and $M = 15$ (cf. [146]). Using Theorem 3.13, we can deduce that the radius of convergence for method (3.350) is given by*

$$R = \frac{1}{\frac{\sqrt{5}+1}{4}M + M_0} = \frac{1}{\frac{15(\sqrt{5}+1)}{4} + 7.5} = .050928802, \tag{3.361}$$

*which is bigger than the corresponding radius*

$$R' = \frac{1}{\frac{\sqrt{5}+1}{4}M + M} = \frac{1}{\frac{15(\sqrt{5}+1)}{4} + 15} = .036852427 \tag{3.362}$$

*obtained by only using the Lipschitz condition (3.309).*

**Example 3.10** *Let $X = Y = \mathbb{R}$, $\Omega = (-\infty, \infty)$ and define function $F$ on $D$ by*

$$F(x) = x^2 - 1. \tag{3.363}$$

*Then, $x^\star = 1$ is a solution of Eq. (2.1), $F$ is Fréchet- differentiable, $F'(x^\star) = 2$ and the divided difference $[x, y; F]$ can be written as*

$$[x, y; F] = x + y \quad (x, y \in \Omega). \tag{3.364}$$

*Hence, for any $x, y, u, v \in \Omega$, we have*

$$|F'(x^\star)^{-1}([x, y; F] - [u, v; F])| = \frac{1}{2}|x + y - (u + v)| \leq \frac{1}{2}(|x - u| + |y - v|), \tag{3.365}$$

*which shows the Lipschitz condition (3.309) and the center–Lipschitz condition (3.310) are true for $M = M_0 = \frac{1}{2}$. Therefore, we can see that all conditions of Theorem 3.13 are satisfied.*

*Let $x_{-1} = 1.5$ and $x_0 = 1.4$. Assume sequence $\{x_n\}$ is generated by CSTM (or STM) with different choices of parameters $a$, $b$ and $c$. Table 3.2 gives the comparison results for $|x_n - x^\star|$ for this example, which shows that STM (or CSTM satisfying both condition (3.300) and (3.301)) is faster than CSTM satisfying only condition (3.300). The computational order of convergence (COC) is also shown in this table. Here, COC is defined by (cf. [764])*

$$\rho \approx \frac{\ln |(x_{n+1} - x^\star)/(x_n - x^\star)|}{\ln |(x_n - x^\star)/(x_{n-1} - x^\star)|}. \tag{3.366}$$

*The last line of Table 3.2 shows that the computational order of convergence for CSTM satisfying only condition (3.300) is near to 1.618, meanwhile STM (or CSTM satisfying both condition (3.300) and (3.301)) has quadratic convergence.*

**Table 3.2: Comparison results of $|x_n - x^\star|$ for Ex. 3.10 using various methods**

| n | SM | CSTM $(a = b = 1, c = .5)$ | CSTM $(a = b = .5, c = 1)$ | STM $(a = 1)$ | STM $(a = .5)$ |
|---|---|---|---|---|---|
| 1 | 6.90e-02 | 4.44e-02 | 5.38e-02 | 1.98e-02 | 3.87e-02 |
| 2 | 1.12e-02 | 4.28c-03 | 5.39e-03 | 5.62e-04 | 1.34e-03 |
| 3 | 3.70e-04 | 4.74e-05 | 7.58e-05 | 5.53e-08 | 9.19e-07 |
| 4 | 2.06e-06 | 5.07e-08 | 1.03e-07 | 4.37e-15 | 6.22e-13 |
| 5 | 3.81e-10 | 6.02e-13 | 1.95e-12 | 3.34e-30 | 2.28e-25 |
| 6 | 3.92e-16 | 7.63e-21 | 5.00e-20 | 1.60e-59 | 3.51e-50 |
| 7 | 7.47e-26 | 1.15e-33 | 2.44e-32 | 4.47e-119 | 7.65e-100 |
| 8 | 1.46e-41 | 2.19e-54 | 3.05e-52 | 2.85e-237 | 3.81e-199 |
| 9 | 5.47e-67 | 6.28e-88 | 1.86e-84 | 1.42e-474 | 9.21e-398 |
| $\rho$ | 1.6188076 | 1.6188739 | 1.6186348 | 2.0077130 | 2.0001332 |

CHAPTER 4

# Steffensen's Method

Steffensen method StM is an alternative method of NM studied in chapter 2 and a variant method of SM presented in chapter 3. We establish in this chapter the convergence of StM under Lipschitz–type conditions.

## 4.1 Convergence

We provide in this section a semilocal convergence analysis of a general Steffensen's method in a Banach space setting. In some interesting special cases, we expand the applicability of this method.

We shall use Steffensen's method StM

$$\begin{aligned} x_{n+1} &= x_n - A_n^{-1} F(x_n) \quad (n \geq 0, \; x_0 \in \mathcal{D}), \\ A_n &:= A(x_n) = [x_n, g(x_n); F], \end{aligned} \tag{4.1}$$

to generate a sequence $\{x_n\}$ approximating a locally unique solution $x^\star$ of (2.1). Here, $g : \mathcal{X} \longrightarrow \mathcal{X}$ is a continuous operator, and $[x, y; F] \in \mathcal{L}(\mathcal{X})$ denotes a divided difference of order one of $F$ at points $x$, $y$, satisfying Definition 3.7. If $g(x) = x$ $(x \in \mathcal{D})$, StM reduces to Newton's method NM given by (2.2). The popular choice for $g$ is given by

$$g(x) = x + \lambda F(x) \quad (x \in \mathcal{D}, \; \lambda \in \mathbb{R}).$$

Other choices for $g$ can be found in (cf. [163], [194]). StM is a usefull alternative to NM in case where operator $F'(x)^{-1}$ $(x \in \mathcal{D})$ is not available or difficult or impossible to compute.

A semilocal convergence analysis for StM under different choices for $g$ and using different hypothese have been given by several authors (cf. [16], [596]). A survey of such results can be also found in (cf. [163], [194]) and the references there.

We need the following definition.

**Definition 4.1** *Let $\alpha \geq 0$, $\beta > 0$, $\eta > 0$, $M_0 \geq 0$, $M_1 > 0$ and*

$$M = \begin{cases} > M_0 & \text{if} \quad M_0 = 0, \\ \geq M_0 & \text{if} \quad M_0 \neq 0, \end{cases} \tag{4.2}$$

*be given constants.*

*Define*

$$L_0 = \gamma\,\eta + M_0, \quad \gamma = \frac{\alpha}{2}\,(2\,\alpha\,M_1 + M), \tag{4.3}$$

$$L = 2\,\alpha\,M_1 + M, \tag{4.4}$$

$$\eta_0 = \begin{cases} \dfrac{2}{M_0 + \sqrt{M_0^2 + 4\,\gamma}} & \text{if} \quad \alpha \neq 0, \\ \dfrac{1}{M_0} & \text{if} \quad \alpha = 0, \end{cases} \tag{4.5}$$

$$\eta_1 = \begin{cases} \dfrac{2\,(\alpha\,M_1 + M - M_0)}{\gamma} & \text{if} \quad \alpha \neq 0, \\ \text{Any positive number} & \text{if} \quad \alpha = 0, \end{cases} \tag{4.6}$$

*$\eta_2$ to be the minimal positive zero of polynomial:*

$$h_1(t) = 2\,\gamma^2\,t^4 + 4\,\gamma\,M_0\,t^3 + 2\,(M_0^2 - 2\,\gamma)\,t^2 - (4\,M_0 + L)\,t + 2 \tag{4.7}$$

*if it exists. Otherwise, let $\eta_2$ by any positive number,*

$$\eta_3 = \begin{cases} \dfrac{8}{L_0 + 4\,M_0 + \sqrt{(L + 4\,M_0)^2 + 64\gamma}} & \text{if} \quad \alpha \neq 0, \\ \dfrac{4}{L + 4\,M_0} & \text{if} \quad \alpha = 0, \end{cases} \tag{4.8}$$

$$\eta_4 = \min\,\{\eta_i,\ i = 0, \cdots, 3\}, \tag{4.9}$$

$$\eta_5 = \min\left\{\eta_2,\ \frac{1}{L + 4\,M_0}\right\}. \tag{4.10}$$

*It is simple algebra to show that*

*(i) if*

$$\eta < \eta_0, \tag{4.11}$$

*then*

$$h_2(t) = \gamma\,t^2 + M_0\,t < 1 \quad (t \in [0, \eta_0]); \tag{4.12}$$

*(ii) if*

$$\eta \leq \eta_1, \tag{4.13}$$

*then*

$$L_0 \leq L; \tag{4.14}$$

*(iii) if*

$$\eta \leq \eta_2, \tag{4.15}$$

*then*

$$h_1(t) \geq 0 \quad (t \in [0, \eta_2]) \tag{4.16}$$

*and*

$$q_A = \overline{L}\,\eta \leq \frac{1}{2}, \tag{4.17}$$

*where*

$$\overline{L} = \frac{1}{8}\,(L + 4\,L_0 + \sqrt{L^2 + 8\,L_0\,L}); \tag{4.18}$$

*(iv) if*

$$\eta < \eta_3, \tag{4.19}$$

*then*

$$h_3(t) = 4\,\gamma\,t^2 + (L + 4\,M_0)\,t - 4 < 0 \quad (t \in [0, \eta_3]). \tag{4.20}$$

*Items (i)–(iv) can be written in a condensed form provided that following conditions holds*

$$\eta \le \eta^\star \begin{cases} \begin{cases} \le \eta_4 & \text{if} \quad \alpha \neq 0,\ \eta_4 \neq \eta_3, \\ < \eta_4 & \text{if} \quad \alpha \neq 0,\ \eta_4 = \eta_3, \end{cases} \\ \begin{cases} \le \eta_5 & \text{if} \quad \alpha = 0,\ M_0 \neq 0,\ \eta_5 = \eta_2, \\ < \dfrac{1}{L + 4\,M_0} & \text{if} \quad \alpha = 0,\ M_0 \neq 0,\ \eta_5 = \dfrac{1}{L + 4\,M_0}, \end{cases} \\ \le \dfrac{2}{L} \quad \text{if} \ \alpha = M_0 = 0. \end{cases}$$

We shall show the following semilocal convergence theorem for StM.

**Theorem 4.1** *Let $F : \mathcal{D} \subseteq \mathcal{X} \longrightarrow \mathcal{X}$ be a Fréchet–differentiable operator, $g : \mathcal{D} \longrightarrow \mathcal{X}$ be a continuous operator, $[x, y; F]$ be a divided difference of order one of $F$ on $\mathcal{D}$ satisfying Definition 3.7 and let $A(x) \in \mathcal{L}(\mathcal{X})$ be given in (4.1). Assume that there exists $x_0 \in \mathcal{D}$ and constants $\alpha \ge 0$, $\beta > 0$, $\mu > 0$, $M_0 \ge 0$, $M$ satisfying (4.2), $M_1 > 0$ and $M_2 > 0$ such that, for all $x, y \in \mathcal{D}$*

$$\Gamma = F'(x_0)^{-1} \in \mathcal{L}(\mathcal{X}), \tag{4.21}$$

$$\| \Gamma\ F(x_0) \| \le \eta, \tag{4.22}$$

$$\| \Gamma\ (F'(x) - F'(y)) \| \le M\ \| x - y \|, \tag{4.23}$$

$$\| \Gamma\ (F'(x) - F'(x_0)) \| \le M_0\ \| x - x_0 \|, \tag{4.24}$$

$$\| \Gamma\ (A(x) - F'(x)) \| \le M_1\ \| x - g(x) \|, \tag{4.25}$$

$$\| \Gamma\ ([x, y; F] - F'(x_0)) \| \le M_2\ (\| x - x_0 \| + \| y - x_0 \|), \tag{4.26}$$

$$\| x - g(x) \| \le \alpha\ \| \Gamma\ F(x) \|, \tag{4.27}$$

$$\eta \le \eta^\star, \tag{4.28}$$

$$\overline{U}(x_0, r) \subseteq \mathcal{D}, \tag{4.29}$$

*where*

$$r = \alpha\,\eta + t^\star \tag{4.30}$$

*and $q_A$, $\eta^\star$, $t^\star$ are given in Definition 4.1, and Lemma 2.11. Then, sequence $\{x_n\}$ $(n \geq 0)$ generated by StM is well defined, remains in $\overline{U}(x_0, r)$ for all $n \geq 0$ and converges to a solution $x^\star$ of equation $F(x) = 0$ in $\overline{U}(x_0, r)$. Moreover, the following estimates hold for all $n \geq 0$*

$$\| x_n - x^\star \| \leq t^\star - t_n, \tag{4.31}$$

*where sequence $\{t_n\}$ $(n \geq 0)$ is given in Lemma 2.11. Furthermore, if*

$$t^\star \leq \frac{1}{2\ M_2}, \tag{4.32}$$

*then, the vector $x^\star$ is the only solution of equation $F(x) = 0$ in $U(x_0, R)$, where*

$$R \in \left[t^\star, \frac{1}{M_2} - t^\star\right]. \tag{4.33}$$

**Proof.** We shall show using induction on $n \geq 0$ the following assertions

$$\| x_{n+1} - x_n \| \leq t_{n+1} - t_n, \tag{4.34}$$

$$\overline{U}(x_{n+1}, t^\star - t_{n+1}) \subseteq \overline{U}(x_n, t^\star - t_n), \tag{4.35}$$

$$\| \Gamma\ F(x_n) \| \leq t_{n+1} - t_n \tag{4.36}$$

and

$$\| g(x_n) - x_0 \| \leq r. \tag{4.37}$$

For every $z \in \overline{U}(x_1, t^\star - t_1)$,

$$\| z - x_0 \| \leq \| z - x_1 \| + \| x_1 - x_0 \| \leq t^\star - t_1 + t_1 - t_0 = t^\star - t_0$$

implies $z \in \overline{U}(x_0, t^\star - t_0)$. We also have by (2.440), (4.22) that

$$\| x_1 - x_0 \| = \| \Gamma\ F(x_0) \| \leq \eta = t_1 - t_0.$$

In view of (4.27), and (4.36), we get

$$\| g(x_0) - x_0 \| \leq \alpha \| \Gamma F(x_0) \| \leq \alpha\eta \leq \alpha\eta + t^\star = r.$$

Hence, estimates (4.34)–(4.37) hold for $n = 0$. Let us assume these estimates hold for all integers $k \leq n$. Then, we have

$$\| x_{n+1} - x_0 \| \leq \sum_{i=1}^{n+1} \| x_i - x_{i-1} \| \leq \sum_{i=1}^{n+1} (t_i - t_{i-1}) = t_{n+1} - t_0 = t_{n+1} \leq t^\star,$$

$$\| x_n + \theta\,(x_{n+1} - x_n) - x_0 \| \leq t_n + \theta\,(t_{n+1} - t_n) \leq t^\star \quad (\theta \in (0, 1))$$

and

$$\begin{aligned} \| g(x_n) - x_0 \| &\leq \| g(x_n) - x_n \| + \| x_n - x_0 \| \\ &\leq \alpha\ \| \Gamma F(x_n) \| + t^\star \\ &\leq \alpha\ (t_{n+1} - t_n) + t^\star \leq \alpha\ \eta + t^\star = r. \end{aligned}$$

For simplicity, denote $x_{\theta,n} = x_n + \theta\,(x_{n+1} - x_n)$. In view of StM, we obtain the identity:

$$F(x_{n+1}) = (F'(x_n) - A_n)\,(x_{n+1} - x_n) + \int_0^1 (F'(x_{\theta,n}) - F'(x_n))\,(x_{n+1} - x_n)\,dt. \tag{4.38}$$

Then, using the induction hypotheses, (4.23)–(4.27), (4.38), we have in turn that

$$\begin{aligned}
&\| \Gamma\, F(x_{n+1}) \| \\
&= \left\| \Gamma \left( (F'(x_n) - A_n)\,(x_{n+1} - x_n) + \int_0^1 (F'(x_{\theta,n}) - F'(x_n))\,(x_{n+1} - x_n)\,dt \right) \right\| \\
&\leq \left\| \Gamma\,(F'(x_n) - A_n)\,(x_{n+1} - x_n) \right\| + \int_0^1 \| \Gamma\,(F'(x_{\theta,n}) - F'(x_n))\,(x_{n+1} - x_n) \| \; dt \\
&\leq M_1 \; \| x_n - g(x_n) \| \, \| x_{n+1} - x_n \| + \frac{\overline{M}}{2} \; \| x_{n+1} - x_n \|^2 \\
&\leq \alpha \; M_1 \; \| \Gamma\, F(x_n) \| \, \| x_{n+1} - x_n \| + \frac{\overline{M}}{2} \; \| x_{n+1} - x_n \|^2 \\
&\leq \left( \alpha\, M_1 + \frac{\overline{M}}{2} \right) (t_{n+1} - t_n)^2 = \frac{1}{2}\,(2\,\alpha\,M_1 + \overline{M})\,(t_{n+1} - t_n)^2 \\
&\leq \frac{1}{2}\,L_1\,(t_{n+1} - t_n)^2
\end{aligned} \tag{4.39}$$

where

$$\overline{M} = \begin{cases} M_0 & \text{if} \quad n = 0 \\ M & \text{if} \quad n > 0 \end{cases} \qquad \text{and} \quad L_1 = \begin{cases} L_0 & \text{if} \quad n = 0 \\ L & \text{if} \quad n > 0. \end{cases} \tag{4.40}$$

Moreover, in view of (4.5), (2.447), (4.24)–(4.28), and the induction hypotheses, we obtain that

$$\begin{aligned}
\| \Gamma\,(A_{n+1} - F'(x_0)) \| &= \| \Gamma\,((A_{n+1} - F'(x_{n+1})) + (F'(x_{n+1}) - F'(x_0))) \| \\
&\leq \| \Gamma\,(A_{n+1} - F'(x_{n+1})) \| + \| \Gamma\,(F'(x_{n+1}) - F'(x_0)) \| \\
&\leq M_1 \; \| g(x_{n+1}) - x_{n+1} \| + M_0 \; \| x_{n+1} - x_0 \| \\
&\leq \alpha\,M_1 \; \| \Gamma\,F(x_{n+1}) \| + M_0 \; \| x_{n+1} - x_0 \| \\
&\leq \frac{\alpha\,M_1\,L_1}{2}\,(t_{n+1} - t_n)^2 + M_0\,t_{n+1} \\
&\leq \left( \frac{\alpha\,M_1\,L_1}{2}\,(t_{n+1} - t_n) + M_0 \right) t_{n+1} \\
&\leq \left( \frac{\alpha\,M_1\,L_1}{2}\,\eta + M_0 \right) t_{n+1} \leq L_0\,t_{n+1} < 1.
\end{aligned} \tag{4.41}$$

It follows from (4.41) and the Banach lemma on invertible operators that $A_{n+1}^{-1}$ exists such that

$$\| A_{n+1}^{-1}\, F'(x_0) \| \leq (1 - L_0\, t_{n+1})^{-1}. \tag{4.42}$$

Using (4.1), we obtain the approximation

$$x_{n+2} - x_{n+1} = -A_{n+1}^{-1}\, F(x_{n+1}) = -(A_{n+1}^{-1}\, F'(x_0))\, (\Gamma\, F(x_{n+1}))$$

Using (4.39), and (4.42), we get that

$$\begin{aligned} \| x_{n+2} - x_{n+1} \| &\leq \| A_{n+1}^{-1}\, F'(x_0) \| \; \| \Gamma\, F(x_{n+1}) \| \\ &\leq \frac{L\, (t_{n+1} - t_n)^2}{2\, (1 - L_0\, t_{n+1})} = t_{n+2} - t_{n+1}, \end{aligned} \tag{4.43}$$

which completes the induction for (4.34). Thus, for every $z \in \overline{U}(x_{n+2}, t^\star - t_{n+2})$, we have

$$\begin{aligned} \| z - x_{n+1} \| &\leq \| z - x_{n+2} \| + \| x_{n+2} - x_{n+1} \| \\ &\leq t^\star - t_{n+2} + t_{n+2} - t_{n+1} = t^\star - t_{n+1}, \end{aligned}$$

which implies $z \in \overline{U}(x_{n+1}, t^\star - t_{n+1})$. That is (4.35) holds for $n+1$ replacing $n$.

Furthermore, it follows from (2.447) and (4.43) that

$$\| \Gamma\;F(x_{n+1}) \| \leq (1 - L\, t_{n+1})\, (t_{n+2} - t_{n+1}) \leq t_{n+2} - t_{n+1}, \tag{4.44}$$

which completes the induction for (4.36). We also have that

$$\begin{aligned} \| g(x_{n+1}) - x_0 \| &\leq \| g(x_{n+1}) - x_{n+1} \| + \| x_{n+1} - x_0 \| \\ &\leq \alpha\, (t_{n+2} - t_{n+1}) + r \\ &\leq \alpha\, (t_1 - t_0) + t^\star = \alpha\, \eta + t^\star = r. \end{aligned}$$

Hence, the induction for (4.34)–(4.37) is completed.

It follows from Lemma 2.11 that sequence $\{t_n\}$ is a Cauchy sequence. In view of (4.34) and (4.35), sequence $\{x_n\}$ $(n \geq 0)$ is a Cauchy sequence too in a Banach space $\mathcal{X}$ and as such it converges to some $x^\star \in \overline{U}(x_0, r)$ (since $\overline{U}(x_0, r)$ is a closed set). By letting $n \longrightarrow \infty$ in (4.39), we obtain $F(x^\star) = 0$. Estimate (4.31) follows from (4.34) by using standard majorization techniques.

To show the uniqueness part, let $y^\star \in U(x_0, R)$ be a solution of equation $F(x) = 0$. Using (4.26), (4.32) and (4.33), we get that

$$\| \Gamma\, ([x^\star, y^\star; F] - F'(x_0)) \| \leq M_2\, (\| x^\star - x_0 \| + \| y^\star - x_0 \|) < M_2\, (t^\star + R) \leq 1. \tag{4.45}$$

It follows by (4.45), and the Banach lemma on invertible operators that linear operator $[x^\star, y^\star; F]$ is invertible. Then, using the estimate $0 = F(x^\star) - F(y^\star) = [x^\star, y^\star; F]\, (x^\star - y^\star)$, we deduce $x^\star = y^\star$. This completes the proof of Theorem 4.1. ⊠

**Remark 4.1** *(a) The number $t^\star$ can be replaced by $t^{\star\star}$ (given in closed form in (2.445)) in (4.29), (4.30), (4.32) and (4.33).*

*(b) Conditions (4.23), (4.24) and (4.26) certainly hold if replaced by the stronger*

$$\| \Gamma ([x, y; F] - F'(z)) \| \leq M_3 (\| x - z \| + \| y - z \|) \tag{4.46}$$

*for all $x, y, z \in \mathcal{D}$ and some $M_3 > 0$. In this case, we can set*

$$M = 2\, M_3 \quad \text{and} \quad M_0 = 2\, M_2 \quad \text{or} \quad M_0 = 2\, M_3.$$

*Note also that $M_2 \leq M_3$.*

*(c) If $g(x) = x$ ($x \in \mathcal{D}$), then, StM reduces to NM. Then, we can set*

$$\alpha = M_1 = 0, \quad L_0 = M_0, \quad M_2 = \frac{M_0}{2} \quad \text{and} \quad L = M.$$

*Then, it is simple algebra to show that the sufficient convergence condition (4.28) reduces to (2.443). The corresponding to (2.443) Newton–Kantorovich hypothesis (cf. [163], [194]) is given by*

$$q_K = L\, \eta \leq \frac{1}{2}. \tag{4.47}$$

*Note that, in general,*

$$M_0 \leq M \quad (\text{i.e., } L_0 \leq L) \tag{4.48}$$

*holds and $\dfrac{M}{M_0}$ can be arbitrarily large. Note also that*

$$q_K \leq \frac{1}{2} \Longrightarrow q_A \leq \frac{1}{2}, \tag{4.49}$$

*but not necessarily vice verca unless if $M_0 = M$. In (cf. [139], [194]), we have provided numerical examples, where (2.443) holds but (4.47) is violated and $L_0 < L$.*

*The Newton–Kantorovich (cf. [163], [194]) majorizing sequence corresponding to (2.440) is*

$$s_0 = 0, \quad s_1 = \eta, \quad s_{n+2} = s_{n+1} + \frac{M\, (s_{n+1} - s_n)^2}{2\, (1 - M\, s_{n+1})} \quad (n \geq 0). \tag{4.50}$$

*A simple inductive argument shows that under (4.47) and if $L_0 < L$, we get that*

$$t_n < s_n \qquad (n > 1), \tag{4.51}$$

$$t_{n+1} - t_n < s_{n+1} - s_n \qquad (n \geq 1), \tag{4.52}$$

$$t^\star - t_n < s^\star - s_n \qquad (n \geq 1) \tag{4.53}$$

*and*

$$t^\star < s^\star, \tag{4.54}$$

*where*

$$s^\star = \lim_{n\to\infty} s_n.$$

*Note also that estimates (2.449), and (2.450) are also finer, since* $q_A < q_K$. *Hence, under the same computational cost, the applicability of NM is expanded under our approach. Note that, in the general case of StM (i.e.,* $g(x) \neq x$, $x \in \mathcal{D}$), *we do not have a related Kantorovich–type theorem to compare our results. The existing sufficient convergence conditions for StM involve stronger hypotheses than ours (cf. [16], [273], [596]).*

(d) *In view of the proof of Theorem 4.1, it follows (see (4.39) and (4.40)) that scalar sequence* $\{r_n\}$ *given by*

$$r_0 = 0, \quad r_1 = \eta, \quad r_2 = r_1 + \frac{L_0\,(r_1 - r_1)^2}{2\,(1 - L_0\,r_1)}$$

$$r_{n+2} = r_{n+1} + \frac{L\,(r_{n+1} - r_n)^2}{2\,(1 - L_0\,r_{n+1})} \qquad (n \geq 1)$$

*is also majorizing sequence for* $\{x_n\}$. *Moreover, for* $L_0 < L$, *we have that*

$$r_n < t_n \qquad (n > 1),$$

$$r_{n+1} - r_n < t_{n+1} - t_n \qquad (n \geq 1),$$

$$r^\star - r_n < t^\star - t_n \qquad (n \geq 1)$$

*and*

$$r^\star \leq t^\star,$$

*where*

$$r^\star = \lim_{n\to\infty} r_n.$$

(e) *If* $\mathcal{X} = \mathbb{R}^k$, *then, divided difference* $[x, y; F]$ *can be given by* $(n \geq 0)$

$$\begin{aligned} J_n :&= [x_n, g(x_n); F] \\ &= (F(x_n + D_n\,e^1) - F(x_n), \cdots, F(x_n + D_n\,e^n) - F(x_n))\; D_n^{-1}, \end{aligned}$$

*where*

$$D_n = diag\,(f_1(x_n), f_2(x_n), \cdots, f_k(x_n)),$$

$$F(x) = (f_1(x), f_2(x), \cdots, f_k(x)),$$

$$f_i : \mathbb{R}^k \longrightarrow \mathbb{R}, \; (i = 1 \cdots k) \quad \text{and} \quad x = (x_1, x_2, \cdots, x_k).$$

*StM in the form*

$$x_{n+1} = x_n - J_n^{-1}\, F(x_n) \qquad (n \geq 0)$$

*avoids the evaluation of $F'(x_n)^{-1}$ which may be too expensive or impossible to invert, but has the same convergence order with NM.*

At the end of this section, we provide a numerical example, where we show that StM can be faster than NM.

**Example 4.1** *We consider the non–linear second order boundary value problem:*

$$\frac{d^2 w(t)}{d\,t^2} = e^{w(t)}, \quad w(0) = w(1) = 0. \tag{4.55}$$

*As in (cf. [39]), we approximate the second derivative by*

$$w'' \approx \frac{w_{i-1} - 2\,w_i + w_{i+1}}{h^2} \qquad (i = 1, \cdots, k-1),$$

$$q_i = i\,h, \quad h = \frac{1}{k}. \tag{4.56}$$

*In view of (4.56), equation (4.55) becomes:*

$$F(w) = A\,w - h^2\,p(w), \tag{4.57}$$

*where $F : \mathbb{R}^{k-1} \longrightarrow \mathbb{R}^{k-1}$,*

$$A = \begin{pmatrix} -2 & 1 & 0 & \cdots & 0 & 0 \\ 1 & -2 & 1 & \cdots & 0 & 0 \\ 0 & 1 & -2 & 1 & 0 & \cdots \\ \vdots & \vdots & \vdots & \vdots & \vdots & \vdots \\ 0 & \cdots & \cdots & \cdots & 1 & -2 \end{pmatrix},$$

$$w = (w_1, w_2, \cdots, w_{k-1})^T$$

*and*

$$p(x) = (e^{w_1}, e^{w_2}, \cdots, e^{w_{k-1}})^T.$$

*Let us solve equation $F(w) = 0$, where, $F$ is given by (4.57) using StM and NM (2.2) for comparison reasons. Let us choose $k = 12$ and initial iterate $w_i(0) = q_i\,(q_i - 1)$, $i = 1, \cdots, 11$. Then, we get the Table*

**Comparison Table between StM and NM on** $\| F(w^{(k)}) \|$

| n | StM : (4.1) | NM: (2.2) |
|---|---|---|
| 1 | 1.4448e-009 | 1.9887e-009 |
| 2 | 1.9664e-017 | 4.5194e-017 |

*The table justifies the claim made in Remark 4.1.*

**Remark 4.2** *(a) Under the hypotheses of Theorem 4.1, we have*

$$M_0 \leq L_0 \quad \text{and} \quad M \leq L \tag{4.58}$$

*(by (4.3), and (4.4)). It follows from (4.17), (2.443) and (4.28) that NM defined by (2.2) converges. Moreover, the predicted estimates by (2.449), and (2.450) show that NM is faster than StM.*

*(b) We can justify the results of the Comparison Table on* $\| F(x_n) \|$ *i.e., on*

*(* $\| F'(x_0)^{-1} F(x_n) \|$*). In view of (4.39), (4.44), we have*

$$\| \Gamma F(x_n) \| \leq \left( M \; \| x_n - g(x_n) \| + \frac{\overline{M}}{2} \; \| x_{n+1} - x_n \| \right) \; \| x_{n+1} - x_n \| \tag{4.59}$$

*and*

$$\| \Gamma F(x_{n+1}) \| \leq (1 - L_0 \; t_{n+1}) \; (t_{n+2} - t_{n+1}), \tag{4.60}$$

*respectively. Moreover, for (2.2), we have*

$$\| F'(y_0)^{-1} F(y_{n+1}) \| \leq \frac{M}{2} \; \| y_{n+1} - y_n \|^2, \tag{4.61}$$

*and*

$$\| F'(y_0)^{-1} F(y_{n+1}) \| \leq (1 - M \; s_{n+1}) \; (s_{n+2} - s_{n+1}). \tag{4.62}$$

*It follows from (4.60) and (4.62) that, if e.g.* $M < L_0$, $t_n \approx s_n$ $(n \geq 0)$*, at least theoretically predicted upper bounds on* $\| \Gamma F(x_{n+1}) \|$ *can be smaller than* $\| \Gamma F(y_{n+1}) \|$*. This observation explains at least theoretically the results of the Comparison Table.*

CHAPTER 5

# Gauss–Newton Method

Gauss–Newton method is also an alternative method for Newton's method. We study in this chapter the convergence of Gauss–Newton method under Lipschitz and average Lipschitz–type conditions.

## 5.1 Convergence

We establish in this section a new semilocal convergence analysis of the Gauss–Newton method GNM for solving nonlinear equation in the Euclidean space. Using our new idea of recurrent functions and a combination of center–Lipschitz, Lipschitz conditions, we provide under the same or weaker hypotheses than before (cf. [226], [469]), a tighter convergence analysis. The results can be extented in case outer or generalized inverses are used.

We are concerned with the problem of finding $x^\star \in \mathbb{R}^i$, minimizing the objective function given by

$$G(x) := \frac{1}{2} \parallel F(x) \parallel^2 = \frac{1}{2} F(x)^T F(x), \tag{5.1}$$

where $\parallel . \parallel$ denotes the Euclidean norm and $F$ is a Fréchet–differentiable function, defined on a convex subset $\mathcal{D}$ of $\mathbb{R}^i$ with value in $\mathbb{R}^j$ $(i \leq j)$. Many problems in applied mathematics and also in engineering are solved by finding such solutions $x^\star$ (cf. [95], [586]).

Except in special cases, the most commonly used solution methods are iterative, when starting from one or several initial approximations a sequence is constructed that converges to the solution of the equation. Iteration methods are also used for solving optimization problems like (5.1).

Iteration sequences converge to an optimal solution of the problem at hand. In particular, here for $x^\star$ to be a local minimum it is necessary to be a zero of the gradien $\nabla G$ of $G$, too:

$$\nabla G(x^\star) = \mathcal{J}^T(x^\star)\, F(x^\star) = 0, \tag{5.2}$$

with

$$\mathcal{J}(x) = F'(x) \qquad (x \in \mathcal{D}). \tag{5.3}$$

The iterative method for computing such zero is so–called Gauss–Newton method GNM, as introduced by Ben–Israel (cf. [226])

$$x_{n+1} = x_n - \mathcal{J}^+(x_n)\, F(x_n) \quad (n \geq 0, \quad x_0 \in \mathcal{D}), \tag{5.4}$$

where $\mathcal{J}^+$ denotes the well known Moore–Penrose–pseudoinverse of $\mathcal{J}$ (cf. [163]) (see also Definition 5.1). There is an extensive literature on convergence results for the GNM. We refer the reader to (cf. [95], [586] and the reference there). In particular, Häußler (cf. [412]) provided a Kantorovich–type semilocal convergence analysis for GNM.

Using the center–Lipschitz conditions (instead of Lipschitz conditions used in (cf. [412])) to find more precise upper bounds on the inverses of the mappings involved and our new idea of recurrent functions, we provide a analysis for GNM with the following advantages (under the same or weaker computational cost and hypotheses)

(a) finer estimates on the distances $\| x_{n+1} - x_n \|$, $\| x_n - x^\star \|$ $(n \geq 0)$;

(b) an at least as precise information on the distances involved.

Numerical examples are provided to show that our results apply, where the corresponding ones in (cf. [226], [469]) do not.

We need the following definition:

**Definition 5.1** *$\mathcal{M}^+$ is the Moore–Penrose–pseudoinverse of matrix $\mathcal{M}$ if the following four axioms hold*

$$(\mathcal{M}^+ \mathcal{M})^T = \mathcal{M}^+ \mathcal{M},$$

$$(\mathcal{M} \mathcal{M}^+)^T = \mathcal{M} \mathcal{M}^+,$$

$$\mathcal{M}^+ \mathcal{M} \mathcal{M}^+ = \mathcal{M}^+$$

*and*

$$\mathcal{M} \mathcal{M}^+ \mathcal{M} = \mathcal{M}.$$

*In the case of a full rank $(m, n)$ matrix $\mathcal{M}$ with rank rank $\mathcal{M} = n$, the pseudo–inverse is given by*

$$\mathcal{M}^+ = (\mathcal{M}^T \mathcal{M})^{-1} \mathcal{M}^T.$$

We need also the following result on majorizing sequences for GNM.

**Lemma 5.1** *Let $\beta > 0$, $\gamma_0 > 0$, $\gamma > 0$, with $\gamma_0 \leq \gamma$, and $\eta \in [0, 1)$ be given. Let*

$$\delta_0 = \frac{\gamma\, \beta + 2\, \eta}{1 - \gamma_0\, \beta}, \tag{5.5}$$

$$\alpha = \frac{-\gamma + \sqrt{\gamma^2 + 8\, \gamma_0\, \gamma}}{4\, \gamma_0}, \tag{5.6}$$

$$\beta^\star = \min \left\{ \frac{\alpha\,(\eta + 1) - \eta - \alpha^2}{\gamma_0}, \frac{2\,(\alpha - \eta)}{\gamma + 2\, \alpha\, \gamma_0} \right\}. \tag{5.7}$$

*Assume that the following hold*

$$\eta < \alpha \quad \text{and} \quad \beta \leq \beta^\star. \tag{5.8}$$

*Then, scalar sequence $\{t_n\}$ $(n \geq 0)$ generated by*

$$t_0 = 0,\ t_1 = \beta,\ t_{n+2} = t_{n+1} + \frac{\gamma\,(t_{n+1} - t_n) + 2\eta}{2\,(1 - \gamma_0\, t_{n+1})}\,(t_{n+1} - t_n) \tag{5.9}$$

*is increasing, bounded from above by*

$$t^{\star\star} = \frac{\eta}{1 - \alpha} \tag{5.10}$$

*and converges to its unique least upper bound $t^\star$ such that*

$$t^\star \in [0, t^{\star\star}]. \tag{5.11}$$

*Moreover, the following estimates hold for all $n \geq 0$*

$$0 \leq t_{n+2} - t_{n+1} \leq \alpha\,(t_{n+1} - t_n) \leq \alpha^{n+1}\,\beta \tag{5.12}$$

*and*

$$t^\star - t_n \leq \frac{\beta\,\alpha^n}{1 - \alpha}. \tag{5.13}$$

**Proof** We note that $\beta^\star > 0$ by the choice of $\eta$ and (5.8). Moreover, we have also $\delta_0 \leq 2\,\alpha$ by (5.8). We shall show using induction on the integer $m$ the following

$$0 < t_{m+2} - t_{m+1} = \frac{\gamma\,(t_{m+1} - t_m) + 2\eta}{2\,(1 - \gamma_0\, t_{m+1})}\,(t_{m+1} - t_m) \leq \alpha\,(t_{m+1} - t_m) \tag{5.14}$$

and

$$\gamma_0\, t_{m+1} < 1. \tag{5.15}$$

If (5.14) and (5.15) hold, then, we have (5.12) holds and

$$\begin{aligned} t_{m+2} &\leq t_{m+1} + \alpha\,(t_{m+1} - t_m) \\ &\leq t_m + \alpha\,(t_m - t_{m-1}) + \alpha\,(t_{m+1} - t_m) \\ &\leq \eta + \alpha\,\beta + \cdots + \alpha^{m+1}\,\beta \\ &= \frac{1 - \alpha^{m+2}}{1 - \alpha}\,\beta < \frac{\eta}{1 - \alpha} = t^{\star\star} \qquad (\text{by } (5.10)). \end{aligned} \tag{5.16}$$

Estimates (5.14) and (5.15) hold for $m = 0$. Indeed, by the initial conditions, (5.8), the choices of $\alpha$ and $\delta_0$, we get that

$$\frac{\gamma\,(t_1 - t_0) + 2\,\eta}{1 - \gamma_0\, t_1} = \frac{\gamma\,\beta + 2\,\eta}{1 - \gamma_0\,\beta} = \delta_0 \leq 2\,\alpha,$$

$$\gamma_0\, t_1 = \gamma_0\,\beta < 1.$$

Let us assume (5.12), (5.14) and (5.15) hold for all $m \leq n+1$. Estimate (5.14) can be re–written as:

$$\gamma\,(t_{m+1} - t_m) + 2\,\eta + \gamma_0\,\delta\, t_{m+1} - 2\,\alpha \leq 0$$

or

$$\gamma\,\alpha^m\,\beta + 2\,\gamma_0\,\alpha\,\frac{1-\alpha^{m+1}}{1-\alpha}\,\beta + 2\,\eta - 2\,\alpha \le 0. \tag{5.17}$$

Estimate (5.17) motivates us to introduce functions $f_m$ on $[0,+\infty)$ $(m \ge 1)$ for $s = \alpha$ by

$$f_m(s) \;=\; \gamma\,s^m\,\beta + 2\,\gamma_0\,s\,(1+s+s^2+\cdots+s^m)\,\beta - 2\,s + 2\,\eta. \tag{5.18}$$

Estimate (5.17) certainly holds if:

$$f_m(\alpha) \le 0 \quad (m \ge 1). \tag{5.19}$$

We need to find a relationship between two consecutive polynomials $f_m$. We have that

$$\begin{aligned} f_{m+1}(s) &= \gamma\,s^{m+1}\,\beta + 2\,\gamma_0\,s\,(1+s+s^2+\cdots+s^m+s^{m+1})\,\beta - 2\,s + 2\eta \\ &= \gamma\,s^m\,\beta - \gamma\,s^m\,\beta + \gamma\,s^{m+1}\,\beta \\ &+ 2\,\gamma_0\,s\,(1+s+s^2+\cdots+s^m)\,\beta + 2\,\gamma_0\,s^{m+2}\beta - 2\,s + 2\eta \\ &= f_m(s) + g(s)\,\beta\,s^m, \end{aligned} \tag{5.20}$$

where

$$g(s) = 2\,\gamma_0\,s^2 + \gamma\,s - \gamma. \tag{5.21}$$

Note that function $g$ has a unique positive root $\alpha$ given by (5.6). In view of (5.20) and (5.21), we have

$$f_m(\alpha) = f_1(\alpha) \qquad (m \ge 1). \tag{5.22}$$

Moreover, define

$$f_\infty(\alpha) = \lim_{m\to\infty} f_m(\alpha) \quad (s \in [0,1),\ m \ge 1). \tag{5.23}$$

Then, we have by (5.22) that

$$f_\infty(\alpha) = f_m(\alpha) \qquad (m \ge 1). \tag{5.24}$$

In view of (5.24), we can show, instead of (5.19), since

$$f_\infty(\alpha) = 2\left(\frac{\gamma_0\,\beta\,\alpha}{1-\alpha} + \eta - \alpha\right)$$

that

$$f_\infty(\alpha) \le 0,$$

which is true by (5.8). That completes the induction.

Estimate (5.13) follows from (5.12) by using standard majorization techniques. Finally, sequence $\{t_n\}$ is non–decreasing, bounded from above by $t^{\star\star}$, and as such it converges to its unique least upper bound $t^\star$. This completes the proof of Lemma 5.1. ⊠

We need the following standard perturbation lemma (cf. [163], [412], [586]).

**Lemma 5.2** *Let $A$ and $B$ be $(m \times n)$ matrices. Assume that*

$$rank\,(A) \leq rank\,(B) = r \leq i \quad (r \geq 1) \tag{5.25}$$

*and*

$$\| A - B \| \| B^+ \| < 1. \tag{5.26}$$

*Then, the following hold*

$$rank\,(A) = r \tag{5.27}$$

*and*

$$\| A^+ \| \leq \frac{\| B^+ \|}{1 - \| B^+ \| \| A - B \|}. \tag{5.28}$$

We can show the semilocal convergence result for GNM.

**Theorem 5.1** *Let $F \in mathcalC^1(\mathcal{D}_0)$, $\mathcal{D}_0 \subseteq \mathcal{D} \subseteq \mathbb{R}^i$ and $\mathcal{D}_0$ be a convex set. Assume there exist $x_0 \in \mathcal{D}_0$ and constants $\beta > 0$, $\beta_0 > 0$ $K > 0$, $K_0 > 0$ and $\eta : \mathcal{D}_0 \longrightarrow \mathbb{R}^+$ such that, for all $x, y \in \mathcal{D}_0$*

$$rank\,(\mathcal{J}(x_0)) = r \leq i \quad (r \geq 1), \tag{5.29}$$

$$rank\,(\mathcal{J}(x)) \leq r, \tag{5.30}$$

$$\| \mathcal{J}^+(x_0)\; F(x_0) \| \leq \beta, \tag{5.31}$$

$$\| \mathcal{J}(x) - \mathcal{J}(y) \| \leq K \;\; \| x - y \|, \tag{5.32}$$

$$\| \mathcal{J}(x) - \mathcal{J}(x_0) \| \leq K_0 \;\; \| x - x_0 \|, \tag{5.33}$$

$$\| \mathcal{J}^+(x_0) \| \leq \beta_0, \tag{5.34}$$

$$\| \mathcal{J}^+(y)\; r(x) \| \leq \eta(x) \;\; \| x - y \| \tag{5.35}$$

*with*

$$r(x) = (\mathcal{I} - \mathcal{J}(x)\; \mathcal{J}^+(x))\; F(x), \tag{5.36}$$

$$\eta(x) \leq \eta < 1, \tag{5.37}$$

$$\overline{U}(x_0, t^\star) \subseteq \mathcal{D}_0, \tag{5.38}$$

*where $t^\star$ is given in (5.11) and hypotheses of Lemma 5.1 hold for*

$$\gamma_0 = \beta_0\; K_0 \quad \text{and} \quad \gamma = \beta_0\; K. \tag{5.39}$$

*Then, the following hold*

$$rank\,(\mathcal{J}(x)) = r \qquad (x \in U(x_0, t^\star)); \tag{5.40}$$

*Sequence $\{x_n\}$ $(n \geq 0)$ generated by GNM is well defined, remains in $\overline{U}(x_0, t^\star)$ for all $n \geq 0$ and converges to a zero $x^\star$ of $\mathcal{J}^+(x)\; F(x)$ in $\overline{U}(x_0, t^\star)$;*

$$\| x_{n+1} - x_n \| \leq t_{n+1} - t_n \tag{5.41}$$

*and*

$$\| x_n - x^\star \| \le t^\star - t_n, \tag{5.42}$$

*where sequence $\{t_n\}$ is given in Lemma 5.1. Moreover, the following equality holds*

$$rank\,(\mathcal{J}(x^\star)) = r \tag{5.43}$$

*and, if $rank\,(\mathcal{J}(x_0)) = i$ and $F(x^\star) = 0$, then, $x^\star$ is unique in $U(x_0, t^{\star\star})$ and also $x^\star$ is the unique zero of $\mathcal{J}^+(x)\ F(x)$ in $U(x_0, t^\star)$ too.*

**Proof** By hypothesis $x_1 \in \overline{U}(x_0, t^\star)$ since $\| x_1 - x_0 \| \le \beta \le t^\star$. Then (5.41) holds for $n = 0$. Assume $x_m \in \overline{U}(x_0, t^\star)$ and (5.41) holds for $m \le n$. Using (5.33) and (5.15), we get that

$$\begin{aligned} \| \mathcal{J}(x_m) - \mathcal{J}(x_0) \| &\le K_0\ \| x_m - x_0 \| \\ &\le K_0\ (t_m - t_0) = K_0\ t_m < \frac{1}{\beta_0}. \end{aligned} \tag{5.44}$$

It follows from (5.44) and Lemma 5.2 that (5.40), (5.43) and

$$\| \mathcal{J}^+(x_m) \| \le \frac{\beta_0}{1 - \beta_0\ K_0\ \| x_m - x_0 \|} \le \frac{\beta_0}{1 - \gamma_0\ t_m} \tag{5.45}$$

hold. Using (5.4), (5.9), (5.32), (5.35)–(5.39), (5.45) and the induction hypotheses, we obtain in turn that

$$\begin{aligned} \| x_{m+1} - x_m \| =& \| \mathcal{J}^+(x_m) \int_0^1 (\mathcal{J}(x_{m-1} + \theta\ (x_m - x_{m-1})) - \mathcal{J}(x_{m-1}))\ (x_m - x_{m-1})\ d\theta \\ &+ \mathcal{J}^+(x_m)\ (\mathcal{I} - \mathcal{J}(x_{m-1})\ \mathcal{J}^+(x_{m-1}))\ F(x_{m-1}) \| \\ \le& \frac{1}{1 - \gamma_0\ t_m} \left( \frac{1}{2}\ \gamma\ \| x_m - x_{m-1} \| + \eta \right)\ \| x_m - x_{m-1} \| \\ \le& \frac{1}{2\ (1 - \gamma_0\ t_m)}\ (\gamma\ (t_m - t_{m-1}) + \eta)\ (t_m - t_{m-1}) = t_{m+1} - t_m, \end{aligned} \tag{5.46}$$

which completes the induction for (5.41). Note also that (5.41) implies that

$$\| x_{k+1} - x_0 \| \le t_{k+1} \quad (k = 1, \cdots, m+1).$$

That is, $x_{m+1} \in \overline{U}(x_0, t^\star)$. In view of Lemma 5.1, sequence $\{x_n\}$ is a Cauchy sequence in $\mathbb{R}^i$, and as such it converges to some $x^\star \in \overline{U}(x_0, t^\star)$ (since $\overline{U}(x_0, t^\star)$ is a closed set).

We claim that $x^\star$ is a zero of $\mathcal{J}^+(x)\ F(x)$. Indeed, we get that

$$\begin{aligned} \| \mathcal{J}^+(x^\star)\ F(x_m) \| \le& \| \mathcal{J}^+(x^\star)\ (\mathcal{I} - \mathcal{J}(x_m)\ \mathcal{J}^+(x_m))\ F(x_m) \| \\ &+ \| \mathcal{J}^+(x^\star) \|\ \| \mathcal{J}(x_m)\ \mathcal{J}^+(x_m)\ F(x_m) \| \\ &\le \eta\ \| x_m - x^\star \| + \| \mathcal{J}^+(x^\star) \|\ \| \mathcal{J}(x_m) \|\ \| x_{m+1} - x_m \|. \end{aligned} \tag{5.47}$$

By using (5.47) and the continuity of mapping $\mathcal{J}(x)$, $F(x)$, we justify the claim.

Finally, estimate (5.42) follows from (5.41) by using standard majorization techniques. The uniqueness part as identical to (cf. [412, Lemma 2.9, p. 122]) is omitted. The proof of Theorem 5.1 is complete. ⊠

We can now state Häu$\beta$ler's result for comparison purposes:

**Theorem 5.2** *(cf. [412]) Under hypotheses (5.29)–(5.37) (excluding (5.33)), further assume that*

$$h_H = \beta\,\gamma \leq \frac{1}{2}\,(1-\eta)^2 \tag{5.48}$$

*and*

$$\overline{U}(x_0, v^\star) \subseteq \mathcal{D}_0, \tag{5.49}$$

*where*

$$v^\star = \lim_{n \longrightarrow \infty} v_n, \tag{5.50}$$

$$v_0 = 0,\ v_1 = \beta,\ v_{n+2} = v_{n+1} + \frac{\gamma\,(v_{n+1} - v_n) + 2\eta}{2\,(1 - \gamma\,v_{n+1})}\,(v_{n+1} - v_n). \tag{5.51}$$

*Then, the conclusions of Theorem 5.1 hold with $v^\star$, $\{v_n\}$ replacing $t^\star$, $\{t_n\}$ $(n \geq 0)$, respectively.*

**Remark 5.1** *Note that in general*

$$\gamma_0 \leq \gamma \tag{5.52}$$

*holds in general and $\dfrac{\gamma}{\gamma_0}$ can be arbitrarily large.*

Using induction on integer, we can easily show that

**Proposition 5.1** *Under only hypotheses of Theorem 5.2 or Theorems 5.1 and 5.2, the following hold for all $n \geq 0$*

$$\|\ x_{n+1} - x_n\ \| \leq t_{n+1} - t_n \leq v_{n+1} - v_n, \tag{5.53}$$

$$t_n \leq v_n \quad (n \geq 2) \tag{5.54}$$

*and*

$$\|\ x_n - x^\star\ \| \leq t^\star - t_n \leq v^\star - v_n. \tag{5.55}$$

*Note also that, if $\gamma_0 < \gamma$, then, strict inequality holds in (5.53) and (5.54) for $n \geq 2$.*

**Remark 5.2** *By Proposition 5.1, the error estimates of Theorem 5.1 can certainly be improved under the same computational cost, since in practice, the computation of $\gamma$ requires that of $\gamma_0$.*

In the next section, we shall show that

(a) conditions of Lemma 5.1 are always weaker than (5.48), when $\gamma_0 < \gamma$ and $i = j$ (i.e., when $\mathcal{J}(x) = F'(x)^{-1}$ $(x \in \mathcal{D}_0)$ in the case of Newton's method), where as they coincide when $\gamma_0 = \gamma$;

(b) conditions of Lemma 5.1 can be weaker than (5.48) when $\gamma_0 < \gamma$.

**Application 5.1** *(Newton's method). That is, $\eta = 0$. Hypothesis*

$$h_G = \beta\,\gamma \leq \frac{(1-\eta)^2}{2} \quad \text{(see } (cf.[398])) \tag{5.56}$$

*reduces to the famous for its simplicity and clarity Newton–Kantorovich hypothesis (cf. [134], [469]) for solving nonlinear equations:*

$$h_K = \gamma\,\beta \leq \frac{1}{2}. \tag{5.57}$$

*Note that in this case, polynomials $f_m$ $(m \geq 1)$ should be in the following form*

$$f_m(s) = \left(\gamma\; s^{m-1} + 2\,\gamma_0\;(1 + s + s^2 + \cdots + s^m)\right)\,\beta - 2 \tag{5.58}$$

*and*

$$f_{m+1}(s) = f_m(s) + g(s)\,s^{m-1}\,\beta. \tag{5.59}$$

*It is simple algebra to show that condition of Lemma 5.1 reduces to the following*

$$h_A = \alpha\,\beta \leq \frac{1}{2}, \tag{5.60}$$

*where*

$$\alpha = \frac{1}{8}\left(\gamma + 4\,\gamma_0 + \sqrt{\gamma^2 + 8\,\gamma_0\,\gamma}\right). \tag{5.61}$$

*In view of (5.57), (5.60) and (5.61), we get that*

$$h_K \leq \frac{1}{2} \Longrightarrow h_A \leq \frac{1}{2}, \tag{5.62}$$

*but not necessarily vice verca unless if $\gamma = \gamma_0$. Moreover, if $\gamma_0 < \gamma$, Condition (5.60) is also weaker than*

$$h_{HSL} = \frac{\gamma_0 + \gamma}{2}\;\beta \leq \frac{1}{2}, \tag{5.63}$$

*provided in (cf. [443]) for nonsingular operators. Note that condition (5.63) was first given by us in (cf. [136], [134]) for the case when linear operator $F'(x_0)$ is invertible.*

We provide examples, where $\gamma_0 < \gamma$, or (5.60) holds but (5.57) is violated.

**Example 5.1** *Let $\mathcal{X} = \mathcal{Y} = \mathbb{R}^2$, equipped with the max–norm and*

$$x_0 = (1,1)^T, \quad U_0 = \{x : \| x - x_0 \| \leq 1 - p\} \quad \left(p \in \left[0, \frac{1}{2}\right)\right).$$

*Define a function $F$ on $U_0$ by*

$$F(x) = (\xi_1^3 - p, \xi_2^3 - p) \qquad (x = (\xi_1, \xi_2)^T). \tag{5.64}$$

*The Fréchet–derivative of operator $F$ is given by*

$$F'(x) = \begin{bmatrix} 3\,\xi_1^2 & 0 \\ 0 & 3\,\xi_2^2 \end{bmatrix}.$$

**Case 1:** $\eta = 0$. *Using hypotheses of Theorem 5.1, we get that*

$$\beta = \frac{1}{3}\,(1-p), \quad \gamma_0 = 3 - p \ \ \textit{and} \ \ \gamma = 2\ (2-p).$$

*The Kantorovich condition (5.57) is violated, since*

$$\frac{4}{3}\,(1-p)\,(2-p) > 1 \quad \left(p \in \left[0, \frac{1}{2}\right)\right).$$

*Hence, there is no guarantee that Newton's method converges to $x^\star = (\sqrt[3]{p}, \sqrt[3]{p})^T$, starting at $x_0$. However, our condition (5.60) is true for all $p \in I = \left[.450339002, \frac{1}{2}\right)$. Hence, the conclusions of our Theorem 5.1 can apply to solve equation (5.64) for all $p \in I$.*

**Case 2:** $0 \neq \eta = 0.01$. *Choose $p = .49$, then, we get that*

$$\gamma_0 = 2.51 < \gamma = 3.02, \qquad \beta = .17,$$

$$\frac{\delta}{2} = .033058514 < \alpha = .53112045$$

*and*

$$\delta_0 = .3347085 < 2\,\alpha.$$

*Note also that condition (5.56) is violated no matter how $\eta$ is chosen in $(0,1)$.*

*Finally, by comparing (5.60) with (5.48), we see that our condition is weaker provided that*

$$a < \frac{\gamma}{(1-\eta)^2}, \tag{5.65}$$

*which can certainly happen. For example, if $\gamma_0 \approx 0$, then $\alpha \approx 0$, in which case (5.65) holds.*

**Application 5.2** *In the case $\mathcal{X} = \mathcal{Y} = \mathbb{R}^j$ ($j$ fixed in $\mathbb{N}$), we can split matrix $F'(x_n)$ into $F'(x_n) = B_n - C_n$ to obtain the inner–outer iteration:*

$$x_{n+1} = x_n - (H_n^{m_n - 1} + \cdots + H_n + \mathcal{I})\ B_n^{-1}\ F(x_n) \quad (n \geq 0), \tag{5.66}$$

$$H_n = B_n^{-1}\ C_n, \tag{5.67}$$

*where $m_n$ is the number of inner iterations. Let us assume $m_n = m$ in iteration (5.66). We can obtain result concerning the estimation of the number of inner iterations under the conditions of Theorem 5.1.*

**Theorem 5.3** *Under the hypotheses of Theorem 5.1, further assume that*

$$\| B_0^{-1} F'(x_0) \| \leq q,$$

$$a_0 \, h^m + m \, b \, h^{m-1} \leq \eta_n, \qquad \sup_n \| H_n \| \leq h < 1,$$

*where*

$$\begin{aligned} a_0 &= \frac{3 - 2\eta + 2\beta \, \gamma^n}{\eta^2}, \\ b &= \frac{2 - \eta}{\eta} \, \frac{q \, (q+1) \, \gamma_0}{[1 - (1-\eta) \, \gamma_0 \, q]^2} \left[ \frac{(1-\eta)^2}{2\gamma} + \frac{1-\eta}{\gamma} + \beta \right]; \end{aligned} \tag{5.68}$$

*the matrix norm has the property:*

$$\| F'(x_0)^{-1} \, R \| \leq \| F'(x_0)^{-1} \, S \|$$

*with $R$ any submatrix of $S$;*

$$\overline{U}(x_0, t^\star) \subseteq \mathcal{D};$$

*and hypotheses of Lemma 5.1 hold.*

*Then, the conclusions of Theorem 5.1 hold true for inexact iteration (5.4).*

**Proof** It follows exactly as in (cf. [398, Corollary 3.3]) and (cf. [166, Theorem 3.7]). Here are the changes (with $\gamma_0$ replacing $\gamma$ in the proof)

$$\| F'(x_0)^{-1} \, F'(x_n) \| \leq 1 + \gamma_0 \, \| x_n - x_0 \|,$$

$$\| F'(x_n)^{-1} \, F'(x_0) \| \leq \frac{1}{1 - \gamma_0 \, \| x_n - x_0 \|},$$

$$\| F'(x_0)^{-1} \, F(x_n) \| \leq \frac{\gamma}{2} \, \| x_n - x_0 \|^2 + \| x_n - x_0 \| + \beta,$$

$$\| F'(x_0)^{-1} \, (B_n - B_{n-1}) \| \leq \gamma \, \| x_n - x_{n-1} \|$$

and

$$\| B_n^{-1} \, F'(x_0)^{-1} \| \leq \frac{q}{1 - \gamma_0 \, \| x_n - x_0 \| \, q}.$$

⊠

**Remark 5.3** *The constant $\overline{b}$ defined in (cf. [398]) (for $\gamma_0 = \gamma$) is larger than $b$, which is an advantage of our approach for the selection of a smaller $\eta$, when $\gamma < \gamma_0$.*

*Note that the hypotheses of Theorem 5.3 are simpler than the hypotheses of (cf. [166, Theorem 3.7]) and weaker than (cf. [398, Corollary 3.3]). Hence, all the above justify the claims made.*

*Note that, in the case $i = j$, the results can be provided in affine–invariant form by simply replacing $F(x)$ by $F'(x_0)^{-1}\, F(x)$ for $x \in \mathcal{D}_0$, and setting $\beta_0 = 1$. The advantages of this approach have been explained in (cf. [163], [326]).*

*Finally, our results immediately extend to the more general case of outer or generalized inverses, by simply replacing perturbation Lemma 5.2 by its analog in (cf. [277, Lemma 2.2, p. 238]) (see also (cf. [95], [163])) and using the same approach as in this section. Note that the crucial majorizing sequence (5.9) remains the same in this new setting. We leave the details in the motivated reader.*

## 5.2 Average–Lipschitz Conditions

We extend in this section the applicability of the Gauss–Newton method for solving singular systems of equations under the notions of average Lipschitz–type conditions introduced recently in (cf. [504]). Using our idea of recurrent functions, we provide a tighter local as well as semilocal convergence analysis for the Gauss–Newton method than in (cf. [504]) who recently extended and improved earlier results (cf. [443], [502], [751]). We also note that our results are obtained under weaker or the same hypotheses as in (cf. [504]).

We are concerned with the problem of approximating a locally unique solution $x^\star$ of (2.1), where $F$ is a Fréchet–differentiable operator defined on an open, nonempty, convex subset $\mathcal{D}$ of $\mathbb{R}^m$ with values in $\mathbb{R}^l$, where $m, l \in \mathbb{N}^\star$.

We shall use the Gauss–Newton method GNM defined by (5.4) to generate a sequence $\{x_n\}$ approximating a solution $x^\star$ of equation

$$F'(x)^+\, F(x) = 0, \tag{5.69}$$

where $F'(x)^+$ denotes the Moore–Penrose inverse of matrix $F'(x)$ ($x \in \mathcal{D}$) (see Definition 5.1). If $m = l$ and $F'(x_n)$ is invertible, then GNM reduces to Newton's method NM. There is an extensive literature on the local as well as the semilocal convergence of GNM under Lipschitz–type conditions. We refer the reader to (cf. [157], [194] and the references therein) for convergence results on Newton–type methods (see also (cf. [136], [151], [165], [231], [502], [596], [782])). In particular, we recommend the paper by Xu and Li (cf. [782]), where GNM is studied under average Lipschitz conditions.

In (cf. [504]), Li, Hu and Wang provided a Kantorovich–type convergence analysis for GNM by inaugurating the notions of a certain type of average Lipschitz conditions. GNM is also studied using the Smale point estimate theory. This way, they unified convergence criteria for GNM. Special cases of their results extend and/or improve important known results (cf. [151], [412]). We are motivated by the elegant work in (cf. [504]) and optimization considerations. In particular, using our new concept of recurrent functions, we provide a tighter convergence analysis for GNM under weaker or the same hypotheses in (cf. [504]) for both the local as well the semilocal case.

Let $\overline{\mathbb{R}} = \mathbb{R} \cup \{+\infty\}$ and $\overline{\mathbb{R}}_+ = [0, +\infty]$. We assume that $L$ and $L_0$ are non–decreasing functions on $[0, R)$, where $R \in \overline{\mathbb{R}}_+$ and

$$L_0(t) \leq L(t) \quad (t \in [0, R)). \tag{5.70}$$

Let $\beta > 0$ and $0 \leq \lambda < 1$ be given parameters. Define a function $g : [0, R] \longrightarrow \overline{\mathbb{R}}$ by

$$g(t) = \beta - t + \int_0^t L_0(u)\,(t-u)\,du. \tag{5.71}$$

Moreover, define a majorizing function $h_\lambda : [0, R] \longrightarrow \overline{\mathbb{R}}$ corresponding to a fixed pair $(\lambda, L)$ by

$$h_\lambda(t) = \beta - (1-\lambda)\,t + \int_0^t L(u)\,(t-u)\,du. \tag{5.72}$$

We have for $t \in [0, R]$ that

$$g'(t) = -1 + \int_0^t L_0(u)\,du, \tag{5.73}$$

$$g''(t) = L_0(t), \tag{5.74}$$

$$h'_\lambda(t) = -(1-\lambda) + \int_0^t L(u)\,du \tag{5.75}$$

and

$$h''_\lambda(t) = L(t). \tag{5.76}$$

It follows from (5.70), (5.73) and (5.75) that

$$g'(t) \leq h'_0(t) \quad (t \in [0, R]). \tag{5.77}$$

Define

$$r_\lambda := \sup\left\{ r \in (0, R) : \int_0^r L(u)\,du \leq 1 - \lambda \right\} \tag{5.78}$$

and

$$b_\lambda := (1-\lambda)\,r_\lambda - \int_0^{r_\lambda} L(u)\,(r_\lambda - u)\,du. \tag{5.79}$$

Set

$$\Delta = \int_0^R L(u)\,du.$$

Then, we have that

$$r_\lambda = \begin{cases} R & \text{if} \quad \Delta < 1 - \lambda, \\ t_\lambda & \text{if} \quad \Delta \geq 1 - \lambda, \end{cases} \tag{5.80}$$

where $t_1 \in [0, R]$ is such that

$$\int_0^{t_\lambda} L(u)\, du = 1 - \lambda$$

and is guaranteed to exist, since $\Delta \geq 1 - \lambda$ in this case.

We also get

$$\begin{cases} b_\lambda \geq \displaystyle\int_0^{r_\lambda} L(u)\, u\, du & \text{if} \quad \Delta < 1 - \lambda, \\ b_\lambda = \displaystyle\int_0^{r_\lambda} L(u)\, u\, du & \text{if} \quad \Delta \geq 1 - \lambda, \end{cases} \tag{5.81}$$

Let us define scalar sequence $\{s_{\lambda,n}\}$ by

$$s_{\lambda,0} = 0, \quad s_{\lambda,n+1} = s_{\lambda,n} - \frac{h_\lambda(s_{\lambda,n})}{g'(s_{\lambda,n})} \quad (n \geq 0). \tag{5.82}$$

Note that if equality holds in (5.70), then

$$g(t) = h_0(t) \qquad (t \in [0, R]) \tag{5.83}$$

and $\{s_{\lambda,n}\}$ reduces to $\{t_{\lambda,n}\}$ introduced in (cf. [504]) and given by

$$t_{\lambda,0} = 0, \quad t_{\lambda,n+1} = t_{\lambda,n} - \frac{h_\lambda(t_{\lambda,n})}{h_0'(s_{\lambda,n})} \quad (n \geq 0). \tag{5.84}$$

We shall show in Lemma 5.4 that, under the same hypothesis (see (5.85)), scalar sequence $\{s_{\lambda,n}\}$ is at least as tight as $\{t_{\lambda,n}\}$. But first, we need a crucial result on majorizing sequences for the GNM.

**Lemma 5.3** *Assume that*

$$\beta \leq b_\lambda. \tag{5.85}$$

*Then, the following hold*

*(i) Function $h_\lambda$ is strictly decreasing on $[0, r_\lambda]$ and has exactly one zero $t^\star_\lambda \in [0, r_\lambda]$ such that*

$$\beta < t^\star_\lambda. \tag{5.86}$$

*(ii) Sequence $\{s_{\lambda,n}\}$ given by (5.82) is strictly increasing and converges to $t^\star_\lambda$.*

**Proof**

(i) This part follows immediately from (5.72), (5.75) and (5.79).

(ii) We shall show this part using induction on $n$. It follows from (5.82) and (5.86) that

$$0 = s_{\lambda,0} < s_{\lambda,1} = \beta < t_\lambda^\star.$$

Assume

$$s_{\lambda,k-1} < s_{\lambda,k} < t_\lambda^\star \quad (k \le n). \tag{5.87}$$

In view of (5.74), $-g'$ is strictly increasing on $[0, R]$ and so, by (5.77), (5.87) and the definition of $r_\lambda$, we have

$$-g'(s_{\lambda,k}) > -g'(t_\lambda^\star) \ge -g'(r_\lambda) \ge -h_\lambda'(r_\lambda) + \lambda \ge 0. \tag{5.88}$$

We also have $h_\lambda(s_{\lambda,k}) > 0$ by (i). That is, it follows from

$$s_{\lambda,k+1} = s_{\lambda,k} - \frac{h_\lambda(s_{\lambda,k})}{g'(s_{\lambda,k})} > s_{\lambda,k}. \tag{5.89}$$

Let us define $p_\lambda$ on $[0, t_\lambda^\star]$ by

$$p_\lambda(t) = t - \frac{h_\lambda(t)}{g'(t)}. \tag{5.90}$$

We have

$$g'(t) < 0 \quad (t \in [0, t_\lambda^\star]) \tag{5.91}$$

except if $\lambda = 0$ and $t = t_\lambda^\star = r_\lambda$. As in (cf. [504]), we set by convention

$$\frac{h_\lambda(t_\lambda^\star)}{g'(t_\lambda^\star)} = \lim_{t \longrightarrow t_\lambda^{\star -}} \frac{h_\lambda(t)}{g'(t)} = 0 \tag{5.92}$$

by Hospital's rule. Therefore, function $p_\lambda$ is well defined and continuous on $[0, t_\lambda^\star]$. It follows from part (i), (5.76) and (5.91) that

$$\begin{aligned} p_\lambda'(t) &= 1 - \frac{h_\lambda'(t)\, g'(t) - h_\lambda(t)\, g''(t)}{(g'(t))^2} \\ &= \frac{-(\lambda + \int_0^t (L(u) - L_0(u))\, du)\, g'(t) + h_\lambda(t)\, L_0(t)}{(g'(t))^2} \\ &> 0 \quad \text{for a.e.} \quad t \in [0, t_\lambda^\star). \end{aligned} \tag{5.93}$$

Using (5.87), (5.89) and (5.93), we get

$$s_{\lambda,k} < s_{\lambda,k+1} = p_\lambda(s_{\lambda,k}) < p_\lambda(t_\lambda^\star) = t_\lambda^\star, \tag{5.94}$$

which completes the induction. Hence, $\{s_{\lambda,n}\}$ is increasing, bounded above by $t_\lambda^\star$ and as such it converges to its unique least upper bound $s^\star \in (0, t_\lambda^\star]$ with $h_\lambda(s^\star) = 0$. Using part (i), we get $s^\star = t_\lambda^\star$.

The proof of Lemma 5.3 is complete. ☒

Next, we compare sequence $\{s_{\lambda,n}\}$ with $\{t_{\lambda,n}\}$.

**Lemma 5.4** *Assume that condition (5.85) is satisfied, then, the following hold for $n \geq 0$*

$$s_{\lambda,n} \leq t_{\lambda,n} \tag{5.95}$$

*and*

$$s_{\lambda,n+1} - s_{\lambda,n} \leq t_{\lambda,n+1} - t_{\lambda,n}. \tag{5.96}$$

*Moreover, if $L_0(t) < L(t)$ for $t \in [0, t_\lambda^\star]$, then (5.95) and (5.96) hold as strict inequalities for $n \geq 1$ and $n \geq 0$, respectively.*

**Proof** It was shown in (cf. [504]) that, under hypothesis (5.85), assertions (i) and (ii) of Lemma 5.3 hold with $\{t_{\lambda,n}\}$ replacing $\{s_{\lambda,n}\}$. We shall show (5.95) and (5.96) using induction. It follows from (5.70), (5.82) and (5.84) that

$$\begin{aligned} s_{\lambda,0} &= t_{\lambda,0}, \quad s_{\lambda,1} = t_{\lambda,1} = \beta, \\ s_{\lambda,1} &= s_{\lambda,0} - \frac{h_\lambda(s_{\lambda,0})}{g'(s_{\lambda,0})} \leq t_{\lambda,0} - \frac{h_\lambda(t_{\lambda,0})}{h_0'(t_{\lambda,0})} = t_{\lambda,1} \end{aligned} \tag{5.97}$$

and

$$s_{\lambda,1} - s_{\lambda,0} = -\frac{h_\lambda(s_{\lambda,0})}{g'(s_{\lambda,0})} \leq -\frac{h_\lambda(t_{\lambda,0})}{h_0'(t_{\lambda,0})} = t_{\lambda,1} - t_{\lambda,0}. \tag{5.98}$$

Hence, (5.95) and (5.96) hold for $n = 0$. Let us assume that (5.95) and (5.96) hold for all $k \leq n$. Then, we have in turn that

$$s_{\lambda,k+1} = s_{\lambda,k} - \frac{h_\lambda(s_{\lambda,k})}{g'(s_{\lambda,k})} \leq t_{\lambda,k} - \frac{h_\lambda(t_{\lambda,k})}{h_0'(t_{\lambda,k})} = t_{\lambda,k+1} \tag{5.99}$$

and

$$s_{\lambda,k+1} - s_{\lambda,k} = -\frac{h_\lambda(s_{\lambda,k})}{g'(s_{\lambda,k})} \leq -\frac{h_\lambda(t_{\lambda,k})}{h_0'(t_{\lambda,k})} = t_{\lambda,k+1} - t_{\lambda,k}, \tag{5.100}$$

which completes the induction for (5.95) and (5.96). Moreover, if $L_0(t) < L(t)$ for $t \in [0, t_\lambda^\star]$, then (5.97)–(5.100) hold as strict inequalities. This completes the proof of Lemma 5.3. ☒

We shall provide a semilocal convergence for GNM using majorizing sequences under the Kantorovich approach. For the remainder of this section, we assume that $F : \mathbb{R}^m \longrightarrow \mathbb{R}^l$ is continuous Fréchet–differentiable,

$$\| F'(y)^+ (\mathcal{I} - F'(x)\, F'(x)^+)\, F(x) \| \leq \kappa \, \| x - y \| \quad (x, y \in \mathcal{D}), \tag{5.101}$$

where $\kappa \in [0, 1)$, $\mathcal{I}$ is the identity matrix,

$$F'(x_0) \neq 0 \tag{5.102}$$

for some $x_0 \in \mathcal{D}$ and

$$\operatorname{rank}(F'(x)) \leq \operatorname{rank}(F'(x_0)) \quad (x \in \mathcal{D}). \tag{5.103}$$

We need the definition of the modified $L$–average Lipschitz condition on $U(x_0, r)$.

**Definition 5.2** *(cf. [504]) Let $r > 0$ be such that $U(x_0, r) \subseteq \mathcal{D}$. Mapping $F'$ satisfies the modified $L$–average Lipschitz condition on $U(x_0, r)$ if, for any $x, y \in U(x_0, r)$ with $\| x - x_0 \| + \| y - x \| < r$,*

$$\| F(x_0)^+ \| \, \| F'(y) - F'(x) \| \leq \int_{\|x-x_0\|}^{\|x-x_0\|+\|y-x\|} L(u)\, du. \tag{5.104}$$

*Condition (5.104) was used in (cf. [504]) as an alternative to $L$–average Lipschitz condition (cf. [443])*

$$\| F(x_0)^+ \, (F'(y) - F'(x)) \| \leq \int_{\|x-x_0\|}^{\|x-x_0\|+\|y-x\|} L(u)\, du, \tag{5.105}$$

*which is a modification of Wang's condition (cf. [751], [753]). Condition (5.104) fits the case when $F'(x_0)$ is not surjective (cf. [504]).*

We also introduce the condition.

**Definition 5.3** *Let $r > 0$ be such that $U(x_0, r) \subseteq \mathcal{D}$. Mapping $F'$ satisfies the modified center $L_0$–average Lipschitz condition on $U(x_0, r)$ if, for any $x \in U(x_0, r)$ with*
*$2 \| x - x_0 \| < r$,*

$$\| F(x_0)^+ \| \, \| F'(y) - F'(x_0) \| \leq \int_{\|x-x_0\|}^{2\,\|x-x_0\|} L_0(u)\, du. \tag{5.106}$$

**Remark 5.4** *If (5.104) holds, then, so do (5.106) and (5.70). Therefore, (5.106) is not an additional hypothesis. We also note that $\dfrac{L}{L_0}$ can be arbitrarily large.*

**Lemma 5.5** *Let*

$$\tilde{r}_0 = \sup\left\{ r \in (0, R) \, : \, \int_0^r L_0(u)\, du \leq 1 \right\}.$$

*Suppose that $0 \leq r \leq \tilde{r}_0$ satisfies $U(x_0, r) \subseteq \mathcal{D}$ and that $F'$ satisfies (5.106) on $U(x_0, r)$.*

*Then, for each $x \in U(x_0, r)$, $\operatorname{rank}(F'(x)) = \operatorname{rank}(F'(x_0))$ and*

$$\| F'(x)^+ \| \leq -g'(\| x - x_0 \|)^{-1} \, \| F'(x_0)^+ \| . \tag{5.107}$$

**Remark 5.5** *If equality holds in (5.70), then Lemma 5.5 reduces to (cf. [504, Lemma 3.2]). Otherwise, (i.e. if strict inequality hold in (5.70)), (5.107) is a more precise estimate than*

$$\| F'(x)^+ \| \leq -h_0'(\| x - x_0 \|)^{-1} \ \| F'(x_0)^+ \| \tag{5.108}$$

*given in (cf. [504]), since*

$$-g'(r)^{-1} < -h_0'(r)^{-1} \qquad (r \in [0, R]). \tag{5.109}$$

*We also have*

$$r_0 \leq \tilde{r}_0 \tag{5.110}$$

*and*

$$\lambda_0 = \kappa \, (1 - \int_0^\beta L(u)\, du) \leq \tilde{\lambda}_0 = \kappa \, (1 - \int_0^\beta L_0(u)\, du), \tag{5.111}$$

*where*

$$\beta = \| F'(x_0)^+ \, F(x_0) \| . \tag{5.112}$$

We also state the result.

**Lemma 5.6** *(cf. [504], [443]) Let $0 \leq c < R$. Define*

$$\chi(t) = \frac{1}{t} \int_0^t L(c+u)\,(t-u)\, du, \qquad 0 \leq t < R - c.$$

*Then, function $\chi$ is increasing on $[0, R-c)$.*

We can show the following semilocal convergence result for GNM. Our approach differs from the corresponding (cf. [504, Theorem 3.1, p. 273]) since we use (5.107) instead (5.108).

**Theorem 5.4** *Let $\lambda \geq \tilde{\lambda}_0$. Assume*

$$\beta \leq b_\lambda \quad \text{and} \quad \overline{U}(x_0, t_\lambda^\star) \subseteq \mathcal{D}; \tag{5.113}$$

*$F'$ satisfies (5.104) and (5.106) on $U(x_0, t_\lambda^\star)$. Then, the sequence $\{x_n\}$ generated by GNM is well defined, remains in $\overline{U}(x_0, t_\lambda^\star)$ for all $n \geq 0$ and converges to a zero $x^\star$ of $F'(.)^+ F(.)$ in $\overline{U}(x_0, t_\lambda^\star)$. Moreover, the following estimates hold*

$$\| x_{n+1} - x_n \| \leq s_{\lambda,n+1} - s_{\lambda,n} \tag{5.114}$$

*and*

$$\| x_n - x^\star \| \leq t_\lambda^\star - s_{\lambda,n}. \tag{5.115}$$

**Proof** We first use mathematical induction to prove that

$$\| x_n - x_{n-1} \| \leq s_{\lambda,n} - s_{\lambda,n-1} \tag{5.116}$$

hold for each $n \geq 0$. We first have

$$\| x_1 - x_0 \| = \| F'(x_0)^+ F(x_0) \| = \beta \leq s_{\lambda,1} - s_{\lambda,0}$$

then, (5.116) holds for $n = 0$. Assume that (5.116) hold for all $n \leq k$. Then

$$\| x_k - x_{k-1} \| \leq s_{\lambda,k} - s_{\lambda,k-1}. \tag{5.117}$$

For $\theta \in [0,1]$, we denote by

$$x_k^\theta = x_{k-1} + \theta\,(x_k - x_{k-1}) \quad \text{and} \quad s_{\lambda,k}^\theta = s_{\lambda,k-1} + \theta\,(s_{\lambda,k} - s_{\lambda,k-1}).$$

Then, for all $\theta \in [0,1]$,

$$\| x_k^\theta - x_0 \| \leq \| x_k^\theta - x_{k-1} \| + \sum_{i=1}^{k-1} \| x_i - x_{i-1} \| \leq s_{\lambda,k}^\theta < t_\lambda^\star \leq r_\lambda \leq r_0. \tag{5.118}$$

In particular,

$$\| x_{k-1} - x_0 \| \leq s_{\lambda,k-1} \quad \text{and} \quad \| x_k - x_0 \| \leq s_{\lambda,k}. \tag{5.119}$$

By Lemma 5.5 and the monotony property of function $-g'(.)^{-1}$, we obtain

$$\begin{aligned} \| F'(x_k)^+ \| &\leq -g'(\| x_k - x_0 \|)^{-1} \; \| F'(x_0)^+ \| \leq \| F'(x_k)^+ \| \\ &\leq -g'(s_{\lambda,k})^{-1} \; \| F'(x_0)^+ \| . \end{aligned} \tag{5.120}$$

We deduce by Lemma 5.6, (5.76) and (5.117) that

$$\begin{aligned} &\int_0^1 \int_{\|x_{k-1}-x_0\|}^{\|x_{k-1}-x_0\|+\theta\,\|x_k-x_{k-1}\|} L(u)\,du \;\; \| x_k - x_{k-1} \| \; d\theta \\ &= \frac{1}{\| x_k - x_{k-1} \|} \int_0^{\|x_k-x_{k-1}\|} L(\| x_{k-1} - x_0 \| + u)\,(\| x_{k-1} - x_k \| - u)\,du \\ &\quad \times \| x_k - x_{k-1} \| \\ &\leq \frac{1}{s_{\lambda,k} - s_{\lambda,k-1}} \int_0^{s_{\lambda,k}-s_{\lambda,k-1}} h_\lambda''(s_{\lambda,k-1} + u)\,(s_{\lambda,k} - s_{\lambda,k-1} - u)\,du \;\; \| x_k - x_{k-1} \| \\ &= (h_\lambda(s_{\lambda,k}) - h_\lambda(s_{\lambda,k-1}) - h_\lambda'(s_{\lambda,k-1})\,(s_{\lambda,k} - s_{\lambda,k-1}))\,\frac{\| x_k - x_{k-1} \|}{s_{\lambda,k} - s_{\lambda,k-1}}. \end{aligned} \tag{5.121}$$

Since,

$$h_\lambda' = g' + \lambda \quad \text{and} \quad -\,h_\lambda(s_{\lambda,k-1}) - g'(s_{\lambda,k-1})\,(s_{\lambda,k} - s_{\lambda,k-1}) = 0$$

then, estimate (5.121) becomes

$$\int_0^1 \int_{\|x_{k-1}-x_0\|}^{\|x_{k-1}-x_0\|+\theta\,\|x_k-x_{k-1}\|} L(u)\,du \;\; \| x_k - x_{k-1} \| \; d\theta \leq h_\lambda(s_{\lambda,k}) - \lambda\,(s_{\lambda,k} - s_{\lambda,k-1}). \tag{5.122}$$

Using (5.70), we obtain the identity

$$
\begin{aligned}
x_{k+1} - x_k &= -F'(x_k)^+ \, F(x_k) \\
&= -F'(x_k)^+ \int_0^1 (F'(x_{k-1} + \theta\,(x_k - x_{k-1})) - F'(x_k))\,(x_k - x_{k-1})\,d\theta \\
&\quad - F'(x_k)^+ \,(\mathcal{I} - F'(x_{k-1})\,F'(x_{k-1})^+)\,F(x_{k-1}).
\end{aligned} \tag{5.123}
$$

By (5.101), (5.104), (5.121) and (5.123), we obtain that

$$
\begin{aligned}
&\| x_{k+1} - x_k \| = \| F'(x_k)^+ \, F(x_k) \| \\
&\leq \int_0^1 \| F'(x_k)^+ \| \, \| F'(x_{k-1} + \theta\,(x_k - x_{k-1})) - F'(x_k) \| \, \| x_k - x_{k-1} \| \;\; d\theta \\
&\quad + \kappa \, \| x_k - x_{k-1} \| \\
&\leq -g'(s_{\lambda,k})^{-1} \int_0^1 \int_{\|x_{k-1}-x_0\|}^{\|x_{k-1}-x_0\|+\theta\,\|x_k - x_{k-1}\|} L(u)\;du \;\; \| x_k - x_{k-1} \| \;\; d\theta \\
&\quad + \kappa \, \| x_k - x_{k-1} \| \\
&\leq -g'(s_{\lambda,k})^{-1}\,h_\lambda(s_{\lambda,k}) + (\kappa + \lambda\,g'(s_{\lambda,k})^{-1})\,(s_{\lambda,k} - s_{\lambda,k-1}).
\end{aligned} \tag{5.124}
$$

Note that $\beta = s_{\lambda,1} \leq s_{\lambda,k}$. By (5.111), we have

$$
\lambda \geq \tilde{\lambda}_0 \geq -g'(s_{\lambda,k})\,\kappa,
$$

then, $\kappa + \lambda\,g'(s_{\lambda,k})^{-1} \leq 0$. Consequently, by (5.124), we get

$$
\| x_{k+1} - x_k \| \leq -g'(s_{\lambda,k})^{-1}\,h_\lambda(s_{\lambda,k}) = s_{\lambda,k+1} - s_{\lambda,k}
$$

and the induction for (5.114) is completed. In view of Lemma 5.6, sequence $\{x_n\}$ $(n \geq 0)$ converges to some $x^\star \in \overline{U}(x_0, t_\lambda^\star)$. Estimate (5.115) follows from (5.114) by using standard majorization techniques. By letting $k \longrightarrow \infty$ in the inequality

$$
\begin{aligned}
\| F'(x^\star)^+ \, F(x_k) \| &\leq \| F'(x^\star)^+ \,(\mathcal{I} - F'(x_k)\,F'(x_k)^+)\,F(x_k) \| + \\
&\quad \| F'(x^\star)^+ \| \, \| F'(x_k)\,F'(x_k)^+\,F(x_k) \| \\
&\leq \kappa \, \| x_k - x^\star \| + \| F'(x^\star)^+ \| \, \| F'(x_k) \| \, \| x_{k+1} - x_k \|
\end{aligned}
$$

we obtain $F'(x^\star)^+ \, F(x^\star) = 0$. This completes the proof of Theorem 5.4. ⊠

**Remark 5.6** *(a) If equality holds in (5.70), then, Theorem 5.4 reduces to (cf. [504, Theorem 3.1]). Otherwise, it follows from Lemmas 5.3, 5.4 and Remark 5.5 that under the same computational cost and hypotheses, we obtain the advantages over the works in (cf. [504]) as stated in the* *introduction of this section. It also follows from (5.124) that* $\{v_{\lambda,n}\}$ *defined by*

$$v_{\lambda,0} = 0, \quad v_{\lambda,1} = \beta,$$

$$v_{\lambda,n+1} = v_{\lambda,n} - \frac{\left(\int_0^1 \int_{v_{\lambda,n-1}}^{v_{\lambda,n-1}+\theta\,(v_{\lambda,n}-v_{\lambda,n-1})} L(u)\,du\,d\theta + \kappa\right)(v_{\lambda,n} - v_{\lambda,n-1})}{g'(v_{\lambda,n})} \tag{5.125}$$

*is a tighter majorizing sequence for GNM than* $\{s_{\lambda,n}\}$, *so that*

$$v_{\lambda,n} \le s_{\lambda,n}, \tag{5.126}$$

$$v_{\lambda,n} - v_{\lambda,n-1} \le s_{\lambda,n} - s_{\lambda,n-1} \tag{5.127}$$

*and*

$$\lim_{n\longrightarrow\infty} v_{\lambda,n} = v_\lambda^\star \le t_\lambda^\star. \tag{5.128}$$

*(see also the proof in Lemma 5.4 and Theorem 5.4). Hence,* $\{v_{\lambda,n}\}$, $v_\lambda^\star$ *can replace* $\{s_{\lambda,n}\}$, $s_\lambda^\star$, *respectively, in Theorem 5.4.*

*At this point, we are wondering if conditions (5.113) can be weakened, since this way the applicability of GNM will be extended. It turns out that, if* $\{v_{\lambda,n}\}$ *is used as a majorizing sequence for GNM, conditions (5.113) can be replaced by weaker ones in many cases. However, before doing that we present the rest of our results, which also constitute improvements of the corresponding (cf. [504]). The proofs are omitted, since they follow from (cf. [504]) by simply using* $g'$ *instead of* $h_0'$. *Note also that Corollary 5.1, Theorem 5.5 and Corollary 5.2 hold with* $\{v_n\}$, $v_0^\star$ *replacing* $\{s_n\}$, $s_0^\star$, *respectively.*

(b) *The condition (5.113) depends upon the choice of* $\lambda$. *The best choice of* $\lambda$ *is* $\tilde{\lambda}_0$ *given by (5.111) and in this case, (5.113) would be implicit. In the following corollary, we consider the simple choice* $\lambda = \kappa$ *such that the condition (5.111) is explicit.*

**Corollary 5.1** *Assume*

$$\beta \le b_\kappa \quad \text{and} \quad \overline{U}(x_0, t_\kappa^\star) \subseteq \mathcal{D} \tag{5.129}$$

*and* $F'$ *satisfies (5.104) and (5.106) on* $U(x_0, t_\lambda^\star)$. *Then, the conclusions of Theorem 5.4 hold in* $\overline{U}(x_0, t_\kappa^\star)$ *with* $\lambda = \kappa$.

We have the following improved version of Theorem 5.4 for $\kappa = 0$.

**Theorem 5.5** *Assume*

$$\beta \le b_0 \quad \text{and} \quad \overline{U}(x_0, t_0^\star) \subseteq \mathcal{D}, \tag{5.130}$$

$$\| F'(y)^+ \,(\mathcal{I} - F'(x)\,F'(x)^+)\,F(x) \,\| = 0 \quad (x, y \in \mathcal{D}) \tag{5.131}$$

*and* $F'$ *satisfies (5.104) and (5.106) on* $U(x_0, t_0^\star)$.

*Then, the conclusions of Theorem 5.4 hold in* $\overline{U}(x_0, t_0^\star)$ *with* $\lambda = 0$. *Furthemore, we have the additional estimate for* $n \geq 1$

$$\| x_{n+1} - x_n \| \leq \frac{s_{0,n+1} - s_{0,n}}{s_{0,n} - s_{0,n-1}} \| x_n - x_{n-1} \| .$$

In the case when $F'(x_0)$ is surjective, we have the following corollary.

**Corollary 5.2** *Assume that* $F'(x_0)$ *is surjective and the conditions of Theorem 5.5 hold. Then, the conclusions of Theorem 5.5 hold for* $t_0^\star$ *replaced by* $s_0^\star$. *Furthemore, we have the additional estimate for* $n \geq 1$

$$\| F'(x_0)^+ F(x_n) \| \leq \frac{s_{0,n+1} - s_{0,n}}{s_{0,n} - s_{0,n-1}} \| F'(x_0)^+ F(x_{n-1}) \| .$$

We shall provide a local convergence analysis for GNM.

Assume there exists $x^\star \in \mathcal{D}$ such that $F(x^\star) = 0$ and $F'(x^\star) \neq 0$,

$$U(x^\star, \tilde{r}_0) \subseteq \mathcal{D}$$

and

$$\operatorname{rank}(F'(x)) \leq \operatorname{rank}(F'(x^\star)) \quad (x \in \mathcal{D}). \tag{5.132}$$

Next, we provide an upper bound on the norm $\| F'(x_0)^+ F(x_0) \|$ needed in the proof of the main result of this section.

**Lemma 5.7** *Let* $0 < r \leq \tilde{r}_0$. *Suppose that* $F'$ *satisfies (5.104) and (5.106) in* $U(x^\star, r)$. *Then, for each* $x_0 \in U(x^\star, r)$, *the following hold*

$$\operatorname{rank}(F'(x_0)) = \operatorname{rank}(F'(x^\star))$$

*and*

$$\| F'(x_0)^+ F(x_0) \| \leq \frac{\| x_0 - x^\star \| + \displaystyle\int_0^{\| x_0 - x^\star \|} L(u)\,(u - \| x_0 - x^\star \|)\, du}{1 - \displaystyle\int_0^{\| x_0 - x^\star \|} L_0(u)\, du}. \tag{5.133}$$

**Proof** Let $x_0 \in U(x^\star, r)$. Using Lemma 5.5, we have $\operatorname{rank}(F'(x_0)) = \operatorname{rank}(F'(x^\star))$ and

$$\| F'(x_0)^+ \| \leq \frac{\| F'(x^\star)^+ \|}{1 - \displaystyle\int_0^{\| x_0 - x^\star \|} L_0(u)\, du}. \tag{5.134}$$

We have also the identity

$$\begin{aligned}
&-F'(x_0)^+ F(x_0) \\
&= F'(x_0)^+ (F(x^\star) - F(x_0) - F'(x_0)(x^\star - x_0)) + F'(x_0)^+ F'(x_0)(x^\star - x_0) \\
&= F'(x_0)^+ \int_0^1 (F'(x_0 - F'(x^\star + \theta (x_0 - x^\star))\,(x^\star - x_0)\, d\theta \\
&+ \Pi_{(Ker\ F'(x_0))^\perp} (x^\star - x_0).
\end{aligned} \tag{5.135}$$

Using (5.104), (5.106), (5.134) and (5.135), we obtain

$$\| F'(x_0)^+ F(x_0) \| \leq \frac{1}{1-\displaystyle\int_0^{\|x_0-x^\star\|} L_0(u)\,du} \int_0^1 \int_{\theta\,\|x_0-x^\star\|}^{\|x_0-x^\star\|} L(u)\ \| x_0 - x^\star \|\ du\,d\theta$$
$$+ \| x_0 - x^\star \|$$
$$= \frac{\| x_0 - x^\star \| + \displaystyle\int_0^{\|x_0-x^\star\|} L(u)\,(u - \| x_0 - x^\star \|)\,du}{1-\displaystyle\int_0^{\|x_0-x^\star\|} L_0(u)\,du}.$$

This completes the proof of Lemma 5.7. ⊠

**Remark 5.7** *If equality holds in (5.70), then, Lemma 5.7 reduces to (cf. [504, Lemma 4.1]). Otherwise, it constitutes an improvement since the estimate*

$$\| F'(x_0)^+ F(x_0) \| \leq \frac{\| x_0 - x^\star \| + \displaystyle\int_0^{\|x_0-x^\star\|} L(u)\,(u - \| x_0 - x^\star \|)\,du}{1-\displaystyle\int_0^{\|x_0-x^\star\|} L(u)\,du} \tag{5.136}$$

*is used in (cf. [504]) and in this case*

$$\frac{1}{1-\displaystyle\int_0^{\|x_0-x^\star\|} L_0(u)\,du} < \frac{1}{1-\displaystyle\int_0^{\|x_0-x^\star\|} L(u)\,du}.$$

*Using (5.133) instead of (5.136), we obtain the following improvements of Lemmas 4.2, 4.3, Theorem 4.1 and Corollaries 4.1–4.3 in (cf. [504]).*

**Lemma 5.8** *Suppose that $F'$ satisfies (5.104) and (5.106) in $U(x^\star, r_0)$. Let $x_0 \in U(x^\star, r_0)$ and let $\overline{L} : \Omega = [0, R - \| x_0 - x^\star \|) \longrightarrow \mathbb{R}$ defined by*

$$\overline{L}(u) = \frac{L(u + \| x_0 - x^\star \|)}{1-\displaystyle\int_0^{\|x_0-x^\star\|} L_0(u)\,du} \qquad (u \in \Omega). \tag{5.137}$$

*Then, the following hold*

*(i)* $r_\kappa \leq \overline{r}_\kappa + \| x_0 - x^\star \| \leq \tilde{r}_0$, *where*

$$\overline{r}_\kappa : = \sup\left\{ r \in (0, R) \ :\ \int_0^1 \overline{L}(u)\,du \leq 1 - \kappa \right\}.$$

*(ii)* $F'$ *satisfies* $\overline{L}$*–average Lipschitz condition in* $U(x_0, r_0 - \| x_0 - x^\star \|)$.

**Lemma 5.9** *Let $\phi_\kappa : [0, r_\kappa] \longrightarrow \mathbb{R}$ defined by*

$$\phi_\kappa(t) = b_\kappa - (2-\kappa)\,t + \kappa\,(r_\kappa - t) \int_0^t L(u)\,du + 2 \int_0^t L(u)\,(t-u)\,du.$$

*Then,* $\phi_\kappa$ *is a strictly decreasing function on* $[0, r_\kappa]$ *and has exact one zero* $\widehat{r}_\kappa \in [0, r_\kappa]$ *satisfying*

$$\frac{b_\kappa}{2-\kappa} < \widehat{r}_\kappa < r_\kappa.$$

**Theorem 5.6** *Suppose that* $F'$ *satisfies (5.104) and (5.106) in* $U(x^\star, r_0)$*. Let* $x_0 \in U(x^\star, \widehat{r}_\kappa)$*, where* $\widehat{r}_\kappa$ *is given in Lemma 5.9. Then,* $\{x_n\}$ *generated by GNM starting at* $x_0$ *converges to a zero of* $F'(.)^+ F(.)$*.*

**Corollary 5.3** *Suppose that* $F'$ *satisfies (5.104) in* $U(x^\star, r_0)$*. Let*

$$x_0 \in U(x^\star, \frac{b_\kappa}{2-\kappa}).$$

*Then,* $\{x_n\}$ *generated by GNM starting at* $x_0$ *converges to a zero of* $F'(.)^+ F(.)$*.*

**Corollary 5.4** *Suppose that* $F'$ *satisfies (5.104) in* $U(x^\star, r_0)$ *and the condition (5.131) holds. Let* $x_0 \in \overline{U}(x^\star, \widehat{r}_0)$*, where* $\widehat{r}_0$ *is the exact one zero of function* $\phi_0 : [0, r_0] \longrightarrow \mathbb{R}$ *defined by*

$$\phi_0(t) = b_0 - 2\,t + 2 \int_0^t L(u)\,(t-u)\,du.$$

*Then,* $\{x_n\}$ *generated by GNM starting at* $x_0$ *converges to a zero of* $F'(.)^+ F(.)$*.*

**Corollary 5.5** *Suppose that* $F'$ *is surjective and satisfies (5.104) in* $U(x^\star, r_0)$*. Let* $x_0 \in \overline{U}(x^\star, \widehat{r}_0)$*, where* $\widehat{r}_0$ *is given in Corollary 5.4. Then,* $\{x_n\}$ *generated by GNM starting at* $x_0$ *converges to a solution of* $F(x) = 0$*.*

**Remark 5.8** *The local results obtained can also be used to solve equation of the form* $F(x) = 0$*, where* $F'$ *satisfies the autonomous differential equation (cf. [157])*

$$F'(x) = T(F(x)), \tag{5.138}$$

*where* $T : \mathcal{Y} \longrightarrow \mathcal{X}$ *is a known continuous operator. Since* $F'(x^\star) = T(F(x^\star)) = T(0)$*, we can apply our results without actually knowing the solution of* $x^\star$ *of equation* $F(x) = 0$*.*

*As an example, let* $\mathcal{X} = \mathcal{Y} = (-\infty, +\infty)$*,* $\mathcal{D} = \overline{U}(0,1)$ *and define function* $F$ *on* $\mathcal{D}$ *by*

$$F(x) = e^x - 1. \tag{5.139}$$

*Then, for* $x^\star = 0$*, we can set* $T(x) = x + 1$ *in (5.138).*

We shall provide a semilocal convergence for GNM using our new concept of recurrent functions. This idea has already produced a finer convergence analysis for iterative methods using invertible operators or outer or generalized inverses (cf. [165], [196]).

We need to define some parameters, sequences and functions.

**Definition 5.4** *Let $x_0 \in \mathcal{D}$, $\kappa \in [0,1)$ and $\lambda \in [0,1)$. Define parameter $\beta$, iteration $\{v_{\lambda,n}\}$, functions $f_{\lambda,n}$, $\varepsilon_{\lambda,n}$, $\mu_{\lambda,n}$ on $[0,1)$ and $\xi_\lambda$ on $I_{\xi_\lambda} = [0,1]^2 \times [1, \frac{1}{1-d_\lambda}]^2$ ($d_\lambda \in [0,1)$) by*

$$v_{\lambda,0} = 0, \quad v_{\lambda,1} = \beta,$$

$$v_{\lambda,n+1} = v_{\lambda,n} + \frac{\delta_{\lambda,n} + \kappa}{1 - \overline{\delta}_{\lambda,n}} (v_{\lambda,n} - v_{\lambda,n-1}) \quad (n \geq 1), \tag{5.140}$$

$$f_{\lambda,n}(d_\lambda) = \int_0^1 \int_{w_{\lambda,n-1}(d_\lambda)}^{z_{\lambda,n-1}(d_\lambda)} L(u)\, du\, d\theta + d_\lambda \int_0^{w_{\lambda,n}(d_\lambda)} L_0(u)\, du + c_\lambda, \tag{5.141}$$

$$\varepsilon_{\lambda,n}(d_\lambda) = \int_0^1 \left( \int_{w_{\lambda,n}(d_\lambda)}^{z_{\lambda,n}(d_\lambda)} L(u)\, du - \int_{w_{\lambda,n-1}(d_\lambda)}^{z_{\lambda,n-1}(d_\lambda)} L(u)\, du \right) d\theta$$
$$+ d_\lambda \int_{w_{\lambda,n}(d_\lambda)}^{w_{\lambda,n+1}(d_\lambda)} L_0(u)\, du, \tag{5.142}$$

$$\mu_{\lambda,n}(d_\lambda) = \int_0^1 \left( \int_{w_{\lambda,n+1}(d_\lambda)}^{z_{\lambda,n+1}(d_\lambda)} L(u)\, du + \int_{w_{\lambda,n-1}(d_\lambda)}^{z_{\lambda,n-1}(d_\lambda)} L(u)\, du \right.$$
$$\left. - 2 \int_{w_{\lambda,n}(d_\lambda)}^{z_{\lambda,n}(d_\lambda)} L(u)\, du \right) d\theta$$
$$+ d_\lambda \left( \int_{w_{\lambda,n+1}(d_\lambda)}^{z_{\lambda,n+1}(d_\lambda)} L_0(u)\, du - \int_{w_{\lambda,n}(d_\lambda)}^{z_{\lambda,n}(d_\lambda)} L_0(u)\, du \right),$$

$$\xi_\lambda(\theta, d_\lambda, a_\lambda, e_\lambda) \tag{5.143}$$
$$= \int_0^1 \left( \int_{(a_\lambda + e_\lambda + e_\lambda\, d_\lambda)\, \beta}^{(a_\lambda + e_\lambda + e_\lambda\, d_\lambda + \theta\, d_\lambda^2)\, \beta} L(u)\, du + \int_{a_\lambda\, \beta}^{(a_\lambda + \theta\, e_\lambda)\, \beta} L(u)\, du \right.$$
$$\left. - 2 \int_{(a_\lambda + e_\lambda)\, \beta}^{(a_\lambda + e_\lambda + \theta\, e_\lambda\, d_\lambda)\, \beta} L(u)\, du \right) d\theta$$
$$+ d_\lambda \left( \int_{(a_\lambda + e_\lambda + e_\lambda\, d_\lambda)\, \beta}^{(a_\lambda + e_\lambda + e_\lambda\, d_\lambda + d_\lambda^2\, e_\lambda)\, \beta} L_0(u)\, du - \int_{(a_\lambda + e_\lambda)\, \beta}^{(a_\lambda + e_\lambda + e_\lambda\, d_\lambda)\, \beta} L_0(u)\, du \right), \tag{5.144}$$

*where*

$$\delta_{\lambda,n} = \int_0^1 \int_{v_{\lambda,n-1}}^{v_{\lambda,n-1} + \theta\, (v_{\lambda,n} - v_{\lambda,n-1})} L(u)\, du\, d\theta, \tag{5.145}$$

$$\overline{\delta}_{\lambda,n} = \int_0^{v_{\lambda,n}} L_0(u)\, du, \tag{5.146}$$

$$z_{\lambda,n}(d_\lambda) = \left( \frac{1 - d_\lambda^n}{1 - d_\lambda} + \theta\, d_\lambda^n \right) \beta, \tag{5.147}$$

$$w_{\lambda,n}(d_\lambda) = \frac{1 - d_\lambda^n}{1 - d_\lambda}\,\beta, \tag{5.148}$$

$$c_\lambda = \kappa - d_\lambda. \tag{5.149}$$

*Define a function $f_{\lambda,\infty}$ on $[0,1)$ by*

$$f_{\lambda,\infty}(d_\lambda) = \lim_{n \longrightarrow \infty} f_{\lambda,n}(d_\lambda). \tag{5.150}$$

**Remark 5.9** *Using (5.141) and (5.150), we get*

$$f_{\lambda,\infty}(d_\lambda) = d_\lambda \int_0^{\frac{\beta}{1-d_\lambda}} L_0(u)\,du + c_\lambda. \tag{5.151}$$

*Then, it follows from (5.141)–(5.149) that the following identities hold*

$$f_{\lambda,n+1}(d_\lambda) = f_{\lambda,n}(d_\lambda) + \varepsilon_{\lambda,n}(d_\lambda), \tag{5.152}$$

$$\varepsilon_{\lambda,n+1}(d_\lambda) = \varepsilon_{\lambda,n}(d_\lambda) + \mu_{\lambda,n}(d_\lambda) \tag{5.153}$$

*and*

$$\mu_{\lambda,n}(d_\lambda) = \xi_\lambda(\theta, d_\lambda, a_\lambda = (1 + d_\lambda + \cdots + d_\lambda^{n-2})\,\beta, e_\lambda = d_\lambda^{n-1}\,\beta). \tag{5.154}$$

We need the following result on majorizing sequences for GNM

**Lemma 5.10** *Let parameters $\beta$, $\kappa$, $\lambda$, iteration $\{v_{\lambda,n}\}$ and functions $f_{\lambda,n}$, $\varepsilon_{\lambda,n}$, $\mu_{\lambda,n}$ and $\xi_\lambda$ be as in Definition 5.4. Assume there exists $\alpha_\lambda \in (0,1)$ such that*

$$\int_0^1 \int_0^{\theta\,\beta} L(u)\,du\,d\theta + \kappa \le \alpha_\lambda \left(1 - \int_0^\beta L_0(u)\,du\right), \tag{5.155}$$

$$c_\lambda = \kappa - \alpha_\lambda < 0, \tag{5.156}$$

$$4[pt]\xi_\lambda(\theta, q_1, q_2, q_3) \ge 0 \quad \text{on} \quad I_{\xi_\lambda}, \tag{5.157}$$

$$4[pt]\varepsilon_{\lambda,1}(\alpha_\lambda) \ge 0 \tag{5.158}$$

*and*

$$f_{\lambda,\infty}(\alpha_\lambda) \le 0. \tag{5.159}$$

*Then, iteration $\{v_{\lambda,n}\}$ given by (5.140) is non–decreasing, bounded from above by*

$$v_\lambda^{\star\star} = \frac{\beta}{1 - \alpha_\lambda} \tag{5.160}$$

*and converges to its unique least bound $v_\lambda^\star$ such that*

$$v_\lambda^\star \in [0, v_\lambda^{\star\star}]. \tag{5.161}$$

*Moreover, the following estimates hold for all $n \ge 0$*

$$0 \le v_{\lambda,n+1} - v_{\lambda,n} \le \alpha_\lambda\,(v_{\lambda,n} - v_{\lambda,n-1}) \le \alpha_\lambda^n\,\beta \tag{5.162}$$

*and*

$$0 \le v_\lambda^\star - v_{\lambda,n} \le \frac{\alpha_\lambda^n\,\beta}{1 - \alpha_\lambda}. \tag{5.163}$$

**Proof** Estimate (5.162) is true if

$$\delta_{\lambda,n} + \kappa \leq \alpha_\lambda \,(1 - \overline{\delta}_{\lambda,n}) \tag{5.164}$$

holds for all $n \geq 1$. It follows from (5.140), (5.155) and (5.156) that estimate (5.164) holds for $n = 1$. Then, we also have that (5.162) holds for $n = 1$ and

$$v_{\lambda,n} \leq \frac{1 - \alpha_\lambda^n}{1 - \alpha_\lambda}\,\beta < v_\lambda^{\star\star}. \tag{5.165}$$

Using the induction hypotheses and (5.165), estimate (5.164) is true if

$$\delta_{\lambda,k} + \alpha_\lambda\,\overline{\delta}_{\lambda,k} + c_\lambda \leq 0 \tag{5.166}$$

or

$$\int_0^1 \int_{w_{\lambda,k-1}(\alpha_\lambda)}^{z_{\lambda,k-1}(\alpha_\lambda)} L(u)\,du\,d\theta + \alpha_\lambda \int_0^{w_{\lambda,k}(\alpha_\lambda)} L_0(u)\,du + c_\lambda \leq 0 \tag{5.167}$$

hold for all $k \leq n$. Estimate (5.167) (for $d_\lambda = \alpha_\lambda$) motivates us to introduce function $f_{\lambda,k}$ given by (5.141) and show instead of (5.167)

$$f_{\lambda,k}(\alpha_\lambda) \leq 0. \tag{5.168}$$

We have by (5.152)–(5.154) (for $d_\lambda = \alpha_\lambda$) and (5.158) that

$$f_{\lambda,k+1}(\alpha_\lambda) \geq f_{\lambda,k}(\alpha_\lambda). \tag{5.169}$$

In view of (5.150), (5.151) and (5.169), estimate (5.168) shall holds if (5.159) is true, since

$$f_{\lambda,k}(\alpha_\lambda) \leq f_{\lambda,\infty}(\alpha_\lambda) \tag{5.170}$$

and the induction is completed. It follows from (5.162) and (5.165) that iteration $\{v_{\lambda,n}\}$ is non–decreasing, bounded from above by $v_\lambda^{\star\star}$ given by (5.160) and as such it converges to $v_\lambda^\star$.

Finally, estimate (5.163) follows from (5.162) by using the standard majorization techniques. This completes the proof of Lemma 5.10. ⊠

We can show the following semilocal convergence result for GNM using recurrent functions, wich is the analog of Theorem 5.4.

**Theorem 5.7** *Let* $\lambda \geq \tilde{\lambda}_0$. *Assume*

$$\overline{U}(x_0, v_\lambda^\star) \subseteq \mathcal{D}; \tag{5.171}$$

*$F'$ satisfies (5.104) and (5.106) on $U(x_0, v_\lambda^\star)$; and Hypotheses of Lemma 5.4 hold.*

*Then, the sequence $\{x_n\}$ generated by GNM is well defined, remains in $\overline{U}(x_0, v_\lambda^\star)$ for all $n \geq 0$ and converges to a zero $x^\star$ of $F'(.)^+\,F(.)$ in $\overline{U}(x_0, v_\lambda^\star)$. Moreover, the following estimates hold*

$$\parallel x_{n+1} - x_n \parallel \leq v_{\lambda,n+1} - v_{\lambda,n} \tag{5.172}$$

*and*

$$\parallel x_n - x^\star \parallel \leq v_\lambda^\star - v_{\lambda,n}. \tag{5.173}$$

**Proof** As in Theorem 5.4, we arrive at the estimate on line above (5.124) (with $v_{\lambda,k}$ replacing $s_{\lambda,k}$), which in view of (5.140) leads to

$$\| x_{k+1} - x_k \| \leq v_{\lambda,k+1} - v_{\lambda,k}. \tag{5.174}$$

Estimates (5.118), (5.119), (5.174) and Lemma 5.10 implies that sequence $\{x_k\}$ is a complete sequence in $\mathbb{R}^m$ and as such it converges to some $x^\star \in \overline{U}(x_0, v_\lambda^\star)$ (since $\overline{U}(x_0, v_\lambda^\star)$ is a closed set). This completes the proof of Theorem 5.7. ⊠

**Remark 5.10** *(a) The point $v_\lambda^{\star\star}$ given in closed form by (5.160) can replace $v_\lambda^\star$ in hypothesis (5.171).*

*(b) Hypotheses of Lemma 5.141 involve only computations at the initial data. These hypotheses differ from (5.113) given in Theorem 5.4. In practice, we shall test to see which of the two are satisfied if any. If both conditions are satisfied, we shall use the more precise error bounds given in Theorem 5.7.*

We show that the conditions of Theorem 5.7 can be weaker than those of Theorem 5.4. We compare the Kantorovich–type conditions with the corresponding ones in this section.

## Semilocal Case

An operator $Q : \mathbb{R}^m \longrightarrow \mathbb{R}^l$ is said to be Lipschitz continuous on $\mathcal{D}_0 \subseteq \mathcal{D}$ with modulus $L > 0$ if

$$\| Q(x) - Q(y) \| \leq L \| x - y \| \quad (x, y \in \mathcal{D}_0) \tag{5.175}$$

and center–Lipschitz continuous on $\mathcal{D}_0$ with modulus $L_0 > 0$ if

$$\| Q(x) - Q(x_0) \| \leq L_0 \| x - x_0 \| \quad (x \in \mathcal{D}_0). \tag{5.176}$$

Let $x_0 \in \mathcal{D}$ and $r > 0$ be such that $U(x_0, r) \subseteq \mathcal{D}$. Clearly, if $\| F'(x_0)^+ \| F'$ is Lipschitz continuous on $U(x_0, r)$ with modulus $L$, then $F'$ satisfies the modified $L$–average Lipschitz condition on $U(x_0, r)$. Similarily, if $\| F'(x_0)^+ \| F'$ is center–Lipschitz continuous on $U(x_0, r)$ with modulus $L_0$, then $F'$ satisfies the modified center $L_0$–average Lipschitz condition on $U(x_0, r)$. Using (5.71), (5.72), (5.78) and (5.79), we get for $t \geq 0$

$$g(t) = \beta - t + \frac{L_0}{2} t^2, \tag{5.177}$$

$$h_\lambda(t) = \beta - (1 - \lambda) t + \frac{L}{2} t^2, \tag{5.178}$$

$$r_\lambda = \frac{1 - \lambda}{L} \tag{5.179}$$

and

$$b_\lambda = \frac{(1-\lambda)^2}{2\,L}. \tag{5.180}$$

Moreover, if

$$\beta \leq b_\lambda, \tag{5.181}$$

then

$$t_\lambda^\star = \frac{1-\lambda-\sqrt{(1-\lambda)^2-2\,\beta\,L}}{L}. \tag{5.182}$$

We have the following improvement of (cf. [504, Theorem 5.1]).

**Theorem 5.8** *Let $\lambda = \tilde{\lambda}_0 = (1-\beta\,L_0)\,\kappa$. Assume*

$$\beta\,L \leq \Delta = \frac{(1-\kappa)^2}{\kappa^2-\kappa+1+\sqrt{2\,\kappa^2-2\,\kappa+1}}; \tag{5.183}$$

$$\overline{U}(x_0, v_\lambda^\star) \subseteq \mathcal{D}; \tag{5.184}$$

*and $\parallel F'(x_0)^+ \parallel$ $F'$ satisfies (5.175) and (5.176) on $U(x_0, t_\lambda^\star)$. Then, sequence $\{x_n\}$ generated by GNM is well defined, remains in $\overline{U}(x_0, t_\lambda^\star)$ for all $n \geq 0$ and converges to a zero $x^\star$ of $F'(.)^+\,F(.)$ in $\overline{U}(x_0, t_\lambda^\star)$. Moreover, the following estimates hold*

$$\parallel x_{n+1} - x_n \parallel \leq s_{\lambda,n+1} - s_{\lambda,n} \tag{5.185}$$

*and*

$$\parallel x_n - x^\star \parallel \leq t_\lambda^\star - s_{\lambda,n}. \tag{5.186}$$

**Proof** Similarly, replace $\{t_{\lambda,n}\}$ by $\{s_{\lambda,n}\}$ in the proof of (cf. [504, Theorem 5.1]). That completes the proof of Theorem 5.8. ⊠

We provide now example, where $\kappa \neq 0$ and the hypotheses of Theorem 5.8 are satisfied, but not earlier ones (cf. [412], [443]).

**Example 5.2** *(cf. [206], [443]) Let $i = j = 2$ and $\mathbb{R}^2$ be equipped with the $\ell_1$–norm. Choose the starting point as*

$$x_0 = (.2505, 0)^T, \quad \mathcal{D} = \{x = (v,w)^T \,:\, -1 < v < 1 \text{ and } -1 < w < 1\}.$$

*Define a function $F$ on $U(x_0, \sigma) \subseteq \mathcal{D}$ ($\sigma = .72$) by*

$$F(x) = (v-w, .5\,(v-w)^2), \qquad x = (v,w)^T \in \mathcal{D}. \tag{5.187}$$

*Then, for each $x = (v,w)^T \in \mathcal{D}$, the Fréchet–derivative of $F$ at $x$ and the Moore–Penrose–pseudoinverse of $F'(x)$ are given by*

$$F'(x) = \begin{bmatrix} 1 & -1 \\ v-w & w-v \end{bmatrix} \tag{5.188}$$

*and*

$$F'(x)^+ = \frac{1}{2\,(1+(v-w)^2)} \begin{bmatrix} 1 & v-w \\ -1 & w-v \end{bmatrix}, \tag{5.189}$$

*respectively. Let* $x = (v_1, w_1)^T \in \mathcal{D}$ *and* $y = (v_2, w_2)^T \in \mathcal{D}$*. By (5.188), we have*

$$\| F'(x) - F'(y) \| = |(v_1 - v_2) - (w_1 - w_2)| \leq \| x - y \| .$$

*That is,* $L = L_0 = 1$*. Using (5.188), (5.189) and (5.101), we obtain that*

$$\| F'(y)^+ \ (\mathcal{I} - F'(x) \ F'(x)^+) \ F(x) \| \leq \frac{2}{5} \ \| x - y \| .$$

*Hence, the constant* $\kappa$ *in hypothesis (5.101) is given by*

$$\kappa = .4.$$

*Using hypotheses of Theorem 5.8, (5.187)–(5.189), (cf. [412, Theorem 2.4]) and [443, Theorem 3.1]) are not applicable. However, our Theorem 5.8 can apply to solve equation (5.187).*

**Remark 5.11** *If* $L = L_0$*, Theorem 5.8 reduces to (cf. [504, Theorem 5.1]). Otherwise, it constitues an improvements (see Lemma 5.4). If* $\kappa = 0$ *(Newton's method for* $h_0$*), we have* $\lambda = 0$ *and* $\Delta = 1/2$*. Then, the sequences* $\{t_{\lambda,n}\}$ *and* $\{v_{\lambda,n}\}$ *reduce to:*

$$t_0 = 0, \quad t_1 = \beta, \quad t_{n+1} = t_n + \frac{L(t_n - t_{n-1})^2}{2(1 - L \ t_n)} \tag{5.190}$$

*and*

$$v_0 = 0, \quad v_1 = \beta, \quad v_{n+1} = v_n + \frac{L \ (v_n - v_{n-1})^2}{2 \ (1 - L_0 \ v_n)}, \tag{5.191}$$

*respectively. The corresponding sufficient convergence conditions are:*

$$h_{LHW} = \beta L \leq \frac{1}{2} \qquad (cf. \ [504]) \tag{5.192}$$

*and*

$$h_{AH} = \beta \, \overline{L} \leq \frac{1}{2}, \tag{5.193}$$

*where*

$$\overline{L} = \frac{1}{8} (L + 4 \, L_0 + \sqrt{L^2 + 8 \, L_0 \, L}) \tag{5.194}$$

*(see Lemma 5.10). Note that*

$$h_{LHW} \leq \frac{1}{2} \Longrightarrow h_{AH} \leq \frac{1}{2}, \tag{5.195}$$

*but not necessarily vice versa unless if* $L = L_0$*. Moreover, since* $\dfrac{L_0}{L}$ *can be arbitrarily small, we have by (5.192) and (5.193) that*

$$\frac{h_{LHW}}{h_{AH}} \longrightarrow \frac{1}{4} \quad \text{as} \quad \frac{L_0}{L} \longrightarrow 0. \tag{5.196}$$

*That is, our approach extends the applicability of GNM by at most four times. The applicability of GNM can be extended even further (see section 2.1).*

*Concerning the error bounds, we have already shown (see Lemma 5.4) that* $\{v_n\}$ *is a tighter majorizing sequence for* $\{x_n\}$ *than* $\{t_n\}$ *(see Example 5.3 (b)).*

**Example 5.3** *(see Example 2.11) Let* $x_0 = (.9, .9)^T$ *and* $\varrho = .7$ *(here* $\varrho$ *is* $\beta$ *in Example 2.11). Using hypotheses of Theorem 5.8, we get that*

$$\beta = .1, \quad L = 2.6, \quad L_0 = 2.3, \quad \overline{L} = 2.39864766,$$

$$h_{LHW} = .26 \quad \text{and} \quad h_{AH} = .239864766.$$

*Then, (5.192) and (5.193) are satisfied. We have also:*

$$F'(.9, .9)^{-1} = .4115226337\ \mathcal{I},$$

*where* $\mathcal{I}$ *is the identity* $2 \times 2$ *matrix. The hypotheses of our Theorem 5.8 and the Kantorovich theorem are satisfied. Then, NM starting at* $x_0$ *converges to* $x^\star = (.8879040017, .8879040017)^T$. *We also can provide the comparison table using the software Maple* 13.

**Comparison table**

| | NM | (5.191) | (5.190) |
|---|---|---|---|
| $n$ | $\parallel x_{n+1} - x_n \parallel$ | $v_{n+1} - v_n$ | $t_{n+1} - t_n$ |
| 0 | .01193415638 | .1 | .1 |
| 1 | .0001618123748 | .01688311688 | .01756756757 |
| 2 | 2.946995467e-8 | .0005067933842 | .0005778355237 |
| 3 | 4.228114294e-11 | 4.57383463010e-7 | 6.265131450e-7 |
| 4 | ~ | 3.725461907e-13 | 7.365175646e-13 |
| 5 | ~ | 2.471607273e-25 | 1.017862116e-24 |
| 6 | ~ | 1.087872853e-49 | 1.944019580e-48 |
| 7 | ~ | 2.107538365e-98 | 7.091269701e-96 |
| 8 | ~ | 7.909885354e-196 | 9.435626465e-191 |
| 9 | ~ | 1.114190851e-390 | 1.670568212e-380 |
| 10 | ~ | 2.210743650e-780 | 5.236621208e-760 |

*The table shows that our error bounds* $v_{n+1} - v_n$ *are tighter than* $t_{n+1} - t_n$.

**Remark 5.12** *If* $L = L_0$, *the error bounds reduce to the classical ones (cf. [504]). Otherwise, the error bounds are tighter since the ratio* $2\,h_{AH}$ *is smaller than* $2\,h_{LHW}$ *(see also the comparison table in Example 5.3).*

Other applications and examples where $L_0 < L$ including the solution of nonlinear Chandrasekhar–type integral equations appearing in radiative transfer are also find in section 2.1.

## Local Case

Let $x^\star \in \mathcal{D}$ be such that $F(x^\star) = 0$, $F'(x^\star) \neq 0$ and (5.132) holds. We present the following improvement of (cf. [504, Theorem 5.2]).

**Theorem 5.9** *Assume that* $\| F'(x^\star)^+ \| \, F'$ *satisfies (5.175) and (5.176) on* $U(x^\star, \frac{1}{L_0})$. *Then, sequence* $\{x_n\}$ *generated by GNM is well defined, remains in* $\overline{U}(x^\star, R_0)$ *for all* $n \geq 0$ *and converges to a zero* $x^\star$ *of* $F'(.)^+ \, F(.)$ *in* $\overline{U}(x^\star, R_0)$, *where*

$$R_0 = \begin{cases} \dfrac{L + 2\,\Delta\,L_0 - \sqrt{(L + 2\,\Delta\,L_0)^2 - 2\,\Delta\,(L^2 + 2\,\Delta\,L_0^2)}}{2\,(L^2 + 2\,\Delta\,L_0^2)} & \text{if} \quad L + 2\,\Delta\,(2\,L_0 - L) \geq 0, \\[2ex] \dfrac{1}{L_0} & \text{if} \quad L + 2\,\Delta\,(2\,L_0 - L) < 0. \end{cases} \tag{5.197}$$

**Proof** Note that the quantity under the radical in non–negative by (5.197). Let

$$\tilde{L} = \frac{L}{1 - L_0 \, \| x_0 - x^\star \|}, \qquad \tilde{r} = \frac{1}{L_0} - \| x_0 - x^\star \|$$

and

$$\tilde{\lambda}_0 = (1 - \beta\,L_0)\,\varrho.$$

Then, by Lemma 5.8, $\| F'(x^\star) \|^+ \; F'$ is $\tilde{L}$–Lipschitz continuous on $\overline{U}(x_0, \tilde{r})$ and

$$\tilde{t}^\star_{\tilde{\lambda}_0} \leq \tilde{t}^\star_\kappa \leq \tilde{r}_\kappa \leq \tilde{r}_0 - \| x_0 - x^\star \| = \tilde{r} \tag{5.198}$$

since

$$\tilde{\lambda}_0 \leq \lambda.$$

That is,

$$U(x^\star, \tilde{t}^\star_\lambda) \subseteq \overline{U}(x_0, \tilde{r}) \subseteq \mathcal{D}.$$

Using Lemme 5.7, we have

$$\beta \leq \frac{(2 - L \, \| x_0 - x^\star \|) \, \| x_0 - x^\star \|}{2\,(1 - L_0 \, \| x_0 - x^\star \|)},$$

so

$$\beta\,\tilde{L} \leq \frac{(2 - L \, \| x_0 - x^\star \|)\,L \, \| x_0 - x^\star \|}{2\,(1 - L_0 \, \| x_0 - x^\star \|)^2} \leq \Delta$$

by the choice of $R_0$ and the function

$$\Upsilon \, : \, (s, t) \longrightarrow \frac{2\,(2 - t)\,t}{2\,(1 - s)^2}.$$

For $s \leq t$, $\Upsilon$ is increasing on $(0, 1)^2$. Therefore, Theorem 5.8 is applicable. This completes the proof of Theorem 5.9. ⊠

**Remark 5.13** *If $L = L_0$, Theorem 5.9 reduces to (cf. [504, Theorem 5.2]). Otherwise, it is constitues an improvement since the radius of convergence*

$$R_1 = \frac{1 - \dfrac{1}{\sqrt{2\,\Delta + 1}}}{L}$$

*given in (cf. [504]) is smaller than $R_0$. Hence, our Theorem 5.9 allows a wider choice of initial guesses $x_0$ and also provides tighter error bounds than (cf. [504, Theorem 5.2]).*

# Chapter 6

# Newton–Type Methods

We present in this chapter Newton–type methods. We establish convergence results using Lipschitz–type conditions.

## 6.1 Convergence with Outer Inverses

We provide in this section new semilocal convergence results for Newton–like method using outer inverses but no Lipschitz conditions in a Banach space setting. The first is the Kantorovich–type approach, whereas the second uses our new concept of recurrent functions. Comparisons are given between the two techniques. Our results are compared favorably with earlier ones using the information and requiring the same computational cost. We are concerned with the problem of approximating a locally unique solution $x^\star$ of equation

$$\mathcal{Q}F(x) = 0, \tag{6.1}$$

where $F$ is a Fréchet–differentiable operator defined on an open convex subset $\mathcal{D}$ of a Banach space $\mathcal{X}$ with values in Banach space $\mathcal{Y}$ and $\mathcal{Q} \in \mathcal{L}(\mathcal{Y}, \mathcal{X})$.

We shall use the Newton–like method NLM

$$x_{n+1} = x_n - A(x_n)^{\#}\, F(x_n) \quad (n \geq 0,\ x_0 \in D) \tag{6.2}$$

to generate a sequence $\{x_n\}$ approximating $x^\star$. Here, $A(x_n) \in \mathcal{L}(\mathcal{X}, \mathcal{Y})$ is an approximation of the Fréchet–derivative $F'(x_n)$ and $A(x_n)^{\#}$ denotes an outer inverse of $A(x_n)$, i.e.,

$$A(x_n)^{\#}\, A(x_n)\, A(x_n)^{\#} = A(x_n)^{\#} \qquad (n \geq 0).$$

This general setting includes generalized Newton methods GNM for undetermined systems, the Gauss–Newton method GNM for nonlinear least–squares problems, a Newton–type method suitable for ill–posed equations and Newton–type method for solving equations provided that $A(x)^{\#} = A(x)^{-1}$ $(x \in \mathcal{D})$ (cf. [95]). Outer inverses and generalized inverses have been used by several authors in connection with NLM. A survey of such results can be found in (cf. [157] (see also (cf. [194], [227], [277])). The Lipschitz condition

$$\| F'(x) - F'(y) \| \leq H \ \| x - y \| \quad (x, y \in \mathcal{D}) \tag{6.3}$$

is the crucial hypothesis in the convergence analysis of NLM. However, they are many examples in the literature, where (6.3) is vilated (cf. [157]) (see also Example 6.1). Here, we expand the applicability of NLM by considering instead of (6.3) condition

$$\| F(x) - F(y) - F'(x)\,(x-y) \| \leq H \ \| x-y \|^2 \quad (x, y \in \mathcal{D}). \tag{6.4}$$

Note that (6.3) implies (6.4) but not necessarily vice versa. Simply, let $\mathcal{X} = \mathcal{Y} = \mathbb{R}$ and consider the function

$$F(x) = \frac{H}{2}\, x^2.$$

It turns out that (6.4) can replace (6.3) in all convergence results (cf. [461], [277]), involving the latter without changing the rest of the hypotheses (see Theorems 6.1, 6.2, 6.3 and Corollary 6.1).

We provide some results on perturbations bounds for outer inverses that can originally be found in (cf. [277]) (see also (cf. [157])). For a comprehensive theory of various inverses in Banach spaces, see (cf. [227], [556]). Let $A$ be a linear operator. Then, $\mathcal{N}(A)$, $\mathcal{R}(A)$ denote the null space and range of $A$, respectively. We need the following Lemmas. The proofs can be found in (cf. [277]).

**Lemma 6.1** *Let $A \in \mathcal{L}(\mathcal{X}, \mathcal{Y})$. If $A^{\#}$ is a bounded outer inverse of $A$, then, the following direct sum decompositions hold*

$$\mathcal{X} = \mathcal{R}(A^{\#}) \oplus \mathcal{N}(A^{\#}\, A)$$

*and*

$$\mathcal{Y} = \mathcal{N}(A^{\#}) \oplus \mathcal{R}(A\, A^{\#}).$$

**Lemma 6.2** *Let $A \in \mathcal{L}(\mathcal{X}, \mathcal{Y})$ and $A^{\#}$ be a bounded outer inverse of $A$. Let $B \in \mathcal{L}(\mathcal{X}, \mathcal{Y})$ be such that $\| A^{\#}\,(B-A) \| < 1$. Then, $B^{\#} := (I + A^{\#}\,(B-A))^{-1}\, A^{\#}$ is a bounded outer inverse of $B$ with $\mathcal{N}(B^{\#}) = \mathcal{N}(A^{\#})$ and $\mathcal{R}(B^{\#}) = \mathcal{R}(A^{\#})$. Moreover, the following hold*

$$\| B^{\#} - A^{\#} \| \leq \frac{\| A^{\#}\,(B-A)\, A^{\#} \|}{1 - \| A^{\#}\,(B-A) \|} \leq \frac{\| A^{\#}\,(B-A) \| \, \| A^{\#} \|}{1 - \| A^{\#}\,(B-A) \|}$$

*and*

$$\| B^{\#}\, A \| \leq \frac{1}{1 - \| A^{\#}\,(B-A) \|}.$$

**Lemma 6.3** *Let $A, B \in \mathcal{L}(\mathcal{X}, \mathcal{Y})$ and $A^{\#}$, $B^{\#}$ are a bounded outer inverses of $A$ and $B$, respectively. Then, $B^{\#}\,(I - A\, A^{\#}) = 0$ if and only if $\mathcal{N}(A^{\#}) \subset \mathcal{N}(B^{\#})$.*

We shall show the following semilocal convergence theorem for NLM.

**Theorem 6.1** *Let $F : \mathcal{D} \subseteq \mathcal{X} \to \mathcal{Y}$ be a Fréchet–differentiable operator. Assume that*

(a) *there exist an approximation $A(x) \in \mathcal{L}(\mathcal{X}, \mathcal{Y})$ of $F'(x)$, an open convex subset $\mathcal{D}_0$ of $\mathcal{D}$, $x_0 \in \mathcal{D}_0$, a bounded outer inverse $A^{\#}$ of $A(x_0) := A$ and constants $\eta > 0$, $K > 0$, $M > 0$, $L > 0$, $\mu \geq 0$, $\ell \geq 0$ such that, for all $x, y \in \mathcal{D}_0$, the following assertions hold*

$$\| A^{\#}\ F(x_0) \| \leq \eta, \tag{6.5}$$

$$\| A^{\#}\ (F(x) - F(y) - F'(y)\,(x-y)) \| \leq \frac{K}{2}\ \| x - y \|^2, \tag{6.6}$$

$$\| A^{\#}\ (F'(x) - A(x)) \| \leq M\ \ \| x - x_0 \| + \mu \tag{6.7}$$

*and*

$$\| A^{\#}\ (A(x) - A(x_0)) \| \leq L\ \ \| x - x_0 \| + \ell; \tag{6.8}$$

(b)

$$b = \mu + \ell < 1, \tag{6.9}$$

$$h\frac{(1-b)^2}{2}, \tag{6.10}$$

*where,*

$$\sigma := \max\,\{K, M + L\},$$
$$\overline{U}(x_0, t^\star) \subseteq \mathcal{D}_0$$

*and*

$$t^\star = \frac{1 - b - \sqrt{(1-b)^2 - h}}{\sigma}.$$

*Then, we have*

(i) *Sequence $\{x_n\}$ $(n \geq 0)$ generated by NLM with*

$$A(x_n)^{\#} = [I + A^{\#}\ (A(x_n) - A(x_0))]^{-1}\ A^{\#}$$

*is well defined, remains in $U(x_0, t^\star)$ for all $n \geq 0$ and converges to a solution $x^\star$ of equation $A^{\#}\ F(x) = 0$;*

(ii) *The solution $x^\star$ is unique in $\tilde{U}(x_0, t^\star) \cap R(A^{\#}, x_0)$, where*

$$\tilde{U}(x_0, t^\star) = \begin{cases} \overline{U}(x_0, t^\star) \cap \mathcal{D}_0 & \text{if} \quad h = \dfrac{(1-b)^2}{2}, \\ U(x_0, t^{\star\star}) \cap \mathcal{D}_0 & \text{if} \quad h < \dfrac{(1-b)^2}{2}, \end{cases}$$

*$t^{\star\star}$ is the large zero of function $f$ given by*

$$f(s) = \frac{\sigma}{2}\,s^2 - (1-b)\,s + \eta$$

*and*

$$R(A^{\#}, x_0) \equiv R(A^{\#}) + x_0 = \{x + x_0 : x \in R(A^{\#})\}.$$

*Moreover, define function $q$ by*

$$q(s) = 1 - L\,s - \ell \tag{6.11}$$

*and sequence $\{t_n\}$ $(n \geq 0)$ by*

$$t_0 = 0, \quad t_1 = \eta, \quad t_{n+2} = t_{n+1} + \frac{f(t_{n+1})}{q(t_{n+1})} \quad (n \geq 0). \tag{6.12}$$

*Then, the following estimates hold for all $n \geq 0$*

$$\| x_{n+1} - x_n \| \leq t_{n+1} - t_n \tag{6.13}$$

*and*

$$\| x_n - x^\star \| \leq t^\star - t_n. \tag{6.14}$$

**Proof** We shall show using induction on $m$, that (6.13) holds. Estimate (6.14) will then follow from (6.13) using standard majorization techniques. By the initial conditions, we have that

$$\| x_1 - x_0 \| \leq t_1 - t_0$$

and (6.13) holds for $m = 0$. Using (6.8), we get that

$$\| A^{\#} \ (A(x_1) - A) \| \leq L \ \| x_1 - x_0 \| + \ell \leq L\ t_1 + \ell \leq L\ t^\star + \ell < 1. \tag{6.15}$$

From perturbation Lemma 6.2 and (6.15), we obtain that $A(x_1)^{\#} := (I + A^{\#}\ (A(x_1) - A))^{-1}\ A^{\#}$ is an outer inverse of $A(x_1)$. Moreover, we have that

$$\| A(x_1)^{\#}\ A \| \leq (1 - L\ \| x_1 - x_0 \| - \ell)^{-1} \leq (1 - L\ t_1 - \ell)^{-1}$$

and $\mathcal{N}(A(x_1)^{\#}) = \mathcal{N}(A^{\#})$. Assume that, for $1 \leq m \leq k$

$$\| x_m - x_{m-1} \| \leq t_m - t_{m-1}$$

and

$$\mathcal{N}(A(x_{m-1})^{\#}) = \mathcal{N}(A^{\#}).$$

Then, we have

$$\| x_m - x_0 \| \leq \sum_{i=1}^{m} (t_i - t_{i-1}) \leq t_m - t_0 = t_m \tag{6.16}$$

and

$$\mathcal{N}(A(x_m)^{\#}) = \mathcal{N}(A(x_{m-1})^{\#}) = \mathcal{N}(A^{\#}). \tag{6.17}$$

Hence, we have by (6.2) and Lemma 6.3 that

$$A(x_m)^{\#}\ (I - A(x_{m-1})\ A(x_{m-1})^{\#}) = 0$$

and

$$\begin{aligned}x_{m+1}-x_m &= -A(x_m)^{\#}\,F(x_m)\\ &= -A(x_m)^{\#}\,(F(x_m)-A(x_{m-1})\,(x_m-x_{m-1})\\ &\quad - A(x_{m-1})\,A(x_{m-1})^{\#}F(x_{m-1}))\\ &= -A(x_m)^{\#}\,(F(x_m)-F(x_{m-1})-A(x_{m-1})\,(x_m-x_{m-1})) \qquad (6.18)\\ &= -A(x_m)^{\#}\,(F(x_m)-F(x_{m-1})-F'(x_{m-1})\,(x_m-x_{m-1})\\ &\quad + (F'(x_{m-1})-A(x_{m-1}))\,(x_m-x_{m-1}))\end{aligned}$$

We have by (6.17) and Lemma 6.3 that

$$A(x_m)^{\#}\;(I-A\;A^{\#})=0.$$

In view of (6.6)–(6.8), (6.16) and (6.18), we have

$$\begin{aligned}&\|x_{m+1}-x_m\|\\ &\le \|\,A(x_m)^{\#}\,A\,\|\,\Big(\|\,A^{\#}(F(x_m)-F(x_{m-1})-F'(x_{m-1})\,(x_m-x_{m-1}))\,\|\\ &\quad +\|\,A^{\#}\,(F'(x_{m-1})-A(x_{m-1}))\,\|\,\|\,x_m-x_{m-1}\,\|\Big)\\ &\le \frac{1}{1-L\,t_m-\ell}\left(\frac{K}{2}\,\|\,x_m-x_{m-1}\,\|^2+(M\,\|\,x_{m-1}-x_0\,\|+\mu)\,\|\,x_m-x_{m-1}\,\|\right)\\ &\le \frac{1}{1-L\,t_m-\ell}\left(\frac{\sigma}{2}\,(t_m-t_{m-1})+M\;t_{m-1}+\mu\right)(t_m-t_{m-1})=t_{m+1}-t_m,\end{aligned} \qquad (6.19)$$

which completes the induction. Hence, we have for any $m$ the following

$$\|\,x_{m+1}-x_m\,\|\le t_{m+1}-t_m,$$

$$\|\,A^{\#}\;(A(x_{m+1})-A)\,\|\le L\;\|\,x_{m+1}-x_0\,\|+\ell\le L\;t_{m+1}+\ell\le L\;t^\star+\ell<1,$$

$$\|\,x_m-x_0\,\|\le t_m\le t^\star$$

and $A(x_{m+1})^{\#}:=(I+A^{\#}\;(A(x_{m+1})-A))^{-1}\;A^{\#}$ is an outer inverse of $A(x)$. It follows that $x_m\in U(x_0,t^\star)$, $m\ge 0$ and $\{x_m\}$ converges to a point $x^\star$ in $\overline{U}(x_0,t^\star)$. The point $x^\star$ is a solution of $A^{\#}\;F(x)=0$. Indeed, by definition

$$A(x_m)^{\#}=(I+A^{\#}\;(A(x_m)-A))^{-1}\;A^{\#}\quad (m\ge 0)$$

and

$$\begin{aligned}0&=\lim_{m\longrightarrow\infty}(I+A^{\#}\,(A(x_m)-A))\,(x_m-x_{m-1})\\ &=\lim_{m\longrightarrow\infty}A^{\#}\,F(x_m)=A^{\#}\,F(x^\star).\end{aligned}$$

Hence, $x^\star$ solves equation $A^\# \, F(x^\star) = 0$.

To show that $x^\star$ is the unique solution of equation (6.1) in $\tilde{U}(x_0, t^\star) \cap R(A^\#, x_0)$, let $y^\star \neq x^\star$ such that $y^\star \in \tilde{U}(x_0, t^\star) \cap R(A^\#, x_0)$ and $A^\# \, F(y^\star) = 0$. Then, $y^\star - x^\star \in \mathcal{R}\,(A^\#)$ and

$$A^\# \, A\,(y^\star - x_k) = A^\# \, A\,(y^\star - x_0) + A^\# \, A\,(x_k - x_0) = y^\star - x_k \quad (k \geq 0).$$

By Lemma 6.2, we have that $\mathcal{R}\,(A(x_m)^\#) = \mathcal{R}\,(A^\#)$, for all $m \geq 0$, so

$$x_{m+1} - x_m = -A(x_m)^\# \, F(x_m) \in \mathcal{R}\,(A(x_m)^\#) = \mathcal{R}\,(A^\#).$$

Furthermore, using Lemma 6.1, we have $\mathcal{R}\,(A^\#) = \mathcal{R}\,(A^\# \, A)$ and $x_{m+1} \in x_m + \mathcal{R}\,(A^\#)$ for all $m \geq 0$. We also have the estimate

$$\begin{aligned} \| \, y^\star - x_1 \, \| &= \| \, y^\star - x_0 + A^\# \, F(x_0) - A^\# \, F(y^\star) \, \| \\ &\leq \| \, A^\# \, (F(y^\star) - F(x_0) - F'(x_0)\,(y^\star - x_0)) \, \| \\ &\quad + \| \, A^\# \, (F'(x_0) - A) \, \| \, \| \, y^\star - x_0 \, \| \\ &\leq \left( \frac{\sigma}{2} \, \| \, y^\star - x_0 \, \| + \ell + \mu \right) \| \, y^\star - x_0 \, \| = \phi(\| \, y^\star - x_0 \, \|), \end{aligned} \tag{6.20}$$

where,

$$\phi(s) = \frac{\sigma}{2} \, s^2 + (\ell + \mu) \, s.$$

Since,

$$\| \, y^\star - x_0 \, \| \leq \| \, y^\star - x_1 \, \| + \| \, x_1 - x_0 \, \| \leq \phi(\| \, y^\star - x_0 \, \|) + \eta,$$

then, $\phi(\| \, y^\star - x_0 \, \|) \geq 0$. Consequently, $y^\star \in \overline{U}(x_0, t^\star)$. We prove by induction that

$$\| \, y^\star - x_m \, \| \leq t^\star - t_m \quad (m \geq 0). \tag{6.21}$$

Inequality (6.21) holds for $m = 0$ since $y^\star \in \overline{U}(x_0, t^\star)$. Suppose that (6.21) holds for $m$. As in (6.18) and (6.19), we have the estimation

$$\begin{aligned} &\|y^\star - x_{m+1} \, \| \\ &\leq \| \, A(x_m)^\# \, A \, \| \, \left( \| \, A^\# \, (F(y^\star) - F(x_m) - F'(x_m)\,(y^\star - x_m)) \, \| \right. \\ &\quad \left. + \| \, A^\# \, (F'(x_m) - A(x_m)) \, \| \, \| \, y^\star - x_m \, \| \right) \\ &\leq (1 - \ell - L\,t_m)^{-1} \left( \frac{K}{2} \, \| \, y^\star - x_m \, \|^2 + (M \, \| \, x_m - x_0 \, \| + \mu) \, \| \, y^\star - x_m \, \| \right) \\ &\leq (1 - \ell - L\,t_m)^{-1} \left( \frac{\sigma}{2} \, (t^\star - t_m) + M\,t_m + \mu \right) (t^\star - t_m) \leq t^\star - t_{m+1}, \end{aligned} \tag{6.22}$$

which complete the induction for (6.21). It follows by (6.22) that $\lim_{m \longrightarrow \infty} x_m = y^\star$. But we showed $\lim_{m \longrightarrow \infty} x_m = x^\star$. Hence, we deduce $x^\star = y^\star$. This completes the proof of Theorem 6.1. ⊠

**Corollary 6.1** *Under the hypotheses of Theorem 6.1 with* $A(x) = F'(x)$ $(x \in \mathcal{D}_0)$*, the method (6.2) with* $A(x_n) : = F'(x_n)$ $(n \geq 0)$ *converges quadratically to a solution* $x^\star \in \tilde{U}(x_0, t^\star) \cap R(A^\#, x_0)$ *of equation* $F'(x_0)^\# F(x) = 0$.

**Proof** Hypotheses (6.5)–(6.10) of Theorem 6.1 become (for $A(x) = F'(x)$, $x \in \mathcal{D}_0$)

$$\| F'(x_0)^\# F(x_0) \| \leq \eta,$$

$$\| F'(x_0)^\# (F(x) - F(y) - F'(y)(x - y)) \| \leq \frac{K}{2} \| x - y \|^2, \tag{6.23}$$

$$\| F'(x_0)^\# (F'(x) - F'(x_0)) \| \leq K_0 \| x - x_0 \|, \tag{6.24}$$

$$K\eta \leq \frac{1}{2}, \tag{6.25}$$

$$U(x_0, t^\star) \subseteq \mathcal{D}_0$$

and

$$t^\star = \frac{1 - \sqrt{1 - K\eta}}{K}.$$

Using the uniqueness part of the proof of Theorem 6.1 and (6.22), we have $y^\star = x^\star$ and

$$\begin{aligned} &\|x^\star - x_{m+1} \| \\ &\leq \| F'(x_m)^\# F'(x_0) \| \| F'(x_0)^\# (F(x^\star) - F(x_m) - F'(x_m)(x^\star - x_m)) \| \\ &\leq (1 - K_0 t^\star)^{-1} \frac{K}{2} \| x^\star - x_m \|^2 . \end{aligned} \tag{6.26}$$

This completes the proof of Corollary 6.1. ⊠

We now state a generalization of an affine invariant version of Mysovskii–type theorem. The proof as similar to one in Theorem 6.1 is omitted (see also (cf. [277])).

**Theorem 6.2** *Let* $F : \mathcal{D} \subseteq \mathcal{X} \to \mathcal{Y}$ *be a Fréchet–differentiable operator. Assume there exist an approximation* $A(x) \in \mathcal{L}(\mathcal{X}, \mathcal{Y})$ *of* $F'(x)$*, an open convex subset* $\mathcal{D}_0$ *of* $\mathcal{D}$*,* $x_0 \in \mathcal{D}_0$*, a bounded outer inverse* $A^\#$ *of* $A(x_0) : = A$ *and constants* $\eta > 0$*,* $K > 0$ *such that, for all* $x, y \in \mathcal{D}_0$*, the following hold*

$$\mathcal{N}(A(x)^\#) = \mathcal{N}(A^\#), \qquad \| A^\# F(x_0) \| \leq \eta,$$

$$\| A^\# (F(x) - F(y) - F'(y)(x - y)) \| \leq \frac{K}{2} \| x - y \|^2,$$

$$h \equiv \frac{1}{2} K \eta \leq 1$$

*and*

$$\overline{U}(x_0, r) \subseteq \mathcal{D}_0,$$

*where*

$$r = \frac{\eta}{1-h}.$$

*Then, sequence* $\{x_n\}$ $(n \geq 0)$ *generated by NLM with* $\mathcal{N}(A(x_k)^{\#}) = \mathcal{N}(A^{\#})$ *is well defined, remains in* $U(x_0, r)$ *for all* $n \geq 0$ *and converges to a solution* $x^\star$ *of equation* $A^{\#}\, F(x) = 0$.

**Remark 6.1** *(i) If (6.6) is replaced by stronger condition*

$$\| \, A^{\#} \, (F'(x) - F'(y)) \, \| \leq K \, \| \, x - y \, \| \quad (x, y \in \mathcal{D}_0), \tag{6.27}$$

*then, our Theorem 6.1 reduces to (cf. [277, Theorem 3.1 p. 141, 142]). Otherwise, it constitutes an improvement since (6.27) implies (6.6), but not vice versa.*

*(ii) Condition (6.24) is only used to show (6.26). However, in this section, we shall show that condition (6.24) in combination with (6.6) can be used to generate more precise majorizing sequences than Theorem 6.1 and weaker sufficient convergence conditions than in Theorem 6.1 and Corollary 6.1 provided that* $K_0 < K$.

*(iii) Condition (6.6) can be replaced by*

$$\| \, A^{\#} \, (F(x) - F(y) - F'(y)\,(x-y)) \, \| \leq K_1 \, \| \, F'(y)\,(x-y) \, \| \, \| \, x - y \, \| \tag{6.28}$$

*and*

$$\| \, F(y) \, \| \leq K_2 \quad (x, y \in \mathcal{D}_0) \tag{6.29}$$

*(see (cf. [461]))*

*or*

$$\| \, A^{\#} \, (F(x) - F(y) - F'(y)\,(x-y)) \, \| \leq K_3 \, \| \, F(x) - F(y) \, \| \, \| \, x - y \, \| \tag{6.30}$$

*and*

$$\| \, F(x) - F(y) \, \| \leq K_4 \quad (x, y \in \mathcal{D}_0). \tag{6.31}$$

*In the case of conditions (6.28) and (6.29), we can set* $K = 2\, K_1\, K_2$, *whereas when (6.30) and (6.31) hold, we let* $K = 2\, K_3\, K_4$.

*(iv) If Lipschitz–type condition of Theorem 6.2 is replaced by stronger*

$$\| \, A^{\#} \, (F'(x + t\,(y - x)) - F'(x)) \, \| \leq K \; t \, \| \, x - y \, \|$$

*for all* $x, y \in \mathcal{D}_0$ *and* $t \in [0, 1]$, *then, Theorem 6.2 reduces to (cf. [277, Theorem 3.2, p. 247]).*

We shall consider the more general equation

$$A^{\#}\,(F(x) + G(x)) = 0, \tag{6.32}$$

where $G : \mathcal{D} \longrightarrow \mathcal{Y}$ is a continuous operator. The corresponding general Newton–like method (GNLM) to (6.32) is given by

$$x_{n+1} = x_n - A(x_n)^{\#} \left(F(x_n) + G(x_n)\right) \quad (n \geq 0, \quad x_0 \in D). \tag{6.33}$$

We use our new concept of recurrent function to study the semilocal convergence of (GNLM). First, we need the following results on the convergnce of majorizing sequences for (GNLM).

**Lemma 6.4** *(cf. [200]) Assume there exist constants $K > 0$, $M > 0$, $\mu \geq 0$, $L > 0$ and $\eta > 0$ such that*

$$2M < K, \tag{6.34}$$

*Quadratic polynomial $f_1$ given by*

$$f_1(s) = 2L\eta s^2 - \left(2(1 - L, \eta) - K\eta\right) s + 2(M\eta + \mu) \tag{6.35}$$

*has a root in $(0, 1)$, denoted by $\frac{\delta}{2}$ and for*

$$\delta_0 = \frac{K\eta + 2\mu}{1 - L\eta}, \tag{6.36}$$

$$\alpha = \frac{2(K - 2M)}{K + \sqrt{K^2 + 8L(K - 2M)}}, \tag{6.37}$$

*the following holds*

$$\delta_0 \leq \delta \leq 2\,\alpha. \tag{6.38}$$

*Then, scalar sequence $\{t_n\}$ $(n \geq 0)$ given by*

$$t_0 = 0, \quad t_1 = \eta,$$

$$t_{n+2} = t_{n+1} + \frac{K\,(t_{n+1} - t_n) + 2\,(M\,t_n + \mu)}{2\,(1 - L\,t_{n+1})}\,(t_{n+1} - t_n) \tag{6.39}$$

*is increasing, bounded from above by*

$$t^{\star\star} = \frac{2\eta}{2 - \delta} \tag{6.40}$$

*and converges to its unique least upper bound $t^\star \in [0, t^{\star\star}]$. Moreover, the following estimates hold for all $n \geq 1$*

$$t_{n+1} - t_n \leq \frac{\delta}{2}\,(t_n - t_{n-1}) \leq \left(\frac{\delta}{2}\right)^n \eta \tag{6.41}$$

*and*

$$t^\star - t_n \leq \frac{2\,\eta}{2 - \delta}\left(\frac{\delta}{2}\right)^n. \tag{6.42}$$

**Remark 6.2** *The hypotheses of Lemma 6.4 have been left as uncluttered as possible. Note that these hypotheses involve only computations only at the initial point* $x_0$. *Next, we shall provide some simpler but stronger hypotheses under which the hypotheses of Lemma 6.4 hold.*

**Lemma 6.5** *(cf. [200]) Let* $K > 0$, $M > 0$, $\mu > 0$ *with* $L > 0$ *and* $\eta > 0$ *be such that*

$$\mu < \alpha, \quad 2\,M < K$$

*and*

$$0 < h_A = a\,\eta \leq \frac{1}{2}, \tag{6.43}$$

*where*

$$a = \frac{1}{4\,(\alpha - \mu)}\ \max\{2\,L\,\alpha^2 + 2\,L\,\alpha + K\,\alpha + 2\,M,\ K + 2\,\alpha\,L\}. \tag{6.44}$$

*Then, the following hold* $f_1$ *has a positive root* $\frac{\delta}{2}$, $\max\{\delta_0,\ \delta\} \leq 2\ \alpha$ *and the conclusions of Lemma 6.4 hold with* $\alpha$ *replacing* $\frac{\delta}{2}$.

We show the semilocal convergence theorem for (GNLM) using outer inverses.

**Theorem 6.3** *Let* $F : \mathcal{D} \subseteq \mathcal{X} \to \mathcal{Y}$ *be a Fréchet–differentiable operator and* $G : \mathcal{D} \to \mathcal{Y}$ *be a continuous operator. Assume that there exist an approximation* $A(x) \in \mathcal{L}(\mathcal{X}, \mathcal{Y})$ *of* $F'(x)$, *an open convex subset* $\mathcal{D}_0$ *of* $\mathcal{D}$, $x_0 \in \mathcal{D}_0$, *a bounded outer inverse* $A^{\#}$ *of* $A(x_0) : = A$ *and constants* $\eta > 0$, $K > 0$, $M > 0$, $L > 0$, $\mu_0, \mu_1 \geq 0$, $\ell \geq 0$ *such that, for all* $x, y \in \mathcal{D}_0$

$$\|\ A^{\#}\ F(x_0)\ \| \leq \eta, \tag{6.45}$$

$$\|\ A^{\#}\ (F(x) - F(y - F'(y)\,(x - y))\ \| \leq \frac{K}{2}\ \|\ x - y\ \|^2, \tag{6.46}$$

$$\|\ A^{\#}\ (F'(x) - A(x))\ \| \leq M\ \|\ x - x_0\ \| + \mu_0, \tag{6.47}$$

$$\|\ A^{\#}\ (G(x) - G(y))\ \| \leq \mu_1\ \|\ x - y\ \|, \tag{6.48}$$

$$\|\ A^{\#}\,(A(x) - A(x_0))]\ \| \leq L\ \|\ x - x_0\ \| + \ell \tag{6.49}$$

*and hypotheses of Lemmas 6.4 or 6.5. Then, sequence* $\{x_n\}$ $(n \geq 0)$ *generated by (GNLM) with*

$$A(x_n)^{\#} = [I + A^{\#}\ (A(x_n) - A(x_0))]^{-1}\ A^{\#} \tag{6.50}$$

*is well defined, remains in* $U(x_0, t^\star)$ *for all* $n \geq 0$ *and converges to a solution* $x^\star$ *of equation* $A^{\#}\,F(x) = 0$. *Moreover, the following estimates hold for all* $n \geq 0$

$$\|\ x_{n+1} - x_n\ \| \leq t_{n+1} - t_n \tag{6.51}$$

*and*

$$\| x_n - x^\star \| \leq t^\star - t_n, \tag{6.52}$$

*where* $\{t_n\}$ *is given by (6.39), with* $\mu = \mu_0 + \mu_1$. *Furthermore, the solution* $x^\star$ *of equation (6.32) is unique in* $\overline{U}(x_0, t^\star)$ *provided that*

$$\left(\frac{K}{2} + M + L\right) t^\star + \mu + \ell < 1. \tag{6.53}$$

**Proof** We shall show, using induction on $m$, that (6.51) holds. Estimate (6.52) will then follow from (6.13) using standard majorization techniques. By the initial conditions, we have

$$\| x_1 - x_0 \| \leq t_1 - t_0$$

and (6.51) holds for $m = 0$. Using (6.49), we get that

$$\| A^{\#} (A(x_1) - A) \| \leq L \| x_1 - x_0 \| + \ell \leq L t_1 + \ell \leq L t^\star + \ell < 1. \tag{6.54}$$

From Lemma 6.2 and (6.54), we obtain that $A(x_1)^{\#} := (I + A^{\#} (A(x_1) - A))^{-1} A^{\#}$ is an outer inverse of $A(x_1)$. Moreover

$$\| A(x_1)^{\#} A \| \leq (1 - L \| x_1 - x_0 \| - \ell)^{-1} \leq (1 - L t_1 - \ell)^{-1}$$

and $\mathcal{N}(A(x_1)^{\#}) = \mathcal{N}(A^{\#})$. Assume that, for $1 \leq m \leq k$

$$\| x_m - x_{m-1} \| \leq t_m - t_{m-1}$$

and

$$\mathcal{N}(A(x_{m-1})^{\#}) = \mathcal{N}(A^{\#}).$$

Then, we have

$$\| x_m - x_0 \| \leq t_m - t_{m-1}$$

and

$$\mathcal{N}(A(x_m)^{\#}) = \mathcal{N}(A(x_{m-1})^{\#}) = \mathcal{N}(A^{\#}).$$

Hence, we have by (6.33) and Lemma 6.3 that

$$A(x_m)^{\#} (I - A(x_{m-1}) A(x_{m-1})^{\#}) = 0$$

and

$$\begin{aligned} x_{m+1} - x_m &= -A(x_m)^{\#} (F(x_m) + G(x_m)) \\ &= -A(x_m)^{\#} \Big( (F(x_m) - F(x_{m-1}) - F'(x_{m-1})(x_m - x_{m-1})) \\ &\quad + (F'(x_{m-1}) - A(x_{m-1}))(x_m - x_{m-1}) + (G(x_m) - G(x_{m-1})) \Big). \end{aligned} \tag{6.55}$$

We have by Lemma 6.3 that

$$A(x_m)^{\#} (I - A A^{\#}) = 0.$$

In view of hypotheses of Theorem and (6.55), we have

$$\begin{aligned}
&\| x_{m+1} - x_m \| \\
&\leq \| A(x_m)^{\#} A \| \ \{(F(x_m) - F(x_{m-1}) - F'(x_{m-1})(x_m - x_{m-1})) \\
&\quad + \| A^{\#} (F'(x_{m-1}) - A(x_{m-1})) \| \| x_m - x_{m-1} \| + \| A^{\#} (G(x_m) - G(x_{m-1})) \| \} \\
&\leq \frac{1}{1 - L\, t_m - \ell} \left( \frac{K}{2} \| x_m - x_{m-1} \|^2 + (M \| x_{m-1} - x_0 \| + \mu) \| x_m - x_{m-1} \| \right) \\
&\leq \frac{1}{1 - L\, t_m - \ell} \left( \frac{K}{2} (t_m - t_{m-1}) + M\, t_{m-1} + \mu \right) (t_m - t_{m-1}) = t_{m+1} - t_m, \qquad (6.56)
\end{aligned}$$

which completes the induction. Hence, we have for any $m$ that

$$\| x_{m+1} - x_m \| \leq t_{m+1} - t_m,$$

$$\| A^{\#} \ (A(x_{m+1}) - A) \| \leq L \ \| x_{m+1} - x_0 \| + \ell \leq L\, t_{m+1} + \ell \leq L\, t^{\star} + \ell < 1,$$

$$\| x_m - x_0 \| \leq \sum_{i=1}^{m} (t_i - t_{i-1}) \leq t_m - t_0 = t_m$$

and $A(x_{m+1})^{\#} := (I + A^{\#} \ (A(x_{m+1}) - A))^{-1} \ A^{\#}$ is an outer inverse of $A(x)$. It follows that $x_m \in U(x_0, t^{\star})$, $m \geq 0$ and $\{x_m\}$ converges to a point $x^{\star}$ in $\overline{U}(x_0, t^{\star})$. The point $x^{\star}$ is a solution of $A^{\#} \ (F(x) + G(x)) = 0$. Indeed, by definition

$$A(x_m)^{\#} = (I + A^{\#} \ (A(x_m) - A))^{-1} \ A^{\#} \quad (m \geq 0)$$

and

$$\begin{aligned}
0 &= \lim_{m \longrightarrow \infty} (I + A^{\#} \ (A(x_m) - A)) \ (x_m - x_{m-1}) \\
&= \lim_{m \longrightarrow \infty} A^{\#} \ (F(x_m) + G(x_m)) = A^{\#} \ (F(x^{\star}) + G(x^{\star})).
\end{aligned}$$

Hence, $x^{\star}$ solves equation $A^{\#} \ (F(x^{\star}) + G(x^{\star})) = 0$.

Finally, to show that $x^{\star}$ is the unique solution of equation (6.1) in $\overline{U}(x_0, t^{\star})$, as in (6.18) and (6.19), we get in turn for $y^{\star} \in \overline{U}(x_0, t^{\star})$ with $A^{\#} \ (F(y^{\star}) + G(y^{\star})) = 0$ the estimation

$$\begin{aligned}
&\| y^{\star} - x_{m+1} \| \\
&\leq \| A(x_m)^{\#} A \| \left\{ (F(y^{\star}) - F(x_m) - F'(x_m)(y^{\star} - x_m)) \right. \\
&\quad \left. + \| A^{\#} (F'(x_m) - A(x_m)) \| \| y^{\star} - x_m \| + \| A^{\#} (G(x_m) - G(y^{\star})) \| \right\} \\
&\leq (1 - \ell - L\, t_m)^{-1} \left( \frac{K}{2} \| y^{\star} - x_m \|^2 + (M \ \| x_m - x_0 \| + \mu) \ \| y^{\star} - x_m \| \right) \qquad (6.57) \\
&\leq (1 - \ell - L\, t_m)^{-1} \left( \frac{K}{2} (t^{\star} - t_m) + M\, t_m + \mu \right) \ \| y^{\star} - x_m \| \\
&\leq (1 - \ell - L\, t^{\star})^{-1} \left( \frac{K}{2} (t^{\star} - t_0) + M\, t^{\star} + \mu \right) \ \| x^{\star} - x_m \| \\
&< \| y^{\star} - x_m \|
\end{aligned}$$

by the uniqueness hypothesis (6.53). It follows by (6.57) that $\lim_{m \longrightarrow \infty} x_m = y^\star$. But we showed $\lim_{m \longrightarrow \infty} x_m = x^\star$. Hence, we deduce $x^\star = y^\star$. The proof of Theorem 6.3 is complete. ⊠

**Remark 6.3** *(i) The point $t^\star$ can be replaced by $t^{\star\star}$, given in closed form by (6.40) in all hypotheses of Theorem 6.3.*

*(ii) If (6.46) is replaced by stronger condition*

$$\| A^{\#} \, (F'(x) - F'(y)) \| \leq K \, \| x - y \| \quad (x, y \in \mathcal{D}_0),$$

*then, our Theorem 6.3 reduces to (cf. [277, Theorem 2.4]).*

*(iii) If $G = 0$, majorizing sequence $\{t_n\}$ given by (6.39) is finer than $\{t_n\}$ given in Theorem 6.1 (see also (cf. [200, Proposition 2.7, p. 123]) for the proof).*

*(iv) If $A(x)^{\#} = A^{\#}$ $(x_0 \in \mathcal{D}_0)$ and $G = 0$ (Newton's method), then, Kantorovich hypothesis (6.25) can be replaced by weaker (if $K_0 < K$)*

$$h_{AH} = \overline{K} \, \eta \leq \frac{1}{2}, \tag{6.58}$$

*where*

$$\overline{K} = \frac{1}{8} \, (K + 4 \, K_0 + \sqrt{K^2 + 8 \, K \, K_0})$$

*and the constant $K_0$ satisfies the center–Lipschitz condition*

$$\| A^{\#} \, (F'(x) - F'(x_0)) \| \leq K_0 \, \| x - x_0 \| \quad (x \in \mathcal{D}_0). \tag{6.59}$$

*Note also that*

$$K_0 \leq K \tag{6.60}$$

*holds in general and $\dfrac{K}{K_0}$ can be arbitrarily large. Hypothesis even weaker than (6.58) has been given in section 2.1.*

We provide now some examples. For simplicity, we set $A(x) = F'(x)$, $A^{\#}\,(x) = A(x)^{-1}$ and $G(x) = 0$ $(x \in \mathcal{D}_0)$.

**Example 6.1** *We consider the integral equation*

$$u(s) = f(s) + \varrho \int_a^b \mathcal{G}(s,t) \, u(t)^{1+\frac{1}{n}} \, dt \quad (n \geq 1).$$

*Here, $f$ is a given continuous function satisfying $f(s) > 0$, $s \in [a, b]$, $\varrho$ is a real number and the kernel $\mathcal{G}$ is continuous and positive in $[a, b] \times [a, b]$. For example, when $\mathcal{G}(s,t)$ is the Green kernel, the corresponding integral equation is equivalent to the boundary value problem*

$$\begin{cases} u'' = \varrho \, u^{1+\frac{1}{n}}, \\ u(a) = f(a), \; u(b) = f(b). \end{cases}$$

*These type of problems have been considered in (cf. [157]). We consider F as follows*

$$F\colon \mathcal{D} \subseteq \mathcal{C}[a,b] \to \mathcal{C}[a,b], \quad \mathcal{D} = \{u \in \mathcal{C}[a,b]\colon u(s) \geq 0,\ s \in [a,b]\}$$

*and*

$$F(u)(s) = u(s) - f(s) - \varrho \int_a^b \mathcal{G}(s,t)\, u(t)^{1+\frac{1}{n}}\, dt. \tag{6.61}$$

$\mathcal{C}[a,b]$ *is equipped with the max–norm. The derivative* $F'$ *is given by*

$$F'(u)\, v(s) = v(s) - \varrho \left(1 + \frac{1}{n}\right) \int_a^b \mathcal{G}(s,t)\, u(t)^{\frac{1}{n}} v(t)\, dt \quad (v \in \mathcal{D}). \tag{6.62}$$

*First, note that* $F'$ *does not satisfy a Lipschitz condition (6.27) in* $\mathcal{D}$*. Let us consider, for instance,* $[a,b] = [0,1]$*,* $\mathcal{G}(s,t) = 1$ *and* $y(t) = 0$*. Then,* $F'(y)\, v(s) = v(s)$ *and*

$$\| F'(x) - F'(y) \| = |\varrho| \left(1 + \frac{1}{n}\right) \int_0^1 x(t)^{\frac{1}{n}}\, dt. \tag{6.63}$$

*We suppose that* $F'$ *is a Lipschitz function, then*

$$\| F'(x) - F'(y) \| \leq K \ \| x - y \| .$$

*Consequently, we obtain*

$$\int_0^1 x(t)^{\frac{1}{n}}\, dt \leq \frac{K}{|\varrho|\,\left(1 + \frac{1}{n}\right)} \max_{x \in [0,1]} x(s) \tag{6.64}$$

*would hold for all* $x \in \mathcal{D}$*. But this is not true. Consider, for example, the functions*

$$x_j(t) = \frac{t}{j} \quad (j \geq 1,\ t \in [0,1]).$$

*If these are substituted into (6.64), we obtain*

$$\frac{1}{j^{1/n}\left(1 + \frac{1}{n}\right)} \leq \frac{K}{j\, |\varrho|\, \left(1 + \frac{1}{n}\right)} \Longleftrightarrow j^{1-\frac{1}{n}} \leq \frac{K}{|\varrho|} \quad (j \geq 1).$$

*This inequality is not true when* $j \to \infty$*. Therefore, condition (6.27) fails in this case. However, condition (6.59) holds. To show this, let* $x_0(t) = f(t)$ *and* $\Xi = \min\limits_{s \in [a,b]} f(s) > 0$*. Then, for all* $v \in \mathcal{D}$*,*

$$\begin{aligned}
&\| (F'(x) - F'(x_0))\, v \| \\
&= |\varrho| \left(1 + \frac{1}{n}\right) \max_{s \in [a,b]} \left| \int_a^b \mathcal{G}(s,t) (x(t)^{\frac{1}{n}} - f(t)^{\frac{1}{n}}) v(t) dt \right| \\
&\leq |\varrho| \left(1 + \frac{1}{n}\right) \max_{s \in [a,b]} \int_a^b \frac{\mathcal{G}(s,t)\, |x(t) - f(t)|}{x(t)^{(n-1)/n} + x(t)^{(n-2)/n} f(t)^{1/n} + \cdots + f(t)^{(n-1)/n}}\, dt \ \| v \| .
\end{aligned}$$

*Hence, we have*

$$\| F'(x) - F'(x_0) \| \leq \frac{|\varrho| \left(1 + \frac{1}{n}\right)}{\Xi^{(n-1)/n}} \max_{s \in [a,b]} \int_a^b \mathcal{G}(s,t)\, dt \ \| x - x_0 \|$$
$$= K_0 \ \| x - x_0 \|,$$

*where*

$$K_0 = \frac{|\varrho| \left(1 + \frac{1}{n}\right)}{\Xi^{(n-1)/n}} \max_{s \in [a,b]} \int_a^b \mathcal{G}(s,t)\, dt,$$

*and so (6.59) is satisfied.*

We finally provide a numerical example to show how the parameters in Theorem 6.3 can be computed.

**Example 6.2** *Let* $\mathcal{X} = \mathcal{Y} = \mathbb{R}^2$, $x_0 = (.495, .495)^T$ *and* $\mathcal{D} = \overline{U}(x_0, \frac{7}{8})$. *Define a function* $F$ *on* $\mathcal{D}$ *by*

$$F(x) = \left( \xi_1^3 - \frac{1}{8}, \xi_2^3 - \frac{1}{8} \right)^T \qquad (x = (\xi_1, \xi_2)^T). \tag{6.65}$$

*Using hypotheses of Theorem 6.3 and Remark 6.3 (iv), we get*

$$L = 5.595, \quad \eta = .0050506751, \quad \ell = \mu_0 = \mu_1 = 0,$$
$$K = 90.0912, \quad M = 0, \quad \delta = .90861451,$$
$$h = .22751069028 < .5.$$

*That is, all hypotheses of Theorem 6.3 are satisfied. Hence, NLM starting at* $x_0$ *converges quadratically to* $x^\star = (.5, .5)^T$.

## 6.2 Convergence of a Moser–Type Method

In this section we are concerned with the problem of approximating a locally unique solution $x^\star$ of (2.1). There are many problems for which NM given by (2.2) is not applicable in its original form. A case of interest occurs when the derivative is not continuously invertible, as for instance, dealing with problems involving small divisors, or other important examples (cf. [157], [277], [541]). Several Newton–type methods have addressed this problem (cf. [88], [792]). Moret in (cf. [541]) unified a large class of such Newton–type methods, where, at each step, the inverse of the derivative, is replaced by a linear operator, which is obtained recursively from the previous one.

Two iterative schemes were provided in (cf. [541]) by

**Scheme 1.** Let the following be given: a Banach space $\mathcal{Z}$, an operator valued mapping $\mathcal{B} : \mathcal{D} \longrightarrow \mathcal{L}(\mathcal{X}, \mathcal{Y})$, $x_0 \in \mathcal{D}$, $\mathcal{S}_{-1} \in \mathcal{L}(\mathcal{Z}, \mathcal{Y})$,

$\mathcal{R}_{-1} \in \mathcal{L}(\mathcal{Z}, \mathcal{X})$.

For $n \geq 0$, let $N_n \in \mathcal{L}(\mathcal{Z}, \mathcal{Z})$ and set

$$\begin{aligned} \mathcal{S}_n &= \mathcal{S}_{n-1}\ N_n + \mathcal{B}(x_n)\ \mathcal{R}_{n-1}, \\ \mathcal{R}_n &= \mathcal{R}_{n-1} + \mathcal{R}_{n-1}\ N_n, \\ x_{n+1} &= x_n + \mathcal{R}_n\ \Delta_n, \end{aligned}$$

$\Delta_n$ being a possible approximate solution of

$$\mathcal{S}_n\ \Delta_n = -F(x_n).$$

That is, $\Delta_n$ satisfies an equation of type

$$\mathcal{S}_n\ \Delta_n = -(F(x_n) + r_n),$$

for a suitable null sequence $\{r_n\} \subset \mathcal{Y}$.

Moret (cf. [541]) provided a semilocal convergence analysis for NM under general conditions on the mapping $\mathcal{B}$, the starting guesses $x_0$, $\mathcal{S}_{-1}$, $\mathcal{R}_{-1}$, the operators $N_n$ and the sequence $\{r_n\}$. Similar results were given for scheme 2.

There are two type of problems common to both schemes: the convergence may be too slow and the schemes may not even be applicable. For example in both schemes a condition of the form:

$$\| \mathcal{I}_{\mathcal{Y}} - \mathcal{T}\,\mathcal{R}_{-1} \| < 1 \tag{6.66}$$

is required, where

$$\mathcal{T} = F'(x_0) \qquad \text{(Scheme 1)} \tag{6.67}$$

or

$$\mathcal{T} = \mathcal{B}(x_0) \qquad \text{(Scheme 2).} \tag{6.68}$$

There are simple numerical examples to show that condition (6.66) is violated. Let scalar function $F$ given by

$$F(x) = \frac{1}{128}\ x^2 - \frac{1}{64}\ \cos x \tag{6.69}$$

and choose $x_0 = 0$. Then, $F'(x_0) = 0$, which shows that (6.66) is violated for $\mathcal{T} = \mathcal{B}(x_0) = F'(x_0)$. Hence, Schemes 1 and 2 cannot be used to approximate $x^\star$. In particular, the classical Newton–Moser method (cf. [157], [277], [541]), obtained from Scheme 1, by setting $\mathcal{Z} = \mathcal{Y}$, $\mathcal{B} = F'$, $\mathcal{S}_{-1} = \mathcal{I}_{\mathcal{Y}}$, $N_n = \mathcal{I}_{\mathcal{Y}} - F'(x_n)\ \mathcal{R}_{n-1}$ and $r_n = 0$ $(n \geq 0)$ cannot be used.

To address these problems, we consider the Newton–type methods of the form NTM

$$x_{n+1} = x_n - (F'(x_n) - \mathcal{A})^{-1}\ F(x_n) \tag{6.70}$$

or more generally NLM

$$x_{n+1} = x_n - (\mathcal{C}(x_n) - \mathcal{A})^{-1}\ F(x_n), \tag{6.71}$$

where $\mathcal{C}(x) \in \mathcal{L}(\mathcal{X}, \mathcal{Y})$ is an approximation of $F'(x)$ $(x \in \mathcal{D})$ and $\mathcal{A}$ a given linear operator. In the case of function $F$ given in (6.69) NTM can be used to approximate $x^\star$, if $\mathcal{A}$ is an invertible operator.

Methods (6.70) and (6.71) can be combined into one, even more general GNTM:

$$x_{n+1} = x_n - \mathcal{A}(x_n)^{-1} \, F(x_n), \tag{6.72}$$

where $\mathcal{A}(x) \in \mathcal{L}(\mathcal{X}, \mathcal{Y})$ $(x \in \mathcal{D})$. Note that, if

$$\mathcal{A}(x) = F'(x) - \mathcal{A} \tag{6.73}$$

or

$$\mathcal{A}(x) = \mathcal{C}(x) - \mathcal{A}, \tag{6.74}$$

we obtain (6.70) and (6.71), respectively.

Sufficient conditions for semilocal convergence of GNTM and estimates on the distances $\| \, x_{n+1} - x_n \, \|$, $\| \, x_n - x^\star \, \|$ have been given by several authors (cf. [88], [277], [600], [792]).

However, in the special case of (6.73) (or (6.74)), we can do better by a direct approach. That is, we can provide (under the same computational cost) weaker sufficient convergence conditions and finer error estimates on $\| \, x_{n+1} - x_n \, \|$, $\| \, x_n - x^\star \, \|$ $(n \geq 0)$.

Let $L_0 > 0$, $L > 0$, $\eta > 0$, $a > 0$, $c > 1$ be given constants. Set

$$b = \frac{c}{2}(L\eta + 2a), \qquad r = \frac{c-1}{c \, L_0}.$$

We need the following result on majorizing sequences for GNTM:

**Lemma 6.6** *Assume that*

$$c \in \left(1, \, \frac{1}{a}\right) \tag{6.75}$$

*and*

$$\eta \leq \frac{(c-1) \, (1 - c \, a)}{c \, \left(L_0 + \dfrac{c-1}{2} \, L\right)}. \tag{6.76}$$

*Then, scalar sequence $\{t_n\}$ $(n \geq 0)$ given by*

$$t_0 = 0, \quad t_1 = \eta, \quad t_{n+2} = t_{n+1} + \frac{L \, (t_{n+1} - t_n) + 2 \, a}{2 \, (1 - L_0 \, t_{n+1})} \, (t_{n+1} - t_n) \tag{6.77}$$

*is non–decreasing, bounded by*

$$t^{\star\star} = \frac{c-1}{c \, L_0} \tag{6.78}$$

*and converges to its unique least upper bound $t^\star$, satisfying:*

$$0 \leq t^\star \leq \frac{\eta}{1-b} \leq t^{\star\star}. \tag{6.79}$$

*Moreover, the following estimates hold for all $n \geq 0$*

$$t_{n+2} - t_{n+1} \leq b \, (t_{n+1} - t_n) \leq b^{n+1} \, \eta. \tag{6.80}$$

**Proof** We shall show using induction on the integer $k$ that

$$0 \leq t_{k+2} - t_{k+1} = \frac{L\ (t_{k+1} - t_k) + 2\ a}{2\ (1 - L_0\ t_{k+1})}\ (t_{k+1} - t_k) \leq b\ (t_{k+1} - t_k) \quad (6.81)$$

and

$$t_{k+1} < r_0. \quad (6.82)$$

Estimates (6.81) and (6.82) hold for $k = 0$, by the initial conditions. Assume (6.81) and (6.82) hold for all $m \leq k$. Then, we have

$$\begin{aligned} 0 \leq t_{k+2} - t_{k+1} &\leq b\ (t_{k+1} - t_k) \\ &\leq b\ (b\ (t_k - t_{k-1})) = b^2\ (t_k - t_{k-1}) \leq \cdots \\ &\leq b^{k+1}\ \eta \end{aligned}$$

and

$$\begin{aligned} t_{k+1} &\leq t_k + b^k\ \eta \\ &\leq t_{k-1} + b^{k-1}\ \eta + b^k\ \eta \\ &\leq t_1 + b^1\ \eta + \cdots + b^k\ \eta \\ &= \frac{1 - b^{k+1}}{1 - b}\ \eta < \frac{\eta}{1 - b} \leq r \qquad \text{(by (6.75) and (6.76))}, \end{aligned}$$

which complete the induction for (6.81) and (6.82).

Finally, sequence $\{t_n\}$ is non–decreasing, bounded above by $t^{\star\star}$ and as such it converges to its unique least upper bound $t^\star$. This completes the proof of Lemma 6.6. ⊠

We shall show the following semilocal convergence theorem for GNTM in the special case, when $\mathcal{A}$ is given by (6.73).

**Theorem 6.4** *Let $F : \mathcal{D} \subseteq \mathcal{X} \longrightarrow \mathcal{Y}$ be a Fréchet–differentiable operator and $\mathcal{A}(x) \in \mathcal{L}(\mathcal{X}, \mathcal{Y})$ be given by (6.73). Assume that there exist an open convex subset $\mathcal{D}$ of $\mathcal{X}$, $x_0 \in \mathcal{D}$, a bounded inverse $\mathcal{A}(x_0)^{-1}$ of $\mathcal{A}(x_0)$ and constants $L_0 > 0$, $L > 0$, $a \geq 0$ and $\eta > 0$ such that, for all $x, y \in \mathcal{D}$*

$$\| \mathcal{A}(x_0)^{-1}\ F(x_0) \| \leq \eta, \quad (6.83)$$

$$\| \mathcal{A}(x_0)^{-1}\ [F'(x) - F'(y)] \| \leq L\ \| x - y \|, \quad (6.84)$$

$$\| \mathcal{A}(x_0)^{-1}\ [F'(x) - \mathcal{A}(x)] \| \leq a, \quad (6.85)$$

$$\| \mathcal{A}(x_0)^{-1}\ [\mathcal{A}(x) - \mathcal{A}(x_0)] \| \leq L_0\ \| x - x_0 \|, \quad (6.86)$$

$$\overline{U}(x_0, t^\star) \subseteq \mathcal{D} \quad (6.87)$$

*and the hypotheses of Lemma 6.6 hold. Then, sequence $\{x_n\}$ $(n \geq 0)$ generated by GNTM is well defined, remains in $\overline{U}(x_0, t^\star)$ for all $n \geq 0$ and converges* *to a solution* $x^\star$ *of equation* $F(x) = 0$ *in* $\overline{U}(x_0, t^\star)$. *Moreover, the following estimates hold for all* $n \geq 0$

$$\| x_{n+1} - x_n \| \leq t_{n+1} - t_n \tag{6.88}$$

*and*

$$\| x_n - x^\star \| \leq t^\star - t_n, \tag{6.89}$$

*where sequence* $\{t_n\}$ $(n \geq 0)$ *and* $t^\star$ *are given in Lemma 6.6. Furthermore, the solution* $x^\star$ *of equation (2.1) is unique in* $\overline{U}(x_0, t^\star)$ *provided that*

$$\left(\frac{L}{2} + a + L_0\right) t^\star + a < 1. \tag{6.90}$$

**Proof** We shall show using induction on $m \geq 0$ that

$$\| x_{m+1} - x_m \| \leq t_{m+1} - t_m \tag{6.91}$$

and

$$\overline{U}(x_{m+1}, t^\star - t_{m+1}) \subseteq \overline{U}(x_m, t^\star - t_m). \tag{6.92}$$

For every $z \in \overline{U}(x_1, t^\star - t_1)$,

$$\| z - x_0 \| \leq \| z - x_1 \| + \| x_1 - x_0 \| \leq t^\star - t_1 + t_1 = t^\star - t_0$$

implies $z \in \overline{U}(x_0, t^\star - t_0)$. We also have

$$\| x_1 - x_0 \| = \| \mathcal{A}(x_0)^{-1} \; F(x_0) \| \leq \eta = t_1 - t_0.$$

That is, (6.91) and (6.92) hold for $m = 0$. Given they hold for $n \leq m$, then, we have

$$\| x_{m+1} - x_0 \| \leq \sum_{i=1}^{m+1} \| x_i - x_{i-1} \| \leq \sum_{i=1}^{m+1} (t_i - t_{i-1}) = t_{m+1} - t_0 = t_{m+1}$$

and

$$\| x_m + \theta \, (x_{m+1} - x_m) - x_0 \| \leq t_m + \theta \, (t_{m+1} - t_m) \leq t^\star \quad (\theta \in (0, 1)).$$

Using (6.82), (6.86), the induction hypotheses, we get that

$$\begin{aligned} \| \mathcal{A}(x_0)^{-1} \; [\mathcal{A}(x_{m+1}) - \mathcal{A}(x_0)] \| &\leq L_0 \;\; \| x_{m+1} - x_0 \| \\ &\leq L_0 \; (t_{m+1} - t_0) \\ &\leq L_0 \; t_{m+1} < 1. \end{aligned} \tag{6.93}$$

It follows from (6.93) and the Banach lemma on invertible operators that $\mathcal{A}(x_{m+1})^{-1}$ exists and

$$\| \mathcal{A}(x_{m+1})^{-1} \; \mathcal{A}(x_0) \| \leq (1 - L \; t_{m+1})^{-1}. \tag{6.94}$$

Using (6.72), we obtain the approximation

$$\begin{aligned} x_{m+2} - x_{m+1} &= -\mathcal{A}(x_{m+1})^{-1}\, F(x_{m+1}) \\ &= -\mathcal{A}(x_{m+1})^{-1}\, \mathcal{A}(x_0)\, \mathcal{A}(x_0)^{-1} \\ &\times \left( \int_0^1 [F'(x_{m+1} + \theta\,(x_m - x_{m+1})) - F'(x_m)]\,(x_{m+1} - x_m)\, d\theta \right. \\ &\left. + (F'(x_m) - \mathcal{A}(x_m))\,(x_{m+1} - x_m) \right) \end{aligned} \tag{6.95}$$

Using (6.84)–(6.86), (6.91) (6.94), (6.95) and the induction hypotheses, we obtain in turn that

$$\begin{aligned} \| x_{m+2} - x_{m+1} \| &\le (1 - L_0\, t_{m+1})^{-1} \left( \frac{L}{2}\, \| x_{m+1} - x_m \|^2 + a\, \| x_{m+1} - x_m \| \right) \\ &\le (1 - L_0\, t_{m+1})^{-1} \left( \frac{L}{2}\,(t_{m+1} - t_m) + a \right) (t_{m+1} - t_m) \\ &= t_{m+2} - t_{m+1}, \end{aligned} \tag{6.96}$$

which shows (6.91) for all $m \ge 0$. Thus, for every $z \in \overline{U}(x_{m+2}, t^\star - t_{m+2})$, we have

$$\begin{aligned} \| z - x_{m+1} \| &\le \| z - x_{m+2} \| + \| x_{m+2} - x_{m+1} \| \\ &\le t^\star - t_{m+2} + t_{m+2} - t_{m+1} = t^\star - t_{m+1}, \end{aligned}$$

which shows (6.92) for all $m \ge 0$.

Lemma 6.6 implies that sequence $\{t_n\}$ is a Cauchy sequence. Moreover, it follows from (6.91) and (6.92) that $\{x_n\}$ $(n \ge 0)$ is also a Cauchy sequence in a Banach space $\mathcal{X}$ and as such it converges to some $x^\star \in \overline{U}(x_0, t^\star)$ (since $\overline{U}(x_0, t^\star)$ is a closed set). By letting $m \longrightarrow \infty$ in (6.96), we obtain $F(x^\star) = 0$. Furthermore, estimate (6.89) is obtained from (6.88) by using standard majorization techniques.

Finally, to show that $x^\star$ is the unique solution of equation (2.1) in $\overline{U}(x_0, t^\star)$, as in (6.95) and (6.96), we get in turn for $y^\star \in \overline{U}(x_0, t^\star)$, with $F(y^\star) = 0$, the estimation:

$$\begin{aligned} \| y^\star - x_{m+1} \| &\le \| \mathcal{A}(x_m)^{-1}\, \mathcal{A}(x_0) \| \\ &\times \left( \int_0^1 \| \mathcal{A}(x_0)^{-1}\, (F'(x_m + \theta\,(y^\star - x_m)) - F'(x_m)) \|\, d\theta \right. \\ &\left. + \| \mathcal{A}(x_0)^{-1}\, [F'(x_m) - \mathcal{A}(x_m)] \| \right) \| y^\star - x_m \| \\ &\le (1 - L_0\, t_{m+1})^{-1} \left( \frac{L}{2}\, \| y^\star - x_m \|^2 + a\, \| y^\star - x_m \| \right) \end{aligned}$$

$$\leq (1 - L_0 \; t_{m+1})^{-1} \left( \frac{L}{2} \; (t^\star - t_m) + a \right) \; \| \; y^\star - x_m \; \|$$

$$\leq (1 - L_0 \; t^\star)^{-1} \left( \frac{L}{2} \; (t^\star - t_0) + a \right) \; \| \; x^\star - x_m \; \|$$

$$< \| \; y^\star - x_m \; \| \tag{6.97}$$

by (6.94). It follows by (6.97) that $\lim_{m \longrightarrow \infty} x_m = y^\star$. But we showed $\lim_{m \longrightarrow \infty} x_m = x^\star$. Hence, we deduce $x^\star = y^\star$. This completes the proof of Theorem 6.4. ⊠

Note that $t^\star$ can be replaced by $t^{\star\star}$ given by (6.78) in the uniqueness hypothesis provided that $\overline{U}(x_0, t^{\star\star}) \subseteq \mathcal{D}$ or in all hypotheses of the theorem.

**Remark 6.4** *According to (cf. [157], [277], [792]), the sufficient convergence condition becomes*

$$\eta \leq \frac{(1-a)^2}{2L} \tag{6.98}$$

*and the majorizing iteration $\{v_n\}$ is given by*

$$v_0 = 0, \quad v_1 = \eta, \quad v_{n+2} = v_{n+1} + \frac{\dfrac{L}{2} \; v_{n+1}^2 - (1-a) \; v_{n+1} + \eta}{1 - L_0 \; v_{n+1}} \quad (n \geq 0). \tag{6.99}$$

*In view of (6.76) and (6.98), our condition is weaker if*

$$\frac{(1-a)^2}{2L} < \frac{(c-1) \; (1 - c \; a)}{c \left( L_0 + \dfrac{c-1}{2} \; L \right)}. \tag{6.100}$$

*Let $L_0 = p \; L$, for $p \in [0, 1]$. Then, (6.100) holds if*

$$\frac{(1-a)^2}{2} < \frac{(c-1) \; (1 - c \; a)}{c \left( p + \dfrac{c-1}{2} \right)}. \tag{6.101}$$

*If $p$ is close enough to zero and e.g. $c = 2$, we have (6.101) holds provided that*

$$(1-a)^2 < 2 \; (1 - 2 \; a)$$

*or*

$$a \in (0, \; \sqrt{2} - 1).$$

*As example, set $p = .1$, $c = 2$, $a = .3$, then condition (6.101) becomes*

$$.245 < .333 \tag{6.102}$$

*Hence, for $\eta \in (.245, \; .333)$, our results can apply, where as the ones in (cf. [157], [277], [792]) cannot guarantee convergence, since (6.98) is violated.*

*Concerning the error estimates if*

$$\frac{(L - 2\ L_0)\ \eta + 2\ (a+1)}{1 - L_0\ \eta}\ (1-\eta) + L\ \eta < 1 + 2\ a, \tag{6.103}$$

*then, we have*

$$v_0 = t_0, \quad v_1 = t_1, \quad v_2 = t_2, \quad t_3 - t_2 < v_3 - v_2.$$

*An inductive argument shows that*

$$t_n < v_n \qquad (n \geq 3),$$

$$t_{n+1} - t_n < v_{n+1} - v_n \qquad (n \geq 2),$$

$$t^\star - t_n < v^\star - v_n \qquad (n \geq 2)$$

*and*

$$t^\star < v^\star.$$

*Estimates (6.103) holds for example, let*

$$L_0 = .00005, \quad L = .0001, \quad \eta = .8, \quad a = .0001$$

*to obtain*

$$.400496017 < 1.0002.$$

*Note also that (6.76) and (6.98) also hold, since*

$$\eta \leq 4999$$

*and*

$$\eta \leq 4999.00005,$$

*respectively. Hence, the claims made in the introduction of this section are now justified.*

*In practice, on will test (6.76), (6.98), (6.103) and then use the combination of the best results.*

Let return back to the numerical examples in the introduction of this section.

**Example 6.3** *Let $\mathcal{D} = [-4,\ 4]$ and choose $c = 2$ and $a = .2$. Then, we have that*

$$L_0 = L = \frac{5}{32}, \qquad \eta = \frac{5}{64} \tag{6.104}$$

*and (6.76) becomes*

$$.15625 < 1.28.$$

*Hence, the conclusions of Theorem 6.4 apply to solve equation $F(x) = 0$.*

## 6.3 Convergence with Slantly Differentiable Operator

In this section, we are concerned with the problem of approximating a locally unique solution $x^\star$ of (2.1). In a number of problems, the operator $F$ in (2.1) is nondifferentiable (cf. [413], [470], [472], [473], [629]). There is an extensive literature on iterative procedures approximating a solution $x^\star$ of equation (2.1). A survey of such results can be found in (cf. [157]) (see also (cf. [27], [139], [166], [651] and the references therein)).

We are motivated by the elegant work by Chen, Nashed and Qi (cf. [279]), where the notions and properties of slanting operators and slant differentiability of operators were inaugurated on a Banach space setting. Using the above notions and Newton–like method:

$$x_{k+1} = x_k - A(x_k)^{-1}\, F(x_k) \quad (k \geq 0, \quad x_0 \in \mathcal{D}), \tag{6.105}$$

they generated a sequence $\{x_k\}$ approximating $x^\star$. Here, operators $A(x)$ ($x \in \mathcal{D}$) are such that $A(x) \in \mathcal{L}(\mathcal{X}, \mathcal{Y})$.

Note that only local convergence results were given in (cf. [279]). We provide local as well as semilocal convergence results for the Newton–like method (6.105). In the local convergence case, our hypotheses are weaker, the radius of convergence larger and the error estimates finer than in (cf. [279]). Some applications and examples are also provided in this section.

We need the notion of slant differentiability:

**Definition 6.1** *An operator $F : \mathcal{D} \subseteq \mathcal{X} \longrightarrow \mathcal{Y}$ is slantly differentiable at $x \in \mathcal{D}$ if there exists an operator $f^\circ : \mathcal{D} \longrightarrow \mathcal{L}(\mathcal{X}, \mathcal{Y})$ such that the family $\{f^\circ(x+h)\}$ of bounded linear mappings is uniformly bounded in the operator norm for $h$ sufficiently small and*

$$\lim_{h \to 0} \frac{F(x+h) - F(x) - f^\circ(x+h)\, h}{\| h \|} = 0. \tag{6.106}$$

*The function $f^\circ$ is called a slanting function for $F$ at $x$.*

**Definition 6.2** *An operator $F : \mathcal{D} \subseteq \mathcal{X} \longrightarrow \mathcal{Y}$ is slantly differentiable in an open domain $\mathcal{D}_0 \subseteq \mathcal{D}$ if there exists an operator $f^\circ : \mathcal{D} \longrightarrow \mathcal{L}(\mathcal{X}, \mathcal{Y})$ such that $f^\circ$ is a slantly operator for $F$ at every point $x \in \mathcal{D}_0$. In this case, $f^\circ$ is called a slanting function for $F$ in $\mathcal{D}_0$.*

**Definition 6.3** *Assume $f^\circ : \mathcal{D} \longrightarrow \mathcal{L}(\mathcal{X}, \mathcal{Y})$ is a slantly operator for $F$ at $x \in \mathcal{D}$.*

*The set*

$$\partial_S\, F(x) : = \{ \lim_{x_k \to x} f^\circ(x_k) \} \tag{6.107}$$

*is the slant derivative of $F$ associated with $f^\circ$ at $x \in \mathcal{D}$. Here, $\lim_{x_k \to x} f^\circ(x_k)$ means the limit of $f^\circ(x_k)$ for any sequence $\{x_k\} \subset \mathcal{D}$ such that $x_k \to x$ and $\lim_{x_k \to x} f^\circ(x_k)$ exists. Note that $\partial_S\, F(x) \neq \emptyset$, since $f^\circ(x) \in \partial_S\, F(x)$.*

We need the result:

**Proposition 6.1** *Assume $F$ is slantly differentiable in a neighborhood of $x$ and $f^\circ$ be a slanting function for $F$ in the neighborhood of $x$. Then, the following are equivalent.*

*(i) There are a neighborhood $N_x$ of $x$ and a positive constant $c$ such that, for any $y \in N_x$, $f^\circ(y)$ is nonsingular and $\| f^\circ(y)^{-1} \| \leq c$.*

*(ii) There are a neighborhood $\widehat{N}_x$ of $x$ and a positive constant $\widehat{c}$ such that, for any $y \in \widehat{N}_x$, every $V \in \partial_S F(y)$ is nonsingular and $\| V^{-1} \| \leq \widehat{c}$.*

We finally need the definitions of smoothing and semismoothing functions of $F$.

**Definition 6.4** *An operator $f : \mathcal{D} \times (0, +\infty) \longrightarrow \mathcal{Y}$ is a smoothing mapping of $F$ if $f$ is continuously differentiable with respect to $x$ and for any $x \in \mathcal{D}$ and any $\epsilon > 0$,*

$$\| F(x) - f(x, \epsilon) \| \leq \mu \ \epsilon, \tag{6.108}$$

*where $\mu$ is a positive constant. The smoothing operator is said to satisfy the slant derivative consistently property at $\widehat{x}$ (in $\mathcal{D}$), if*

$$\lim_{\epsilon \to 0^+} f_x(x, \epsilon) = f^\circ(x) \in \mathcal{L}(\mathcal{X}, \mathcal{Y}) \tag{6.109}$$

*exists for $x$ in a neighborhood of $\widehat{x}$ (in $\mathcal{D}$) and $f^\circ$ serves as a slanting mapping for $F$ at $\widehat{x}$ (in $\mathcal{D}$).*

**Definition 6.5** *An operator $F : \mathcal{D} \subseteq \mathcal{X} \longrightarrow \mathcal{Y}$ is semismooth at $x$ if there is a slantly operator $f^\circ$ for $F$ in a neighborhood $N_x$ of $x$ such that $f^\circ$ and the associated slant derivative satisfy the conditions:*

*(i)* $\lim_{t \to 0^+} f^\circ(x + t\ h)\, h$ *exists for every $h \in \mathcal{X}$ and*

$$\lim_{h \to 0} \frac{\lim_{t \to 0^+} f^\circ(x + t\ h)\, h - f^\circ(x + h)\ h}{\| h \|} = 0. \tag{6.110}$$

*(ii)*

$$f^\circ(x + h)\ h - V\ h = o(\| h \|) \qquad (V \in \partial_S F(x + h)). \tag{6.111}$$

We provide three local convergence results for the Newton–like method (6.105).

**Theorem 6.5** *Assume that $F : \mathcal{D} \subseteq \mathcal{X} \longrightarrow \mathcal{Y}$ be slantly differentiable operator at a solution $x^\star$ of equation $F(x) = 0$. Let $f^\circ$ be a slanting operator for $F$ at $x^\star$, $A(x) \in \mathcal{L}(\mathcal{X}, \mathcal{Y})$ ($x \in \mathcal{D}$) with $A(x^\star)^{-1} \in \mathcal{L}(\mathcal{Y}, \mathcal{X})$,*

$$\| A(x^\star)^{-1}\ (A(x^\star + h) - A(x^\star)) \| \leq L \ \| h \|, \tag{6.112}$$

$$\| A(x^\star)^{-1} \ (A(x^\star + h) - f^\circ(x^\star + h)) \| \leq M \ \ \| h \| + M_0 \tag{6.113}$$

*as* $\| h \| \to 0$*;*

*By the definition of slant differentiability, there exist* $c > 0$ *and* $\delta = \delta(c) > 0$ *such that, for all* $\| h \| \leq \delta$, $\| f^\circ(x^\star + h) \| \leq c$ *and*

$$\frac{\| F(x^\star + h) - F(x^\star) - f^\circ(x^\star + h) \|}{\| h \|} \leq c; \tag{6.114}$$

$$M_0 + c < 1; \tag{6.115}$$

*Set*

$$r = \min \left\{ \delta, \ \frac{1}{L}, \ \frac{1 - c - M_0}{M + L} \right\}, \tag{6.116}$$

$$U(x^\star, r) \subseteq \mathcal{D}. \tag{6.117}$$

*Then, sequence* $\{x_n\}$ *generated by Newton–like method (6.105) is well defined, remains in* $U(x^\star, r)$ *for all* $n \geq 0$ *and converges to* $x^\star$*, provided that* $x_0 \in U(x^\star, r)$*. Moreover, the following estimates hold for all* $k \geq 0$

$$\| x_{k+1} - x^\star \| \leq q \ \ \| x_k - x^\star \|, \tag{6.118}$$

*where*

$$0 \leq q = \frac{c + M_0 + M \ \ \| x_0 - x^\star \|}{1 - L \ \ \| x_0 - x^\star \|} < 1. \tag{6.119}$$

**Proof** By hypothesis, $x_0 \in U(x^\star, r)$. Hence, $x_1$ is well defined. Assume $x_m \in U(x^\star, r)$ for all $m \leq k$. Using (6.112) for $h = x_m - x^\star$ and (6.116), we obtain that

$$\| A(x^\star)^{-1} \ (A(x^\star + (x_m - x^\star)) - A(x^\star)) \| \leq L \ \ \| x_m - x^\star \| < L \ r \leq 1. \tag{6.120}$$

It follows from (6.120) and the Banach lemma on invertible operators that $A(x_m)^{-1}$ exists on $U(x^\star, r)$ and

$$\| A(x_m)^{-1} \ A(x^\star) \| \leq (1 - L \ \ \| x_m - x^\star \|)^{-1} < (1 - L \ r)^{-1}. \tag{6.121}$$

Using (6.105), (6.113), (6.114), (6.116) and (6.121), we obtain in turn that

$$\begin{aligned}
&\| x_{k+1} - x^\star \| \\
&= \| x_k - x^\star - A(x_k)^{-1} \ F(x_k) + A(x_k)^{-1} \ F(x^\star) \| \\
&\leq \| A(x_k)^{-1} \ A(x^\star) \| \ \ ( \| A(x^\star)^{-1} \ (F(x_k) - F(x^\star) - f^\circ(x_k) \, (x_k - x^\star)) \| \\
&\quad + \| A(x^\star)^{-1} \ (A(x_k) - f^\circ(x_k)) \ (x_k - x^\star) \| \ ) \\
&\leq \frac{1}{1 - L \ \ \| x_k - x^\star \|} \ (c + M_0 + M \ \ \| x_k - x^\star \|) \ \ \| x_k - x^\star \| \\
&\leq q \ \ \| x_k - x^\star \| < \| x_k - x^\star \| < r,
\end{aligned} \tag{6.122}$$

which imply $x_{k+1} \in U(x^\star, r)$ and $\lim\limits_{k \to \infty} x_k = x^\star$. That completes the proof of Theorem 6.5. ⊠

**Theorem 6.6** *Assume $F : \mathcal{D} \subseteq \mathcal{X} \longrightarrow \mathcal{Y}$ be slantly differentiable operator at a solution $x^\star$ of equation $F(x) = 0$; Let $f^\circ$ be a slanting operator for $F$ at $x^\star$, $0 < \| f^\circ(x^\star)^{-1} \| \leq a$, $A(x) \in \mathcal{L}(\mathcal{X}, \mathcal{Y})$ $(x \in \mathcal{D})$,*

$$\| A(x^\star + h) - A(x^\star) \| \leq L \ \| h \|,$$
$$\| A(x^\star + h) - f^\circ(x^\star + h) \| \longrightarrow 0 \tag{6.123}$$

*as $\| h \| \to 0$;*

*By Definition 6.1 and the perturbation Banach lemma, there exist a ball $U(x^\star, \delta_0)$ such that $U(x^\star, \delta_0) \subseteq \mathcal{D}$ and positive constants $a_0 \geq a$, $\rho_1$, $\rho_2 \in (0, 1)$, $\rho_1 + \rho_2 < 1$ such that, for any $x \in U(x^\star, r)$*

$$\| A(x^\star)^{-1} \| \leq a_0, \tag{6.124}$$

$$\| F(x) - F(x^\star) - f^\circ(x) \ (x - x^\star) \| \leq \frac{\rho_1}{a_0} \ \| x - x^\star \| \tag{6.125}$$

*and*

$$\| A(x) - f^\circ(x) \| \leq \frac{\rho_2}{a_0}. \tag{6.126}$$

*Set*

$$r = \min \left\{ \delta_0, \, \delta_1, \, \frac{1 - (\rho_1 + \rho_2)}{a_0 \ L} \right\}. \tag{6.127}$$

$$U(x^\star, r) \subseteq \mathcal{D}. \tag{6.128}$$

*Then, sequence $\{x_n\}$ generated by Newton–like method (6.105) is well defined, remains in $U(x^\star, r)$ for all $n \geq 0$ and converges to $x^\star$, provided that $x_0 \in U(x^\star, r)$. Moreover, the following estimates hold for all $k \geq 0$*

$$\| x_{k+1} - x^\star \| \leq q_0 \ \| x_k - x^\star \|, \tag{6.129}$$

*where*

$$q_0 = \frac{\rho_1 + \rho_2}{1 - a_0 \ L \ \| x_0 - x^\star \|}. \tag{6.130}$$

*Furthermore, the convergence is superlinear.*

**Proof** Estimates (6.121) now becomes

$$\| A(x_m)^{-1} \| \leq \frac{a_0}{1 - a_0 \ L \ \| x_m - x^\star \|}, \tag{6.131}$$

whereas estimate (6.122) gives

$$\begin{aligned} \| x_{k+1} - x^\star \| &\leq \frac{\rho_1 + \rho_2}{1 - a_0 \ L \ \| x_k - x^\star \|} \ \| x_k - x^\star \| \\ &\leq q_0 \ \| x_k - x^\star \| < \| x_k - x^\star \| < r, \end{aligned} \tag{6.132}$$

which imply $x_{k+1} \in U(x^\star, r)$ and $\lim_{k \to \infty} x_k = x^\star$.

Finally, using Definition 6.1 and (6.123), we deduce that

$$\| x_{k+1} - x^\star \| = O(\| x_k - x^\star \|). \tag{6.133}$$

This completes the proof of Theorem 6.6. ⊠

In view of Theorem 6.6 and Proposition 6.1, we obtain the special cases:

**Proposition 6.2** *Assume $F : \mathcal{D} \subseteq \mathcal{X} \longrightarrow \mathcal{Y}$ be slantly differentiable operator at a solution $x^\star$ of equation $F(x) = 0$. Let $f^\circ$ be a slanting operator for $F$ at $x^\star$ and $\| f^\circ(x^\star)^{-1} \| \leq a$, where $a$ is a positive constant. Then, the following statements hold in $U(x^\star, r)$.*

*(i) Newton–like method*

$$x_{k+1} = x_k - f^\circ(x_k)^{-1}\; F(x_k)$$

*superlinearly converges to $x^\star$.*

*(ii) If $f : \mathcal{D} \times (0, +\infty) \longrightarrow \mathcal{Y}$ is smoothing function, which satisfies the slant consistently property (6.109) in $U(x^\star, r)$, then, the smoothing NM defined by*

$$x_{k+1} = x_k - f'(x_k, \epsilon_k)^{-1}\; F(x_k)$$

*superlinearly converges to $x^\star$.*

*(iii) If $F$ is semismooth at $x^\star$, then the semismooth NM*

$$x_{k+1} = x_k - V_k^{-1}\; F(x_k) \qquad (V_k \in \partial_S F(x_k))$$

*superlinearly converges to $x^\star$.*

**Remark 6.5** *Theorem 6.5 cannot directly be compared with Theorem 3.4 in (cf. [279]) since different, but easier to verify conditions are used in our Theorem 6.6. However, our Theorem 6.6 compares favorably to Theorem 3.4 in (cf. [279]), where stronger condition:*

$$\| f^\circ(x) \| \leq a \tag{6.134}$$

*is used. There are also other reasons:*

**Application 6.1** *In the special case where $f^\circ = F'$, $A = F'$, we get $M_0 = M = 0$ and introducing Lipschitz condition*

$$\| F'(x) - F'(y) \| \leq K \;\; \| x - y \| \qquad (x \in \mathcal{D}). \tag{6.135}$$

*We can set $c = \dfrac{K}{2}$. Our estimate (6.132) gives*

$$\| x_{k+1} - x^\star \| \leq \frac{a\;K \;\; \| x_k - x^\star \|^2}{2(1 - L\;a \;\; \| x_k - x^\star \|)}, \tag{6.136}$$

*which implies*

$$r_A = \frac{2}{a\ (2L + K)}, \tag{6.137}$$

*whereas the corresponding one in (cf. [279, Theorem 3.4]) becomes*

$$\| x_{k+1} - x^\star \| \le \frac{a\ K\ \ \| x_k - x^\star \|^2}{2(1 - K\ \ \| x_k - x^\star \|)}, \tag{6.138}$$

*which leads to*

$$r_R = \frac{2}{3\ a\ K}. \tag{6.139}$$

*Note that*

$$L \le K \tag{6.140}$$

*holds in general and* $\frac{K}{L}$ *can be arbitrarily large. Using (6.137), (6.139) and (6.140), we obtain*

$$r_R \le r_A. \tag{6.141}$$

*Moreover, if strict inequality holds in (6.140), then, so does in (6.141). Furthermore, our error estimates (6.136) are finer than (6.138). The radius* $r_R$ *was obtained by Rheinboldt (cf. [638]).*

We provide two semilocal convergence results for the Newton–like method (6.105).

**Theorem 6.7** *Assume that* $F\ :\ \mathcal{D} \subseteq \mathcal{X} \longrightarrow \mathcal{Y}$ *be slantly differentiable operator on an open domain* $\mathcal{D}_0 \subseteq \mathcal{D}$. *Let* $f^\circ$ *be a slanting operator for* $F$ *on* $\mathcal{D}_0$, $A(x) \in \mathcal{L}(\mathcal{X}, \mathcal{Y})$ $(x \in \mathcal{D})$ *and there exist* $x_0 \in \mathcal{D}_0$, $a > 0$, $L > 0$ *such that*

$$\| f^\circ(x_0)^{-1} \| \le a, \tag{6.142}$$

$$\| A(x_0)^{-1}\ (A(x_0 + h) - A(x_0)) \| \le L\ \ \| h \|, \tag{6.143}$$

$$\| A(x_0)^{-1}\ (A(x_0 + h) - f^\circ(x_0 + h) \| \longrightarrow 0 \tag{6.144}$$

*as* $\| h \| \rightarrow 0$;

*By Definition 6.1 and the perturbation Banach lemma, there exist* $\overline{U}(x_0, \delta_0) \subseteq \mathcal{D}_0$ *and positive constants* $a_0 \ge a$, $\rho_1$, $\rho_2 \in (0,1)$, $\rho_1 + 2\rho_2 < 1$ *such that, for all* $x, y \in \overline{U}(x_0, \delta)$, $A(x_0)$ *is nonsingular,*

$$\| A(x_0)^{-1} \|\ \le a_0, \tag{6.145}$$

$$\| F(x) - F(y) - f^\circ(x)\ (x - y) \|\ \le \frac{\rho_1}{a_0}\ \ \| x - y \| \tag{6.146}$$

*and*

$$\| A(x) - f^\circ(x) \| \le \frac{\rho_2}{a_0}. \tag{6.147}$$

*For*

$$\eta_0 = \frac{2}{a_0\ L} \left( \frac{1 - (\rho_1 + 2\ \rho_2)}{1 + \sqrt{\rho_1 + 2\rho_2}} \right)^2, \tag{6.148}$$

$$\eta_1 = \frac{(1-(\rho_1+2\,\rho_2))\,\delta}{2-\delta\,a_0\,L}, \tag{6.149}$$

$$\delta\,a_0\,L < 2, \tag{6.150}$$

$$\|\,A(x_0)^{-1}\,F(x_0)\,\| < \eta < \min\{\eta_0,\,\eta_1\}, \tag{6.151}$$

$$r_- = \frac{2\eta}{1+a_0\,L\,\eta-(\rho_1+2\rho_2)+\sqrt{(1+a_0\,L\,\eta-(\rho_1+2\,\rho_2))^2-4\,a_0\,L\,\eta}}, \tag{6.152}$$

$$\overline{U}(x_0,r) \subseteq \mathcal{D}_0, \tag{6.153}$$

*where*

$$r \in [r_-,\,\delta]. \tag{6.154}$$

*Then, sequence $\{x_n\}$ generated by Newton–like method (6.105) is well defined, remains in $U(x_0,r)$ for all $n \geq 0$ and superlinearly converges to a unique solution $x^\star$ in $U(x_0,r)$ of equation $F(x)=0$. Moreover, the following estimates hold*

$$\|\,x_{k+1}-x_k\,\| \;\leq b\;\|\,x_k-x_{k-1}\,\| \quad (k\geq 1), \tag{6.155}$$

$$\|\,x_k-x_0\,\| \;\leq \frac{1-b^k}{1-b}\,\eta \leq \frac{\eta}{1-b} \quad (k\geq 0) \tag{6.156}$$

*and*

$$\|\,x_k-x^\star\,\| \leq \frac{b^k}{1-b}\,\eta \quad (k\geq 0), \tag{6.157}$$

*where*

$$b = \frac{a_0}{1-a_0\,L\,r}. \tag{6.158}$$

**Proof** Using (6.150), (6.151), we see $\eta_1 \geq 0$, $b \in [0,1)$, $r_- \geq 0$, $[r_-,\delta] \neq \emptyset$ and $x_1 \in U(x_0,r)$, (since $\eta \leq r$). Assume $x_m \in U(x_0,r)$ for all $m \leq k$. We obtain using (6.143) and the choice of $r$

$$\begin{aligned}\|\,A(x_0)^{-1}\,\|\;\|\,A(x_m)-A(x_0)\,\| &= \|\,A(x_0)^{-1}\,\|\;\|\,A(x_0+(x_m-x_0))-A(x_0)\,\|\\ &\leq a_0\,L\;\|\,x_m-x_0\,\| \leq a_0\,L\,r < 1.\end{aligned} \tag{6.159}$$

It follows from (6.159) and the Banach lemma on invertible operators that $A(x_m)^{-1}$ exists with

$$\|\,A(x_m)^{-1}\,\| \leq \frac{a_0}{1-a_0\,L\;\|\,x_m-x^\star\,\|} \leq \frac{a_0}{1-a_0\,L\,r}. \tag{6.160}$$

In view of (6.105), we obtain the approximation

$$\begin{aligned}F(x_{m+1}) &= (F(x_{m+1})-F(x_m)-f^\circ(x_{m+1})\,(x_{m+1}-x_m))\\ &\quad+(f^\circ(x_{m+1})-A(x_{m+1}))\,(x_{m+1}-x_m)+(f^\circ(x_m)-A(x_m))\,(x_{m+1}-x_m)\\ &\quad-f^\circ(x_m)\,(x_{m+1}-x_m).\end{aligned} \tag{6.161}$$

Using (6.105), (6.146), (6.147), (6.160), we obtain in turn that

$$\begin{aligned}
\| x_{m+2} - x_{m+1} \| &= \| A(x_{m+1})^{-1} \; F(x_{m+1}) \| \\
&\leq \| A(x_{m+1})^{-1} \| \; \| F(x_{m+1}) \| \\
&\leq \| A(x_{m+1})^{-1} \| \\
&\quad \times ( \| F(x_{m+1}) - F(x_m) - f^\circ(x_{m+1}) \; (x_{m+1} - x_m) \| \\
&\quad + \| (f^\circ(x_{m+1}) - A(x_{m+1})) \; (x_{m+1} - x_m) \| \\
&\quad + \| (f^\circ(x_m) - A(x_m)) \; (x_{m+1} - x_m) \| + \| f^\circ(x_m) \; (x_{m+1} - x_m) \| ) \\
&\leq \frac{a_0}{1 - a_0 \; L \; r} \left( \frac{\rho_1}{a_0} + \frac{\rho_2}{a_0} + \frac{\rho_2}{a_0} \right) \; \| x_{m+1} - x_m \| \\
&\leq b \; \| x_{m+1} - x_m \|,
\end{aligned} \tag{6.162}$$

which implies (6.155). We also get that

$$\begin{aligned}
\| x_{m+1} - x_0 \| &\leq \| x_{m+1} - x_m \| + \| x_m - x_{m-1} \| + \cdots + \| x_1 - x_0 \| \\
&\leq (b^m + b^{m-1} + \cdots + 1) \; \| x_1 - x_0 \| \\
&\leq \frac{1 - b^{m+1}}{1 - b} \; \| x_1 - x_0 \| \leq \frac{\eta}{1 - b} \leq r,
\end{aligned} \tag{6.163}$$

which shows $x_{m+1} \in \overline{U}(x_0, r)$ and (6.156). We also have for all $m \geq k$

$$\begin{aligned}
&\| x_{k+m} - x_k \| \\
&\leq \| x_{k+m} - x_{k+m-1} \| + \| x_{k+m-1} - x_{k+m-2} \| + \cdots + \| x_{k+1} - x_k \| \\
&\leq (b^{k+m-1} + b^{k+m-2} + \cdots + b^k) \; \| x_1 - x_0 \| \\
&= b^k \; \frac{1 - b^m}{1 - b} \; \eta.
\end{aligned} \tag{6.164}$$

It follows from (6.164), that sequence $\{x_n\}$ is a Cauchy sequence in a Banach space $\mathcal{X}$ and as such it converges to some $x^\star \in \overline{U}(x_0, r)$. By letting $m \longrightarrow \infty$ in (6.164), we obtain (6.157). Moreover, by letting $m \longrightarrow \infty$ in (6.162), we obtain $F(x^\star) = 0$.

Finally, to show the uniqueness, let $y^\star \in \overline{U}(x_0, r)$ be a solution of equation $F(x) = 0$. By replacing $y$ by $y^\star$ in (6.146), we obtain in turn as in (6.162) that

$$\begin{aligned}
\| x_{m+1} - y^\star \| &= \| x_m - y^\star - A(x_m)^{-1} \; F(x_m) + A(x_m)^{-1} \; F(y^\star) \| \\
&\leq \| A(x_m)^{-1} \| \; \Bigg( \| F(x_m) - F(y^\star) - f^\circ(x_m) \; (x_m - y^\star) \| \\
&\quad + \| (A(x_m) - f^\circ(x_m)) \; (x_m - y^\star) \| \Bigg)
\end{aligned}$$

$$\leq \frac{a_0}{1 - a_0\, L\, r} \left( \frac{\rho_1}{a_0} + \frac{\rho_2}{a_0} \right) \parallel x_m - y^\star \parallel$$

$$< b \parallel x_m - y^\star \parallel \leq \parallel x_m - y^\star \parallel, \tag{6.165}$$

which implies $\lim\limits_{k\to\infty} x_k = y^\star$. But we showed $\lim\limits_{k\to\infty} x_k = x^\star$. Hence, we deduce $x^\star = y^\star$. The superlinear convergence of sequence $\{x_n\}$ follows from Definition 6.2, (6.144) since $\parallel x_{k+1} - x^\star \parallel = O(\parallel x_k - x^\star \parallel)$ (by (6.165)). This completes the proof of Theorem 6.7. ⊠

In view of Theorem 6.7 and Proposition 6.1 as in Proposition 6.2, we obtain the special cases:

**Theorem 6.8** *Assume $F : \mathcal{D} \subset \mathcal{X} \longrightarrow \mathcal{Y}$ be slantly differentiable operator on a open domain $\mathcal{D}_0 \subseteq \mathcal{D}$.*

*Let $x_0 \in \mathcal{D}_0$ and $f^\circ$ be a slanting operator for $F$ on $\mathcal{D}_0$ such that $\parallel f^\circ(x_0)^{-1} \parallel \leq a$, where $a > 0$. Then, the conclusions of Proposition 6.2 hold in $U(x_0, r)$.*

Note that the results of this section cannot be compared with the ones in (cf. [279]) since only local results were provided there.

# 6.4 A Intermediate Newton Method

We revisit in this section a one–step intermediate Newton iterative scheme that was used by Uko and Velásquez in (cf. [725]) for the constructive solution of nonlinear equations of the type $f(u) + g(u) = 0$ . By utilizing weaker hypotheses of the Zabrejko-Nguen kind and a modified majorizing sequence we perform a semilocal convergence analysis which yields finer error bounds and more precise information on the location of the solution that the ones obtained in (cf. [725]). We also give two generalizations of the well-known Kantorovich theorem on the solvability of nonlinear equations and the convergence of Newton's method. Illustrative examples are provided.

Let $X$ and $Y$ be Banach spaces, let $u_0 \in X$, $D$ be a non-empty closed ball $B[u_0, T]$ in $X$ and $F\colon D \longmapsto Y$ be a continuous function. We denote by $D_0 = B(u_0, T)$. We are interested in the solvability of (2.1) the function $F$ has a splitting

$$F = f + g,$$

where $f, g\colon D \longmapsto Y$ are continuous functions that are Fréchet differentiable on $D_0$ and that therefore we can reformulate equation (2.1) in the form

$$f(u) + g(u) = 0. \tag{6.166}$$

In (cf. [725]), it was shown that if $f$ and $g$ satisfy Zabrejko-Nguen type conditions, then, the iterates obtained from the equation

$$u_{m+1} = u_m - [f'(u_m) + g'(u_0)]^{-1}[f(u_m) + g(u_m)] \quad (m \geq 0) \tag{6.167}$$

converge to a solution of problem (6.166). When $g = 0$, the scheme (6.167) becomes the classical Newton scheme for the equation $f(u) = 0$ and when $f = 0$, it becomes the modified Newton scheme for the equation $g(u) = 0$. Therefore this scheme provides an interesting unified setting for the study of both Newton's method and the modified Newton's method. The iterative scheme (6.167) can also be viewed as an intermediate scheme between the method of Newton, defined by the iterations

$$u_{m+1} = u_m - F'(u_m)^{-1}F(u_m) \quad (m \geq 0) \tag{6.168}$$

and the modified Newton scheme, defined by the iterations

$$u_{m+1} = u_m - F'(u_0)^{-1}F(u_m) \quad (m \geq 0).$$

It is well–known that although Newton's method usually requires fewer iterations than the modified Newton method to achieve a specified level of accuracy, the later is less expensive to implement than Newton's method. This led several authors (cf. [29, 131, 725]) to propose intermediate Newton methods which converge faster than the modified method and are cheaper to implement than Newton's method. The iterative scheme (6.167) is an interesting intermediate Newton scheme of this kind and is particularly useful in situations in which the Jacobian derivative of $f$ is relatively easy to compute.

The Kantorovich Theorem is a fundamental tool in nonlinear analysis for proving the existence and uniqueness of solutions of nonlinear equations arising in various fields (cf. [163, 209, 469]). The generalized Kantorovich Theorems that we present in this section are extensions of the Kantorovich and Argyros results and should ultimately lead to an enlargement of the class of nonlinear problems that can be solved with the Kantorovich technique and/or a weakening of the solvability conditions for some of the previously solved problems.

We define the majorant sequences that we will use and give their main properties.

**Proposition 6.3** *Let $a \geq 0$, let $\alpha(t)$, $\alpha_0(t)$ and $\beta_0(t)$ be non-negative non-decreasing functions defined on an interval $[0, T]$. For any $0 \leq t \leq T$, let*

$$\kappa(t) = \int_0^t \alpha(s)\,ds, \qquad \kappa_0(t) = \int_0^t \alpha_0(s)\,ds,$$

$$\sigma(t) = a + \int_0^t \kappa(s)\,ds - t, \qquad \sigma_0(t) = a + \int_0^t \kappa_0(s)\,ds - t,$$

$$\pi_0(t) = \int_0^t \beta_0(s)\,ds, \qquad \tau_0(t) = \int_0^t \pi_0(s)\,ds.$$

*Let $\mu(t) \equiv \sigma(t) + \tau_0(t)$, $t_0 = 0$, $t_1 = a$ and for $m \geq 1$, let*

$$t_{m+1} = t_m - \frac{\mu(t_m) - \mu(t_{m-1}) - (t_m - t_{m-1})\sigma'(t_{m-1}))}{\sigma_0'(t_m)}. \tag{6.169}$$

*Suppose that $\kappa_0(T) < 1$ and $t_m \leq T$ for all $m \geq 1$. Then, sequence $\{t_m\}$ is well defined and convergent and for $m \geq 1$, we have that*

$$t_{m-1} \leq t_m \leq t_* \equiv \lim_{n\to\infty} t_m \leq T. \tag{6.170}$$

**Proof** Since $\sigma_0'(t_m) = \kappa_0(t_m) - 1 \leq \kappa_0(T) - 1 < 0$ for all $m \geq 1$ and $\sigma(t)$ is convex, we see that, if $t_{m-1} \leq t_m$, then

$$\begin{aligned}&\mu(t_m) - \mu(t_{m-1}) - (t_m - t_{m-1})\sigma'(t_{m-1})\\ &= \tau_0(t_m) - \tau_0(t_{m-1}) + \sigma(t_m) - \sigma(t_{m-1}) - (t_m - t_{m-1})\sigma'(t_{m-1}) \geq 0.\end{aligned}$$

Therefore, it follows from an easy induction argument that the sequence $\{t_m\}$ is well defined, monotone increasing and bounded above by $T$ and as such it converges to its unique least upper bound $t_*$. ⊠

**Remark 6.6** *According to (cf. [725, Proposition 1]) the condition $\kappa_0(T) < 1$ can be replaced by the requirement that the function $\mu(t) = \sigma(t) + \tau_0(t)$ have a unique zero $t_*$ in $[0, T]$.*

In the next series of results, we will consider specific choices of $\alpha(t)$, $\alpha_0(t)$ and $\beta(t)$ for which the hypotheses of Proposition 6.3 hold.

**Proposition 6.4** *Let $a \geq 0$, $L_0 \geq 0$, $0 \leq M_0 \leq M$, $0 \leq \theta < 2$, $T = 2a/(2-\theta)$ and suppose that*

$$\frac{M\theta}{2} + (2L_0 + \theta M_0)(\frac{\theta}{2}) \leq M, \tag{6.171}$$

$$(M + 2L_0 + M_0\theta)a \leq \theta. \tag{6.172}$$

*Then, sequence defined by*

$$t_0 = 0,\ t_1 = a,\ t_{m+1} = t_m + \frac{M(t_m - t_{m-1})^2 + L_0(t_m^2 - t_{m-1}^2)}{2(1 - M_0 t_m)} \quad (m \geq 1) \tag{6.173}$$

*is well defined and converges to a real number $t_*$ that satisfies condition (6.170) and the inequality*

$$(M_0 + L_0)t_* \leq 1. \tag{6.174}$$

**Proof** If we set $\alpha(t) \equiv Mt$, $\alpha_0(t) \equiv M_0 t$ and $\beta_0(t) \equiv L_0 t$, then, sequence in (6.169) reduces to the form (6.173). Therefore it suffices to verify the hypotheses of Proposition 6.3. If $a = 0$ or $\theta = 0$ or $M^2 + L_0^2 = 0$, then, $t_m = t_1$ for all $m \geq 1$ and the conclusions of the Proposition hold trivially. So we assume that $a > 0$ and $M^2 + L_0^2 > 0$ and $\theta > 0$. We show by induction that the inequalities

$$t_{m-1} \leq t_m \leq T, \tag{6.175}$$

$$M(t_m - t_{m-1}) + 2L_0 t_m + M_0\theta t_m \leq \theta, \tag{6.176}$$

$$M_0 t_m < 1 \tag{6.177}$$

hold for all $m \geq 1$. It follows from (6.172) that they hold when $m = 1$. Suppose, by induction, that $k \geq 1$ and that these inequalities hold for all $m \leq k$. Then, it follows from (6.176), (6.173) and (6.177) that the inequalities

$$\begin{aligned} M(t_m - t_{m-1}) + L_0(t_m + t_{m-1}) + M_0\theta t_m &\leq M(t_m - t_{m-1}) + 2L_0 t_m + M_0\theta t_m \\ &\leq \theta, \end{aligned}$$

$$t_{m+1} - t_m = (t_m - t_{m-1})\frac{M(t_m - t_{m-1}) + L_0(t_m + t_{m-1})}{2(1 - M_0 t_m)} \leq \frac{\theta}{2}(t_m - t_{m-1})$$

hold for all $m \leq k$. Hence, we have

$$0 \leq t_{k+1} - t_k \leq \frac{\theta}{2}(t_k - t_{k-1}) \leq \cdots \leq \left(\frac{\theta}{2}\right)^k a, \tag{6.178}$$

$$\begin{aligned} t_{k+1} &= t_1 + (t_2 - t_1) + \cdots + (t_{k+1} - t_k), \\ &\leq \left[1 + \cdots + \left(\frac{\theta}{2}\right)^k\right] a = \frac{a[1 - (\frac{\theta}{2})^{k-1}]}{1 - \frac{\theta}{2}} < \frac{a}{1 - \frac{\theta}{2}} = T. \end{aligned} \tag{6.179}$$

Therefore (6.175) holds when we replace $m$ with $k+1$. It follows from (6.179), (6.172) and (6.171) that

$$\begin{aligned} &M(t_{k+1} - t_k) + 2L_0 t_{k+1} + M_0\theta t_{k+1} \\ &\leq Ma\left(\frac{\theta}{2}\right)^k + \frac{(2L_0 + M_0\theta)a[1 - (\frac{\theta}{2})^{k+1}]}{1 - \frac{\theta}{2}} \\ &= T\left[(2L_0 + \theta M_0) + \left(\frac{\theta}{2}\right)^k \left(M - M\left(\frac{\theta}{2}\right) - (2L_0 + \theta M_0)\left(\frac{\theta}{2}\right)\right)\right] \\ &\leq T\left[(2L_0 + \theta M_0) + M - M\left(\frac{\theta}{2}\right) - (2L_0 + \theta M_0)\left(\frac{\theta}{2}\right)\right] \\ &= T(M + 2L_0 + M_0\theta)\left(1 - \left(\frac{\theta}{2}\right)\right) \\ &= (M + 2L_0 + M_0\theta)a \leq \theta. \end{aligned}$$

This shows that the inequality (6.176) holds when $m = k+1$ and implies that

$$t_{k+1} \leq \frac{Mt_k + \theta}{M + 2L_0 + M_0\theta}. \tag{6.180}$$

If $M_0 = 0$, then, (6.177) holds trivially when we replace $m$ with $k+1$. If $M_0 > 0$, then, it follows from (6.180) and the induction hypotheses that

$$t_{k+1} < \frac{(M/M_0) + \theta}{M + 2L_0 + M_0\theta} \leq \frac{(M/M_0) + \theta}{M + M_0\theta} = \frac{1}{M_0}.$$

Therefore, (6.177) also holds when we replace $m$ with $k+1$. We conclude, by induction, that (6.175)–(6.177) hold for all $m \geq 1$.

It follows from (6.173) and a straightforward induction argument that the sequence $\{t_m\}$ is monotone increasing and bounded above by $T$ and as such it converges to its unique least upper bound $t_*$. To conclude the proof, we rewrite (6.176) in the form $M(t_m - t_{m-1}) + 2L_0 t_m \leq \theta(1 - M_0 t_m)$ and let $m$ tend to infinity. We obtain the inequality

$$2L_0 t_* \leq \theta(1 - M_0 t_*) \leq 2(1 - M_0 t_*),$$

which implies (6.174). ⊠

A setback of this result is the fact that $\theta$ has to be found by trial and error and there is no guarantee that the choice we make leads to minimal conditions on $M$, $M_0$, $L_0$ and $a$ that ensure the convergence of the sequence $\{t_m\}$. The next result resolves this problem by identifying the best possible value of $\theta$ provided that $M_0$ and $M$ are not allowed to vanish at the same time.

**Proposition 6.5** *Let* $0 \leq L_0$, $0 \leq \gamma$, $0 \leq M_0 \leq M \leq \gamma M_0$ *and set*

$$\theta_*(M, M_0, L_0) \equiv \begin{cases} \frac{4M}{M+2L_0+\sqrt{[2L_0+M]^2+8MM_0}}, & \text{if } M + L_0 > 0, \\ 0, & \text{if } M + L_0 = 0. \end{cases} \tag{6.181}$$

*Then, the following assrtion holds*

$$0 \leq \theta_*(M, M_0, L_0) < 2. \tag{6.182}$$

*Moreover, there exists* $\theta \in [0, 2)$ *satisfying conditions (6.171)–(6.172) if and only if*

$$[2L_0 + M + M_0\theta_*(M, M_0, L_0)]a \leq \theta_*(M, M_0, L_0). \tag{6.183}$$

**Proof** If $M + L_0 = 0$, then, condition (6.182) holds trivially. If $M + L_0 > 0$, then $\theta_*(M, M_0, L_0) \leq \frac{4M}{2M+4L_0} \leq 2$. Since $M$ and $L_0$ cannot both vanish at the same time, one of these inequalities is strict, which shows (6.182) also holds in this case. If $M + L_0 = 0$, then, $M = L_0 = M_0 = 0$. Therefore, conditions (6.171)–(6.172) and (6.183) are redundant and so the equivalence of these conditions holds trivially. Therefore, without loss of generality, we may assume that $M + L_0 > 0$. In this case it is easy to verify – by solving a quadratic inequality – that $\theta_*(M, M_0, L_0)$ is by definition the largest value of $\theta$ for which the inequality (6.171) holds.

Suppose that (6.183) holds. Then, if we set $\theta = \theta_*(M, M_0, L_0)$, we see that condition (6.171) holds as an equality and that (6.172) reduces to condition (6.183).

Suppose now that conditions (6.171) and (6.172) hold for some $\theta \in [0, 2)$. Then, condition (6.172) implies that $M_0 a \leq \frac{M_0\theta}{M+M_0\theta+2L_0} < 1$. It follows that condition (6.172) can be expressed in the equivalent form

$$0 \leq (M + 2L_0)a/(1 - M_0 a) \leq \theta.$$

Since $\theta_*(M, M_0, L_0)$ is the largest value of $\theta$ for which the inequality (6.171) holds we conclude that $0 \leq (M + 2L_0)a/(1 - M_0 a) \leq \theta \leq \theta_*(M, M_0, L_0)$.

This shows that (6.183) holds and that completes the proof of the Proposition. ⊠

The next result gives an alternative set of hypotheses that guarantee the convergence of the scalar majorant sequence (6.173).

**Proposition 6.6** *Suppose that* $a \geq 0$, $0 \leq M_0 \leq M$, $0 \leq L_0 \leq M$, $0 \leq \theta < 2$, $T = 2a/(2-\theta)$ *and*

$$\frac{M\theta}{2} + (L_0 + \theta M_0)\left(\frac{\theta}{2}\right) \leq M, \tag{6.184}$$

$$(M + L_0 + M_0\theta)a \leq \theta. \tag{6.185}$$

*Then, sequence* $\{t_m\}$ *defined in (6.173) is well defined and converges to a real number* $t_*$ *that satisfies conditions (6.170)and (6.174).*

**Proof** As in the proof of Proposition 6.4, we may assume without loss of generality that $a > 0$ and and $M^2 + L_0^2 > 0$ and $\theta > 0$. We show by induction that the inequalities

$$t_{m-1} \leq t_m \leq T, \tag{6.186}$$

$$M(t_m - t_{m-1}) + L_0(t_m + t_{m-1}) + M_0\theta t_m \leq \theta, \tag{6.187}$$

$$M_0 t_m < 1 \tag{6.188}$$

hold for all $m \geq 1$. It follows from (6.185) that they hold when $m = 1$. Suppose, by induction, that $k \geq 1$ and that these inequalities hold for all $m \leq k$. Then, it follows from (6.173) and (6.188) that the inequalities

$$t_{m+1} - t_m \leq \frac{\theta}{2}(t_m - t_{m-1})$$

hold for all $m \leq k$. By reasoning as in the proof of Proposition 6.4 we see that (6.178)–(6.179) hold and (6.186) holds when we replace $k$ with $k+1$. Using (6.179), (6.172) and (6.171), we see that

$$\begin{aligned}
&M(t_{k+1} - t_k) + L_0(t_{k+1} + t_k) + M_0\theta t_{k+1} \\
&\leq Ma\left(\frac{\theta}{2}\right)^k + \frac{a[2L_0 + M_0\theta - (L_0 + M_0\theta)(\frac{\theta}{2})^{k+1} - L_0(\frac{\theta}{2})^k]}{1 - \frac{\theta}{2}} \\
&= T\left[(2L_0 + \theta M_0) + \left(\frac{\theta}{2}\right)^k\left(M - M\left(\frac{\theta}{2}\right) - L_0 - (L_0 + \theta M_0)\left(\frac{\theta}{2}\right)\right)\right] \\
&\leq T\left[(2L_0 + \theta M_0) + \left(M - M\left(\frac{\theta}{2}\right) - L_0 - (L_0 + \theta M_0)\left(\frac{\theta}{2}\right)\right)\right] \\
&= T(M + L_0 + M_0\theta)\left(1 - \left(\frac{\theta}{2}\right)\right) \\
&= (M + L_0 + M_0\theta)a \leq \theta.
\end{aligned}$$

This shows that the inequality (6.187) holds when $m = k+1$ and implies that

$$t_{k+1} \leq \frac{(M - L_0)t_k + \theta}{M + L_0 + M_0\theta}. \tag{6.189}$$

If $M_0 = 0$, then, (6.188) holds trivially when we replace $m$ with $k+1$. If $M_0 > 0$, then, it follows from (6.189) and the induction hypotheses that

$$t_{k+1} \leq \frac{(M + L_0)t_k + \theta}{M + L_0 + M_0\theta} < \frac{(M + L_0)/M_0 + \theta}{M + L_0 + M_0\theta} = \frac{1}{M_0}.$$

Therefore, (6.188) also holds when we replace $m$ with $k+1$. We conclude, by induction, that (6.186)–(6.188) hold for all $m \geq 1$. The conclusion now follows as in the proof of Proposition 6.4. ⊠

The next result identifies the best possible value of $\theta$ that can be used in Proposition 6.6 whenever $M$ and $M_0$ are not allowed to vanish at the same time.

**Proposition 6.7** *Let $0 \leq \gamma$, $0 \leq M_0 \leq M \leq \gamma M_0$, $0 \leq L_0 \leq M$ and let*

$$\Theta_*(M, M_0, L_0) \equiv \begin{cases} \frac{4(M-L_0)}{M+L_0+\sqrt{[L_0+M]^2+8M_0(M-L_0)}}, & \text{if } M + L_0 > 0 \\ 0, & \text{if } M + L_0 = 0. \end{cases} \tag{6.190}$$

*Then, the following assertion holds*

$$0 \leq \Theta_*(M, M_0, L_0) < 2. \tag{6.191}$$

*Moreover, there exists $\theta \in [0, 2)$ satisfying conditions (6.184)–(6.185) if and only if*

$$[L_0 + M + M_0\Theta_*(M, M_0, L_0)]a \leq \Theta_*(M, M_0, L_0). \tag{6.192}$$

**Proof** The proof is similar to the demonstration of the analogous result in Proposition 6.5. ⊠

We will use now the scalar majorant sequences to derive Zabrejko-Nguen and Kantorovich type existence results for problem (6.166) and convergence results for the intermediate Newton scheme (6.167).

The following result will be used in the sequel. The proof can be found in (cf. [806, Proposition 1]).

**Lemma 6.7** *Let $v$ be a function defined on the closed ball $B[u_0, T]$ in $X$ with values in $Y$. Suppose that there exists a non-decreasing function $\theta(t)$ defined on the closed interval $[0, T]$ such that, for all $0 \leq t \leq T$, we have*

$$\|v(x) - v(y)\| \leq \theta(t)\|x - y\| \quad (x, y \in B(u_0, t)).$$

*Then, whenever $0 \leq t \leq s \leq T$, $x \in B[u_0, t]$ and $y \in B[x, s - t]$, we have*

$$\|v(x) - v(y)\| \leq \int_t^s \theta(s)\, ds.$$

We next prove the convergence of the intermediate Newton-type scheme (6.167) under weaker Zabrejko-Nguen-type hypotheses than the ones used in (cf. [725]).

**Theorem 6.9** *Suppose that $J_0 = f'(u_0) + g'(u_0)$ is invertible and that there exists $a \geq 0$ such that*

$$\|J_0^{-1}[f(u_0) + g(u_0)]\| \leq a. \tag{6.193}$$

*Suppose, further, that there exist functions $\alpha(t)$, $\alpha_0(t)$ and $\beta_0(t)$ satisfying the hypotheses of Proposition 6.3 and whenever $0 \leq t \leq T$ and $x, y \in B(u_0, t)$, we have*

$$\|J_0^{-1}[f'(x) - f'(y)]\| \leq \alpha(t)\|x - y\|, \tag{6.194}$$

$$\|J_0^{-1}[f'(x) - f'(u_0)]\| \leq \kappa_0(t), \tag{6.195}$$

$$\|J_0^{-1}[g'(x) - g'(u_0)]\| \leq \pi_0(t). \tag{6.196}$$

*Let $t_*$ and $\{t_m\}$ be defined as in Proposition 6.3 and suppose that $t_* \leq T$ and that*

$$\kappa_0(T) + \pi_0(T) < 1. \tag{6.197}$$

*Then, the intermediate Newton iterates in (6.167) are well defined and converge to a unique solution $u$ of equation (6.166) in $B[u_0, T]$ with error estimates*

$$\|u_m - u_{m-1}\| \leq t_m - t_{m-1}, \tag{6.198}$$

$$\|u_m - u_0\| \leq t_m, \tag{6.199}$$

$$\|u - u_m\| \leq t_* - t_m. \tag{6.200}$$

**Proof** Let $\sigma(t)$, $\tau_0(t)$ and $\mu(t)$ be defined as in Proposition 6.3.

If $a = 0$, then, $u = u_0$ solves equation (6.166) and since $u_m = u_0$ and $t_m = t_0$ for all $m \geq 1$, the estimates (6.198)–(6.200) hold trivially. In the rest of the proof we assume $a > 0$. Since $\|u_1 - u_0\| = a \leq t_1 - t_0$, we see that (6.198)–(6.199) hold when $m = 1$.

Suppose now, by induction, that $m \geq 1$ and that the $u_m$ are well defined and satisfy (6.198)–(6.199). Then, on letting $J_m \equiv f'(u_m) + g'(u_0) = J_0(I + A)$ or, equivalently, $A = J_0^{-1}[f'(u_m) - f'(u_0)]$ and applying Lemma 6.7, we see that $\|A\| \leq \kappa_0(t_m) \leq \kappa_0(T) < 1$. Therefore, it follows from the Banach Lemma on invertible operators that $(I + A)^{-1}$ exists and $\|(I + A)^{-1}\| \leq 1/[1 - \kappa_0(t_m)] = -1/\sigma_0'(t_m)$. Therefore, $J_m$ is invertible and $\|J_m^{-1}J_0\| = \|(I + A)^{-1}\| \leq -1/\sigma_0'(t_m)$. Now, we have

$$\begin{aligned}
&\|J_0^{-1}[f(u_m) + g(u_m)]\| \\
&\leq \|J_0^{-1}[f(u_m) - f(u_{m-1}) - f'(u_{m-1})(u_m - u_{m-1})]\| \\
&\quad + \|J_0^{-1}[g(u_m) - g(u_{m-1}) - g'(u_0)(u_m - u_{m-1})]\|
\end{aligned}$$

$$\leq \left\| \int_0^1 J_0^{-1}[f'(u_m + s(u_m - u_{m-1})) - f'(u_{m-1})](u_m - u_{m-1})\, ds \right\|$$
$$+ \left\| \int_0^1 J_0^{-1}[g'(u_m + s(u_m - u_{m-1})) - g'(u_0)](u_m - u_{m-1})\, ds \right\|.$$

Therefore, an application of (6.194)–(6.196), Lemma 6.7 and the induction hypotheses shows that

$$\begin{aligned}
&\|J_0^{-1}[f(u_m) + g(u_m)]\| \\
&\leq \int_0^1 \left[ \int_{t_{m-1}}^{t_m + s(t_m - t_{m-1})} \alpha(w)\, dw \right] (t_m - t_{m-1})\, ds \\
&\quad + \int_0^1 \pi_0(t_m + s(t_m - t_{m-1}))(t_m - t_{m-1})\, ds \\
&= \int_0^1 [\kappa(t_m + s(t_m - t_{m-1})) - \kappa(t_{m-1})](t_m - t_{m-1})\, ds + \int_{t_{m-1}}^{t_m} \pi_0(s)\, ds \\
&= \int_{t_{m-1}}^{t_m} \kappa(s)\, ds - \kappa(t_{m-1})(t_m - t_{m-1}) + \tau_0(t_m) - \tau_0(t_{m-1}) \\
&= (t_m - t_{m-1})(1 - \kappa(t_{m-1})) + \sigma(t_m) - \sigma(t_{m-1}) + \tau_0(t_m) - \tau_0(t_{m-1}) \\
&= \mu(t_m) - \mu(t_{m-1}) - (t_m - t_{m-1})\sigma'(t_{m-1})).
\end{aligned}$$

Hence, we have

$$\begin{aligned}
\|u_{m+1} - u_m\| &= \|J_m^{-1}[f(u_m) + g(u_m)]\| \leq \|J_m^{-1} J_0\| \|J_0^{-1}[f(u_m) + g(u_m)]\| \\
&\leq -\frac{\mu(t_m) - \mu(t_{m-1}) - (t_m - t_{m-1})\sigma'(t_{m-1})}{\sigma_0'(t_m)} \\
&= t_{m+1} - t_m, \\
\|u_{m+1} - u_0\| &\leq \|u_{m+1} - u_m\| + \|u_m - u_0\| \leq t_{m+1} - t_m + t_m = t_{m+1}.
\end{aligned}$$

It follows that (6.198) and (6.199) also hold when $m$ is replaced with $m+1$ and hence, by induction, that they hold for all positive integral values of $m$.

This implies that

$$\|u_{m+q} - u_m\| \leq \sum_{k=m+1}^{m+q} \|u_k - u_{k-1}\| \leq \sum_{k=m+1}^{m+q} (t_k - t_{k-1}) = t_{m+q} - t_m.$$

Since $\{t_m\}$ is a Cauchy sequence, it follows that $\{u_m\}$ is also a Cauchy sequence converging to some $u \in B[u_0, T]$. On letting $q$ tend to infinity, we see that (6.200) holds. It follows from (6.167) that

$$[f'(u_m) + g'(u_0)](u_{m+1} - u_m) + f(u_m) + g(u_m) = 0$$

and on letting $m$ tend to infinity we see that $u$ solves equation (6.166).

To prove the uniqueness, we suppose that $v$ is another solution of equation (6.166) in $B[u_0, T]$. Then, on setting $F = f + g$ and making use of (6.195) and (6.196), we see that

$$\begin{aligned}
\|u - v\| &= \|J_0^{-1}[F(u) - F(v) - F'(u_0)(u - v)]\| \\
&\le \left\| \int_0^1 J_0^{-1}[F'(su + (1-s)v) - F'(u_0)](u - v)\, ds \right\| \\
&\le \int_0^1 \kappa_0(s\|u - u_0\| + (1-s)\|v - u_0\|)\|u - v\|\, ds \\
&\quad + \int_0^1 \pi_0(s\|u - u_0\| + (1-s)\|v - u_0\|)\|u - v\|\, ds \\
&\le \int_0^1 (\kappa_0(T) + \pi_0(T))\|u - v\|\, ds \\
&= (\kappa_0(T) + \pi_0(T))\|u - v\|.
\end{aligned}$$

Therefore, it follows immediately from condition (6.197) that $u = v$. This completes the proof. ⊠

**Remark 6.7** *If we introduce the following stronger condition that was used in (cf. [725])*

$$\|J_0^{-1}[g'(x) - g'(y)]\| \le \beta(t)\|x - y\| \quad (x, y \in B[u_0, t]),$$

*then, in order to compare the iterates* $\{t_m\}$ *with the analogous ones used in (cf. [725]), we set* $\pi(t) = \int_0^t \beta(s)\, ds$, $\tau(t) = \int_0^t \pi(s)\, ds$ *and* $\lambda(t) \equiv \sigma(t) + \tau(t)$. *It is evident that*

$$\alpha_0(t) \le \alpha(t) \text{ and } \beta_0(t) \le \beta(t) \quad (t \in [0, T]). \tag{6.201}$$

*Consider the iterations defined by* $s_0 = 0$, $s_1 = a$ *and for* $m \ge 1$, *let*

$$s_{m+1} = s_m - \frac{\lambda(s_m) - \lambda(s_{m-1}) - (t_m - t_{m-1})\sigma'(t_{m-1})}{\sigma'(t_m)}. \tag{6.202}$$

*Suppose that* $s_* = \lim_{m\to\infty} s_m$ *exists. If* $\alpha_0(t) = \alpha(t)$ *and* $\beta_0(t) = \beta(t)$ *for all* $t \in [0, T]$, *then,* $s_m = t_m$ *for all* $m$ *and* $s_* = t_*$ *and in this case our results coincide with those of (cf. [725]). Examples have been given in (cf. [131, 136]) with* $\alpha(t) \gg \alpha_0(t)$ *and* $\beta(t) \gg \beta_0(t)$. *Therefore, one or more of the inequalities in (6.201) can be strict and whenever this happens, a straightforward induction argument shows that whenever* $m \ge 2$ *we have*

$$\begin{aligned}
t_m &< s_m, \\
t_{m+1} - t_m &< s_{m+1} - s_m, \\
t_* - t_m &\le s_* - s_m, \\
t_* &\le s_*.
\end{aligned}$$

*In this case Theorem 6.9 constitutes an improvement over the results of (cf. [725]) because the error estimates are more precise and more precise information is given on the solution of equation (6.166). It is also interesting to observe that these improvements are obtained under weaker hypotheses and at the same computational cost as the results of (cf. [725]) because the evaluation of $\alpha$ and $\beta$ requires the evaluation of $\alpha_0$ and $\beta_0$.*

We now present two generalizations of Argyros' extension (cf. [136]) of the Kantorovich theorem on the solvability of nonlinear equations and the convergence of Newton's method. In the sequel, $t_*$ and majorant sequence $\{t_m\}$ will be defined as in Proposition 6.4.

**Theorem 6.10** *Suppose that $J_0 = f'(u_0)+g'(u_0)$ is invertible and there exist $u \geq 0$, $L_0 \geq 0$, $0 \leq 0 < 2$ and $0 \leq M_0 \leq M$ such that (6.171), (6.172), (6.193) hold and*

$$\|J_0^{-1}[f'(x) - f'(y)]\| \leq M\|x - y\| \quad (x, y \in D_0), \tag{6.203}$$

$$\|J_0^{-1}[f'(x) - f'(u_0)]\| \leq M_0\|x - u_0\| \quad (x \in D_0), \tag{6.204}$$

$$\|J_0^{-1}[g'(x) - g'(u_0)]\| \leq L_0\|x - u_0\| \quad (x \in D_0). \tag{6.205}$$

*Then, the intermediate Newton iterates in (6.167) are well defined and converge to a solution $u$ of equation (6.166) in $B[u_0, t_*]$ and the error estimates (6.198)–(6.200) hold. If $(M_0 + L_0)(t_* + T) < 2$, then, this solution is unique in $B[u_0, T]$.*

**Proof** The existence result follows from Theorem 6.9 and Proposition 6.4. To prove uniqueness, we suppose that $v$ is another solution of equation (6.166) in $B[u_0, t_*]$. Then, on setting $F = f + g$ and using the Lipschitz conditions (6.204) and (6.205), we see that

$$\begin{aligned}
\|u - v\| &= \|J_0^{-1}[F(u) - F(v) - F'(u_0)(u - v)]\| \\
&\leq \left\| \int_0^1 J_0^{-1}[F'(su + (1 - s)v) - F'(u_0)](u - v)\, ds \right\| \\
&\leq \int_0^1 (M_0 + L_0)[s\|u - u_0\| + (1 - s)\|v - u_0\|]\|u - v\|\, ds \\
&= \frac{1}{2}(M_0 + L_0)[\|u - u_0\| + \|v - u_0\|]\|u - v\| \\
&\leq \frac{1}{2}(M_0 + L_0)(T + t_*)\|u - v\|.
\end{aligned}$$

Then, it follows immediately from the hypotheses that $u = v$. This completes the proof. ⊠

**Theorem 6.11** *Suppose that $J_0 = f'(u_0) + g'(u_0)$ is invertible and there exist $a \geq 0$, $0 \leq \theta < 2$, $0 \leq M_0 \leq M$ and $0 \leq L_0 \leq M$ such that (6.184), (6.185), (6.193), (6.203), (6.204) and (6.205) hold. Then, the intermediate*

*Newton iterates in (6.167) are well defined and converge to solution $u$ of equation (6.166) in $B[u_0, t_*]$ and the error estimates (6.198)–(6.200) hold. If $(M_0 + L_0)(t_* + T) < 2$, then, this solution is unique in $B[u_0, T]$.*

**Proof** This result follows immediately from Theorem 6.9 and Proposition 6.6. ⊠

**Remark 6.8** *If we set $L_0 = 0$ and $g = 0$ in Theorem 6.10 or 6.11, we recover Argyros' generalization (cf. [136]) of the Kantorovich Theorem and if further set $\theta = 1$ and $M = M_0$, we recover the Kantorovich theorem (cf. [469]).*

**Remark 6.9** *If we set $M_0 = 0$ in condition (6.204), then, $f'(x) = f'(u_0)$ for all $x \in D_0$, which implies that $f'(x) = f'(y)$ for all $x, y \in D_0$. Evidently, it makes sense to set $M = 0$ also, in this case. Therefore, in the study of the solvability of equation (6.166), there is no loss of generality in assuming that whenever $M_0 = 0$, then, $M = 0$. The simplest way of making sure that this happens is by imposing a condition of the form $0 \leq M_0 \leq M \leq \gamma M_0$ – as was done in Propositions 6.5 and 6.7 – where $\gamma$ is any positive constant. This condition will also be employed in the next two results.*

**Remark 6.10** *A setback of Theorems 6.10 and 6.11 is the fact that $\theta$ has to be found by trial and error and there is no guarantee that the choice we make leads to minimal conditions on $M$, $M_0$, $L_0$ and $a$ that ensure the solvability of problem (6.166). The two next results resolve this problem by identifying the best possible values of $\theta$.*

**Theorem 6.12** *Suppose that $J_0 = f'(u_0) + g'(u_0)$ is invertible and there exist $a \geq 0$, $\gamma \geq 0$, $L_0 \geq 0$ and $0 \leq M_0 \leq M \leq \gamma M_0$ such that (6.193) and (6.203)–(6.205) hold. Suppose, further, that*

$$[2L_0 + M + M_0\theta_*(M, M_0, L_0)]a \leq \theta_*(M, M_0, L_0), \tag{6.206}$$

*where $\theta_*(M, M_0, L_0)$ is defined in (6.181). Then, the intermediate Newton iterates in (6.167) are well defined and converge to a solution $u$ of equation (6.166) in $B[u_0, t_*]$ and the error estimates (6.198)–(6.200) hold. If $(M_0 + L_0)(t_* + T) < 2$, then, this solution is unique in $B[u_0, T]$.*

**Proof** This result follows immediately from Theorem 6.10 and Proposition 6.5. ⊠

**Theorem 6.13** *Suppose that $J_0 = f'(u_0) + g'(u_0)$ is invertible and there exist $a \geq 0$, $\gamma \geq 0$, $0 \leq L_0 \leq M_0$ and $0 \leq M_0 \leq M$ such that (6.193), (6.203), (6.204) and (6.205) hold. Suppose, further, that*

$$[L_0 + M + M_0\Theta_*(M, M_0, L_0)]a \leq \Theta_*(M, M_0, L_0), \tag{6.207}$$

*where $\Theta_*(M, M_0, L_0)$ is defined in (6.190). Then, the intermediate Newton iterates in (6.167) are well defined and converge to a solution $u$ of equation (6.166) in $B[u_0, t_*]$ and the error estimates (6.198)–(6.200) hold. If $(M_0 + L_0)(t_* + T) < 2$, then, this solution is unique in $B[u_0, T]$.*

**Proof** This result follows immediately from Theorem 6.11 and Proposition 6.7. ⊠

**Remark 6.11** *Theorems 6.12 and 6.13 are generalizations of the well-known Kantorovich theorem (cf. [469]) on the solvability of nonlinear equations and the convergence of Newton's method. If the conditions (6.193) and (6.203)–(6.205) hold with* $g = 0$, $f = F$ *and* $L_0 = 0$, *then, it follows from the Kantorovich theorem that, if*

$$2Ma \leq 1, \tag{6.208}$$

*then, Newton iterates defined by (6.168) converge to a solution of equation (6.166). In this case it was shown in (cf. [136]) that the same conclusion holds under the weaker condition*

$$(M + M_0)a \leq 1. \tag{6.209}$$

*In fact, this condition is precisely condition (6.172) of Theorem 6.10 with* $L_0 = 0$ *and* $\theta = 1$. *However, in this case the Newton scheme (6.168) coincides with the intermediate Newton scheme (6.167) and it follows from Theorem 6.13 that the convergence of the Newton iterates to a solution of equation (6.166) holds under the condition*

$$[M + M_0\Theta_*(M, M_0, 0)]a \leq \Theta_*(M, M_0, 0), \tag{6.210}$$

*which is weaker than (6.209). Hypothesis even weaker than (6.210) can be found in section 2.1.*

**Example 6.4** *Let* $X = Y = \mathbb{R}$, $D = [8/9, 10/9]$, $g = 0$, $f(x) = x^3 + x/3 - 0.47118$ *and* $u_0 = 1$. *Then, it is easy to verify that* $M = 2$, $M_0 = 1.9$, $L_0 = 0$, $\theta_*(M, M_0, L_0) = 1.0171451$ *and* $a = 0.258646$. *In this case condition (6.210) holds, but the Kantorovich condition (6.208) and the Argyros condition (6.209) do not hold.*

**Example 6.5** *Let* $X = Y = \mathbb{R}$, $u_0 = 1$, $T = 1 - c$, $D = [c, 2 - c]$ *for some* $c \in [0, 1/2)$, $g = 0$ *and* $f(x) = x^3 - c$. *We see that* $a = (1 - c)/3$, $M_0 = 3 - c$, $L_0 = 0$ *and* $M = 2(2 - c)$. *The Kantorovich condition (6.208) does not hold since* $2Ma = 4(1-c)(2-c)/3 > 1$ *whenever* $c \in [0, 1/2)$. *In (cf. [136]), it was shown that Newton's method converges for this problem whenever* $c \in [0.4505, 1/2)$. *We can get a slight improvement of this result by using our generalized Kantorovich Theorems. For instance, it follows from (6.210) and a brief calculation that the slightly weaker condition* $c \in [0.450339002, 1/2)$ *is sufficient for the convergence of Newton's method for this problem.*

# CHAPTER 7

# Inexact Methods

Sometimes NM presented in chapter 2 has some disadvantages. For example, it requires the exact computation of the Jacobians involved and the computational cost is great, especially when the system is large. To give an answer to part of this problem, we present in this section an inexact methods.

## 7.1 Residual Control Conditions

A local convergence analysis of inexact Newton–type methods using a new type of residual control was recently presented by Li and Shen. Here, we introduce the center Hölder condition on the operator involved and use it in combination with the Hölder condition to provide a new local convergence analysis with the following advantages: larger radius of convergence and tighter error bounds on the distances involved. These results are obtained under the same hypotheses and computational cost. Numerical examples further validating the theoretical results are also provided.

We are concerned with the problem of approximating a solution $x^\star$ of (2.1). We studied in chapter 2 NM. However, NM has two disadvantages. The first one is that it requires the exact computation of the Jacobians involved, whereas the second one is finding the exact solutions of linear equations

$$F'(x_n)(x_{n+1} - x_n) = -F(x_n). \tag{7.1}$$

The computational cost of these two disadvantages is great, especially when the system is large. Newton–type methods have been used to deal with the first disadvantage (cf. [139], [157], [250], [376], [798]). Inexact Newton–type methods handle the second disadvantage (cf. [101], [157], [308], [501], [801]).

The Inexact Newton–type methods have usually the form INTM: For $n = 0$ and a given initial guess $x_0$ until convergence do:

1. For the residual $r_n$ and the iteration $x_n$, find the step $s_n$ satisfying

$$B_n s_n = -F(x_n) + r_n.$$

2. $x_{n+1} = x_n + s_n$.
3. Set $n + 1$ for $n$ and return to Step 1.

Here, $\{B_n\}$ is a sequence of invertible operators from $X$ to $Y$ and $\{r_n\}$ is a sequence belongs in $Y$ and in general depends on $\{x_n\}$. If $r_n = 0$ for all $n \geq 0$, we obtain the Newton–type method NTM

$$x_{n+1} = x_n - B_n^{-1} F(x_n) \quad (x_0 \in D, \ \ n \geq 0). \tag{7.2}$$

In particular, if $B_n = F'(x_n)$ for all $n \geq 0$, NTM reduces to NM.

A local as well as a semilocal convergence for INTM, NTM and NM under various conditions has been given by several authors (cf. [101], [139], [166], [446], [501], [801]). A survey of such results can also be found in (cf. [157], [469]) and the references there. In this section we are motivated by the desire to obtain larger radius of convergence and the recent elegant work by Li and Shen (cf. [501]), who improved earlier results in this area (cf. [543]). In particular, they provided a semilocal convergence analysis for INTM using the control of the residual condition

$$\|P_n r_n\| \leq \theta_n \|P_n F(x_n)\|^{1+\beta} \quad (n \geq 0) \tag{7.3}$$

for sequence $\{P_n\}$ of invertible operators from $Y$ to $X$, some scalar sequence $\{\theta_n\}$ and parameter $\beta \geq 0$. Here, we introduce the needed center-Hölder condition on $F'$ and use it to find tighter upper bounds on the norms of $\|F'(x_n)^{-1} F'(x^\star)\|$ than in (cf. [268], [269], [501]) (using the less precise Hölder condition). This approach leads to a local convergence analysis with the following advantages (under the same computational cost) over the works in (cf. [157], [268], [269]), in case the center Hölder constant is smaller than the Hölder constant:

(a) Larger radius of convergence,
(b) Tighter upper bounds on the distances $\|x_n - x^\star\|$ $(n \geq 0)$.

Advantage (a) allows a wider choice of initial guesses for $x_0$, whereas advantage (b) requires fewer computations in order to obtain a desired error tolerance $\varepsilon > 0$. These improvements are important in computational mathematics (cf. [157]).

Throughout this section, we shall assume $x^\star$ is a simple zero of operator $F$. This is, $F'(x^\star)^{-1} \in \mathcal{L}(Y, X)$. Furthermore, let $0 \leq \beta \leq p \leq 1$ and further assume that $F'(x^\star)^{-1} F'$ satisfies in $U(x^\star, R)$ the Hölder condition

$$\|F'(x^\star)^{-1}(F'(x) - F'(y))\| \leq L_p \|x - y\|^p \quad (x, y \in U(x^\star, R)). \tag{7.4}$$

Let $R_0$ be the maximum of all radii $R > 0$ such that

$$\|F'(x^\star)^{-1}(F'(x) - F'(y))\| \leq L_0 < 1 \quad (x, y \in U(x^\star, R)). \tag{7.5}$$

Define for $L_p \neq 0$

$$R_p = \left(\frac{1}{L_p}\right)^{\frac{1}{p}} \quad (p \geq \beta, \ p \neq 0). \tag{7.6}$$

In view of (7.4) and (7.5), there exist $L_p^0$, $p \in [0,1]$ such that center–Hölder condition

$$\|F'(x^\star)^{-1}(F'(x) - F'(x^\star))\| \le L_p^0 \|x - x^\star\|^p \quad (x \in U(x^\star, R)), \tag{7.7}$$

holds. Let $R_0^0$ be the maximum of all radii $R > 0$ such that

$$\|F'(x^\star)^{-1}(F'(x) - F'(x^\star))\| \le L_0^0 < 1 \quad (x \in U(x^\star, R)). \tag{7.8}$$

Define for $L_p^0 \neq 0$

$$R_p^0 = \left(\frac{1}{L_p^0}\right)^{\frac{1}{p}} \quad (p \ge \beta,\ p \neq 0). \tag{7.9}$$

The existence of $L_p$ and $L_0$ implies that of $L_p^0$ and $L_0^0$. Therefore, (7.7) and (7.8) do not constitute additional hypotheses, since in practice the computation of $L_p$, $L_0$, require that of $L_p^0$ and $L_0^0$, respectively. Clearly,

$$L_p^0 \le L_p \quad (p \in [0,1]) \tag{7.10}$$

holds in general and $\dfrac{L_p}{L_p^0}$ can be arbitrarily large. Moreover, we have that

$$R_p \le R_p^0 \,. \tag{7.11}$$

Note that, if strict inequality holds in (7.10), then, so does in (7.11). Furthermore, we need to define parameter

$$\alpha_p = L_p - (1+p)L_p^0 \,. \tag{7.12}$$

We have the following lemma, which can be proved using Banach's lemma on invertible operators with standard arguments in (cf. [157], [469]).

**Lemma 7.1** *Let $x \in U(x^\star, R_p^0)$. Then, $F'(x)$ is invertible and satisfies that*

$$\|F'(x)^{-1}F'(x^\star)\| \le (1 - L_p^0\|x - x^\star\|^p)^{-1}. \tag{7.13}$$

**Remark 7.1** *If $L_p^0 = L_p$ in (7.10), then, Lemma 2.1 reduces to the corresponding one in (cf. [387, p. 546]). Otherwise, (7.10) constitutes a sharper estimate. This modification leads to the advantages (a) and (b) already stated in the introduction of this section.*

*Convergence analysis of INTM is closely dependent on controls of the error $E_n := F'(x_n) - B_n$ and of the residual $r_n$ in step $n$. Throughout the whole section, we always assume that residual controls (7.3) are satisfied for some nonnegative-valued sequence $\{\theta_n\}$ and that $\{B_n\}$ satisfy*

$$\|B_n^{-1}F'(x_n)\| \le \omega_1 \quad (n \ge 0) \tag{7.14}$$

*for some positive constant $\omega_1$. Moreover, we define*

$$\nu = \sup_{n\ge 0} \theta_n \|(P_nB_n)^{-1}\| \|P_nB_n\|^{1+\beta}. \tag{7.15}$$

In the first theorem below, we will consider error controls for $\{E_n\}$ which are dependent on the solution $x^\star$ of (2.1). We also need to define a functions $h$ and $h_0$ on $(0, R_p^0)$ by

$$h(r) = (\omega_1 h_0(r))^{1+\beta}, \tag{7.16}$$

$$h_0(r) = \frac{1 + \left(\dfrac{L_p}{p+1} - L_p^0\right) r^p}{1 - L_p^0 r^p}. \tag{7.17}$$

It follows from (7.17) that

$$h_0'(r) = \frac{pL_p r^{p-1}}{(p+1)(1 - L_p^0 r^p)^2} > 0. \tag{7.18}$$

Hence, functions $h$ and $h_0$ are increasing on $(0, R_p^0)$. We assume $p$ is a given number in $[0, 1]$. Consequently, the Hölder constants are also fixed.

**Theorem 7.1** *Suppose that $F(x^\star) = 0$ and errors $\{E_n\}$ satisfy the following controls: there exists a nonnegative constant $\omega_2$ such that*

$$\|B_n^{-1} E_n\| \leq \omega_2 \|x_n - x^\star\|^\beta \quad (n \geq 0). \tag{7.19}$$

*Let $0 < R \leq R_p^0$ satisfy*

$$\frac{\omega_1 L_p R^p}{(p+1)(1 - L_p^0 R^p)} + \omega_2 R^\beta + \nu \left( \frac{\omega_1 (p + 1 + \alpha_p R^p)}{(p+1)(1 - L_p^0 R^p)} \right)^{\beta+1} R^\beta \begin{cases} < 1, & \beta = p = 0, \\ \leq 1, & 0 < \beta \leq p \leq 1 \end{cases} \tag{7.20}$$

*and*

$$U(x^\star, R) \subseteq D. \tag{7.21}$$

*Then, for each $x_0 \in U(x^\star, R)$, sequence $\{x_n\}$ generated by INTM converges to $x^\star$ with order $\beta + 1$. Moreover, the following estimates hold*

$$\|x_{n+1} - x^\star\| \leq q \|x_n - x^\star\|^{1+\beta} \quad (n \geq 0), \tag{7.22}$$

*where*

$$q = \frac{\omega_1 L_p \|x_0 - x^\star\|^{p-\beta}}{(p+1)(1 - L_p^0 \|x_0 - x^\star\|^p)} + \omega_2 + \nu \left( \frac{\omega_1 (p + 1 + \alpha_p \|x_0 - x^\star\|^p)}{(p+1)(1 - L_p^0 \|x_0 - x^\star\|^p)} \right)^{\beta+1}. \tag{7.23}$$

**Proof** Let $x_0 \in U(x^\star, R)$ and $x_0 \neq x^\star$. Since $q\|x_0 - x^\star\|^\beta < 1$ by (7.20) and (7.23), it suffices to verify that (7.22) holds. Let $n \geq 0$ be an integer. We claim that, if $x_m \in U(x^\star, R)(m \leq n)$, then

$$\|x_{m+1} - x^\star\| \leq \left( \frac{\omega_1 L_p \|x_m - x^\star\|^{p-\beta}}{(p+1)(1 - L_p^0 \|x_m - x^\star\|^p)} + \omega_2 + \nu \left( \frac{\omega_1 (p + 1 + \alpha_p \|x_m - x^\star\|^p)}{(p+1)(1 - L_p^0 \|x_m - x^\star\|^p)} \right)^{\beta+1} \right) \|x_m - x^\star\|^{1+\beta} \tag{7.24}$$

and

$$\|x_{m+1} - x^\star\| < \|x_m - x^\star\|. \tag{7.25}$$

In fact, writing $x_\tau = x^\star + \tau(x_m - x^\star)$ for $0 \le \tau \le 1$, we have by INTM that

$$\begin{aligned} x_{m+1} - x^\star &= x_m - x^\star - B_m^{-1}(F(x_m) - F(x^\star)) + B_m^{-1}r_m \\ &= B_m^{-1}F'(x_m)F'(x_m)^{-1}F'(x^\star)\int_0^1 F'(x^\star)^{-1}[F'(x_m) - F'(x_\tau)](x_m - x^\star)d\tau \\ &\quad - B_m^{-1}E_n(x_m - x^\star) + B_m^{-1}r_m. \end{aligned} \tag{7.26}$$

Since $x_m \in U(x^\star, R)$, (7.4), (7.13) and (7.14) imply that

$$\begin{aligned} &\left\| B_m^{-1}F'(x_m)F'(x_m)^{-1}F'(x^\star)\int_0^1 F'(x^\star)^{-1}[F'(x_m) - F'(x_\tau)]d\tau \right\| \\ &\le \frac{\omega_1 L_p \|x_m - x^\star\|^p}{(p+1)(1 - L_p^0\|x_m - x^\star\|^p)}. \end{aligned} \tag{7.27}$$

Furthermore, by (7.4), we have

$$\begin{aligned} &\|F'(x_m)^{-1}F(x_m)\| \\ &\le \left(1 + \|F'(x_m)^{-1}F'(x^\star)\| \int_0^1 \|F'(x^\star)^{-1}(F'(x_\tau) - F'(x_m))\|d\tau\right)\|x_m - x^\star\| \\ &\le \frac{p + 1 + \alpha_p\|x_m - x^\star\|^p}{(p+1)(1 - L_p^0\|x_m - x^\star\|^p)}\|x_m - x^\star\|, \end{aligned} \tag{7.28}$$

which together with (7.14) yields

$$\begin{aligned} \|B_m^{-1}F(x_m)\| &\le \|B_m^{-1}F'(x_m)\|\|F'(x_m)^{-1}F(x_m)\| \\ &\le \frac{\omega_1(p + 1 + \alpha_p\|x_m - x^\star\|^p)}{(p+1)(1 - L_p^0\|x_m - x^\star\|^p)}\|x_m - x^\star\|. \end{aligned} \tag{7.29}$$

Thus, by (7.3), we have

$$\begin{aligned} \|B_m^{-1}r_m\| &\le \theta_n\|B_m^{-1}P_m^{-1}\|\|P_mF(x_m)\|^{\beta+1} \le \nu\|B_m^{-1}F(x_m)\|^{\beta+1} \\ &\le \nu\Big(\frac{\omega_1(p + 1 + \alpha_p\|x_m - x^\star\|^p)}{(p+1)(1 - L_p^0\|x_m - x^\star\|^p)}\Big)^{\beta+1}\|x_m - x^\star\|^{\beta+1} \end{aligned} \tag{7.30}$$

thanks to the definition of $\nu$ in (7.15). On the other hand, it follows from (7.19) that

$$\|B_m^{-1}E_m(x_m - x^\star)\| \le \omega_2\|x_m - x^\star\|^{\beta+1}. \tag{7.31}$$

Combining (7.26), (7.27), (7.30) and (7.31), one sees that (7.24) holds. Consequently,

$$\begin{aligned}&\|x_{m+1}-x^\star\|\\&<\left(\frac{\omega_1 L_p R^{p-\beta}}{(p+1)(1-L_p^0 R^p)}+\omega_2+\nu\left(\frac{\omega_1(p+1+\alpha_p R^p)}{(p+1)(1-L_p^0 R^p)}\right)^{\beta+1}\right)R^\beta\|x_m-x^\star\|\\&\leq\|x_m-x^\star\|.\end{aligned} \tag{7.32}$$

thanks to the assumption that $x_m \in U(x^\star, R)$, the increasing of function $h(r)$ and (7.20). Hence, (7.25) holds and the claim is proved.

Below we will proceed by mathematical induction. Clearly, (7.22) holds for $n = 0$. Assume now that (7.22) holds for all $n \leq m-1$ and $m \geq 1$ is a fixed integer. Then, by (7.25),

$$\|x_m - x^\star\| < \|x_{m-1} - x^\star\| < \cdots < \|x_0 - x^\star\| \tag{7.33}$$

and $x_m \in U(x^\star, R)$. Thus (7.24) can be applied to conclude that

$$\begin{aligned}\|x_{m+1}-x^\star\| \leq \Bigg(&\frac{\omega_1 L_p \|x_m-x^\star\|^{p-\beta}}{(p+1)(1-L_p^0\|x_m-x^\star\|^p)}+\omega_2\\&+\nu\left(\frac{\omega_1(p+1+\alpha_p\|x_m-x^\star\|^p)}{(p+1)(1-L_p^0\|x_m-x^\star\|^p)}\right)^{\beta+1}\Bigg)\|x_m-x^\star\|^{1+\beta}.\end{aligned} \tag{7.34}$$

By (7.33), we have

$$\frac{\omega_1 L_p \|x_m-x^\star\|^{p-\beta}}{(p+1)(1-L_p^0\|x_m-x^\star\|^p)}+\omega_2+\nu\left(\frac{\omega_1(p+1+\alpha_p\|x_m-x^\star\|^p)}{(p+1)(1-L_p^0\|x_m-x^\star\|^p)}\right)^{\beta+1}\leq q. \tag{7.35}$$

Hence, (7.22) holds for $n = m$, which completes the proof of the Theorem. This completes the proof. ⊠

Note that error control for $\{E_n\}$ in Theorem 7.1 depends on the information at the solution $x^\star$, which makes the algorithm infeasible from the point of view of practical application because the solution $x^\star$ is often unknown. In the next theorem, we will consider another error control for $\{E_n\}$, which is independent of the solution $x^\star$.

**Remark 7.2** *Suppose that (7.20) is satisfied. Let $x_0 \in U(x^\star, R)$ and $m \geq 0$ is an integer. Then, the following implication holds*

$$\|B_n^{-1}E_n\| \leq \widehat{\omega}_2\|x_n - x^\star\|^\beta \quad \textit{holds for all } n \leq m \Rightarrow x_{m+1} \in U(x^\star, R). \tag{7.36}$$

*(This can be seen from the proof of Theorem 7.1)*

**Theorem 7.2** *Suppose that $F(x^\star) = 0$ and errors $\{E_n\}$ satisfy the following controls: there exists a nonnegative constant $\omega_2$ such that*

$$\|B_n^{-1}E_n\| \leq \widehat{\omega}_2\|F(x_n)\|^\beta \quad (n \geq 0). \tag{7.37}$$

*Let* $0 < R \le R_p^0$ *satisfy*

$$\frac{\omega_1 L_p R^p}{(p+1)(1-L_p^0 R^p)} + \widehat{\omega}_2 R^\beta \|F'(x^\star)\|^\beta \left(1 + \frac{L_p^0 R^p}{p+1}\right)^\beta$$

$$+\nu\left(\frac{\omega_1(p+1+\alpha_p R^p)}{(p+1)(1-L_p^0 R^p)}\right)^{\beta+1} R^\beta \left\{ \begin{array}{ll} <1, & \beta = p = 0, \\ \le 1, & 0 < \beta \le p \le 1, \end{array}\right. \tag{7.38}$$

*and*

$$U(x^\star, R) \subseteq D. \tag{7.39}$$

*Then, for each* $x_0 \in U(x^\star, R)$, *sequence* $\{x_n\}$ *generated by INTM converges to* $x^\star$ *with order* $\beta + 1$. *Moreover, (7.22) holds for* $q$ *replaced by*

$$q_1 = \frac{\omega_1 L_p \|x_0 - x^\star\|^{p-\beta}}{(p+1)(1 - L_p^0 \|x_0 - x^\star\|^p)} + \widehat{\omega}_2 \|F'(x^\star)\|^\beta \left(1 + \frac{L_p^0 \|x_0 - x^\star\|^p}{p+1}\right)^\beta$$

$$+\nu\left(\frac{\omega_1(p+1+\alpha_p\|x_0 - x^\star\|^p)}{(p+1)(1-L_p^0\|x_0 - x^\star\|^p)}\right)^{\beta+1}. \tag{7.40}$$

**Proof** Let

$$\omega_2 = \widehat{\omega}_2 \|F'(x^\star)\|^\beta (1 + \frac{L_p^0 R^p}{p+1})^\beta. \tag{7.41}$$

Then, (7.20) is satisfied. Let $n \ge 0$ is an integer. Next, we shall show that, if $x_n \in U(x^\star, R)$, then

$$\|B_n^{-1} E_n\| \le \omega_2 \|x_n - x^\star\|^\beta. \tag{7.42}$$

To do this, recall that $x_\tau = x^\star + \tau(x_n - x^\star)$. It follows from (7.7) that

$$\|F'(x^\star)^{-1} F(x_n)\| \le \int_0^1 \|F'(x^\star)^{-1} F'(x_\tau)\| d\tau \|x_n - x^\star\|$$

$$\le \left(1 + \frac{L_p^0 R^p}{p+1}\right) \|x_n - x^\star\|. \tag{7.43}$$

Thus, by (7.37) and (7.43),

$$\|B_n^{-1} E_n\| \le \widehat{\omega}_2 \|F(x_n)\|^\beta \le \widehat{\omega}_2 \|F'(x^\star)\|^\beta \left(1 + \frac{L_p^0 R^p}{p+1}\right)^\beta \|x_n - x^\star\|^\beta = \omega_2 \|x_n - x^\star\|^\beta \tag{7.44}$$

and (7.42) is proved. Note that $x_0 \in U(x^\star, R)$. Thus, with help of Remark 7.2, one can easily use mathematical induction to verify that (7.42) holds for all $n \ge 0$. Consequently, Theorem 7.1 is applicable and the proof is complete. ⊠

In particular, for the case when $B_n = F'(x_n)$ for all $n \ge 0$, one sees that (7.14) is satisfied with $\omega_1 = 1$, while (7.19) and (7.37) coincide and are satisfied with $\omega_2 = \widehat{\omega}_2 = 0$ for any $0 \le \beta \le p \le 1$. Therefore, applying Theorem 7.2, we have the following corollary for the inexact Newton method, which includes the corresponding result of (cf. [543]) (for $p = 1$, $\beta = 0$ and $L_p^0 = L_p$) as a special case.

**Corollary 7.1** *Suppose that $F(x^\star) = 0$. Let*

$$\nu = \sup_{n\geq 0} \theta_n \|(P_n F'(x_n))^{-1}\| \|P_n F'(x_n)\|^{\beta+1}$$

*and*

$$0 < R \leq R_p^0$$

*satisfying*

$$\frac{L_p R^p}{(p+1)(1-L_p^0 R^p)} + \nu \left( \frac{(p+1+\alpha_p R^p)}{(p+1)(1-L_p^0 R^p)} \right)^{\beta+1} R^\beta \begin{cases} <1, & \beta = p = 0, \\ \leq 1, & 0 < \beta \leq p \leq 1, \end{cases} \tag{7.45}$$

*and*

$$U(x^\star, R) \subseteq D. \tag{7.46}$$

*Then, for each $x_0 \in U(x^\star, R)$, sequence $\{x_n\}$ generated by INTM converges to $x^\star$ with order $\beta + 1$ and (7.22) holds with $q$ replaced by*

$$q_2 = \frac{L_p \|x_0 - x^\star\|^{p-\beta}}{(p+1)(1-L_p^0\|x_0 - x^\star\|^p)} + \nu \left( \frac{(p+1+\alpha_p\|x_0 - x^\star\|^p)}{(p+1)(1-L_p^0\|x_0 - x^\star\|^p)} \right)^{\beta+1}. \tag{7.47}$$

We present now separately some corollaries of Theorem 7.2 for the interesting case when $p = \beta = 0$ (this means that the Hölder condition is not necessary satisfied). Note that in this case (7.19) and (7.37) coincide with $\omega_2 = \widehat{\omega}_2$.

**Corollary 7.2** *Suppose that $F(x^\star) = 0$. Let $p = \beta = 0$ and $\nu$ given by (7.15). Suppose that*

$$\frac{\omega_1 L_0}{1 - L_0^0} + \omega_2 + \frac{\nu\omega_1(1 - L_0 + L_0^0)}{1 - L_0^0} < 1. \tag{7.48}$$

*Let $0 < R \leq R_0^0$ such that*

$$U(x^\star, R) \subseteq D. \tag{7.49}$$

*Then, for each $x_0 \in U(x^\star, R)$, sequence $\{x_n\}$ generated by INTM converges linearly to $x^\star$ and (7.22) holds with $\beta = 0$ and $q$ replaced by*

$$q_3 = \frac{\omega_1 L_0}{1 - L_0^0} + \omega_2 + \frac{\nu\omega_1(1 - L_0 + L_0^0)}{1 - L_0^0}. \tag{7.50}$$

**Corollary 7.3** *Suppose that $F(x^\star) = 0$. Let $p = \beta = 0$ and set $\nu = \sup_{n\geq 0} \theta_n cond(P_n F'(x_n))$. Suppose that*

$$\frac{L_0}{1 - L_0^0} + \frac{\nu(1 - L_0 + L_0^0)}{1 - L_0^0} < 1 \tag{7.51}$$

*and let $0 < R \leq R_0^0$ such that*

$$U(x^\star, R) \subseteq D. \tag{7.52}$$

*Then, for each $x_0 \in U(x^\star, R)$, sequence $\{x_n\}$ generated by INTM converges linearly to $x^\star$ and (7.22) holds with $\beta = 0$ and $q$ replaced by*

$$q_4 = \frac{L_0}{1 - L_0^0} + \frac{\nu(1 - L_0 + L_0^0)}{1 - L_0^0}. \tag{7.53}$$

**Remark 7.3** *The results obtained above reduce to the corresponding ones in (cf. [501]) if $L_p = L_p^0$. Otherwise, they constitute an improvement with advantages (a) and (b) as stated in the introduction of this section.*

There are cases when center–Hölder condition (7.8) only holds. Then, we have a result corresponding to Theorem 7.1 that can handle this case:

**Theorem 7.3** *Assume that operator $F$ has a simple zero $x^\star$, center–Hölder condition (7.8) holds and errors $\{E_n\}$ satisfy the controls (7.19). Let $0 < R \le R_p^0$ satisfy*

$$\frac{(p+2)\omega_1 L_p^0 R^p}{(p+1)(1 - L_p^0 R^p)} + \omega_2 R^\beta + \nu\left(\frac{\omega_1(p+1+L_p^0 R^p)}{(p+1)(1 - L_p^0 R^p)}\right)^{\beta+1} R^\beta \begin{cases} < 1, & \beta = p = 0, \\ \le 1, & 0 < \beta \le p \le 1, \end{cases} \tag{7.54}$$

*and*

$$U(x^\star, R) \subseteq D. \tag{7.55}$$

*Then, for each $x_0 \in U(x^\star, R)$, sequence $\{x_n\}$ generated by INTM converges to $x^\star$ with order $\beta + 1$. Moreover, the following estimates hold*

$$\|x_{n+1} - x^\star\| \le q_5 \|x_n - x^\star\|^{1+\beta} \quad (n \ge 0), \tag{7.56}$$

*where*

$$q_5 = \frac{(p+2)\omega_1 L_p^0 \|x_0 - x^\star\|^{p-\beta}}{(p+1)(1 - L_p^0\|x_0 - x^\star\|^p)} + \omega_2 + \nu\left(\frac{\omega_1(p+1+L_p^0\|x_0 - x^\star\|^p)}{(p+1)(1 - L_p^0\|x_0 - x^\star\|^p)}\right)^{\beta+1}. \tag{7.57}$$

**Proof** We follow the proof of Theorem 7.1, but this time we use (7.8) and the estimate

$$\begin{aligned} \left\| \int_0^1 F'(x^\star)^{-1}[F'(x_n) - F'(x_\tau)]d\tau \right\| &\le \left\| \int_0^1 F'(x^\star)^{-1}[F'(x_n) - F'(x^\star)]d\tau \right\| \\ &\quad + \left\| \int_0^1 F'(x^\star)^{-1}[F'(x^\star) - F'(x_\tau)]d\tau \right\| \\ &\le L_p^0 \|x_n - x^\star\|^p + \frac{L_p^0}{p+1}\|x_n - x^\star\|^p \\ &= \frac{p+2}{p+1} L_p^0 \|x_n - x^\star\|^p \end{aligned} \tag{7.58}$$

in (7.27)–(7.30) to arrive at (7.54) and (7.56) instead of (7.20) and (7.22). This completes the proof of Theorem 7.3. ⊠

Similarly, we have the following results corresponding to Theorem 7.2 and Corollary 7.1 using only center–Hölder condition (7.8):

**Theorem 7.4** *Assume that operator $F$ has a simple zero $x^\star$, center–Hölder condition (7.8) holds and errors $\{E_n\}$ satisfy the controls (7.37). Let $0 < R \leq R_p^0$ satisfy*

$$\frac{(p+2)\omega_1 L_p^0 R^p}{(p+1)(1-L_p^0 R^p)} + \widehat{\omega}_2 R^\beta \|F'(x^\star)\|^\beta \left(1 + \frac{L_p^0 R^p}{p+1}\right)^\beta$$
$$+ \nu \left(\frac{\omega_1(p+1+L_p^0 R^p)}{(p+1)(1-L_p^0 R^p)}\right)^{\beta+1} R^\beta \begin{cases} < 1, & \beta = p = 0, \\ \leq 1, & 0 < \beta \leq p \leq 1, \end{cases} \tag{7.59}$$

*and*

$$U(x^\star, R) \subseteq D. \tag{7.60}$$

*Then, for each $x_0 \in U(x^\star, R)$, sequence $\{x_n\}$ generated by INTM converges to $x^\star$ with order $\beta+1$. Moreover, (7.22) holds for $q$ replaced by*

$$q_6 = \frac{(p+2)\omega_1 L_p^0 \|x_0 - x^\star\|^{p-\beta}}{(p+1)(1-L_p^0\|x_0 - x^\star\|^p)} + \widehat{\omega}_2 \|F'(x^\star)\|^\beta \left(1 + \frac{L_p^0 \|x_0 - x^\star\|^p}{p+1}\right)^\beta$$
$$+ \nu \left(\frac{\omega_1(p+1+L_p^0\|x_0 - x^\star\|^p)}{(p+1)(1-L_p^0\|x_0 - x^\star\|^p)}\right)^{\beta+1}. \tag{7.61}$$

**Corollary 7.4** *Suppose that $F(x^\star) = 0$ and let*

$$\nu = \sup_{n \geq 0} \theta_n \|(P_n F'(x_n))^{-1}\| \|P_n F'(x_n)\|^{\beta+1}.$$

*Let $0 < R \leq R_p^0$ satisfy*

$$\frac{(p+2)L_p^0 R^p}{(p+1)(1-L_p^0 R^p)} + \nu \left(\frac{(p+1+L_p^0 R^p)}{(p+1)(1-L_p^0 R^p)}\right)^{\beta+1} R^\beta \begin{cases} < 1, & \beta = p = 0, \\ \leq 1, & 0 < \beta \leq p \leq 1, \end{cases} \tag{7.62}$$

*and*

$$U(x^\star, R) \subseteq D. \tag{7.63}$$

*Then, for each $x_0 \in U(x^\star, R)$, sequence $\{x_n\}$ generated by INTM converges to $x^\star$ with order $\beta+1$ and (7.22) holds with $q$ replaced by*

$$q_7 = \frac{(p+2)L_p^0 \|x_0 - x^\star\|^{p-\beta}}{(p+1)(1-L_p^0\|x_0 - x^\star\|^p)} + \nu \left(\frac{(p+1+L_p^0\|x_0 - x^\star\|^p)}{(p+1)(1-L_p^0\|x_0 - x^\star\|^p)}\right)^{\beta+1}. \tag{7.64}$$

We give now some examples to show the application of our results.

**Example 7.1** *Let $X = Y = \mathbb{R}$, $D = [1,6]$. Set $E_n = 0$, $B_n = F'(x_n)^{-1}$, $\theta_n = 0$, $P_n = I$ and $\beta = \frac{1}{2}$. Then, we obtain $\omega_2 = 0 = \nu$ and $\omega_1 = 1$. Define scalar function $F$ on $D$ by*

$$F(x) = -\frac{1}{15}x^{\frac{3}{2}} + \frac{1}{10}x. \tag{7.65}$$

*Then, we obtain the solution $x^\star = \frac{9}{4}$, $F'(x) = -\frac{1}{10}x^{\frac{1}{2}} + \frac{1}{10}$ and $\|F'(x^\star)\| = \frac{1}{20}$. Moreover, for all $x, y \in \left(\frac{5}{4}, \frac{13}{4}\right) = U\left(\frac{9}{4}, 1\right)$, $F'(x)$ is Hölder continuous with exponent $\frac{1}{2}$ in $U\left(\frac{9}{4}, 1\right)$*

$$\begin{aligned}\|F'(x) - F'(y)\| &= \frac{1}{10}|x^{\frac{1}{2}} - y^{\frac{1}{2}}| \\ &\le \frac{1}{10}|x^{\frac{1}{2}} - y^{\frac{1}{2}}|^{\frac{1}{2}}|x^{\frac{1}{2}} + y^{\frac{1}{2}}|^{\frac{1}{2}} = \frac{1}{10}|x-y|^{\frac{1}{2}},\end{aligned} \tag{7.66}$$

$$\begin{aligned}\|F'(x) - F'(x^\star)\| &= \frac{1}{10}\left|x^{\frac{1}{2}} - \frac{3}{2}\right| \\ &\le \frac{1}{20}\left|x^{\frac{1}{2}} - \frac{3}{2}\right|^{\frac{1}{2}}\left|x^{\frac{1}{2}} + \frac{3}{2}\right|^{\frac{1}{2}} = \frac{1}{20}\left|x - \frac{3}{2}\right|^{\frac{1}{2}},\end{aligned} \tag{7.67}$$

*in which the inequality of (7.67) is deduced from the following assertion*

$$4\left|x^{\frac{1}{2}} - \frac{3}{2}\right| \le x^{\frac{1}{2}} + \frac{3}{2} \quad (x \in D). \tag{7.68}$$

*That is, we have $p = \frac{1}{2}$ and*

$$L_p^0 = 1 < 2 = L_p. \tag{7.69}$$

*Hence, we get*

$$R_p^0 = 1 > R_p = \left(\frac{1}{2}\right)^2 = \frac{1}{4} \tag{7.70}$$

*and all conditions of Theorem 7.1 are true. Applying Theorem 7.1, we obtain that estimates (7.22) hold, where $q$ is defined by*

$$q = \frac{4}{3\left(1 - \left|x_0 - \frac{9}{4}\right|^{\frac{1}{2}}\right)}. \tag{7.71}$$

*On the other hand, if we apply the corresponding result of Li and Shen's paper (cf. [501, Theorem 2.1, p. 546]), we obtain that estimates (7.22) hold, where $q$ is replaced by*

$$q' = \frac{4}{3\left(1 - 2\left|x_0 - \frac{9}{4}\right|^{\frac{1}{2}}\right)}. \tag{7.72}$$

*Clearly, we have for any $x_0 \in U\left(\frac{9}{4}, \frac{1}{4}\right)$ and $x_0 \neq \frac{9}{4}$:*

$$q < q'. \tag{7.73}$$

*In Table 7.1, we give the comparison results of error bounds for the estimates (7.22) for Li and Shen's paper with ours. We set $x_0 = 2.05$ and thus $\left|x_0 - \frac{9}{4}\right|^{\frac{1}{2}} = .447213595$. Hence $q = 2.412022659$ and $q' = 12.62951461$.*

**Table 7.1:** The comparison results: the error bounds for (7.22)

| $n$ | [501] | ours |
|---|---|---|
| 0 | 1.129618127 | .215737865 |
| 1 | .025372913 | .004845795 |
| 2 | 9.62218E-06 | 1.83767E-06 |
| 3 | 1.4107E-12 | 2.6942E-13 |
| 4 | 1.18193E-22 | 2.25729E-23 |

**Example 7.2** *Let $X = Y = C[0,1]$ with the max–norm. Let also $D = \{x \in C[0,1] : \|x\| \leq 1\}$ and define*

$$F(x)(s) = x(s) - \int_0^1 stx^3(t)dt. \tag{7.74}$$

*Then, $x^\star = x^\star(s) = 0$ for $0 \leq s \leq 1$ is zero of operator $F$ and*

$$[F'(x)w](s) = w(s) - 3\int_0^1 stx^2(t)w(t)dt \tag{7.75}$$

*holds for all $x, w \in D$. Using (7.75), we have*

$$F'(x^\star) = I_X \tag{7.76}$$

*and*

$$[(F'(x) - F'(y))w](s) = -3\int_0^1 st(x^2(t) - y^2(t))w(t)dt \quad (x, y, w \in D). \tag{7.77}$$

*In view of (7.77), we obtain*

$$\begin{aligned}\|F'(x) - F'(y)\| &\leq 3\int_0^1 t\|x^2(t) - y^2(t)\|dt \\ &\leq 3\int_0^1 t(\|x\| + \|y\|)\|x - y\|dt \leq 3\|x - y\| \quad (x, y \in D)\end{aligned} \tag{7.78}$$

*and*

$$\begin{aligned}\|F'(x)-F'(x^\star)\| &\le 3\int_0^1 t\|x^2(t)\|dt\\ &\le 3\int_0^1 t\|x\|\|x-x^\star\|dt\\ &\le \frac{3}{2}\|x-x^\star\| \quad (x\in D).\end{aligned} \tag{7.79}$$

*Then, we can easily obtain from (7.76), (7.78) and (7.79) that*

$$L_1^0=\frac{3}{2}<3=L_1. \tag{7.80}$$

We finally provide an example, where only center–Hölder condition holds.

**Example 7.3** *Let $X=Y=\mathbb{R}$ and $D=X$. Define a function $F$ on $D$ by*

$$F(x)=\int_0^x\left(1+x\sin\frac{\pi}{x}\right)dx. \tag{7.81}$$

*Then, we get*

$$F'(x)=\begin{cases}1+x\sin\frac{\pi}{x}, & x\neq 0,\\ 1, & x=0.\end{cases} \tag{7.82}$$

*Note that $x^\star=0$ is zero of $F$ and $F'$ satisfies*

$$\|F'(x^\star)^{-1}[F'(x)-F'(x^\star)]\|=|x\sin\frac{\pi}{x}|\le|x|=\|x-x^\star\| \quad (x\in D). \tag{7.83}$$

*That is, $L_1^0=1$.*

*We shall show $L_1$ does not exist. Assume that there exists a constant $L_1>0$ such that*

$$\|F'(x^\star)^{-1}[F'(x)-F'(y)]\|\le L_1\|x-y\| \quad (x,y\in D). \tag{7.84}$$

*Suppose $m\ge 1$ is a fixed integer. Then, for $x=\frac{1}{m}$ and $y=\frac{2}{2m+1}$, we have*

$$\begin{aligned}\|F'(x^\star)^{-1}[F'(x)-F'(y)]\| &= \left|x\sin\frac{\pi}{x}-y\sin\frac{\pi}{y}\right|\\ &=\left|\frac{1}{m}\sin(m\pi)-\frac{2}{2m+1}\sin\frac{(2m+1)\pi}{2}\right|\\ &=\frac{2}{2m+1}\\ &\le L_1\|x-y\|=L_1\left|\frac{1}{m}-\frac{2}{2m+1}\right|\\ &=L_1\frac{1}{m(2m+1)},\end{aligned} \tag{7.85}$$

*which means that*

$$L_1 \geq 2m \quad (m \geq 1). \tag{7.86}$$

*This is impossible since $L_1 > 0$ is a constant.*

## 7.2 Average Lipschitz Conditions

In this section, we are concerned with the problem of approximating a locally unique solution $x^\star$ of (2.1). We present in this section the convergence of inexact Newton method INMB given by

$$For \ \ n = 0 \ \ step\, 1 \ \ until \ \ convergence \ \ do,$$

$$Find \ the \ step \ \Delta_n \ with \ satisfies \ \mathcal{B}_n \, \Delta_n = -F(x_n) + r_n$$

where

$$\frac{\| \mathcal{P}_n \, r_n \|}{\| \mathcal{P}_n \, F(x_n) \|} \leq \eta_n \leq 1.$$

Set

$$x_{n+1} = x_n + \Delta_n \quad (n \geq 0). \tag{7.87}$$

Here, $\mathcal{P}_n$ is an invertible operator and $\mathcal{B}_n^{-1} \in \mathcal{L}(\mathcal{Y}, \mathcal{X})$ for each $n \geq 1$.

The INMB was considered by Morini in (cf. [543]), whereas, if $\mathcal{P}_n = I$ $(n \geq 0)$, the method has been studied extensively in (cf. [88], [102], [163], [166], [313], [758]) under various Lipschitz–type hypotheses. A survey of such results can be found in (cf. [157]) (see, also (cf. [139], [275], [758])). The advantages of introducing operators $\mathcal{P}_n$ have been explained in (cf. [543]). In case $\mathcal{B}_n = F'(x_n)$ $(n \geq 0)$, we will denote INMB by INMF.

In this section, we are motivated by the work in (cf. [275], [758]), where, radius Lipschitz–type conditions are used (see (7.90)) to provide a local as well as a semilocal convergence for NM. We use weaker and needed center–Lipschitz–type conditions (see (7.89)) to find upper bounds on the distances $\| F'(x)^{-1} \, F'(x^\star) \|$ $(x \in \mathcal{D})$ instead of the stronger (7.90) used in (cf. [275], [758]) for NM. It turns out that this approach leads to a local convergence analysis not only for NM, but also for INMB and INMF with the following advantages, and under the same computational cost (see Remark 7.4):

(a) larger convergence radii,

(b) finer estimates for the distances $\| x_n - x^\star \|$ $(n \geq 0)$.

We provide four local convergence results for INMB and INMF.

**Theorem 7.5** *Let $F : \mathcal{D} \subseteq \mathcal{X} \longrightarrow \mathcal{Y}$ be a Fréchet–differentiable operator. Assume there exist $x^\star \in \mathcal{D}$, satisfying equation (2.1) such that $F'(x^\star)^{-1} \in \mathcal{L}(\mathcal{Y}, \mathcal{X})$, $r_0 > 0$ with*

$$U(x^\star, r_0) \subseteq \mathcal{D}, \tag{7.88}$$

*positive integrable functions $L_0$ and $L$, satisfying center–Lipschitz condition and radius Lipschitz–type condition:*

$$\| F'(x^\star)^{-1} \, (F'(x) - F'(x^\star)) \, \| \le \int_0^{\rho(x)} L_0(t)\, dt, \tag{7.89}$$

$$\| F'(x^\star)^{-1} \, [F'(x) - F'(x_\theta)] \, \| \le \int_{\theta\, \rho(x)}^{\rho(x)} L(t)\ dt, \tag{7.90}$$

*respectively, for all $x \in U(x^\star, r_0)$, $x_\theta = x^\star + \theta\, (x - x^\star)$, $\rho(x) = \| \, x - x^\star \, \|$, $\theta \in [0,1]$,*

$$v_n = \theta_n \ \| \, (\mathcal{P}_n\, F'(x_n))^{-1} \, \| \, \| \, \mathcal{P}_n\, F'(x_n) \, \| = \theta_n \ \mathrm{cond}\ (\mathcal{P}_n\, F'(x_n)) \le v < 1 \tag{7.91}$$

*and*

$$(1 - v) \int_0^{r_0} L_0(t)\, dt + (1 + v) \int_0^{r_0} L(t)\, dt \le 1 - v. \tag{7.92}$$

*Then, sequence $\{x_n\}$ $(n \ge 0)$ generated by INMF is well defined, remains in $U(x^\star, r_0)$ for all $n \ge 0$ and converges to $x^\star$, provided that $x_0 \in U(x^\star, r_0)$. Moreover, the following estimate holds*

$$\| \, x_{n+1} \quad x^\star \, \| \le \alpha \ \| \, x_n \quad x^\star \, \|, \tag{7.93}$$

*where*

$$\alpha = (1 + v) \ \frac{\displaystyle\int_0^{\rho(x_0)} L(t)\ dt}{1 - \displaystyle\int_0^{\rho(x_0)} L_0(t)\ dt} + v \le 1. \tag{7.94}$$

**Proof** By hypothesis $x_0 \in U(x^\star, r_0)$ and $\alpha \in [0,1)$, since, by (7.92), the positivity of $L_0$, $L$, we have that

$$\alpha < (1 + v) \ \frac{\displaystyle\int_0^{r_0} L(t)\ dt}{1 - \displaystyle\int_0^{r_0} L_0(t)\ dt} + v \le 1.$$

Let us assume $x_m \in U(x^\star, r_0)$, $m \le n$, we shall show (7.93) and $x_{m+1} \in U(x^\star, r_0)$ hold, for all $m$. In view of (7.89), (7.92) and the induction hypotheses, we get that

$$\| \, F'(x^\star)^{-1} \ (F(x_m) - F(x_0)) \, \| \le \int_0^{\rho(x_m)} L_0(t)\ dt \le \int_0^{r_0} L_0(t)\ dt < 1. \tag{7.95}$$

It follows from the Banach lemma on invertible operators that $F'(x_m)^{-1}$ exists and

$$\| \, F'(x_m)^{-1}\, F'(x^\star) \, \| \le \left(1 - \int_0^{\rho(x_m)} L_0(t)\ dt\right)^{-1} \le \left(1 - \int_0^{r_0} L_0(t)\, dt\right)^{-1}. \tag{7.96}$$

Using INMF, we obtain the approximation

$$\begin{aligned} x_{m+1} - x^\star &= x_m - x^\star - F'(x_m)^{-1}\,(F(x_m) - F(x^\star) - r_m) \\ &= F'(x_m)^{-1}\,F'(x^\star)\int_0^1 F'(x^\star)^{-1}\,(F(x_m) - F(x_\theta))\,(x_m - x^\star)\,d\theta \\ &\quad + F'(x_m)\,\mathcal{P}_m^{-1}\,\mathcal{P}_m\,r_m. \end{aligned} \tag{7.97}$$

By (7.90), (7.91), (7.94), (7.96), (7.97) and the induction hypotheses, we obtain in turn that

$$\begin{aligned} &\| x_{m+1} - x^\star \| \\ &\leq \| F'(x_m)^{-1}\,F'(x^\star) \| \int_0^1 \| F'(x^\star)^{-1}\,(F(x_m) - F(x_\theta)) \|\,\| x_m - x^\star \|\,d\theta \\ &\quad + \theta_m\,\| (\mathcal{P}_m\,F'(x_m))^{-1} \|\,\| \mathcal{P}_m\,F(x_m) \| \\ &\leq \frac{1}{1 - \displaystyle\int_0^{\rho(x_m)} L_0(t)\,dt} \int_0^1 \int_{\theta\,\rho(x_m)}^{\rho(x_m)} L(t)\,dt\,\rho(x_m)\,d\theta \\ &\quad + \theta_m\,\| (\mathcal{P}_m\,F'(x_m))^{-1} \|\,\| \mathcal{P}_m\,F'(x_m)\,F'(x_m)^{-1}\,F(x_m) \| \\ &\leq \frac{\displaystyle\int_0^{\rho(x_m)} L(t)\,t\,dt}{1 - \displaystyle\int_0^{\rho(x_m)} L_0(t)\,dt} + \theta_m\,\mathrm{cond}\,(\mathcal{P}_m\,F'(x_m)) \left( \| x_m - x^\star \| + \frac{\displaystyle\int_0^{\rho(x_m)} L(t)\,dt}{1 - \displaystyle\int_0^{\rho(x_m)} L_0(t)\,dt} \right) \\ &\leq (1 + v_m)\,\frac{\displaystyle\int_0^{\rho(x_m)} L(t)\,t\,dt}{1 - \displaystyle\int_0^{\rho(x_m)} L_0(t)\,dt} + v_m\,\rho(x_m) \\ &\leq \left( (1 + v_m)\,\frac{\displaystyle\int_0^{\rho(x_m)} L(t)\,t\,dt}{1 - \displaystyle\int_0^{\rho(x_m)} L_0(t)\,dt} + v_m \right) \rho(x_m) \leq \alpha\,\| x_m - x^\star \| < \| x_m - x^\star \|, \end{aligned} \tag{7.98}$$

which implies $x_{m+1} \in U(x^\star, r_0)$ and $\lim_{m \longrightarrow \infty} x_m = x^\star$. This completes the proof of Theorem 7.5. ⊠

**Proposition 7.1** *Under hypotheses (7.89)–(7.91) of Theorem 7.5 and further, assume that function*

$$L_c(t) = t^{1-c}\,L(t) \tag{7.99}$$

*is nondecreasing for some $c \in [0, 1]$; there exists $r_1 > 0$, such that*

$$U(x^\star, r_1) \subseteq \mathcal{D}, \tag{7.100}$$

$$\frac{(1+v)\displaystyle\int_0^{r_1} L(t)\ t\ dt}{r_1\left(1-\displaystyle\int_0^{r_1} L_0(t)\ dt\right)} + v \leq 1. \tag{7.101}$$

*Then, sequence $\{x_n\}$ $(n \geq 0)$ generated by INMF is well defined, remains in $U(x^\star, r_1)$ for all $n \geq 0$ and converges to $x^\star$, provided that $x_0 \in \overline{U}(x^\star, r_1)$. Moreover, the following estimate holds*

$$\| x_{n+1} - x^\star \| \leq \beta_n \ \| x_n - x^\star \|, \tag{7.102}$$

*where*

$$\beta_n = (1+v)\ \frac{\displaystyle\int_0^{\rho(x_0)} L(t)\ t\ dt\ \rho(x_n)^c}{\rho(x_0)^{1+c}\left(1-\displaystyle\int_0^{\rho(x_0)} L_0(t)\ dt\right)} + v$$

$$\leq \beta = (1+v)\ \frac{\displaystyle\int_0^{\rho(x_0)} L(t)\ t\ dt}{\rho(x_0)\left(1-\displaystyle\int_0^{\rho(x_0)} L_0(t)\ dt\right)} + v < 1. \tag{7.103}$$

**Proof** We follow the proof of Theorem 7.5 until (7.97). Define function $f_{d,c}$ $(d \geq 0)$ by

$$f_{d,c}(s) = \frac{1}{s^{c+d}} \int_0^s t^d\ L(t)\ dt. \tag{7.104}$$

In view of Lemma 2.2 in (cf. [758]), function $f_{d,c}$ is non–decreasing. It follows in turn that

$$\| x_{m+1} - x^\star \| \leq (1+v_m)\ \frac{\displaystyle\int_0^{\rho(x_m)} L(t)\ t\ dt}{1-\displaystyle\int_0^{\rho(x_m)} L_0(t)\ dt} + v_m\ \rho(x_m)$$

$$= (1+v_m)\ \frac{f_{1,c}(\rho(x_m))}{1-\displaystyle\int_0^{\rho(x_m)} L_0(t)\ dt}\ \rho(x_m)^{1+c} + v_m\ \rho(x_m)$$

$$\leq 1+v_m)\ \frac{f_{1,c}(\rho(x_0))}{1-\displaystyle\int_0^{\rho(x_0)} L_0(t)\ dt}\ \rho(x_m)^{1+c} + v_m\ \rho(x_m)$$

$$\leq \beta_m\ \| x_m - x^\star \| \leq \beta\ \| x_m - x^\star \| < \| x_m - x^\star \|, \tag{7.105}$$

which implies $x_{m+1} \in U(x^\star, r_1)$ and $\lim\limits_{m \longrightarrow \infty} x_m = x^\star$. This completes the proof of Proposition 7.1. ⊠

**Theorem 7.6** *Under hypotheses (7.88)–(7.91) of Theorem 7.5 (for $r_2 > 0$, replacing $r_0$) and further, assume that*

$$\| \mathcal{B}(x)^{-1} \; F'(x) \| \leq w_1, \tag{7.106}$$

$$\| \mathcal{B}(x)^{-1} \; (F'(x) - \mathcal{B}(x)) \| \leq w_2, \tag{7.107}$$

*hold for all $x \in U(x^\star, r_2)$ and*

$$(1+v) \; w_1 \int_0^{r_2} L(t) \; dt + (1-w_2-w_1 \; v) \int_0^{r_2} L_0(t) \; dt \leq 1-w_2-w_1 \; v. \tag{7.108}$$

*Then, sequence $\{x_n\}$ $(n \geq 0)$ generated by INMB is well defined, remains in $U(x^\star, r_2)$ for all $n \geq 0$ and converges to $x^\star$, provided that $x_0 \in U(x^\star, r_2)$. Moreover, the following estimate holds*

$$\| x_{n+1} - x^\star \| \leq \gamma \; \| x_n - x^\star \|, \tag{7.109}$$

*where*

$$\gamma = (1+v) \; \frac{w_1 \displaystyle\int_0^{\rho(x_0)} L(t) \; dt}{1 - \displaystyle\int_0^{\rho(x_0)} L_0(t) \; dt} + w_2 + w_1 \; v < 1. \tag{7.110}$$

**Proof** Using the properties of functions $L_0$, $L$, (7.98) and (7.110), we obtain $\gamma \in (0,1)$. By INMB, if $x_m \in U(x^\star, r_2)$, we have the approximation

$$\begin{aligned} x_{m+1} - x^\star &= x_m - x^\star - \mathcal{B}_m^{-1} \; (F(x_m) - F(x^\star)) + \mathcal{B}_m^{-1} \; r_m \\ &= -\mathcal{B}_m^{-1} F'(x_m) \int_0^1 F'(x_m)^{-1} F'(x^\star) F'(x^\star)^{-1} (F'(x_m) - F'(x_\theta))(x_m - x^\star) d\theta \\ &\quad + \mathcal{B}_m^{-1} \; (F'(x_m) - \mathcal{B}_m) \; (x_m - x^\star) + \mathcal{B}_m^{-1} \; \mathcal{P}_m^{-1} \; \mathcal{P}_m \; r_m. \end{aligned} \tag{7.111}$$

In view of (7.89)–(7.91), (7.96), (7.106) and the induction hypotheses, we obtain in turn that

$$\begin{aligned} &\| x_{m+1} - x^\star \| \\ &\leq \| \mathcal{B}_m^{-1} \; F'(x_m) \| \int_0^1 \| F'(x_m)^{-1} \; F'(x^\star) \| \| F'(x^\star)^{-1} \; (F'(x_m) - F'(x_\theta)) \| \\ &\times \| x_m - x^\star \| \; d\theta + \| \mathcal{B}_m^{-1} \; (F'(x_m) - \mathcal{B}_m) \| \| x_m - x^\star \| \\ &+ \theta_m \; \| (\mathcal{P}_m \, \mathcal{B}_m)^{-1} \| \| \mathcal{B}_m \, F(x_m) \| \end{aligned}$$

$$\leq \frac{w_1}{1 - \displaystyle\int_0^{\rho(x_m)} L_0(t)\,dt} \int_0^1 \int_{\theta\,\rho(x_m)}^{\rho(x_m)} L(t)\,dt\,\rho(x_m)\;d\theta + w_2\,\rho(x_m)$$

$$+\,\theta_m\ \|\ \mathcal{P}_m^{-1}\,F'(x_m)\ \|\,\|\ (\mathcal{P}_m\,F'(x_m))^{-1}\ \|\,\|\ \mathcal{P}_m\,F'(x_m)\ \|\ \|\ F'(x_m)^{-1}\,F(x_m)\ \|$$

$$\leq \frac{w_1 \displaystyle\int_0^{\rho(x_m)} L(t)\,t\,dt}{1 - \displaystyle\int_0^{\rho(x_m)} L_0(t)\,dt} + w_2\,\rho(x_m) + w_1\,v_m \left( \rho(x_m) + \frac{\displaystyle\int_0^{\rho(x_m)} L(t)\,t\,dt}{1 - \displaystyle\int_0^{\rho(x_m)} L_0(t)\,dt} \right)$$

$$\leq (1+v_m)\,\frac{w_1 \displaystyle\int_0^{\rho(x_m)} L(t)\,t\,dt}{1 - \displaystyle\int_0^{\rho(x_m)} L_0(t)\,dt} + (w_2 + w_1\,v_m)\,\rho(x_m)$$

$$\leq \left( (1+v_m)\,\frac{w_1 \displaystyle\int_0^{\rho(x_m)} L(t)\,dt}{1 - \displaystyle\int_0^{\rho(x_m)} L_0(t)\,dt} + w_2 + w_1\,v_m \right) \rho(x_m) \leq \gamma\ \|\ x_m - x^\star\ \| < \|\ x_m - x^\star\ \|, \tag{7.112}$$

which implies $x_{m+1} \in \overline{U}(x^\star, r_2)$ and $\lim\limits_{m \longrightarrow \infty} x_m = x^\star$. This completes the proof of Theorem 7.6. ⊠

**Proposition 7.2** *Under hypotheses (7.88)–(7.91) (for $r_3 > 0$, replacing $r_0$), (7.99), (7.106), (7.107) (for $r_3$, replacing $r_2$) and further, assume $r_3$ satisfies*

$$(1+v)\,\frac{w_1 \displaystyle\int_0^{r_3} L(t)\ t\ dt}{r_3 \left(1 - \displaystyle\int_0^{r_3} L_0(t)\ dt\right)} + w_2 + w_1\ v \leq 1. \tag{7.113}$$

*Then, sequence $\{x_n\}$ $(n \geq 0)$ generated by INMB is well defined, remains in $U(x^\star, r_3)$ for all $n \geq 0$ and converges to $x^\star$. Moreover, the following estimate holds*

$$\|\ x_{n+1} - x^\star\ \| \leq \delta\ \|\ x_n - x^\star\ \|, \tag{7.114}$$

*where,*

$$\delta = (1+v)\,\frac{w_1 \displaystyle\int_0^{\rho(x_0)} L(t)\,t\,dt}{\rho(x_0) \left(1 - \displaystyle\int_0^{\rho(x_0)} L_0(t)\,dt\right)} + w_2 + w_1\,v < 1. \tag{7.115}$$

**Proof** Using the properties of functions $L$, $L_0$, (7.113) and (7.115), we deduce $\delta \in (0, 1)$. If $x_m \in U(x^\star, r_3)$, as in Proposition 7.1, using (7.112), we get in turn that

$$\| x_{m+1} - x^\star \| \leq (1 + v_m) \, \frac{w_1 \displaystyle\int_0^{\rho(x_m)} L(t)\, t\, dt}{1 - \displaystyle\int_0^{\rho(x_m)} L_0(t)\, dt} + (w_2 + w_1\, v_m)\, \rho(x_m)$$

$$\leq (1 + v) \, \frac{w_1\, f_{1,c}(\rho(x_m))}{1 - \displaystyle\int_0^{\rho(x_m)} L_0(t)\, dt} \, \rho(x_m)^{1+c} + (w_2 + w_1\, v)\, \rho(x_m)$$

$$\leq (1 + v) \, \frac{w_1\, f_{1,c}(\rho(x_0))}{1 - \displaystyle\int_0^{\rho(x_0)} L_0(t)\, dt} \, \rho(x_m)^{1+c} + (w_2 + w_1\, v)\, \rho(x_m)$$

$$\leq \delta \, \| x_m - x^\star \| < \| x_m - x^\star \|,$$

which implies $x_{m+1} \in U(x^\star, r_3)$ and $\lim_{m \longrightarrow \infty} x_m = x^\star$. This completes the proof of Proposition 7.2. ⊠

**Remark 7.4** *Note that in general*

$$L_0(t) \leq L(t) \qquad (t \leq 0), \tag{7.116}$$

*and* $\dfrac{L(t)}{L_0(t)}$ *can be arbitrarily large. In order for us to compare our results with relevant ones already in the literature, let us consider the case* $\mathcal{B}_n = F'(x_n)$ *(*$n \geq 0$*),* $v = w_1 = w_2 = 0$*. Then, the convergence radii in (cf. [758] and [275]) (for* $c = 1$*) are given by*

$$\int_0^r L(t)\, dt \leq \frac{1}{2} \tag{7.117}$$

*and*

$$\frac{1}{\overline{r}} \int_0^{\overline{r}} (\overline{r} + t)\, L(t)\, dt \leq 1, \tag{7.118}$$

*respectively. In view of (7.92), (7.101), (7.116), (7.117) and (7.118), we obtain that*

$$r \leq r_0 \tag{7.119}$$

*and*

$$\overline{r} \leq r_1. \tag{7.120}$$

*In case strict inequality holds in (7.116), then, so does in (7.119) and (7.120). Moreover, the error estimates (7.93) and (7.102) are finer (smaller) than the corresponding ones in (cf. [275] and [758]). Note that these advantages are obtained under the same computational cost since, in practice, the computation of function* $L$ *requires that of* $L_0$*.*

## 7.3 Two–step Methods

Using our new concept of recurrent functions, we approximate a locally unique solution of a nonlinear equation by an inexact two–step Newton–like algorithm in a Banach space setting. Our semilocal analysis provides tighter error bounds than before and in many interesting cases, weaker sufficient convergence conditions.

In (cf. [69], [135], [163]), we introduced and studied the inexact two–step Newton–like method ITSNLM:

$$y_n = x_n - A(x_n)^{-1}\, F(x_n),$$

$$x_{n+1} = y_n - z_n \quad (n \geq 0,\ x_0 \in \mathcal{D}) \tag{7.121}$$

to generate a sequence $\{x_n\}$ approximating $x^\star$ the locally unique solution of (2.1). Here, $A(x) \in \mathcal{L}(\mathcal{X}, \mathcal{Y})$ $(x \in \mathcal{D})$ the space of bounded linear operators from $\mathcal{X}$ into $\mathcal{Y}$ and $\{z_n\}$ is a sequence in $\mathcal{X}$ depending on $\{x_n\}$ and earlier iterates. If $z_n = 0$ and $A(x) = F'(x)$ $(x \in \mathcal{D})$, we obtain NM where as if $z_n = F'(y_n)^{-1}\, F(y_n)$, we get the two–step Newton method. For $z_n = 0$ and $A(x) = [x, x_+; F]$ (or $[x_-, x; F]$), where $[.,.; F]$ is a divided difference of order one for operator $F$, we obtain the Secant method (cf. [163], [194]). Moreover, if $z_n = L_n\, F''(x_n)\,(y_n - x_n)^2$ for some $L_n \in \mathcal{L}(\mathcal{X}, \mathcal{Y})$ $(n \geq 0)$, we obtain Chebyshev–type methods (cf. [163], [259], [400], [416]). Many other studies with different choices of $\{z_n\}$ were given in (cf. [69], [135], [163]). Several authors have also examined the convergence for ITSNLM but for special choices of sequences $\{z_n\}$ (cf. [41], [767]). As far as we know most studies on inexact methods (with the exception of our works in (cf. [204])) involve a choice of sequence $\{z_n\}$ given by $z_n = L_n\, F(x_n)$ $(n \geq 0)$. Hence, our ITSNLM unifies all earlier inexact methods.

In this section, we provided a unifying semilocal convergence analysis for ITSNLM under general conditions on the operators involved, using our new concept of recurrent functions. By semilocal analysis, we mean that we use information from a neighborhood of the initial guess $x_0$. The analysis is called local if the information is received from a neighborhood of the solution $x^\star$. The local convergence of ITSNLM is not examined in this section, although it can be provided in a similar way to the semilocal one. The most popular approach for providing sufficient conditions for the semilocal convergence Newton–type methods as well as error estimates on the distances involved is the so–called Kantorovich approach KA (cf. [163], [194], [469], [613], [614]). The KA involves recurrent "majorizing" sequences, but the sufficient convergence conditions are not also necessary. The results in (cf. [41], [139]) are based on the KA. In this section, we use our new concept of recurrent functions. This concept is more flexible than the KA, since it yields a more precise upper bound on the norm, e.g., for NM $\parallel F'(x)\, F'(x_0) \parallel \leq (1 - K_0 \parallel x - x_0 \parallel)^{-1}$, which is more precise than $\parallel F'(x)\, F'(x_0) \parallel \leq (1 - K \parallel x - x_0 \parallel)^{-1}$ if $K_0 < K$ for $x$ in some neighborhood of $x_0$. This modification leads to more precise majorizing sequences and to (2.443) instead of (7.177) (see, also Remarks 7.6, 7.7 and Example 7.6). We consider our approach more flexible than the KA, since it allows for a more precise upper error bound on the norm $\| A(x_n)^{-1} A(x_0) \|$ $(n \geq 0)$, leading to more precise majorizing sequences. Moreover, the information on the uniqueness of the solution is more precise. We are motivated by optimization considerations. We provide a new analysis with new sufficient convergence conditions (which can be weaker than before for the specialized choices of sequence $\{z_n\}$), but tighter error bounds than in (cf. [69], [135], [163]). In particular, our results in (cf. [204]) are a special cases of the ones obtained here. Therefore, in practice we shall test all set of conditions hoping that at least one of them will be satisfied. The same approach shall apply for special choices of sequence $\{z_n\}$ given by us or others in (cf. [259], [767]). For the special case of NM, our sufficient convergence conditions provide tighter error bounds under weaker hypotheses than the celebrated Kantorovich theorem for solving equations. Our results also cover the case when $F$ is a nondifferentiable not considered in earlier studies. The results obtained here can be extended to hold for ITSNLM involving outer or generalized inverses along the works of Nashed, Chen (cf. [277]), [200]).

The semilocal convergence analysis of ITSNLM under weak conditions is provided in this section. First, we need the following result on majorizing sequences for ITSNLM.

**Lemma 7.2** *Let $a \geq 0$, $b \geq 0$, $c > 0$, $\eta \geq 0$, $s_0 \geq 0$, $K_0 > 0$, $K > 0$, $M \geq 0$, $L \geq 0$, $\mu \geq 0$ and $\ell \in [0, 1)$ be given constants.*
*Define constants $\Xi$, $\alpha$, $\beta$, $\gamma$ and $\delta$ by*

$$\Xi = K\,(1 + a^2\,\eta^{2b}) + \sqrt{(K\,(1 + a^2\,\eta^{2b}))^2 + 8\,L\,(1 + a\,\eta^b)\,(K\,(1 + a^2\,\eta^{2b}) - 2\,M\,(1 + a\,\eta^b))}$$

$$\alpha = \frac{2\,K\,(1 + a^2\,\eta^{2b}) - 2\,M\,(1 + a\,\eta^b)}{\Xi} < 1, \tag{7.122}$$

$$\beta = K\,a^2\,\eta^{1+2b} + K\,\eta + 2\,\mu + 2\,a\,c\,\eta^b + 2\,\alpha\,L\,(\eta + a\,\eta^{1+b}), \tag{7.123}$$

$$\gamma = K\,a^2\,\alpha\,\eta^{1+2b} + K\,\alpha\,\eta + 2\,(M + L\,\alpha)\,\frac{\eta + a\,\eta^{1+b}}{1 - \alpha} + 2\,\mu + 2\,a\,c\,\eta^b, \tag{7.124}$$

$$\delta = \max\{\beta, \gamma\} \tag{7.125}$$

*and scalar sequences $\{s_n\}$, $\{t_n\}$ by*

$$t_0 = 0,\ s_0 = \eta,\ t_{n+1} = s_n + a\,(s_n - t_n)^{1+b},$$
$$s_{n+1} = t_{n+1} + \frac{\varrho_n}{2\,(1 - \ell - L\,t_{n+1})}, \tag{7.126}$$

*where*

$$\varrho_n = K(t_{n+1} - s_n)^2 + K(s_n - t_n)^2 + 2(M\,t_n + \mu)(s_n - t_n) + 2(1 + K_0\,s_n)(t_{n+1} - s_n).$$

*Assume that*

$$2\,M\,(1+a\,\eta^b) < K\,(1+a^2\,\eta^{2\,b}) \tag{7.127}$$

*and*

$$\delta \leq 2\,\alpha\,(1-\ell). \tag{7.128}$$

*Then, scalar sequence* $\{s_n\}$ $(n \geq 0)$ *is increasing, bounded from above by*

$$s^{\star\star} = \left(\frac{1}{1-\alpha} + \frac{a\,\eta^b}{1-\alpha^{1+b}} + \alpha\right)\eta \tag{7.129}$$

*and converges to its unique least upper bound* $s^\star$ *satisfying* $s^\star \in [0, s^{\star\star}]$. *Moreover, the following estimates hold for all* $n \geq 0$

$$0 \leq s_{n+1} - t_{n+1} \leq \alpha\,(s_n - t_n). \tag{7.130}$$

**Proof** It follows from (7.122) that $\alpha \in (0,1)$. We shall show using induction on the integer $k$ that

$$0 \leq \frac{K\,a^2\,(s_k-t_k)^{1+2\,b} + K\,(s_k-t_k) + 2\,(M\,t_k+\mu) + 2\,a\,(1+K_0\,s_k)\,(s_k-t_k)^b}{2\,(1-\ell-L\,t_{k+1})} \leq \alpha. \tag{7.131}$$

Estimate (7.130) will follow from (7.126) and (7.131). Using the definition of $\beta$, (7.125) and (7.128), we conclude that (7.130) and (7.131) hold for $k=0$. Let us assume (7.130) and (7.131) hold for all $n \leq k$. We have in turn that

$$\begin{aligned}
t_{k+1} &= s_k + a\,(s_k-t_k)^{1+b} \\
&\leq t_k + \alpha^k\,\eta + a\,(s_k-t_k)^{1+b} \\
&\leq s_{k-1} + a\,(s_{k-1}-t_{k-1})^{1+b} + \alpha^k\,\eta + a\,(s_k-t_k)^{1+b} \\
&\leq \alpha^{k-1}\,\eta + s_{k-2} + a\,(s_{k-2}-t_{k-2})^{1+b} + a\,(s_{k-1}-t_{k-1})^{1+b} + \alpha^k\,\eta + a\,(s_k-t_k)^{1+b} \\
&\leq s_1 + (\alpha^2+\alpha^3+\cdots+\alpha^k)\,\eta + a\,((s_1-t_1)^{1+b}+\cdots+(s_k-t_k)^{1+b}) \\
&\leq t_1 + \alpha\,\eta + (\alpha^2+\alpha^3+\cdots+\alpha^k)\,\eta + a\,((s_1-t_1)^{1+b}+\cdots+(s_k-t_k)^{1+b}) \\
&\leq s_0 + a\,(s_0-t_0)^{1+b} + (\alpha+\alpha^2+\cdots+\alpha^k)\,\eta + a\,((s_1-t_1)^{1+b}+\cdots+(s_k-t_k)^{1+b}) \\
&\leq \eta + (\alpha+\alpha^2+\cdots+\alpha^k)\,\eta + a\,((s_0-t_0)^{1+b}+(s_1-t_1)^{1+b}+\cdots+(s_k-t_k)^{1+b}) \\
&= \frac{1-\alpha^{k+1}}{1-\alpha}\,\eta + a\,(\eta^{1+b}+(\alpha\,\eta)^{1+b}+\cdots+(\alpha^k\,\eta)^{1+b}) \\
&= \frac{1-\alpha^{k+1}}{1-\alpha}\,\eta + a\,(1+\alpha^{1+b}+(\alpha^{1+b})^2+\cdots+(\alpha^{1+b})^k)\,\eta^{1+b} \\
&= \frac{1-\alpha^{k+1}}{1-\alpha}\,\eta + a\,\frac{1-\alpha^{(1+b)\,(k+1)}}{1-\alpha^{1+b}}\,\eta^{1+b} \\
&< \frac{\eta}{1-\alpha} + \frac{a}{1-\alpha^{1+b}}\,\eta^{1+b} < s^{\star\star}
\end{aligned} \tag{7.132}$$

and

$$s_{k+1} \leq t_{k+1} + \alpha\,(s_k-t_k) \leq \frac{\eta}{1-\alpha} + \frac{a\,\eta^{1+b}}{1-\alpha^{1+b}} + \alpha^{k+1}\,\eta \leq s^{\star\star}. \tag{7.133}$$

In view of the induction hypotheses, (7.132) and (7.133), estimate (7.131) shall be true if

$$K a^2 (\alpha^k \eta)^{1+2b} + K \alpha^k \eta + 2ac(\alpha^k \eta)^b + 2\alpha L\left(\frac{1-\alpha^{k+1}}{1-\alpha}\eta + \frac{1-\alpha^{(1+b)(k+1)}}{1-\alpha^{1+b}}\eta^{1+b}\right)$$
$$+ 2\left(M\left(\frac{1-\alpha^k}{1-\alpha}\,\eta + a\,\frac{1-\alpha^{(1+b)k}}{1-\alpha^{1+b}}\,\eta^{1+b}\right) + \mu\right) - 2\,(1-\ell)\,\alpha \leq 0. \tag{7.134}$$

But, by (7.122), $\alpha \in [0,1)$ and

$$(\alpha^k\,\eta)^{1+2\,b} \leq \alpha^{k\,(1+2\,b)}\,\eta^{1+2\,b} \leq \alpha^k\,\alpha^{2\,k\,b}\,\eta^{1+2\,b} \leq \alpha^k\,\eta^{1+2\,b}.$$

Estimate (7.134) motivates us to introduce functions $f_k$ on $[0,1)$ $(k \geq 1)$ for $t = \alpha$ in the following form

$$f_k(t) = K a^2 \eta^{1+2b} t^k + K\eta t^k + 2ac\eta^b + 2\left(M\left(\frac{1-t^k}{1-t}\eta + a\,\frac{1-t^k}{1-t}\eta^{1+b}\right) + \mu\right)$$
$$+ 2\,t\,L\left(\frac{1-t^{k+1}}{1-t}\,\eta + a\,\frac{1-t^{k+1}}{1-t}\,\eta^{1+b}\right) - 2\,(1-\ell)\,t. \tag{7.135}$$

We need a relationship between two consecutive functions $f_k$. We have that

$$\begin{aligned} f_{k+1}(t) &= K\,a^2\,\eta^{1+2\,b}\,t^{k+1} + K\,\eta\,t^{k+1} + 2\,a\,c\,\eta^b \\ &\quad + 2\left(M\left(\frac{1-t^{k+1}}{1-t}\,\eta + a\,\frac{1-t^{k+1}}{1-t}\,\eta^{1+b}\right) + \mu\right) \\ &\quad + 2\,t\,L\left(\frac{1-t^{k+2}}{1-t}\,\eta + a\,\frac{1-t^{k+2}}{1-t}\,\eta^{1+b}\right) - 2\,(1-\ell)\,t \\ &\quad - K\,a^2\,\eta^{1+2\,b}\,t^k - K\,\eta\,t^k - 2\,a\,c\,\eta^b \\ &\quad - 2\left(M\left(\frac{1-t^k}{1-t}\,\eta + a\,\frac{1-t^k}{1-t}\,\eta^{1+b}\right) + \mu\right) \\ &\quad - 2\,t\,L\left(\frac{1-t^{k+1}}{1-t}\,\eta + a\,\frac{1-t^{k+1}}{1-t}\,\eta^{1+b}\right) + 2\,(1-\ell)\,t + f_k(t) \\ &= f_k(t) + K\,a^2\,\eta^{1+2\,b}\,t^{k+1} + K\,\eta\,t^{k+1} - K\,a^2\,\eta^{1+2b}\,t^k - K\,t^k\,\eta \\ &\quad + 2\,M(\eta\,t^k + a\,\eta^{1+b}\,t^k) + 2\,L\,t\,(\eta\,t^{k+1} + a\,\eta^{1+b}\,t^{k+1}). \end{aligned}$$

Hence, we get

$$f_{k+1}(t) = f_k(t) + g(t)\,t^k\,\eta, \tag{7.136}$$

where

$$g(t) = 2L(1+a\eta^b)t^2 + K(1+a^2\eta^{2b})t + 2M(1+a\eta^b) - K(1+a^2\eta^{2\,b}). \tag{7.137}$$

It follows from (7.127) and (7.137) that function $g$ has a unique zero $\alpha$ given by (7.122). In view of (7.136), we have

$$f_k(\alpha) = f_1(\alpha) \qquad (k \geq 1). \tag{7.138}$$

Consequently, estimate (7.134) holds if

$$f_k(\alpha) \leq 0 \qquad (k \geq 1) \tag{7.139}$$

or

$$f_1(\alpha) \leq 0. \tag{7.140}$$

But (7.140) holds by the choice of $\gamma$ and (7.128). Moreover, define

$$f_\infty(\alpha) = \lim_{k \to \infty} f_k(\alpha).$$

Then, we have

$$f_\infty(\alpha) = \lim_{k \to \infty} f_k(\alpha) \leq 0.$$

That completes the induction. Finally, sequence $\{s_n\}$ is non–decreasing, bounded from above by $s^{\star\star}$ and as such it converges to $s^\star$. This completes the proof of Lemma 7.2. ⊠

We shall show the main semilocal convergence result for ITSNLM.

**Theorem 7.7** *Let $F : \mathcal{D} \subseteq \mathcal{X} \longrightarrow \mathcal{Y}$ be a Fréchet–differentiable operator. Assume there exist $x_0 \in \mathcal{D}$, a sequence $\{z_n\} \subseteq \mathcal{X}$, an operator $A(x) \in \mathcal{L}(\mathcal{X}, \mathcal{Y})$ $(x \in \mathcal{D})$ and constants $a \geq 0$, $b \geq 0$, $K \geq K_0 > 0$, $M \geq 0$, $L \geq 0$, $\ell \in [0, 1)$, $\mu \geq 0$ $\eta \geq 0$, $s_0 \geq \eta$ such that, for all $x, y \in \mathcal{D}$*

$$A(x_0)^{-1} \in \mathcal{L}(\mathcal{Y}, \mathcal{X}), \tag{7.141}$$

$$\| A(x_0)^{-1} \ F(x_0) \| \leq \eta, \tag{7.142}$$

$$\| A(x_0)^{-1} \ (F'(x) - F'(x_0)) \| \leq K_0 \ \| x - x_0 \|, \tag{7.143}$$

$$\| A(x_0)^{-1} \ (F'(x) - F'(y)) \| \leq K \ \| x - y \|, \tag{7.144}$$

$$\| A(x_0)^{-1} \ (A(x) - F'(x)) \| \leq M \ \| x - x_0 \| + \mu, \tag{7.145}$$

$$\| A(x_0)^{-1} \ (A(x) - A(x_0)) \| \leq L \ \| x - x_0 \| + \ell, \tag{7.146}$$

$$\| x_{n+1} - y_n \| = \| z_n \| \leq a \ \| A(x_n)^{-1} \ F(x_n) \|^{1+b}, \tag{7.147}$$

$$\overline{U}(x_0, s^\star) \subseteq \mathcal{D} \tag{7.148}$$

*and hypotheses (7.127)–(7.128) of Lemma 7.2 hold, where $\{s_n\}$, $\delta$, $\alpha$, $s^\star$, $s^{\star\star}$ are given in Lemma 7.2 with*

$$c = 1 + K_0 \, s^{\star\star}. \tag{7.149}$$

*Then, sequence* $\{y_n\}$ $(n \geq 0)$ *generated by ITSNLM is well defined, remains in* $\overline{U}(x_0, s^\star)$ *for all* $n \geq 0$ *and converges to a solution* $x^\star \in \overline{U}(x_0, s^\star)$ *of equation* $F(x) = 0$. *Moreover, the following estimates hold*

$$\| y_n - x_n \| \leq s_n - t_n, \tag{7.150}$$

$$\| x_{n+1} - y_n \| \leq t_{n+1} - s_n, \tag{7.151}$$

$$\| x_{n+1} - x_n \| \leq t_{n+1} - t_n, \tag{7.152}$$

$$\| y_{n+1} - y_n \| \leq s_{n+1} - s_n, \tag{7.153}$$

$$\| y_n - x^\star \| \leq s^\star - s_n \tag{7.154}$$

*and*

$$\| x_n - x^\star \| \leq s^\star - t_n. \tag{7.155}$$

*Furthermore, if there exists* $R \geq s^\star$ *such that*

$$\overline{U}(x_0, R) \subseteq \mathcal{D} \tag{7.156}$$

*and*

$$K\,(s^\star + R) < 2\,(1 - \mu), \tag{7.157}$$

*then,* $x^\star$ *is the unique solution of (2.1) in* $\overline{U}(x_0, R)$.

**Proof** We shall use mathematical induction to show (7.150)–(7.155) hold for all $n \geq 1$. Estimate (7.150) holds for $n = 0$ by (7.126) and (7.142). We also have that $y_0 \in \overline{U}(x_0, s^\star)$ since $s^\star \geq \eta$. It also follows from (7.126), (7.132) and (7.133) that

$$t_0 \leq s_0 \leq t_1 \leq s_1 \leq s^\star.$$

We get

$$\| x_1 - y_0 \| = \| z_0 \| \leq a\, \| y_0 - x_0 \|^{1+b} \leq a\,(s_0 - t_0)^{1+b} = t_1 - s_0$$

and

$$\| x_1 - x_0 \| \leq \| x_1 - y_0 \| + \| y_0 - x_0 \| \leq t_1 - s_0 + s_0 - t_0 = t_1 - t_0 \leq s^\star. \tag{7.158}$$

That is, $x_1 \in \overline{U}(x_0, s^\star)$ and (7.151), (7.152) hold for $n = 0$. Let us assume $x_{k+1} \in \overline{U}(x_0, s^\star)$ and (7.151), (7.152) hold for $k + 1 < n$. Using (7.143) for $x = x_{k+1}$, we get that

$$\begin{aligned} \| A(x_0)^{-1}\,(A(x_{k+1}) - A(x_0)) \| &\leq L\, \| x_{k+1} - x_0 \| + \ell \\ &\leq L\, t_{k+1} + \ell \leq L\, s^\star + \ell < 1 \quad \text{(by (7.139))}. \end{aligned} \tag{7.159}$$

It follows from (7.159) and the Banach Lemma on invertible operators that $A(x_{k+1})^{-1}$ exists, so that

$$\| A(x_{k+1})^{-1}\, F'(x_0) \| \leq \frac{1}{1 - L\, \| x_{k+1} - x_0 \| - \ell} \leq \frac{1}{1 - L\, t_{k+1} - \ell}. \tag{7.160}$$

In view of ITSNLM, we obtain the approximation

$$\begin{aligned} F(x_{k+1}) &= F(x_{k+1}) - F(y_k) - F'(y_k)\,(x_{k+1} - y_k) \\ &\quad + F(y_k) + F'(y_k)\,(x_{k+1} - y_k) - F(x_k) - A(x_k)\,(y_k - x_k) \\ &\quad - F'(x_k)\,(y_k - x_k) + F'(x_k)\,(y_k - x_k) \qquad (7.161) \\ &= (F(x_{k+1}) - F(y_k) - F'(y_k)\,(x_{k+1} - y_k)) \\ &\quad + (F(y_k) - F(x_k) - F'(x_k)\,(y_k - x_k)) \\ &\quad + (F'(x_k) - A(x_k))\,(y_k - x_k) + F'(y_k)\,(x_{k+1} - y_k) \\ &= \int_0^1 (F'(y_k + \theta\,(x_{k+1} - y_k)) - F'(y_k))\,(x_{k+1} - y_k)\,d\theta \\ &\quad + \int_0^1 (F'(x_k + \theta\,(y_k - x_k)) - F'(x_k))\,(y_k - x_k)\,d\theta \\ &\quad + (F'(x_k) - A(x_k))\,(y_k - x_k) + (F'(x_k) - F'(x_0) + F'(x_0))\,(x_{k+1} - y_k) \end{aligned}$$

Using (7.143)–(7.145), the triangle inequality and the induction hypotheses, we get

$$\begin{aligned} &\| A(x_0)^{-1}\,F(x_{k+1})\,\| \\ &\le \frac{K}{2}\,\|\,x_{k+1} - y_k\,\|^2 + \frac{K}{2}\,\|\,y_k - x_k\,\|^2 + (M\,\|\,x_k - x_0\,\| + \mu)\,\|\,y_k - x_k\,\| \\ &\quad + (K_0\,\|\,y_k - x_0\,\| + 1)\,\|\,x_{k+1} - y_k\,\| \\ &\le \frac{K}{2}\,(t_{k+1} - s_k)^2 + \frac{K}{2}\,(s_k - t_k)^2 + (M\,t_k + \mu)\,(s_k - t_k) + (1 + K_0\,s_k)\,(t_{k+1} - s_k) \\ &\le \frac{K}{2}\,(t_{k+1} - s_k)^2 + \frac{K}{2}\,(s_k - t_k)^2 + (M\,t_k + \mu)\,(s_k - t_k) + 2\,c\,(t_{k+1} - s_k) \\ &\quad \text{(by (7.133) and (7.149)).} \qquad (7.162) \end{aligned}$$

In view of ITSNLM, (7.126), (7.160) and (7.162), we get that

$$\begin{aligned} &\|y_{k+1} - x_{k+1}\,\| = \|\,(A(x_{k+1})^{-1}\,A(x_0))\,(A(x_0)^{-1}\,F(x_{k+1}))\,\| \\ &\le \|\,A(x_{k+1})^{-1}\,A(x_0)\,\|\,\|\,A(x_0)^{-1}\,F(x_{k+1})\,\| \\ &\le \frac{K\,(t_{k+1} - s_k)^2 + K\,(s_k - t_k)^2 + 2\,(M\,t_k + \mu)\,(s_k - t_k) + 2\,(1 + K_0\,s_k)\,(t_{k+1} - s_k)}{2\,(1 - \ell - L\,t_{k+1})} \\ &= s_{k+1} - t_{k+1}, \qquad (7.163) \end{aligned}$$

which shows (7.150) for all $k \ge 1$. We also have that

$$\begin{aligned} \|\,x_{k+1} - y_k\,\| &= \|\,z_k\,\| \le a\,\|\,y_k - x_k\,\|^{1+b} \le a\,(s_k - t_k)^{1+b} = t_{k+1} - s_k, \\ \|\,x_{k+1} - x_k\,\| &\le \|\,x_{k+1} - y_k\,\| + \|\,y_k - x_k\,\| \\ &\le t_{k+1} - s_k + s_k - t_k = t_{k+1} - t_k, \qquad (7.164) \end{aligned}$$

$$\begin{aligned}\| x_{k+1} - x_0 \| &\leq \sum_{i=0}^{k} \| x_{i+1} - x_i \| \leq \sum_{i=0}^{k} (t_{i+1} - t_i) = t_{k+1} \leq s^\star, \\ \| y_{k+1} - x_k \| &\leq \| y_{k+1} - x_{k+1} \| + \| x_{k+1} - x_k \| \\ &\leq s_{k+1} - t_{k+1} + t_{k+1} - t_k = s_{k+1} - t_k \end{aligned} \tag{7.165}$$

and

$$\| y_{k+1} - x_0 \| \leq \| y_{k+1} - x_{k+1} \| + \| x_{k+1} - x_0 \| \leq s_{k+1} - t_{k+1} + t_{k+1} - t_0 \leq s^\star,$$

which complete the induction. In view of Lemma 7.2, sequence $\{s_n\}$ is a Cauchy sequence. It then follows from (7.150)–(7.153) that $\{y_n\}$ $(n \geq 0)$ is a Cauchy sequence too in a Banach space $\mathcal{X}$ and as such it converges to some $x^\star \in \overline{U}(x_0, s^\star)$ (since $\overline{U}(x_0, s^\star)$ is a closed set). By letting $k \longrightarrow \infty$ in (7.162) and noticing that $s_k \leq s^{\star\star}$, we obtain $F(x^\star) = 0$. Estimates (7.154) and (7.155) follow from (7.150)–(7.153) by using standard majorization techniques.

We shall finally show the uniqueness part. Let $y^\star \in \overline{U}(x_0, R)$ be a solution of $F(x) = 0$ and set $\mathcal{Q} = \int_0^1 F'(y^\star + \theta\,(x^\star - y^\star))\, d\theta$. Using (7.144), (7.145), (7.156), (7.157) and (7.157), we have in turn that

$$\begin{aligned} &\| A(x_0)^{-1}\, (\mathcal{Q} - A(x_0)) \| \\ &\leq \left\| A(x_0)^{-1} \int_0^1 (F'(y^\star + \theta\ (x^\star - y^\star)) - F'(x_0))\ d\theta \right\| \\ &\quad + \| A(x_0)^{-1}\, (F'(x_0) - A(x_0)) \| \\ &\leq K \int_0^1 \| y^\star + \theta\,(x^\star - y^\star) - x_0 \|\ d\theta + \mu \\ &\leq K \int_0^1 (\theta\ \| x^\star - x_0 \| + (1-\theta)\ \| y^\star - x_0 \|)\, d\theta + \mu \\ &\leq \frac{K}{2}\ (s^\star + R) + \mu < 1. \end{aligned} \tag{7.166}$$

It follows from (7.166) and the Banach Lemma on invertible operators that $\mathcal{Q}^{-1}$ exists. By (7.165) and the identity $0 = F(x^\star) - F(y^\star) = \mathcal{Q}\,(x^\star - y^\star)$, we conclude $x^\star = y^\star$. This completes the proof of Theorem 7.7. ⊠

**Remark 7.5** *(a) Note that $s^{\star\star}$ given in closed form by (7.129) can replace $s^\star$ in condition (7.148).*

*(b) If we assume*

$$\| A(x_0)^{-1}\ F'(x) \| \leq c_0 \quad (x \in \mathcal{D}), \tag{7.167}$$

*then, in view of (7.149) and (7.163), $c_0$ can replace $c$ in all the above results.*

(c) *It follows from (7.163) that tighter than $\{s_n\}$ majorizing sequence $\{\overline{s}_n\}$ given by*

$$\overline{t}_0 = 0,\ \overline{s}_0 \geq \eta,\ \overline{t}_{n+1} = \overline{s}_n + a\,(\overline{s}_n - \overline{t}_n)^{1+b},$$
$$\overline{s}_{n+1} = \overline{t}_{n+1} + \frac{\overline{\varrho_n}}{2\,(1 - L\,\overline{t}_{n+1} - \ell)}, \tag{7.168}$$

*where*

$$\overline{\varrho_n} = K\,(\overline{t}_{n+1} - \overline{s}_n)^2 + K\,(\overline{s}_n - \overline{t}_n)^2 + 2\,(M\,\overline{t}_n + \mu)\,(\overline{s}_n - \overline{t}_n) + 2\,(1 + K_0\,\overline{s}_n)\,(\overline{t}_{n+1} - \overline{s}_n)$$

*can be used in Theorem 7.7.*

(d) *The sufficient convergence conditions (see e.g. (7.127) and (7.128)) introduced here are based on our new idea of recurrent functions and they differ from by the corresponding ones given us in (cf. [69], [135]), where a Kantorovich–type analysis was used. In practice, we will test these conditions and apply the ones that are satisfied (if any). In the case that both set of conditions are satisfied and ($A(x) = F'(x)$ ($x \in \mathcal{D}$), $\ell = \mu = 0$, $L = K_0$, $M = 0$) we shall use the error bounds of this paper since they are always at least as tight. Indeed, note that $K$ was used instead of $K_0$ for the computations of the upper bounds on $\| F'(x_n)^{-1}\, F'(x_0) \|$ ($n \geq 0$). However,*

$$K_0 \leq K \tag{7.169}$$

*holds in general and $\dfrac{K}{K_0}$ can be arbitrarily large. If strict inequality holds in (7.169), the error bounds of this paper are tighter than the ones in (cf. [579]). If $K_0 = K$, the error estimates in this section and in (cf. [579]) are the same (see also Remark 7.6 and Example 7.6).*

(e) *Note that in case (for special choices of sequence $\{z_n\}$), (see also the introduction, Lemma 2.11, Theorem 7.8 and Remark 7.6), our method ITSNLM reduces to earlier ones, then, we proceed as in (d) above.*

(f) *According to the proof of Theorem 7.7, sequence $\{z_n\}$ does not have to be included in $\mathcal{D}$. However, a popular choice for $z_n$ is given by*

$$z_n = x_n + \xi_n\,(y_n - x_n) \quad (\xi_n \in [0,1]).$$

We provide numerical examples and special cases.

**Example 7.4 Case $z_n \neq 0$ and $A(x) = F'(x)$ ($x \in \mathcal{D}$).** *Let $\mathcal{X} = \mathcal{Y} = \mathcal{C}[a,b]$, $\mathcal{D} = U(1,1)$ and define an operator $\mathcal{P}$ on $\mathcal{D}$ by*

$$\mathcal{P}(x(s)) = \lambda\, x(s) \int_0^1 \mathcal{T}(s,t)\, x(t)\, dt - x(s) + y(s). \tag{7.170}$$

*Note that every zero of $\mathcal{P}$ satisfies the equation*

$$x(s) = y(s) + \lambda\, x(s) \int_0^1 \mathcal{T}(s,t)\, x(t)\, dt. \tag{7.171}$$

*Assume that $\lambda$ is a real number and the kernel $\mathcal{T}(s,t)$ is a continuous function in two variables $s$, $t$, satisfying, for all $(s,t) \in [0,1]^2$,*

*(i)* $0 < \mathcal{T}(s,t) < 1$,
*(ii)* $\mathcal{T}(s,t) + \mathcal{T}(t,s) = 1$

*The space $\mathcal{X}$ is equipped with the max–norm. That is,*

$$\| x \| = \max_{0 \le s \le 1} |x(s)|.$$

*Let us assume for simplicity that*

$$\mathcal{T}(s,t) = \frac{s}{s+t} \qquad ((s,t) \in [0,1]^2). \tag{7.172}$$

*Choose $x_0(s) = y(s) = 1$ for all $s \in [0,1]$, $\lambda = .25$ and*

$$z_n = \frac{1}{100} F''(x_n)\,(y_n - x_n)^2, \tag{7.173}$$

*where $F''$ is the second Fréchet–derivative of operator $F$ (cf. [163]). Note that function $\mathcal{T}$ given by (7.172) satisfies conditions (i) and (ii). Then, using (7.140)–(7.149), (7.122)–(7.125), (7.127) and (7.128), we obtain*

$$\| \mathcal{P}'(x_0(s))^{-1} \| \le 1.53039421,$$

$$K_0 = K = 2\,|\lambda| \max_{0 \le s \le 1} \left| \int_0^1 \frac{s}{s+t} dt \right| \|\mathcal{P}'(x_0(s))^{-1}\| = 2\,|\lambda| \ln 2\, \|\mathcal{P}'(x_0(s))^{-1}\|$$
$$= .530394215,$$

$$\eta = \| \mathcal{P}'(x_0(s))^{-1}\, \mathcal{P}(x_0(s)) \| \ge |\lambda| \ln 2\, \| \mathcal{P}'(x_0(s))^{-1} \| = .265197107,$$

$$b = 1, \quad a = \frac{1}{100} \| \mathcal{P}''(x) \| = \frac{2 \ln 2\, |\lambda|}{100} = .0034657359 \quad (x \in \mathcal{D}),$$

$$\alpha = .499423497, \quad s^{\star\star} = .663453567, \quad c = 1.351891934,$$

$$\gamma = .283591402, \quad \beta = \delta = .283770148$$

*and*

$$\delta = .283770148 \le 2\,\alpha = .998846994.$$

*Moreover, with $s^{\star\star}$ replacing $s^\star$ in (7.157), we get*

$$s^{\star\star} \le R < \frac{2}{K_0} - s^{\star\star} = 3.107326625. \tag{7.174}$$

*That is, all hypotheses of Theorem 7.7 are satisfied. Hence, sequence $\{x_n\}$ converges to a unique solution $x^\star$ in $\mathcal{D}$ (by (7.174)) of equation (7.171), so that error estimates (7.150)–(7.155) hold with $\{s_n\}$, $s^\star$ or $\{\overline{s}_n\}$, $\overline{s}^\star = \lim_{n\to\infty} \overline{s}_n$, respectively.*

**Example 7.5 Case $z_n = 0$ and $A(x) = F'(x)$, ($x \in \mathcal{D}$) (Newton's method)**. *In this case, we set $a = 0$ to obtain NM given by (2.2) and*

$$t_0 = 0, \quad t_1 = \eta, \quad t_{n+2} = t_{n+1} + \frac{K(t_{n+1} - t_n)^2}{2(1 - K_0 t_{n+1})} \quad (n \geq 0). \tag{7.175}$$

*Lemma 7.2 and Theorem 7.7 reduce to Lemma 2.11 and Theorem 7.8, respectively:*

**Theorem 7.8** *(cf. [196]) Let $F : \mathcal{D} \subseteq \mathcal{X} \longrightarrow \mathcal{Y}$ be a Fréchet–differentiable operator. Assume there exist $x_0 \in \mathcal{D}$ and constants $K_0 > 0$, $K > 0$, $\eta \geq 0$, such that for all $x, y \in \mathcal{D}$, we have hypotheses (7.141)–(7.143) hold,*

$$\overline{U}(x_0, t^\star) \subseteq \mathcal{D}$$

*and hypothesis (2.443) of Lemma 2.11 holds. Then, sequence $\{x_n\}$ $(n \geq 0)$ generated by (2.2) is well defined, remains in $\overline{U}(x_0, t^\star)$ for all $n \geq 0$ and converges to a solution $x^\star \in \overline{U}(x_0, t^\star)$ of equation $F(x) = 0$. Moreover, the following estimates hold*

$$\| x_{n+1} - x_n \| \leq t_{n+1} - t_n \tag{7.176}$$

*and*

$$\| x_n - x^\star \| \leq t^\star - t_n,$$

*where $\{t_n\}$ and $t^\star$ are given in Lemma 2.11. Furthermore, if there exists $R \geq t^\star$ such that*

$$\overline{U}(x_0, R) \subseteq \mathcal{D}$$

*and*

$$K_0\,(t^\star + R) < 2,$$

*then, $x^\star$ is the unique solution of equation (2.1) in $\overline{U}(x_0, R)$.*

**Remark 7.6** *(a) The famous for its simplicity and clarity Kantorovich condition corresponding to (2.443) is given by*

$$h_K = K\eta \leq \frac{1}{2} \qquad \text{(cf. [163], [194], [469])}. \tag{7.177}$$

*The majorizing sequence corresponding to (7.177) is*

$$\overline{t}_0 = 0, \quad \overline{t}_1 = \eta, \quad \overline{t}_{n+2} = \overline{t}_{n+1} + \frac{K\,(\overline{t}_{n+1} - \overline{t}_n)^2}{2\,(1 - K\,\overline{t}_{n+1})} \quad (n \geq 0). \tag{7.178}$$

*If $K_0 = K$, Lemma 2.11 and Theorem 7.8 reduce to the corresponding ones given by Kantorovich and others (cf. [469]). Otherwise (i.e. $K_0 < K$), the sufficient convergence conditions are always weaker and the error estimates are tighter (cf. [139], [196]). Numerical examples for this case have been already given in (cf. [163], [194], [200]) and section 2.1.*

(b) *A simple induction argument shows*

$$t_n \le \bar{t}_n, \tag{7.179}$$

$$t_{n+1} - t_n \le \bar{t}_{n+1} - \bar{t}_n \tag{7.180}$$

*and*

$$t^\star \le \bar{t}^\star = \lim_{n \longrightarrow \infty} \bar{t}_n,$$

*where* $\{t_n\}$ *and* $t^\star$ *are given in Lemma 2.11. But if* $K_0 < K$, *inequalities (7.179) and (7.180) are strict.*

Next, we provide a numerical example for NM (2.2), where $K_0 < K$, (2.443) holds, but (7.177) fails in the first part. We also compare in the second part of this example our error bounds $t_{n+1} - t_n$ and $\bar{t}_{n+1} - \bar{t}_n$ in the case where the both conditions (2.443) and (7.177) hold.

**Example 7.6** *Consider Example 2.11.*

(a) *Let* $x_0 = (1,1)^T$. *We have the same as (a) of Example 2.11.*

(b) *Let* $x_0 = (.9, .9)^T$ *and* $\beta = .7$. *Using hypotheses of Theorem 7.8, we get that*

$$\eta = .1, \quad K = 2.6, \quad K_0 = 2.3, \quad \overline{K} = 2.39864766,$$

$$h_K = .26 \quad \text{and} \quad h_A = .239864766.$$

*Then, (7.177) and (2.443) are satisfied. The hypotheses of our Theorem 7.8 and the Kantorovich theorem are satisfied. Then NM (2.2) converges to* $x^\star = (.8879040017, .8879040017)^T$, *starting at* $x_0$. *We also can provide the comparison table using the software Maple* 13.

**Comparison table.**

| $n$ | (2.2) $\| x_{n+1} - x_n \|$ | (7.176) $t_{n+1} - t_n$ | (7.180) $\bar{t}_{n+1} - \bar{t}_n$ |
|---|---|---|---|
| 0 | .01193415638 | .1 | .1 |
| 1 | .0001618123748 | .01688311688 | .01756756757 |
| 2 | 2.946995467e-8 | .0005067933842 | .0005778355237 |
| 3 | 4.228114294e-11 | 4.57383463010e-7 | 6.265131450e-7 |
| 4 | ∼ | 3.725461907e-13 | 7.365175646e-13 |
| 5 | ∼ | 2.471607273e-25 | 1.017862116e-24 |
| 6 | ∼ | 1.087872853e-49 | 1.944019580e-48 |
| 7 | ∼ | 2.107538365e-98 | 7.091269701e-96 |
| 8 | ∼ | 7.909885354e-196 | 9.435626465e-191 |
| 9 | ∼ | 1.114190851e-390 | 1.670568212e-380 |
| 10 | ∼ | 2.210743650e-780 | 5.236621208e-760 |

*The table shows that our error bounds (7.176) are tighter than* $\bar{t}_{n+1} - \bar{t}_n$ *given by Kantorovich's Theorem (cf. [163], [194], [469]).*

**Remark 7.7** *The results obtained in this section extend in the case (2.388), where $F$ is as in the introduction and $G : \mathcal{D} \longrightarrow \mathcal{Y}$ is a continuous operator, satisfying*

$$\| A(x_0)^{-1} (G(x) - G(y)) \| \leq N \| x - y \| \quad ((x, y) \in \mathcal{D}^2). \tag{7.181}$$

*Condition (7.181) implies the continuity but not necessarily the differentiability of operator $G$. The iteration corresponding to (2.388) is given by*

$$\begin{aligned} y_n &= x_n - A(x_n)^{-1} (F(x_n) + G(x_n)) \quad (n \geq 0, \quad x_0 \in \mathcal{D}), \\ x_{n+1} &= y_n - z_n. \end{aligned} \tag{7.182}$$

*The identity corresponding to (7.161) is given by*

$$\begin{aligned} & F(x_{k+1}) + G(x_{k+1}) \\ & = F(x_{k+1}) - F(y_k) - F'(y_k)(x_{k+1} - y_k) \\ & \quad + F(y_k) + F'(y_k)(x_{k+1} - y_k) - F(x_k) - A(x_k)(y_k - x_k) \\ & \quad - F'(x_k)(y_k - x_k) + F'(x_k)(y_k - x_k) + G(x_{k+1}) - G(x_k) \\ & = (F(x_{k+1}) - F(y_k) - F'(y_k)(x_{k+1} - y_k)) \\ & \quad + (F(y_k) - F(x_k) - F'(x_k)(y_k - x_k)) \\ & \quad + (F'(x_k) - A(x_k))(y_k - x_k) + F'(y_k)(x_{k+1} - y_k) \\ & \quad + G(x_{k+1}) - G(x_k) \\ & = \int_0^1 (F'(y_k + \theta\, (x_{k+1} - y_k)) - F'(y_k))\, (x_{k+1} - y_k)\, d\theta \\ & \quad + \int_0^1 (F'(x_k + \theta\, (y_k - x_k)) - F'(x_k))\, (y_k - x_k)\, d\theta \\ & \quad + (F'(x_k) - A(x_k))(y_k - x_k) \\ & \quad + (F'(x_k) - F'(x_0) + F'(x_0))(x_{k+1} - y_k) + G(x_{k+1}) - G(x_k) \end{aligned}$$

*Using (7.143)–(7.145), the triangle inequality and the induction hypotheses, we get*

$$\begin{aligned} & \|A(x_0)^{-1} (F(x_{k+1}) + G(x_{k+1}))\| \\ & \leq \frac{K}{2} \| x_{k+1} - y_k \|^2 + \frac{K}{2} \| y_k - x_k \|^2 + (M \| x_k - x_0 \| + \mu) \| y_k - x_k \| \\ & \quad + (K_0 \| y_k - x_0 \| + 1) \| x_{k+1} - y_k \| + 2\, N \| x_{k+1} - x_k \| \\ & \leq \frac{K}{2} (t_{k+1} - s_k)^2 + \frac{K}{2} (s_k - t_k)^2 + (M\, t_k + \mu)(s_k - t_k) \\ & \quad + (1 + K_0\, s_k)(t_{k+1} - s_k) + 2\, N\, (t_{k+1} - t_k) \end{aligned}$$

$$\leq \frac{K}{2}\,(t_{k+1}-s_k)^2+\frac{K}{2}\,(s_k-t_k)^2+(M\,t_k+\mu)\,(s_k-t_k)+2c\,(t_{k+1}-s_k)$$
$$+\,2N\,(t_{k+1}-t_k).$$

*But, since*

$$\begin{aligned}\|\, x_{k+1}-x_k \,\| &= \|\, y_k-x_k+z_k \,\|\\ &\leq \|\, y_k-x_k \,\| + \|\, z_k \,\|\\ &\leq s_k-t_k+a\,(s_k-t_k)^{1+b}=(1+a\,(s_k-t_k)^b)\,(s_k-t_k),\end{aligned}$$

*the majorizing sequence should given by*

$$t_0=0,\ s_0=\eta,\ t_{n+1}=s_n+a\,(s_n-t_n)^{1+b},$$
$$s_{n+1}=t_{n+1}+\frac{\overline{\overline{\varrho_n}}}{2\,(1-L_0\,t_{n+1})}$$

*with*

$$\overline{\overline{\varrho_n}}=K\,(t_{n+1}-s_n)^2+K\,(s_n-t_n)^2+2\,c\,(t_{n+1}-s_n)+$$
$$2\,N\,(t_{n+1}-t_n)+(2\,M\,t_k+\mu)\,(s_k-t_k),$$

*whereas the term* $2\,a\,c\,\eta^b$ *in (7.123) and (7.124) should be*

$$2\,a\left(1+K_0\,s^{\star\star}+\frac{N}{a\,\eta^b}\,(1+a\,\eta^b)\right)\eta^b$$

*if* $a\neq 0$ *and* $\eta\neq 0$ *and* $2\,N$ *if* $a=0$. *(similar changes for majorizing sequence* $\{\overline{s}_n\}$*). Then, with the above changes, the conclusions of all the results obtained here hold with equation (2.1) replaced by (2.388) (with the exception of the uniqueness part in Theorems 7.7 and 7.8).*

## 7.4 Zabrejko–Zincenko–Type Conditions

In this section, we provide a semilocal convergence analysis for Newton–type methods using our idea of recurrent functions in a Banach space setting. As in (cf. [281], [806], [813]), we use Zabrejko–Zincenko conditions. In particular, we show that the convergence domains given in (cf. [163]) can be extended under the same computational cost.

We shall use the Newton–type method NTM:

$$\begin{aligned}y_{n+1}&=y_n-A(y_n)^{-1}\ P(y_n)\quad (n\geq 0,\ y_0\in\mathcal{D}),\\ P(x)&=F(x)+G(x)\ (x\in\mathcal{D}),\end{aligned}\tag{7.183}$$

where $A(x)\in\mathcal{L}(\mathcal{X},\mathcal{Y})$ to generate a sequence approximating $x^\star$ an unique solution of (2.388).

(a) If

$$A(x) = F'(x) \qquad (x \in \mathcal{D}), \tag{7.184}$$

we obtain the Zabrejko–Nguen iteration (cf. [806]) given by

$$y_{n+1} = y_n - F'(y_n)^{-1}\ P(y_n) \quad (n \geq 0,\ y_0 \in \mathcal{D}). \tag{7.185}$$

(b) If

$$A(x) = F'(x) + [x, y; G] \qquad (x \in \mathcal{D}), \tag{7.186}$$

where $[x, y; F]$ is a divided difference of order one for operator $G$, then, we obtain an iteration faster than (7.185), first considered by Cătinaş (cf. [261]) in the following form

$$y_{n+1} = y_n - (F'(y_n) + [y_n, y_{n-1}; G])^{-1}\ P(y_n) \quad (n \geq 0,\ y_{-1}, y_0 \in \mathcal{D}). \tag{7.187}$$

(c) If

$$A(x) = F'(x), \quad G(x) = 0 \qquad (x \in \mathcal{D}), \tag{7.188}$$

NTM reduces to Newton's method given by (2.2).

Several other choices are possible (cf. [163], [195]). A local as well as a semilocal convergence analysis for all these methods has been provided by many authors under Lipschitz–type conditions (cf. [86], [139], [281], [813]). A survey of such results can be found in (cf. [163]), and the references there. We also refer the reader to the elegant related works by Proinov (cf. [613], [614]), and Ezquerro/Hernández (cf. [355]) whose works are also improved here in at least the Newton's method case.

Let $x^0 \in \mathcal{D}$ and $R > 0$ be such that

$$U(x^0, R) \subseteq \mathcal{D}. \tag{7.189}$$

Chen and Yamamoto (cf. [281]) provided a semilocal convergence for NTM for $y^0 \in U(x^0, R)$ under the conditions $(\mathcal{C})$ given by

$A(x^0)$ exists and for any $x$, $y \in \overline{U}(x^0, R)$, we have

$$\| A(x^0)^{-1}\ (A(x) - A(x^0))\ \| \leq \overline{\nu_0}(\| x - x^0 \|) + a,$$

$$\| A(x^0)^{-1}\ (F'(x + t\,(x - y)) - A(x))\ \|$$

$$\leq \overline{\nu}(\| x - x^0 \| + t\ \| y - x \|) - \overline{\nu_0}(\| x - x^0 \|) + b \quad (t \in [0, 1])$$

and

$$\| A(x^0)^{-1}\ (G(x) - G(y))\ \| \leq \omega(r)\ \ \| x - y \|,$$

where $\overline{\nu}(r + t) - \overline{\nu_0}(r)$, $t \geq 0$ and $\omega(r)$ are non–decreasing, non–negative functions with

$$\omega(0) = \overline{\nu_0}(0) = \overline{\nu}(0) = 0,$$

$\overline{\nu_0}(r)$ is differentiable, $\overline{\nu_0}'(r) > 0$ at every point of $[0, R]$ and the constants $a$, $b$ satisfy $a \geq 0$, $b \geq 0$ and $a + b < 1$. Set

$$\| A(x^0)^{-1} \, P(x^0) \| \leq \eta,$$

$$\phi(r) = \eta - r + \int_0^1 \overline{\nu}(t) \, dt,$$

$$\psi(r) = \int_0^r \omega(t) \, dt,$$

$$\chi(r) = \phi(r) + \psi(r) + (a + b) \, r.$$

Further, assume

$$\chi(R) \leq 0$$

and define scalar sequence $\{s_n\}$ by

$$s_0 \in [0, R], \qquad s_{n+1} = s_n + \frac{u(s_n)}{p(s_n)} \quad (n \geq 0),$$

where

$$u(r) = \chi(r) - \chi^\star, \qquad p(r) = 1 - \overline{\nu_0}(r) - a,$$

$\chi^\star$ is the minimal value of $\chi(r)$ in $[0, R]$ and $s^\star$ denotes the minimal point. Moreover, $t^\star$ denotes the unique zero of $\chi$ in $(0, s^\star]$. Under these assumptions, there exists a unique solution $x^\star$ in $\in \overline{U}(x^0, t^\star)$ such that

$$\| y_{n+1} - y_n \| \leq s_{n+1} - s_n$$

and

$$\| x^\star - y_n \| \leq s^\star - s_n.$$

We shall use the more general set of conditions $(\mathcal{H})$ given by

$(\mathcal{H}_1)$ $A(x^0)$ exists and, for any $x$, $y \in \overline{U}(x^0, R)$ $(0 < r \leq R)$:

$$\| A(x^0)^{-1} \, (A(x) - A(x^0)) \| \leq \nu_0(\| x - x^0 \|) + a,$$

$(\mathcal{H}_2)$

$$\begin{aligned} &\| A(x^0)^{-1} \, (F'(x + t\,(x - y)) - A(x)) \| \\ &\leq \nu(t \, \| y - x \|) + \omega_0(\| x - x^0 \|) + b \quad (t \in [0, 1]), \end{aligned}$$

$(\mathcal{H}_3)$

$$\| A(x^0)^{-1} \, (G(x) - G(y)) \| \leq \omega(r) \;\; \| x - y \|$$

and

$(\mathcal{H}_4)$

$$\overline{U}(x^0, R) \subseteq \mathcal{D},$$

where $\nu_0$, $\nu$ and $\omega_0$ are non–decreasing, non–negative functions on $[0, R]$ with

$$\nu_0(0) = \nu(0) = \omega_0(0) = \omega(0) = 0$$

and constants $a$ and $b$ are non–negative. Some more hypotheses are given in Lemma 7.3.

A semilocal convergence analysis is provided under the $(\mathcal{H})$ conditions. We also compare the two sets of hypotheses. Numerical examples are also provided to show how we can solve equations in cases not covered before (cf. [86], [813]).

We need to define some parameters, functions and sequences.

**Definition 7.1** *Let $y_0 \in U(x^0, r)$. Define parameters $r_0$, $r_1$, iteration $\{r_n\}$, functions $f_n$, $h_n$, $p_n$ on $[0,1)$ and $q \in I_q = [0,1)^2 \times [0, r_1 - r_0] \times [r_1 - r_0, \frac{r_1 - r_0}{1-s}]^3$ $(s \in [0,1))$ by*

$$r_0 \geq \| y_0 - x^0 \|, \quad r_1 > r_0 + \| A(y_0)^{-1} \, (F(y_0) + G(y_0)) \|,$$

$$r_{n+1} = r_n + \frac{\displaystyle\int_0^1 \nu(t\,(r_n - r_{n-1}))\,dt + \omega_0(r_n) + \omega(r_n) + b}{1 - a - \nu_0(r_n)} \, (r_n - r_{n-1}), \qquad (7.190)$$

$$f_n(s) = \int_0^1 \nu(t\,s^{n-1}\,(r_1 - r_0))\,dt + \omega_1((1 + s + \cdots + s^{n-1})\,(r_1 - r_0)) + c, \quad (7.191)$$

$$h_n(s) = \int_0^1 (\nu(t\,s^n\,(r_1 - r_0)) - \nu(t\,s^{n-1}\,(r_1 - r_0))\,dt +$$

$$\omega_1((1 + s + \cdots + s^n)\,(r_1 - r_0)) - \omega_1((1 + s + \cdots + s^{n-1})\,(r_1 - r_0)), \qquad (7.192)$$

*where*

$$c = b - \alpha\,(1 - a) \qquad \text{for some } \alpha \in (0,1), \qquad (7.193)$$

$$\omega_1(s) = \omega_0(s) + \omega(s) + \alpha\,\nu_0(s), \qquad (7.194)$$

$$\begin{aligned} p_n(s) = & \int_0^1 (\nu(t\,s^{n+1}\,(r_1 - r_0)) + \nu(t\,s^{n-1}\,(r_1 - r_0)) - 2\,\nu(t\,s^n\,(r_1 - r_0))\,dt \\ & + \omega_1((1 + s + \cdots + s^{n+1})\,(r_1 - r_0)) + \omega_1((1 + s + \cdots + s^{n-1})\,(r_1 - r_0)) \\ & - 2\,\omega_1((1 + s + \cdots + s^n)\,(r_1 - r_0)), \end{aligned} \qquad (7.195)$$

$$\begin{aligned} q(t, s, \lambda, \beta_0, \beta, \gamma_0) = & \int_0^1 (\nu(t\,\lambda\,s^2) + \nu(t\,\lambda) - 2\,\nu(t\,\lambda\,s))\,dt \\ & + \omega_1(\beta_0 + \beta + \gamma_0) + \omega_1(\beta) - 2\,\omega_1(\beta_0 + \beta). \end{aligned} \qquad (7.196)$$

*Define a function $f_\infty$ on $[0,1)$ by*

$$f_\infty(s) = \lim_{n \to \infty} f_n(s). \qquad (7.197)$$

*It follows from (7.191) and (7.197) that*

$$f_\infty(s) = \omega_1(\frac{r_1 - r_0}{1 - s}) + c. \tag{7.198}$$

*It can also easily be seen from (7.191), (7.192), (7.195) and (7.196) that the following hold*

$$f_{n+1}(s) = f_n(s) + h_n(s), \tag{7.199}$$

$$h_{n+1}(s) = h_n(s) + p_n(s), \tag{7.200}$$

*and*

$$q(t, s, s^{n-1}(r_1 - r_0), s^n(r_1 - r_0), (1 + s + \cdots + s^{n-1})(r_1 - r_0), s^{n+1}(r_1 - r_0)) = p_n(s). \tag{7.201}$$

We need the following result on majorizing sequences for NTM.

**Lemma 7.3** *Let constants $a$, $b$, parameters $r_0$, $r_1$, functions $\nu_0$, $\nu$, $\omega_0$, $\omega$ be as in the Introduction, and parameters $\alpha$, $c$, functions $\omega_1$, $f_n$, $h_n$, $p_n$, $q$ be as in Definition 7.1. Assume there exists $\alpha \in (0, 1)$ such that*

$$\nu_0(r_1) + a < 1, \tag{7.202}$$

$$\frac{\int_0^1 \nu(t(r_1 - r_0))\, dt + \omega_0(r_1) + \omega(r_1) + b}{1 - a - \nu_0(r_1)} \le \alpha, \tag{7.203}$$

$$c < 0, \tag{7.204}$$

$$q(t, s, \lambda, \beta_0, \beta, \gamma_0) \ge 0 \quad \text{on } I_q, \tag{7.205}$$

$$h_1(\alpha) \ge 0 \tag{7.206}$$

*and*

$$f_\infty(\alpha) \le 0. \tag{7.207}$$

*Then, scalar sequence $\{r_n\}$ $(n \ge 0)$ given by (7.190) is non–decreasing, bounded from above by*

$$r^{\star\star} = \frac{r_1 - r_0}{1 - \alpha} \tag{7.208}$$

*and converges to its unique least upper bound $r^\star$ satisfying $r^\star \in [0, r^{\star\star}]$. Moreover, the following estimates hold for all $n \ge 0$*

$$0 \le r_{n+1} - r_n \le \alpha\,(r_n - r_{n-1}) \le \alpha^n\,(r_1 - r_0) \tag{7.209}$$

*and*

$$r^\star - r_n \le \frac{r_1 - r_0}{1 - \alpha}\,\alpha^n. \tag{7.210}$$

**Proof** Estimate (7.209) is true if

$$0 \le \frac{\int_0^1 \nu(t\,(r_n - r_{n-1}))\,dt + \omega_0(r_n) + \omega(r_n)}{1 - a - \nu_0(r_n)} \le \alpha \tag{7.211}$$

holds for all $n \ge 1$. In view of (7.190), (7.202), (7.203), estimate (7.211) holds true for $n = 1$. We also have by (7.211) that $0 \le r_2 - r_1 \le \alpha\,(r_1 - r_0)$.

Let us assume that (7.209) and (7.211) hold for all $k \le n$. Then, we have that

$$r_n \le \frac{1-\alpha^n}{1-\alpha}\,(r_1 - r_0). \tag{7.212}$$

Using the induction hypotheses and (7.212), estimate (7.211) is true if

$$\int_0^1 \nu(t\,\alpha^{n-1}\,(r_1 - r_0))\,dt + \omega_1(\frac{1-\alpha^n}{1-\alpha}\,(r_1 - r_0)) + c \le 0, \tag{7.213}$$

where $c$, $\omega_1$ are given by (7.193) and (7.194), respectively. Estimate (7.213) (for $s = \alpha$) motivates us to introduce function $f_n$ given by (7.191) and show instead of (7.213):

$$f_n(\alpha) \le 0 \qquad (n \ge 1). \tag{7.214}$$

We have by (7.199)–(7.201) (for $s = \alpha$) and (7.206) that

$$f_{n+1}(\alpha) \ge f_n(\alpha) \qquad (n \ge 1). \tag{7.215}$$

In view of (7.197), (7.198) and (7.215), estimate (7.214) shall holds if (7.207) is true since

$$f_n(\alpha) \le f_\infty(\alpha) \qquad (n \ge 1). \tag{7.216}$$

The induction is completed. It follows that iteration $\{r_n\}$ is non–decreasing, bounded above by $r^{\star\star}$ (given by (7.208)) and as such it converges to $r^\star$.

Finally, estimate (7.201) follows from (7.209) by using standard majorization techniques. This completes the proof of Lemma 7.3. ⊠

The hypotheses $(\mathcal{H})$ and those of Lemma 7.3 will be called $(\mathcal{A})$.

We can show the main semilocal convergence result for NTM.

**Theorem 7.9** *Assume hypotheses $(\mathcal{A})$ hold. Then, sequence $\{y_n\}$ $(n \ge 0)$ generated by NTM is well defined, remains in $\overline{U}(x^0, r^\star)$ for all $n \ge 0$ and converges to a solution $x^\star \in \overline{U}(x^0, r^\star)$ of equation $F(x)+G(x) = 0$. Moreover, the following estimates hold for all $n \ge 0$*

$$\| \, y_{n+1} - y_n \, \| \le r_{n+1} - r_n \tag{7.217}$$

*and*

$$\| \, y_n - x^\star \, \| \le r^\star - r_n, \tag{7.218}$$

*where sequence $\{r_n\}$ $(n \geq 0)$ and $r^\star$ are given in Lemma 7.3.*

*Furthermore, if there exists*

$$R_0 \in [r^\star, R] \tag{7.219}$$

*such that*

$$\int_0^1 \nu(t\,(R_0 + r_0))\,dt + \omega_0(r^\star) + \omega(R_0) + \nu_0(r^\star) + a + b \leq 1, \tag{7.220}$$

*then, the solution $x^\star$ of equation (2.388) is unique in $U(x^0, R_0)$.*

**Proof** We shall show using induction:

$$\| y_n - y_{n-1} \| \leq r_n - r_{n-1} \tag{7.221}$$

and

$$\| y_n - x^0 \| \leq r_n. \tag{7.222}$$

Estimates (7.221) and (7.222) hold for $n = 1$, by (7.185) and (7.190). Assume (7.221) and (7.222) hold for $n \leq k$. Using $(\mathcal{H}_1)$ and (7.202), we get that

$$\begin{aligned} \| A(x_0)^{-1}\ [A(y_1) - A(x^0)] \| &\leq \nu_0(\| y_1 - x^0 \|) + a \\ &\leq \nu_0(r_1) + a < 1. \end{aligned} \tag{7.223}$$

It follows from (7.223) and the Banach lemma on invertible operators that $A(y_1)^{-1}$ exists and

$$\| A(y_1)^{-1}\ A(x^0) \| \leq (1 - a - \nu_0(r_1))^{-1}. \tag{7.224}$$

We also showed in Lemma 7.3 that

$$\nu_0(r_k) + a < 1. \tag{7.225}$$

It follows as in (7.223) with $r_k$, $y_k$ replacing $r_1$, $y_1$, respectively, that $A(y_k)^{-1}$ exists and

$$\| A(y_k)^{-1}\ A(x^0) \| \leq (1 - a - \nu_0(r_k))^{-1}. \tag{7.226}$$

Using (7.183), $(\mathcal{H}_2)$, $(\mathcal{H}_3)$, (7.190), (7.221), (7.222) and (7.226), we obtain in turn that

$$\begin{aligned} &\|y_{k+1} - y_k \| \\ &= \| A(y_k)^{-1}(F(y_k) + G(y_k)) \| \end{aligned}$$

$$\leq \| A(y_k)^{-1}A(x^0) \| \, \|A(x^0)^{-1}\big(F(y_k)+G(y_k)-A(y_{k-1})(y_k-y_{k-1})$$
$$-F(y_{k-1})-G(y_{k-1})\big)\|$$
$$\leq (1-a-\nu_0(r_k))^{-1}\Big(\int_0^1 \| A(x^0)^{-1}[F'(y_{k-1}+\theta\,(y_k-y_{k-1}))-A(y_{k-1})] \|$$
$$\times \| y_k-y_{k-1} \| \; d\theta + \| A(x^0)^{-1}(G(y_k)-G(y_{k-1})) \| \Big)$$
$$\leq (1-a-\nu_0(r_k))^{-1}\Big(\int_0^1 \nu(t \| y_k-y_{k-1} \|)dt$$
$$+\,\omega_0(\| y_k-x^0 \|)+\omega(\| y_k-x^0 \|)+b\Big) \| y_k-y_{k-1} \|$$
$$\leq (1-a-\nu_0(r_k))^{-1}\Big(\int_0^1 \nu(t\,(r_k-r_{k-1}))\,dt+\omega_0(r_k)+\omega(r_k)+b\Big)$$
$$\times (r_k-r_{k-1}) = r_{k+1}-r_k. \tag{7.227}$$

Moreover, we have

$$\| y_{k+1}-x^0 \| \leq \| y_{k+1}-y_k \| + \| y_k-x^0 \| \leq (r_{k+1}-r_k)+r_k = r_{k+1} \leq r^\star. \tag{7.228}$$

The induction for (7.221) and (7.222) is completed. In view of Lemma 7.3, (7.221) and (7.222), sequence $\{y_n\}$ $(n \geq 0)$ is a Cauchy sequence in a Banach space $\mathcal{X}$, and as such it converges to some $x^\star \in \overline{U}(x^0, r^\star)$ (since $\overline{U}(x^0, r^\star)$ is a closed set). Estimate (7.218) follows from (7.217) by using standard majorization techniques. Using (7.227), we obtain

$$\|A(x^0)^{-1}\,(F(x_k)+G(x_k))\, \|$$
$$\leq \Big(\int_0^1 \nu(t\,(r_k-r_{k-1}))\,dt+\omega_0(r_k)+\omega(r_k)+b\Big)\,(r_k-r_{k-1}). \tag{7.229}$$

By letting $k \longrightarrow \infty$ in (7.229), we obtain $F(x^\star)+G(x^\star)=0$.

Finally to show that $x^\star$ is the unique solution of equation (2.388) in $\overline{U}(x^0, R_0)$, let $y^\star \in \overline{U}(x^0, R_0)$ with $F(y^\star)+G(y^\star)=0$.

Using the approximation

$$y^\star - y_{k+1} = y^\star - y_k + A(y_k)^{-1}\,(F(y_k)+G(y_k)) - A(y_k)^{-1}\,(F(y^\star)+G(y^\star)) \tag{7.230}$$

as in (7.227), we obtain in turn that

$$\|y^\star - y_{k+1} \|$$
$$\leq (1-a-\nu_0(r_k))^{-1}\Big(\int_0^1 \| A(x^0)^{-1}\,[F'(y_k+\theta\,(y^\star-y_k))$$
$$-A(y_k)] \| \, \| y^\star-y_k \| \; d\theta + \| A(x^0)^{-1}\,(G(y^\star)-G(y_k)) \| \Big)$$

$$\leq (1 - a - \nu_0(r_k))^{-1} \left( \int_0^1 \nu(t \parallel y^\star - y_k \parallel) \, dt + \omega_0(\parallel y_k - x^0 \parallel) \right. \tag{7.231}$$

$$\left. + \, \omega(\parallel y_k - x^0 \parallel) + b \right) \parallel y^\star - y_k \parallel$$

$$< (1 - a - \nu_0(r^\star))^{-1} \left( \int_0^1 \nu(t \, (R_0 + r_0)) \, dt + \omega_0(r^\star) \right.$$

$$\left. + \, \omega(R_0) + b \right) \parallel y_k - y^\star \parallel \leq \parallel y_k - y^\star \parallel \quad \text{(by (7.220))}.$$

It follows by (7.231) that $\lim_{k \longrightarrow \infty} y_k = y^\star$. But we showed $\lim_{k \longrightarrow \infty} y_k = x^\star$. Hence, we deduce $x^\star = y^\star$. This completes the proof of Theorem 7.9. ⊠

**Application 7.1** *(Newton's method). Let $A(x) = F'(x)$ and $G(x) = 0$ ($x \in \mathcal{D}$).*
*Then, in the case of the ($\mathcal{C}$) conditions, we have that*

$$y_0 = x^0 = x_0, \quad s_0 = 0, \quad s_1 = \eta, \quad a = b = 0,$$

$$\overline{\nu_0}(r) = L_0 r, \quad \overline{\nu}(r) = L r, \quad \omega(r) = 0,$$

$$\chi(r) = \phi(r) = \frac{L}{2} r^2 - r + \eta, \quad \chi^\star = \frac{2L\eta - 1}{2L},$$

$$s^\star = \frac{1}{L}, \qquad t^\star = \frac{1 - \sqrt{1 - 2L\eta}}{L}, \quad R \in [t^\star, s^\star]$$

$$u(r) = \frac{1}{2L} (L \, r - 1)^2, \quad p(r) = 1 - L_0 \, r,$$

$$h_{CY} = L\eta \leq \frac{1}{2} \tag{7.232}$$

*and*

$$s_0 = 0, \quad s_1 = \eta,$$

$$s_{n+1} = s_n + \frac{(L \, s_n - 1)^2}{2L \, (1 - L_0 \, s_n)} = s_n + \frac{L \, (s_n - s_{n-1})^2}{2 \, (1 - L_0 \, s_n)} \qquad (n \geq 1). \tag{7.233}$$

*Moreover, in the case of the ($\mathcal{A}$) conditions, we have that*

$$y_0 = x^0 = x_0, \quad r_0 = 0, \quad r_1 = \eta, \quad a = b = 0,$$

$$\nu_0(r) = L_0 \, r, \quad \nu(r) = L \, r, \quad \omega(r) = \omega_0(r) = 0,$$

$$\alpha = \frac{2L}{L + \sqrt{L^2 + 8 \, L_0 \, L}}$$

$$h_{AH} = \overline{L} \, \eta \leq \frac{1}{2}, \tag{7.234}$$

*where* $\overline{L}$ *is given in Lemma 2.11 and*

$$s_0 = 0, \quad s_1 = \eta, \quad s_{n+1} = s_n + \frac{L\,(s_n - s_{n-1})^2}{2\,(1 - L_0\, s_n)} \qquad (n \geq 1). \tag{7.235}$$

*Note that*

$$L_0 \leq L \tag{7.236}$$

*holds in general and* $\dfrac{L}{L_0}$ *can be arbitrarily large. Let us now compare the results. It follows from (7.232), (7.234) and (7.236)*

$$h_{CY} \leq \frac{1}{2} \Longrightarrow h_{AH} \leq \frac{1}{2}, \tag{7.237}$$

*but not necessarily vice versa unless if* $L_0 = L$.

**Remark 7.8** *The convergence domains approach in (cf. [281]) do not necessarily produce the weakest sufficient convergence conditions not even in the simplest possible case of a Newton–like method which is NM (2.2). The recurrent functions approach produces sufficient convergence (7.234) that can always replace (7.232). Hence, the applicability of Newton's method has been extended, under the same hypotheses and computational cost as in (cf. [281], [469]). Note that the results in (cf. [355], [613], [614]) are also improved in at least the Newton's method case, since their conditions also lead to (7.232) instead of (7.234).*

*In the Newton's method case, although the majorizing sequences* $\{s_n\}$ *and* $\{r_n\}$ *coincide the convergence domains approach fails to take advantage of the relationship between* $L_0$ *and* $L$ *since* $L_0$ *does not appear in (7.232). The same is happening in the general case, since function* $\chi$ *does not depends on* $\overline{\nu_0}$. *However, our approach depends on* $\nu_0$ *for the derivation of the sufficient convergence conditions. Under our method the ratio "*$\alpha$*" of convergence for* $\{s_n\}$ *is known, but this is not true for iteration* $\{r_n\}$.

We also refer the reader to chapter 2 for some applications, examples and sufficient convergence conditions that are even weaker than (7.234).

## Chapter 8

# Werner's Method

The local convergence of Werner's method was studied by several authors under Lipschitz–type conditions on the first and second Fréchet–derivatives given in non–affine invariant form. We present in this chapter convergence analysis of this method using our new concept of recurrent functions.

## 8.1 Convergence Analysis

We are concerned in this section with the problem of approximating a locally unique solution $x^\star$ of (2.1). We revisit in this section Werner's method (cf. [600], [766]) given by

$$x_{n+1} = x_n - A_n^{-1}F(x_n), \quad A_n = F'\left(\frac{x_n + y_n}{2}\right)$$

$$y_{n+1} = x_{n+1} - A_n^{-1}F(x_{n+1}) \quad (n \geq 0,\ x_0, y_0 \in \mathcal{D}). \tag{8.1}$$

The local convergence of Werner's method (8.1) was given in (cf. [600], [766]) under Lipschitz conditions on the first and second Fréchet–derivatives given in non–affine invariant form (see (8.53) and (8.54)). The order of convergence of Werner's method (8.1) is $1 + \sqrt{2}$. The derivation of this method and its importance has well been explained in (cf. [600], [766]) (see also (cf. [139])). The two–step method uses one inverse and two function evaluations. Note that if $x_0 = y_0$, then, (8.1) becomes Newton's method (cf. [78], [792]).

We provide a semilocal convergence analysis using our new idea of recurrent functions. Our Lipschitz hypotheses are provided in affine invariant form. As far as we know the semilocal analysis of Werner's method has not been studied in this setting. We are mostly interested in finding weak sufficient convergence conditions, so as to extend the applicability of the method. Our new approach can also be used on other one–step or two–step iterative methods (cf. [78], [139], [163], [792]).

It is convenient for us to define some auxiliary functions appearing in connection to majorizing sequences for Werner's method (8.1). Let $\ell_0 > 0$, $\ell > 0$, $\alpha \geq 0$, $\eta > 0$, $\overline{\eta} \geq \eta$ and $\beta = 1 + \alpha$ be given constants. It is convenient for us to define function $\overline{f}_1$ on $[0, +\infty)$ by

$$\overline{f}_1(t) = \ell t^\beta + 4\ell_0 t - 2. \tag{8.2}$$

We have that

$$\overline{f}_1(0) = -2 < 0. \tag{8.3}$$

There exists sufficiently large $u > 0$ such that

$$\overline{f}_1(t) > 0, \quad t > u. \tag{8.4}$$

It follows from (8.3), (8.4) and the intermediate value theorem that there exists $v \in (0, u)$, such that

$$\overline{f}_1(v) = 0. \tag{8.5}$$

The number $v$ is the unique positive zero in $(0, +\infty)$ of function $\overline{f}_1$ since

$$\overline{f}_1'(t) = \ell\beta t^\alpha + 4\ell_0 > 0 \quad (t \geq 0). \tag{8.6}$$

That is function $\overline{f}_1$ is increasing and as such it crosses the positive axis only once. Moreover, define a function $g$ on $[0, +\infty)$ by

$$g(t) = 2\ell_0 t^3 + 2\ell_0 t^2 + \ell\eta^\alpha t - \ell\eta^\alpha. \tag{8.7}$$

We have as above that

$$g(0) = -\ell\,\eta^\alpha < 0 \tag{8.8}$$

and

$$g(t) > 0 \qquad (t > \zeta) \tag{8.9}$$

for sufficiently large $\zeta > 0$. Hence, as above, there exists $\delta_+ \in (0, \zeta)$ such that

$$g(\delta_+) = 0. \tag{8.10}$$

The number $\delta_+$ is the unique positive zero of function $g$ on $(0, +\infty)$, since

$$g'(t) = 6\,\ell_0\,t^2 + 4\,\ell_0\,t + \ell\,\eta^\alpha > 0 \quad (t \geq 0). \tag{8.11}$$

Set

$$\delta_0 = \frac{\ell\eta^\beta}{1 - \ell_0(\eta + \overline{\eta})}, \qquad \ell_0\,(\eta + \overline{\eta}) \neq 1, \tag{8.12}$$

$$v_\infty = 1 - 2\,\ell_0\,\eta \tag{8.13}$$

and

$$\delta_1 = \max\left\{\frac{\delta_0}{2}, \delta_+\right\}. \tag{8.14}$$

We can show the following result on majorizing sequences for Werner's method (8.1).

**Lemma 8.1** *Let $\ell_0 > 0$, $\ell > 0$, $\alpha \geq 0$, $\eta > 0$, $\overline{\eta} \geq \eta$ and $\beta = 1 + \alpha$ be given constants. Assume*

$$\ell_0\,(\eta + \overline{\eta}) < 1, \quad \eta \leq v \tag{8.15}$$

*and*

$$\delta_1 \leq v_\infty, \tag{8.16}$$

*where* $v$, $\delta_1$, $\delta_+$, $v_\infty$ *were defined by (8.5), (8.14), (8.10) and (8.13), respectively. We also choose*

$$\delta \in [\delta_1,\ v_\infty]. \tag{8.17}$$

*Then, sequence* $\{t_n\}$ $(n \geq 0)$ *generated by*

$$t_0 = 0, \quad t_1 = \eta, \quad t_{n+2} = t_{n+1} + \frac{\ell\ (t_{n+1} - t_n)^{1+\beta}}{2\ (1 - \ell_0\ (t_{n+1} + s_{n+1}))} \tag{8.18}$$

*with*

$$s_0 = 0, \quad s_1 = \overline{\eta}, \quad s_{n+2} = t_{n+2} + \frac{\ell\ (t_{n+2} - t_{n+1})^{1+\beta}}{2\ (1 - \ell_0\ (t_{n+1} + s_{n+1}))}, \tag{8.19}$$

*is non–decreasing, bounded above by*

$$t^{\star\star} = \frac{2\ \eta}{2 - \delta} \tag{8.20}$$

*and converges to its unique least upper bound* $t^\star$ *with*

$$t^\star \in [0, t^{\star\star}]. \tag{8.21}$$

*Moreover, the following estimates hold for all* $n \geq 0$

$$t_n \leq s_n, \tag{8.22}$$

$$0 < t_{n+2} - t_{n+1} \leq \frac{\delta}{2}\ (t_{n+1} - t_n) \leq \left(\frac{\delta}{2}\right)^{n+1} \eta \tag{8.23}$$

*and*

$$0 < s_{n+2} - t_{n+2} \leq \frac{\delta}{2}\ (s_{n+1} - t_{n+1}) \leq \left(\frac{\delta}{2}\right)^{n+2} \eta. \tag{8.24}$$

**Proof** We shall show using induction on $m$ that

$$\begin{aligned} 0 \leq t_{m+2} - t_{m+1} &= \frac{\ell\ (t_{m+1} - t_m)^{\beta}}{2\ (1 - \ell_0\ (t_{m+1} + s_{m+1}))}\ (t_{m+1} - t_m) \\ &\leq \frac{\delta}{2}\ (t_{m+1} - t_m), \end{aligned} \tag{8.25}$$

$$\begin{aligned} 0 \leq s_{m+2} - t_{m+2} &= \frac{\ell\ (t_{m+2} - t_{m+1})^{\beta}}{2\ (1 - \ell_0\ (t_{m+1} + s_{m+1}))}\ (t_{m+2} - t_{m+1}) \\ &\leq \frac{\delta}{2}\ (t_{m+2} - t_{m+1}) \end{aligned} \tag{8.26}$$

and

$$\ell_0\ (t_{m+1} + s_{m+1}) < 1. \tag{8.27}$$

Estimates (8.25)–(8.27) for $m = 0$ will hold if

$$\frac{\ell\ (t_1 - t_0)^\beta}{1 - \ell_0\ (t_1 + s_1)} = \frac{\ell\ \eta^\beta}{1 - \ell_0\ (\eta + \overline{\eta})} = \delta_0 \leq \delta, \tag{8.28}$$

$$\frac{\ell\ (t_2 - t_1)^\beta}{1 - \ell_0\ (t_1 + s_1)} \leq \frac{\ell\left(\dfrac{\delta}{2}\ \eta\right)^\beta}{1 - \ell_0\ (\eta + \overline{\eta})} = \overline{\delta}_0 \leq \delta_0 \leq \delta \tag{8.29}$$

and

$$\ell_0\ (t_1 + s_1) = \ell_0\ (\eta + \overline{\eta}) < 1, \tag{8.30}$$

respectively, which are true by (8.17) and (8.15). Let us assume (8.22)–(8.27) hold for all $n \leq m+1$. Then, we get from (8.25) and (8.26) that

$$t_{m+2} \leq \frac{1 - \left(\dfrac{\delta}{2}\right)^{m+2}}{1 - \dfrac{\delta}{2}}\eta < \frac{2\eta}{2 - \eta} = t^{\star\star} \tag{8.31}$$

and

$$s_{m+2} \leq t_{m+2} + \left(\frac{\delta}{2}\right)^{m+2}\ \eta \leq \left\{\frac{1 - \left(\dfrac{\delta}{2}\right)^{m+2}}{1 - \dfrac{\delta}{2}} + \left(\frac{\delta}{2}\right)^{m+2}\right\}\ \eta. \tag{8.32}$$

We shall only show (8.25), since (8.26), will follows (as (8.29) follows from (8.28)). Using the induction hypotheses, (8.25) certainly holds if

$$\ell\ (t_{m+1} - t_m)^\beta + \ell_0\ \delta\ (t_{m+1} + s_{m+1}) - \delta \leq 0$$

or

$$\ell\left\{\left(\frac{\delta}{2}\right)^m\ \eta\right\}^\beta + \ell_0\ \delta\ \left\{\frac{1 - \left(\dfrac{\delta}{2}\right)^{m+1}}{1 - \dfrac{\delta}{2}} + \frac{1 - \left(\dfrac{\delta}{2}\right)^{m+1}}{1 - \dfrac{\delta}{2}} + \left(\frac{\delta}{2}\right)^{m+1}\right\}\ \eta - \delta \leq 0$$

or, since $\beta \geq 1$,

$$\ell\ \left(\frac{\delta}{2}\right)^m\ \eta^\beta + \ell_0\ \delta\ \left\{2\ \frac{1 - \left(\dfrac{\delta}{2}\right)^{m+1}}{1 - \dfrac{\delta}{2}} + \left(\frac{\delta}{2}\right)^{m+1}\right\}\ \eta - \delta \leq 0. \tag{8.33}$$

We are motivated from (8.33) to define functions $f_m$ ($m \geq 1$) on $[0, +\infty)$ for $v = \dfrac{\delta}{2}$ and show instead of (8.33) that

$$f_m(v) \quad = \quad \ell\ v^{m-1}\ \eta^\beta + 2\ \ell_0\ (2\ (1 + v + \cdots + v^m) + v^{m+1})\ \eta - 2 \leq 0. \tag{8.34}$$

We need a relationship between two consecutive functions $f_m$. We get that

$$\begin{aligned} f_{m+1}(v) &= \ell\ v^m\ \eta^\beta + 2\ \ell_0\ (2\ (1+v+\cdots+v^{m+1}) + v^{m+2})\ \eta - 2 \\ &= \ell\ v^m\ \eta^\beta + \ell\ v^{m-1}\ \eta^\beta - \ell\ v^{m-1}\ \eta^\beta \\ &\quad + 2\ \ell_0(2\ (1+v+\cdots+v^m) + v^{m+1} + v^{m+1} + v^{m+2})\ \eta - 2 \\ &= f_m(v) + \ell\ v^m\ \eta^\beta - \ell\ v^{m-1}\ \eta^\beta + 2\ \ell_0\ (v^{m+1} + v^{m+2})\ \eta \\ &= f_m(v) + g(v)\ v^{m-1}\ \eta, \end{aligned} \tag{8.35}$$

where function $g$ is given by (8.7). We have by (8.34) that

$$f_1(0) = \ell\ \eta^\beta + 4\ \ell_0\ \eta - 2 < 0, \tag{8.36}$$

$$f_m(0) = 4\ \ell_0\ \eta - 2 < 0 \quad (m > 1) \tag{8.37}$$

and for sufficiently large $v > 0$

$$f_m(v) > 0. \tag{8.38}$$

It follows from (8.36)–(8.38) and the intermediate value theorem that there exists $v_m > 0$ such that $f_m(v_m) = 0$. Moreover, each $v_m$ is the unique positive zero of $f_m$ since $f_m'(v) > 0$ for $v \in [0, +\infty)$. We shall show

$$f_m(v) \leq 0 \quad (v \in [0, v_m],\ m \geq 1). \tag{8.39}$$

If there exists $m \geq 0$ such that $v_{m+1} \geq \frac{\delta}{2}$, then, using (8.7) and (8.35), we get that

$$f_{m+1}(v_{m+1}) = f_m(v_{m+1}) + g(v_{m+1})\ v_{m+1}^{m-1}\ \eta$$

or

$$f_m(v_{m+1}) \leq 0$$

since $f_{m+1}(v_{m+1}) = 0$ and $g(v_{m+1})\ v_{m+1}^{m-1}\ \eta \geq 0$, which imply $v_{m+1} \leq v_m$. We can certainly choose the last of the $v_m$'s denoted by $v_\infty$ (obtained from (8.33) by letting $m \longrightarrow \infty$ and given in (8.13)) to be $v_{m+1}$. It follows sequence $\{v_m\}$ is non–increasing, bounded below by zero and as such it converges to its unique maximum lowest bound $v^\star$ satisfying $v^\star \geq v_\infty$. Then, estimate (8.39) certainly holds if

$$\frac{\delta}{2} \leq v_\infty,$$

which is true by hypothesis (8.16).

Finally, sequences $\{t_n\}$, $\{s_n\}$ are non–decreasing, bounded above by $t^{\star\star}$ given by (8.21). Hence, they converge to their common, and unique least upper bound $t^\star$ satisfying (8.21).

This also completes the proof of Lemma 8.1. ⊠

We can also provide a second majorizing result. Let us define function $h_m$ $(m \geq 1)$ as $f_m$ by

$$h_m(s) = \ell\ s^{m-1}\ \eta^\beta + 4\ \ell_0\ (1 + s + \cdots + s^m)\ \eta - 2, \tag{8.40}$$

$$\overline{\delta}_+ = \frac{-\ell\ \eta^\beta + \sqrt{\ell^2\ \eta^{2\,\alpha} + 16\ \ell_0\ \ell\ \eta^\alpha}}{8\ \ell_0}, \tag{8.41}$$

$$\overline{\delta}_0 = \frac{\ell\ \eta^\beta}{1 - \ell_0\ (\eta + \overline{\overline{\eta}})}, \qquad \ell_0\ (\eta + \overline{\overline{\eta}}) \neq 1, \tag{8.42}$$

$$\overline{\delta}_1 = \max\left\{\frac{\overline{\delta}_0}{2},\ \overline{\delta}_+\right\} \tag{8.43}$$

and

$$\overline{v}_\infty = v_\infty. \tag{8.44}$$

Then, with the above changes and simply following the proof of Lemma 8.1, we can provide another result on majorizing sequences for Werner' s method (8.1), using a different approach than in Lemma 8.1.

**Lemma 8.2** *Let $\ell_0 > 0$, $\ell > 0$, $\alpha \geq 0$, $\eta > 0$, $0 < \overline{\overline{\eta}} \leq \eta$ and $\beta = 1 + \alpha$ be given constants. Assume that*

$$\ell_0\ (\eta + \overline{\overline{\eta}}) < 1 \tag{8.45}$$

*and*

$$\overline{\delta}_1 \leq \overline{v}_\infty, \tag{8.46}$$

*where $\overline{\delta}_1$, $\overline{v}_\infty$, $\overline{\delta}_+$ are given by (8.43), (8.44) and (8.41), respectively. We also choose*

$$\delta \in [\overline{\delta}_1,\ \overline{v}_\infty]. \tag{8.47}$$

*Then, scalar sequence $\{v_n\}$ $(n \geq 0)$ given by*

$$v_0 = 0, \quad v_1 = \eta, \quad v_{n+2} = v_{n+1} + \frac{\ell\ (v_{n+1} - v_n)^{1+\beta}}{2\ (1 - \ell_0\ (v_{n+1} + \overline{s}_{n+1}))}, \tag{8.48}$$

*with*

$$\overline{s}_0 = 0, \quad \overline{s}_1 = \overline{\overline{\eta}}, \quad \overline{s}_{n+2} = v_{n+2} + \frac{\ell\ (v_{n+2} - v_{n+1})^{1+\beta}}{2\ (1 - \ell_0\ (v_{n+1} + \overline{s}_{n+1}))}, \tag{8.49}$$

*is non–decreasing, bounded above by $t^{\star\star}$ and converges to its unique least upper bound $t^\star$ with $t^\star \in [0, t^{\star\star}]$, where $t^{\star\star}$ is given by (8.20). Moreover, the following estimates hold for all $n \geq 0$*

$$\overline{s}_n \leq v_n, \tag{8.50}$$

$$0 < v_{n+2} - v_{n+1} \leq \frac{\delta}{2}\ (v_{n+1} - v_n) \leq \left(\frac{\delta}{2}\right)^{n+1} \eta \tag{8.51}$$

*and*

$$0 < v_{n+2} - \overline{s}_{n+2} \leq \frac{\delta}{2}\,(v_{n+1} - \overline{s}_{n+1}) \leq \left(\frac{\delta}{2}\right)^{n+2}\,\eta. \tag{8.52}$$

We also need a lemma due to Werner (cf. [600, Lemma 1, p. 335]).

**Lemma 8.3** *Let $G : \mathcal{D} \subseteq \mathcal{X} \longrightarrow \mathcal{Y}$ be a twice Fréchet differentiable operator. Assume that there exist a positive constants $L_1$, $L_{2,\alpha}$ and $\alpha \in [0,1]$ such that*

$$\| G'(x) - G'(y) \| \leq L_1 \ \| x - y \| \tag{8.53}$$

*and*

$$\| G''(x) - G''(y) \| \leq L_{2,\alpha} \ \| x - y \|^\alpha \quad (x, y \in \mathcal{D}). \tag{8.54}$$

*Then, the following estimates hold for all $x, y, z \in \mathcal{D}$*

$$\| G(x) - G(y) - G'(z)\,(x-y) \| \leq L_1 \int_0^1 \| (1-t)\,y + t\,x - z \| \ dt \ \| x - y \| \tag{8.55}$$

*and for $\theta \in [0,1]$, $x, y \in \mathcal{D}$, $z_\theta = \theta\,x + (1-\theta)\,y$*

$$\begin{aligned} \| G(x) - G(y) - G'(z_\theta)\,(x-y) \| \ \leq \ & \left(\frac{1}{4} + \left(\theta - \frac{1}{2}\right)^2\right) \frac{L_{2,\alpha} \ \| x - y \|^{2+\alpha}}{(\alpha+1)\,(\alpha+2)} \\ & + L_1 \left|\theta - \frac{1}{2}\right| \ \| x - y \|^2 . \end{aligned} \tag{8.56}$$

We can show the following semilocal convergence result for Werner's method (8.1).

**Theorem 8.1** *Let $F : \mathcal{D} \subseteq \mathcal{X} \longrightarrow \mathcal{Y}$ be a twice Fréchet differentiable operator. Assume there exist points $x_0, y_0 \in \mathcal{D}$, $L_0 > 0$, $\alpha \in [0,1]$ and $L_{2,\alpha} > 0$ such that, for all $x, y \in \mathcal{D}$*

$$A_0^{-1} \in \mathcal{L}(\mathcal{Y}, \mathcal{X}), \tag{8.57}$$

$$\left\| A_0^{-1} \left[ F'(x) - F'\left(\frac{x_0 + y_0}{2}\right)\right]\right\| \leq L_0 \left\| x - \frac{x_0 + y_0}{2} \right\|, \tag{8.58}$$

$$\| A_0^{-1} \, [F''(x) - F''(y)] \| \leq L_{2,\alpha} \ \| x - y \|^\alpha, \tag{8.59}$$

$$y_0 \in \overline{U}(x_0, t^\star) = \{x \in \mathcal{X}, \ \| x - x_0 \| \leq t^\star\} \subseteq \mathcal{D}, \tag{8.60}$$

$$\| A_0^{-1} \ F(x_0) \| \leq \eta, \tag{8.61}$$

$$\| A_0^{-1} \ F(x_1) \| \leq \overline{\eta}, \tag{8.62}$$

*where*

$$x_1 = x_0 - F'\left(\frac{x_0 + y_0}{2}\right)^{-1} F(x_0) \tag{8.63}$$

*and conditions of Lemma 8.1 hold with*

$$\ell_0 = \frac{L_0}{2}, \qquad \ell = \frac{L_{2,\alpha}}{2\,\beta\,(1+\beta)}. \tag{8.64}$$

*Then, the sequence $\{x_n\}$ defined by Werner's method (8.1) is well defined, remains in $\overline{U}(x_0, t^\star)$ for all $n \geq 0$ and converges to a unique solution $x^\star$ of equation $F(x) = 0$ in $U(x_0, t^\star)$. Moreover, the following estimate holds for all $n \geq 0$*

$$\| x_n - x^\star \| \leq t^\star - t_n, \tag{8.65}$$

*where sequence $\{t_n\}$ $(n \geq 0)$ is given in Lemma 8.1.*

**Proof** We shall show using induction on the integer $m$ that

$$\| x_{m+1} - x_m \| \leq t_{m+1} - t_m \tag{8.66}$$

and

$$\| y_{m+1} - x_{m+1} \| \leq s_{m+1} - t_{m+1}. \tag{8.67}$$

Estimates (8.66) and (8.67) hold for $m = 0$ by the initial conditions. Let us assume (8.66), (8.67) hold true and $x_m, y_m \in \overline{U}(x_0, t^\star)$ for all $n \leq m + 1$. Using (8.59), we obtain that

$$\begin{aligned} \| A_0^{-1} \, (A_0 - A_n) \| &\leq L_0 \left\| \frac{x_n + y_n}{2} - \frac{x_0 + y_0}{2} \right\| \\ &\leq \frac{L_0}{2} \, ( \| x_n - x_0 \| + \| y_n - y_0 \| ) \\ &\leq \frac{L_0}{2} \, ((t_n - t_0) + (s_n - t_0)) \\ &= \ell_0 \, (t_n + s_n) < 1 \quad \text{(by (8.27))}. \end{aligned} \tag{8.68}$$

It follows from (8.68) and the Banach lemma of invertible operators that $A_n^{-1}$ exists so that

$$\| A_n^{-1} \, A_0 \| \leq \frac{1}{1 - \ell_0 \, (t_n + s_n)}. \tag{8.69}$$

In view of (8.1), we obtain the approximations:

$$F(x_{m+1}) = F(x_{m+1}) - F(x_m) - F'\left(\frac{x_m + y_m}{2}\right)(x_{m+1} - x_m), \tag{8.70}$$

$$F(x_{m+2}) = F(x_{m+2}) - F(x_{m+1}) - F'\left(\frac{x_m + y_m}{2}\right)(y_{m+1} - x_{m+1}). \tag{8.71}$$

By composing both sides of (8.70), (8.71) by $A_0^{-1}$, using Lemma 8.3 for $\theta = \frac{1}{2}$, $G = A_0^{-1} \, F$, we obtain that

$$\| A_0^{-1} \, F(x_{m+1}) \| \leq \frac{L_{2,\alpha}}{4 \, (\alpha + 1) \, (\alpha + 2)} \, \| x_{m+1} - x_m \|^{2+\alpha} \leq \ell \, (t_{m+1} - t_m)^{1+\beta} \tag{8.72}$$

and

$$\| A_0^{-1} \, F(x_{m+2}) \| \leq \frac{L_{2,\alpha}}{4 \, (\alpha + 1) \, (\alpha + 2)} \, \| x_{m+2} - x_{m+1} \|^{2+\alpha} \leq \ell \, (t_{m+2} - t_{m+1})^{1+\beta}, \tag{8.73}$$

respectively. Using (8.1), (8.69), (8.18), (8.19), (8.72) and (8.73), we obtain that

$$\begin{aligned}\| x_{m+2} - x_{m+1} \| &\leq \| A_{m+1}^{-1} A_0 \| \, \| A_0^{-1} F(x_{m+1}) \| \\ &\leq \frac{\ell \, (t_{m+1} - t_m)^{1+\beta}}{2 \, (1 - \ell_0 \, (t_{m+1} + s_{m+1}))} \\ &= t_{m+2} - t_{m+1}\end{aligned}$$

and

$$\begin{aligned}\| y_{m+2} - x_{m+2} \| &\leq \| A_{m+1}^{-1} A_0 \| \, \| A_0^{-1} F(x_{m+2}) \| \\ &\leq \frac{\ell \, (t_{m+2} - t_{m+1})^{1+\beta}}{2 \, (1 - \ell_0 \, (t_{m+1} + s_{m+1}))} \\ &= s_{m+2} - t_{m+2},\end{aligned}$$

which complete the induction for (8.66) and (8.67). By Lemma 8.1, (8.66) and (8.67), sequence $\{x_n\}$ $(n \geq 0)$ is a Cauchy sequence in a Banach space $\mathcal{X}$ and as such it converges to some $x^\star \in \overline{U}(x_0, t^\star)$ (since $\overline{U}(x_0, t^\star)$ is a closed set). By letting $m \longrightarrow \infty$ in (8.72), we obtain $F(x^\star) = 0$.

Finally, to show the uniqueness, let $y^\star \in U(x_0, t^\star)$ be a solution of equation $F(x) = 0$. Let: $\mathcal{M} = \displaystyle\int_0^1 F'(x^\star + t\,(y^\star - x^\star))\,dt$. Using (8.59), we obtain in turn that

$$\begin{aligned}&\| A_0^{-1} (A_0 - \mathcal{M}) \| \\ &\leq L_0 \int_0^1 \left\| \frac{x_0 + y_0}{2} - (x^\star + t\,(y^\star - x^\star)) \right\| \, dt \\ &\leq L_0 \int_0^1 \left( (1-t) \left\| \frac{(x_0 - x^\star) + (y_0 - x^\star)}{2} \right\| + t \left\| \frac{(y_0 - y^\star) + (x_0 - y^\star)}{2} \right\| \right) dt \\ &\leq \frac{L_0}{4} \left( \| x_0 - x^\star \| + \| y_0 - x^\star \| + \| y^\star - x_0 \| + \| y^\star - y_0 \| \right) \\ &< \frac{L_0}{4} \, 4 \, t^\star = L_0 \, t^\star \leq 1 \quad \text{(by (8.27))}. \qquad (8.74)\end{aligned}$$

In view of (8.74) and the Banach lemma on invertible operators, $\mathcal{M}^{-1}$ exists. It follows from the identity: $0 = F(x^\star) - F(y^\star) = \mathcal{M}\,(x^\star - y^\star)$, that $x^\star = y^\star$. This completes the proof of Theorem 8.1. ⊠

**Remark 8.1** *(a) The most appropriate choices for $\delta$ in Lemmas 8.1 and 8.2 seem to be $\delta = \delta_1$ and $\delta = \overline{\delta}_1$, respectively.*

*(b) Note that the conclusions of Theorem 8.1 hold if Lemma 8.1 is replaced by Lemma 8.2 and (8.19) by (8.49).*

*(c) The limit point $t^\star$ (see Theorem 8.1) can be replaced by $t^{\star\star}$ given in closed form by (8.20).*

Chapter 9

# Halley's Method

We present in this Chapter local convergence analysis for Halley's method.

## 9.1 Local Convergence

We provide local convergence results for Halley's method in order to approximate a locally unique zero of an operator in a Banach space setting using convex majorants. Kantorovich–type and Smale–type results are considered as applications and special cases.

In this Section, we are concerned with the problem of approximating a locally unique zero $x^\star$ of operator $F$, where $F$ is a nonlinear operator defined on an open convex subset $D$ of a Banach space $X$ with values in a Banach space $Y$. Halley's method (cubically convergent)

$$x_{n+1} = x_n - \Gamma_F(x_n)F'(x_n)^{-1}F(x_n) \quad (n \geq 0,\ x_0 \in D), \tag{9.1}$$

where,

$$\Gamma_F(x) = (I - L_F(x))^{-1}$$

and

$$L_F(x) = \frac{1}{2}F'(x)^{-1}F''(x)F'(x)^{-1}F(x)$$

is also undoubtedly the most popular iterative method as Newton's method for generating sequences approximating $x^\star$ of (2.1).

We suppose that operator $F$ has a zero $x^\star$. An interesting problem is to find the radius of convergence for Halley's method (9.1). An open ball $U(x^\star, r) \subset X$ is called a convergence ball of an iterative method, is the sequence generated by this iterative method starting from any initial values in it converges. The convergence ball of an iterative method is very important because it shows the extent of difficulty for choosing initial guesses for iterative methods.

Recently, we provided a local convergence analysis for Newton's method (cf. [170]) using the following majorant conditions:

$$||F'(x^\star)^{-1}[F'(y) - F'(x)]|| \leq h'(\|y - x\| + \|x - x^\star\|) - h'(\|x - x^\star\|), \tag{9.2}$$

$$||F'(x^\star)^{-1}[F'(y) - F'(x^\star)]|| \leq h'_\star(\|y - x^\star\|) - h'_\star(0) \tag{9.3}$$

for all $x, y \in U(x^\star, R)$, $R > 0$, where $\|y-x\|+\|x-x^\star\| < R$ and $h$, $h_\star : [0, R) \to \mathbb{R}$ are continuously differentiable, convex, strictly increasing functions with $h'_\star(0) = -1$ and $h_\star(R_0) = 0$ for some $R_0 \in (0, R)$. Note that function $h_\star$ is a special case of $h$,

$$h'_\star(t) \leq h'(t), \quad (t \in [0, R])$$

holds in general and $\frac{h'}{h'_\star}$ can be arbitrarily large. In the Lipschitz and center–Lipschitz case (9.2) and (9.3) become

$$||F'(x^\star)^{-1}[F'(y) - F'(x)]|| \leq \alpha\|y - x\| \tag{9.4}$$

and

$$||F'(x^\star)^{-1}[F'(y) - F'(x^\star)]|| \leq \alpha_\star\|y - x^\star\| \quad (x, y \in U(x^\star, R), \tag{9.5}$$

respectively, provided that

$$h(t) = \frac{\alpha}{2}t^2 - t + \alpha_0, \quad \alpha > 0, \quad \alpha_0 > 0 \tag{9.6}$$

and

$$h_\star(t) = \frac{\alpha_\star}{2}t^2 - t + \alpha_1, \quad \alpha_\star > 0, \quad \alpha_1 > 0. \tag{9.7}$$

Functions $h$, $h_\star$ given by (9.6) and (9.7) are continuously differentiable, convex, strictly increasing on $(0, R_0)$ where $R_0 = \frac{1}{\alpha_\star}$ and $h'(0) = h'_\star(0) = -1$. Using (9.2) and (9.3) we provided a local convergence analysis for Newton's method (cf. [163], [209]) with the following advantages over the corresponding ones in (cf. [139], [367], [365], [569], [638], [710]) (using only (9.2))

(a) tighter error bounds on the distances $\|x_n - x^\star\|$;

(b) an at least as large radius of convergence.

These advantages are due to the fact that for the upper bounds on the norm $\|F'(x)^{-1}F'(x^\star)\|$ we use (9.3) which is more precise (and cheaper) than (9.2) (used in (cf. [367], [365], [569], [638], [710])). Note that under (9.2) for $h$ given in (9.6) the radius of convergence given in (cf. [367], [365]) coincides with the one provided by Traub-Wosniakowski (cf. [710]) and Rheinboldt (cf. [638])

$$r_{TR} = \frac{2}{3\alpha}. \tag{9.8}$$

However, our radius of convergence is given by

$$r_{AH} = \frac{2}{2\alpha_\star + \alpha}. \tag{9.9}$$

Note that

$$\frac{r_{AH}}{r_{TR}} = \frac{3}{2\frac{\alpha_\star}{\alpha} + 1} \to 3 \quad as \quad \frac{\alpha_\star}{\alpha} \to 0. \tag{9.10}$$

Hence, our radius $r_{AH}$ can be at most three times larger than $r_{TR}$. Concerning the error estimates, we have in our case

$$\|x_{n+1} - x^\star\| \le \frac{\alpha\|x_n - x^\star\|^2}{2(1 - \alpha_\star\|x_n - x^\star\|)} \tag{9.11}$$

whereas

$$\|x_{n+1} - x^\star\| \le \frac{\alpha\|x_n - x^\star\|^2}{2(1 - \alpha\|x_n - x^\star\|)} \tag{9.12}$$

is given in (cf. [163], [209]). Therefore, if $\alpha = \alpha_\star$, the estimates (9.11) and (9.12) coincide. Otherwise (i.e. if $\alpha_\star < \alpha$) ours are tighter.

In what follows, we extend our ideas to study the local convergence of Halley's method (9.2). Suppose that $F$ is a twice Fréchet–differentiable operator and there exists $x^\star \in D$ such that $F'(x^\star)^{-1} \in L(Y, X)$, the space of bounded linear operators from $Y$ into $X$. In addition, let $R > 0$ be such that $U(x^\star, R) \subseteq D$ and $h$, $h_\star : [0, R) \to \mathbb{R}$ be twice continuously differentiable function, we say the operator $F''$ satisfies the majrant conditions if

$$||F'(x^\star)^{-1}[F''(y) - F''(x)]|| \le h''(\|y - x\| + \|x - x^\star\|) - h''(\|x - x^\star\|), \tag{9.13}$$

$$||F'(x^\star)^{-1}[F''(y) - F''(x^\star)]|| \le h''_\star(\|y - x^\star\|) - h''_\star(0) \tag{9.14}$$

for all $x, y \in U(x^\star, R)$, where $\|y - x\| + \|x - x^\star\| < R$ and the following conditions hold

$(C_1)$ $h''(0) \ge 0$, $h''_\star(0) > 0$, $h'_\star(0) = -1$;
$(C_2)$ $h''$, $h''_\star$ are strictly increasing in $[0, R)$, $h''_\star$ is convex on $[0, R)$.
$(C_3)$ $h'_\star$ has zeros in $(0, R)$. Denote by $R_0$ the minimal zero of $h'_\star$ in $(0, R)$.
$(C_4)$ $\|F'(x^\star)^{-1}F''(x^\star)\| \le h''_\star(0)$.

Note that, if (9.13) holds, then, so does (9.14). That is, (9.14) is not an additional hypothesis. We also have

$$h''_\star(t) \le h''(t) \quad (t \in [0, R]) \tag{9.15}$$

holds in general and $\dfrac{h''}{h''_\star}$ can be arbitrarily large. We show the cubic convergence of Halley's method under conditions $(C_1)$–$(C_4)$. It is convenient for us to define some functions and parameters. Let us define function $g$ on $[0, R]$ by

$$g(r) = h''_\star(r)(2 + h'_\star(r))r - 2(h'_\star(r))^2.$$

By the (C) conditions $g$ is a continuous function on $[0, R]$ with

$$g(0) = -2(h'_\star(0))^2 = -2 < 0$$

and

$$g(R_0) = h''_\star(R_0)(2 + h'_\star(R_0))R_0 - 2(h'_\star(R_0))^2 = 2h''_\star(R_0)R_0 > 0.$$

It follows from the intermediate value theorem that there exists a zero of function $g$ on $(0, R_0)$. Denote by $R_1$ the minimal such zero of $g$ on $(0, R_0)$. Then, we have

$$q(r) = \frac{h''_\star(r)(2 + h'_\star(r))r}{2(h'_\star(r))^2} < 1,$$

$$G(r) = \frac{1}{1 - q(r)} > 0$$

and

$$P(r) = -\frac{1}{2}G(r)(\frac{1}{h'_\star(r)})[\frac{1}{r} - \frac{(h''_\star(r))^2}{2h'_\star(r)}] > 0$$

are well defined on $[0, R_1]$ since by $(C_1)$ and $(C_2)$, $-1 < h'_\star(r) < 0$. Similarly, we show that there exists a minimal zero $R^\star \in (0, R_1)$ of function

$$Q(r) = P(r)r^2 - 1$$

and

$$Q(r) < 0$$

is true for $r \in [0, R^\star)$. Moreover, replace function $h_\star$ by $h$ in conditions $(C_1)$–$(C_4)$ and in the definition of $g$, $q$, $G$, $P$ and $Q$. Denote resulting functions by $g_0$, $q_0$, $G_0$, $P_0$ and $Q_0$, respectively. Furthermore, denote by $r_0$, $r_1$ and $r^\star$ the parameters corresponding to $R_0$, $R_1$ and $R^\star$, respectively.

We need a result describing the properties of the "$h$" functions.

**Lemma 9.1** *Let $R > 0$ and $h_\star : [0, R) \to \mathbb{R}$ be a twice continuously differentiable function which satisfies the conditions $(C_1)$–$(C_3)$ (for $h_\star$). Then, the following hold*

*(a) $h'_\star$ is strictly convex and strictly increasing on $[0, R)$.*

*(b) $-1 < h'_\star(r) < 0$ for $r \in (0, R_0)$.*

**Proof**

(a) It follows from $(C_2)$ and $h''_\star(0) > 0$.

(b) By $(C_1)$, (a) and the definition of $R_0$ we get (b).

This completes the proof of the lemma. ⊠

We need a Banach–type estimate on the inverse of $F'(x)$.

**Lemma 9.2** *Suppose $\|x - x^\star\| \le r < R_0$. If $h_\star : [0, R_0) \to \mathbb{R}$ is twice continuously differentiable and the majorizing function of $F$ at $x^\star$. Then, $F'(x)^{-1} \in L(Y, X)$ and*

$$\|F'(x)^{-1}F'(x^\star)\| \le -\frac{1}{h'_\star(\|x - x^\star\|)} \le -\frac{1}{h'_\star(r)}. \tag{9.16}$$

**Proof** Let $t \in [0, R_0)$ and $x \in \overline{U}(x^\star, R_0)$. We have the identity

$$F'(x) = F'(x^\star) + \int_0^1 [F''(x^\star + \theta(x - x^\star)) - F''(x^\star)](x - x^\star)d\theta + F''(x^\star)(x - x^\star). \tag{9.17}$$

Using (9.14), $(C_1)$, $(C_4)$ and Lemma 9.1 (b), we get in turn that

$$\begin{aligned}
&\|F'(x^\star)^{-1}(F'(x) - F'(x^\star))\| \\
&\leq \int_0^1 \|F'(x^\star)^{-1}[F''(x^\star + \theta(x - x^\star)) - F''(x^\star)]\|\|x - x^\star\|d\theta \\
&+ \|F'(x^\star)^{-1}F''(x^\star)\|\|x - x^\star\| \qquad (9.18) \\
&\leq \int_0^1 [h''_\star(\theta\|x - x^\star\|) - h''_\star(0)]\|x - x^\star\|d\theta + h''_\star(0)\|x - x^\star\| \\
&= h'_\star(\|x - x^\star\|) - h'_\star(0) \leq h'_\star(r) - h'_\star(0) < 1.
\end{aligned}$$

It follows from (9.18) and the Banach lemma on invertible operators that $F'(x)^{-1} \in L(Y, X)$ so that (9.16) holds. This completes the proof of the lemma. ⊠

We also need an upper bound on $F''(x)$.

**Lemma 9.3** *Suppose $\|x - x^\star\| \leq r < R_0$. If $h_\star : [0, R_0) \to \mathbb{R}$ is twice continuously differentiable and majorizing function to $F$ at $x^\star$. Then, we have*

$$\|F'(x^\star)^{-1}F''(x)\| \leq h''_\star(\|x - x^\star\|) \leq h''_\star(r). \tag{9.19}$$

**Proof** Using (9.14) and $(C_4)$, we obtain that

$$\begin{aligned}
\|F'(x^\star)^{-1}F''(x)\| &\leq \|F'(x^\star)^{-1}[F''(x) - F''(x^\star)]\| + \|F'(x^\star)^{-1}F''(x^\star)\| \\
&\leq h''_\star(\|x - x^\star\|) - h''_\star(0) + h''_\star(0) \\
&= h''_\star(\|x - x^\star\|) \leq h''_\star(r)
\end{aligned}$$

since, $h''_\star$ is strictly increasing. This completes the proof of the lemma. ⊠

We provide the main local convergence result for Halley's method (9.1).

**Theorem 9.1** *Under conditions $(C_1)$–$(C_4)$, sequence $\{x_n\}$ generated by Halley's method (9.1) is well defined, remains in $U(x^\star, R^\star)$ for all $n \geq 0$ and converges to the unique zero $x^\star$ of operator $F$ in $U(x^\star, R_0)$, provided that $x_0 \in U(x^\star, R^\star)$. Moreover, the following estimates hold*

$$\|x_{n+1} - x^\star\| \leq P(R^\star)\|x_n - x^\star\|^3. \tag{9.20}$$

*In particular, Halley's method (9.1) converges Q–cubically to $x^\star$.*

**Proof** Using (9.1) and $F(x^\star)=0$, we get the identity

$$x_{k+1}-x^\star = \Gamma_F(x_k)F'(x_k)^{-1}F'(x^\star)F'(x^\star)^{-1}[-F'(x_k)(x^\star-x_k)-F(x_k)]$$
$$+\Gamma_F(x_k)L_F(x_k)(x^\star-x_k)$$
$$=\Gamma_F(x_k)F'(x_k)^{-1}F'(x_\star)F'(x_\star)^{-1}\int_0^1(1-\theta)[F''(x_k+\theta(x^\star-x_k))$$
$$-F''(x_k)](x^\star-x_k)^2d\theta$$
$$-\frac{1}{2}\Gamma_F(x_k)F'(x_k)^{-1}F'(x^\star)F'(x^\star)^{-1}F''(x_k)[F'(x_k)^{-1}F'(x^\star)F'(x^\star)^{-1}$$
$$\times\int_0^1(1-\theta)F''(x_k+\theta(x^\star-x_k))(x^\star-x_k)^2d\theta](x^\star-x_k) \tag{9.21}$$

By (9.19) and the convexity of $h''$, we get

$$h''(\theta\|x_k-x^\star\|+\|x_k-x^\star\|)-h''(\|x_k-x^\star\|)\le[h''(\theta R^\star+R^\star)-h''(R^\star)]\frac{\theta\|x^\star-x_k\|}{\theta R^\star}. \tag{9.22}$$

Using (9.16), (9.19) and the definition of operator $L_F$, we have, for $x\in\overline{U}(x^\star,R^\star)$

$$\|L_F(x)\|$$
$$\le\frac{1}{2}\|F'(x)^{-1}F'(x^\star)\|^2\|F'(x^\star)^{-1}F''(x)\|\left\|\int_0^1F'(x^\star)^{-1}F'(x^\star+\theta(x-x^\star))d\theta(x-x^\star)\right\|$$
$$\le\frac{h''(\|x-x^\star\|)}{2(h'(\|x-x^\star\|))^2}(2+h'_\star(\|x-x^\star\|))\|x-x^\star\|$$
$$\le g(\|x-x^\star\|)\le g(R^\star)<1. \tag{9.23}$$

It follows from (9.23) and the Banach lemma on invertible operators that $\Gamma_F\in L(Y,X)$ and

$$\|\Gamma_F\|\le\frac{1}{1-g(\|x-x^\star\|)}=G(\|x-x^\star\|). \tag{9.24}$$

Using (9.16), (9.19), (9.21), (9.22) and (9.24) (for $x=x_k$), we get in turn

$$\|x_{k+1}-x^\star\|$$
$$\le\frac{-G(\|x_k-x^\star\|)\int_0^1(1-\theta)d\theta\|x_k-x^\star\|^3}{R^\star h'_\star(\|x_k-x^\star\|)}$$
$$+\frac{G(\|x_k-x^\star\|)(h''_\star(\|x_k-x^\star\|))^2\|x_k-x^\star\|^3}{4(h'_\star(\|x_k-x^\star\|))^2} \tag{9.25}$$
$$=P(R^\star)\|x_k-x^\star\|^3=\frac{\|x_k-x^\star\|^3}{R^{\star 2}}\le\|x_k-x^\star\|<R^\star\quad(k\ge0),$$

which shows (9.20), $x_{k+1}\in U(x^\star,R^\star)$ and $\lim_{k\to\infty}x_k=x^\star$.

Finally, to show the uniqueness part, let $y^\star \in U(x^\star, R_0)$ be zero of operator $F$. Define a linear operator

$$L = \int_0^1 F'(y^\star + \theta(x^\star - y^\star))d\theta. \tag{9.26}$$

Then, as in (9.16), we get

$$\|F'(x^\star)^{-1}(L - F'(x^\star))\| \leq h'_\star(\|y^\star - x^\star\|) - h'_\star(0) < 1 \tag{9.27}$$

since $z = y^\star + \theta(x^\star - y^\star) \in \overline{U}(x^\star, R_0)$. Using (9.27) and the Banach lemma, we get $L^{-1} \in L(Y, X)$. Then, in view of the identity $0 = F(x^\star) - F(y^\star) = L(x^\star - y^\star)$, we get $x^\star = y^\star$. This completes the proof of the theorem. ⊠

According to the discussion above Lemma 9.1 if function $h_\star$ is replaced by $h$ in conditions $(C_1)$–$(C_4)$, Theorem 9.1 gives:

**Theorem 9.2** *Under conditions $(C_1)$–$(C_4)$ with function $h$ replacing $h_\star$, sequence $\{x_n\}$ generated by Halley's method (9.1) is well defined, remains in $U(x^\star, r^\star)$ for all $n \geq 0$ and converges to the unique zero $x^\star$ of operator $F$ in $U(x^\star, r_0)$, provided that $x_0 \in U(x^\star, r^\star)$. Moreover, the following estimates hold*

$$\|x_{n+1} - x^\star\| \leq P_0(r^\star)\|x_n - x^\star\|^3. \tag{9.28}$$

*In particular, Halley's method (9.1) converges Q–cubically to $x^\star$.*

**Remark 9.1** *The results obtained here can be used to find zeros $x^\star$ of operators $F$ satisfying autonomous differentiable equations (cf. [163], [209])*

$$F'(x) = T(F(x)), \tag{9.29}$$

*where $T : Y \to X$ is a known differentiable operator. As*

$$F''(x^\star) = F'(x^\star)T'((F(x^\star))) = T(F(x^\star))T'(F(x^\star)) = T(0)T'(0), \tag{9.30}$$

*we do not need to know $x^\star$ to verify (9.13), (9.14) and $(C_4)$. Equations where operators $F$ satisfy (9.29) appear in many branches of applied sciences, such as minimization problems (cf. [163], [209])*

We consider two special cases.

**1. Kantorovich–type.** We define functions $h$, $h_\star$ by

$$h(r) = \frac{a}{6}r^3 + \frac{b}{2}r^2 - r + a_0 \quad (a_0 > 0) \tag{9.31}$$

and

$$h_\star(r) = \frac{a_\star}{6}r^3 + \frac{b}{2}r^2 - r + a_1 \quad (a_1 > 0) \tag{9.32}$$

for $a > 0$, $b > 0$ satisfying the conditions, for all $x, y \in U(x^\star, R)$,

$$\|F'(x^\star)^{-1}[F''(y) - F''(x)]\| \leq a\|y - x\|, \tag{9.33}$$

$$\|F'(x^\star)^{-1}[F''(x) - F''(x^\star)]\| \leq a_\star\|x - x^\star\| \tag{9.34}$$

and

$$\|F'(x^\star)^{-1}F''(x^\star)\| \leq b. \tag{9.35}$$

Functions $h$, $h_\star$ satisfy conditions $(C_1)$–$(C_4)$, with

$$R_0 = \frac{2}{b + \sqrt{b^2 + 2a_\star}}. \tag{9.36}$$

We provide a numerical example for case 1.

**Example 9.1** *Returning back to Remark 9.1 with $D = U(0, 1)$, we get*

$$a = e, \quad a_\star = e - 1 \ \text{and} \ b = 1.$$

*Then, we get*

$$R_0 = 0.6438496678987803, \quad r_0 = 0.5654448141543782,$$

$$R_1 = 0.3203645228203808, \quad r_1 = 0.2878878975673239,$$

$$R^\star = 0.2829778199426982, \quad r^\star = 0.2572415738179739.$$

*Choosing $x_0 = 0.25 \in U(0, 1)$, we can produce sequence $\{x_n\}$ by using Halley's method (9.1). The comparison results for the error bound by using (9.20) and (9.28) are given in Table 9.1. The table shows that tighter error bounds on the distances $\|x_n - x^\star\|$ can be obtained by using both functions $h$ and $h_\star$ instead of by using only function $h$.*

**Table 9.1:** The comparison results for the error bound of Halley's method

| n | The right hand of (9.20) | The right hand of (9.28) |
|---|---|---|
| 0 | 1.95126e-01 | 2.36123e-01 |
| 1 | 2.70579e-08 | 3.27429e-08 |
| 2 | 7.35104e-29 | 8.89552e-39 |
| 3 | 1.31477e-49 | 1.59101e-49 |

**2. Smale–type.** We study the local convergence of Halley's method (9.1) assuming that $F$ is analytic and satisfies

$$\|F'(x^\star)^{-1}F^{(n)}(x^\star)\| \leq n!\gamma_\star^{n-1} \ (n \geq 2), \tag{9.37}$$

where

$$\gamma_\star = \sup_{n>1} \|\frac{F'(x^\star)^{-1}F^{(n)}(x^\star)}{n!}\|^{\frac{1}{n-1}}. \tag{9.38}$$

Note that Smale used (1) (for $x^\star$ replaced by $x_0$) to study the convergence of (NM) (2.2) (cf. [683]). Define functions $h, h_\star$ by

$$h(r) = \frac{\gamma_0 r^2}{1-\gamma_0 r} - r + \beta_0, \ \ \beta_0 > 0, \ 0 \le r < \frac{1}{\gamma_0}, \tag{9.39}$$

$$h_\star(r) = \frac{\gamma_1 r^2}{1-\gamma_1 r} - r + \beta_1, \ \ \beta_1 > 0, \ 0 \le r < \frac{1}{\gamma_1}. \tag{9.40}$$

Conditions (9.13), (9.14) using (9.39), (9.40) become, respectively,

$$\begin{aligned} &\|F'(x^\star)^{-1}[F''(y) - F''(x)]\| \\ &\le 2\gamma_0\left(\frac{1}{(1-\gamma_0\|y-x\| - \gamma_0\|x-x^\star\|)^3} - \frac{1}{(1-\gamma_0\|x-x^\star\|)^3}\right) \end{aligned} \tag{9.41}$$

and

$$\|F'(x^\star)^{-1}[F''(x) - F''(x^\star)]\| \le 2\gamma_1\left(\frac{1}{(1-\gamma_1\|x-x^\star\|)^3} - 1\right). \tag{9.42}$$

We have $\gamma_1 \le \gamma_0$ and $\gamma_1 = \gamma_0 = \gamma_\star$ is certainly a choice for $\gamma_1$ and $\gamma_0$. Condition (9.41) (for $x^\star, \gamma$ replaced by $x_0, \gamma$ is the so–called $\gamma-$condition introduced by Wang and Han to study the semilocal convergence of Newton's method (cf. [751]) (see also our works in (cf. [163], [209])). It turns out that there is an easier way ro verify condition (9.13) (or (9.14)).

**Proposition 9.1** *Let $F : D \subset X \to Y$ be a thrice continuously differentiable operator. Let $h : [0, R) \to \mathbb{R}$ be a thrice continuous differentiable function with convex $h''$. Then, $F$ satisfies (9.13) if and only if*

$$\|F'(x^\star)^{-1}F'''(x)\| \le h'''(\|x - x^\star\|). \tag{9.43}$$

**Proof** If $F$ satisfies (9.43), then, we get in turn that

$$\begin{aligned} \|F'(x^\star)^{-1}[F''(y) - F''(x)]\| &\le \int_0^1 \|F'(x^\star)^{-1}F'''(x + \theta(y-x))(y-x)\|d\theta \\ &\le \int_0^1 h'''(\|x-x^\star\| + \theta\|y-x\|)\|y-x\|d\theta \\ &= h''(\|y-x\| + \|x-x^\star\|) - h''(\|x-x^\star\|), \end{aligned}$$

which shows (9.13). If $F$ satisfies (9.13), then, (9.43) certainly holds. This completes the proof of the proposition. ⊠

**Remark 9.2** *If $h$ is given by (9.39), then, (9.43) reduces to*

$$\|F'(x^\star)^{-1}F'''(x)\| \le \frac{6\gamma^2}{(1-\gamma\|x-x^\star\|)^4}, \tag{9.44}$$

*where we set $\gamma_0 = \gamma_1 = \gamma$ for simplicity. As an example, note that if $F$ is an analytic operator and $h$ is given by (9.39), we have*

$$\begin{aligned} \|F'(x^\star)^{-1}F'''(x)\| &= \|\sum_{n=0}^{\infty} \frac{1}{n!} F'(x^\star)^{-1}F^{(n+3)}(x^\star)(x-x^\star)^n\| \\ &\le \gamma^2 \sum_{n=0}^{\infty}(n+3)(n+2)(n+1)(\gamma\|x-x^\star\|)^n \\ &= \frac{6\gamma^2}{(1-\gamma\|x-x^\star\|)^4}, \quad for \ \ \gamma\|x-x^\star\| < 1, \end{aligned}$$

*which implies that $F$ satisfies the $\gamma-$condition in this case.*

## CHAPTER 10

# Methods for Variational Inequalities

In this chapter we provide convergence results for variational inequalities under Lipschitz–type conditions on the derivatives and Lipschitz–like assumption on set–valued maps.

## 10.1 Subquadratic convergent method

We present a new approach of the convergence of Chebyshev–type iterative method in Banach space for solving variational inclusions under different assumptions used in (cf. [389, 391, 153]). We relax Lipschitz, Hölder or center–Hölder type conditions by introducing $\omega$–type–conditioned second order Fréchet derivative. Under this condition, we show that the sequence is locally superquadratically convergent if some Aubin continuity property is satisfied. In particular, we recover a quadratic and a cubic convergence.

For $X$ and $Y$ two Banach spaces, $f : X \to Y$ denotes a twice (Fréchet) differentiable function while $G : X \longrightarrow 2^Y$ stands for a set–valued mapping with closed graph. We are concerned with the problem of approximating a solution $x^*$ of the generalized equation given by

$$0 \in f(x) + G(x) \tag{10.1}$$

and we consider the following iterative method for solving (10.1)

$$0 \in A(x_{k+1}, x_k) + G(x_{k+1}), \tag{10.2}$$

where

$$A(y, x) = f(x) + \nabla f(x)(y - x) + \frac{1}{2}\nabla^2 f(x)(y - x)^2 \quad (x, y \in X). \tag{10.3}$$

Algorithm (10.2) is based on the second–degree Taylor polynomial expansion $A$ of $f$. The cubically convergence of method (10.2) is presented in (cf. [389]) when the set–valued mapping $[A(\cdot, x^*) + G(\cdot)]^{-1}$ is Aubin continuous around $(0, x^*)$ (or pseudo–Lipschitz at $(0, x^*)$), and the function $f$ is $\mathcal{C}^2$ and the second Fréchet derivative of $f$ is $L$–Lipschitz in some neighborhood $V$ of $x^*$

$$\| \nabla^2 f(x) - \nabla^2 f(y) \| \leq L \| x - y \| \quad (x, y \in V). \tag{10.4}$$

Geoffroy and Piétrus (cf. [391]) showed that the sequence (10.2) is locally superquadratic convergent to the solution $x^*$ whenever $\nabla^2 f$ satisfies some $\alpha$–Hölder–type condition on some neighborhood $V$ of $x^*$ with constant $K$ ($\alpha,\ K > 0$)

$$\| \nabla^2 f(x) - \nabla^2 f(y) \| \leq K \ \| x - y \|^\alpha \quad (x, y \in V). \tag{10.5}$$

The stability of (10.2) is investigated in (cf. [390]) with respect to some perturbations; more precisely, if we consider the perturbed equation $y \in f(x) + G(x)$ ($y$ is some parameter in $Y$) then the attraction region does not depend on small perturbations of the parameter $y$.

Argyros (cf. [153]) provided a finer local superquadratic convergence of algorithm (10.2) using $\alpha$–center–Hölder condition on some neighborhood $V$ of $x^*$ with constant $K_0$ ($\alpha,\ K_0 > 0$)

$$\| \nabla^2 f(x) - \nabla^2 f(x^*) \| \leq K_0 \ \| x - x^* \|^\alpha \quad (x \in V). \tag{10.6}$$

In this section, we use different conditions to the previous one to study the convergence of (10.2). We relax these usual Lipschitz and Hölder conditions by $\omega$–conditioned second derivative. This condition is used in (cf. [354, 418]) to study Newton's method for solving nonlinear equations ($G = \{0\}$ in (10.1)). The main conditions required are

$$\| \nabla^2 f(x) - \nabla^2 f(y) \| \leq \omega(\| x - y \|) \quad (x, y \in V), \tag{10.7}$$

$$\begin{gathered} \| \nabla^2 f(x) - \nabla^2 f(y) \| \leq \sigma(\| x - y \|) \ \| x - y \|^\theta, \\ \text{for all } x,\ y \text{ in } V \text{ and } \theta \text{ is fixed in } (0, 1], \end{gathered} \tag{10.8}$$

$$\| \nabla^2 f(x) - \nabla^2 f(x^*) \| \leq \mu(\| x - x^* \|), \quad (x, y \in V), \tag{10.9}$$

$$\| \nabla^2 f(x) - \nabla^2 f(x^*) \| \leq \vartheta(\| x - x^* \|) \ \| x - x^* \|^\theta \quad (x, y \in V), \tag{10.10}$$

where $\omega,\ \sigma,\ \mu,\ \vartheta\ :\ \mathbb{R}_+ \longrightarrow \mathbb{R}_+$ are a continuous nondecreasing functions. When the condition (10.7) is satisfied, we say that $\nabla^2 f$ is $\omega$–conditioned. The condition (10.9) is called $\mu$–center–condition on the second derivative $\nabla^2 f$. Similar conditions to (10.7) and (10.9) on the Fréchet derivative $\nabla f$ are used in (cf. [178]) to study of Newton's methods for solving (10.1). The inspiration for considering (10.8) comes from (cf. [806, 87]).

**Example 10.1** *Such a study of this section can be of interest, for example, to variational inequalities for saddle points (cf. [660]). Let $A$ and $B$ be nonempty, closed and convex subsets of $\mathbb{R}^n$ and $\mathbb{R}^m$, respectively, and let $\mathcal{L}$ : $\mathbb{R}^n \times \mathbb{R}^m \longrightarrow \mathbb{R}$ be some $C^1$ convex–concave on $A \times B$. The point $(\bar{x},\bar{y}) \in A \times B$ is a saddle point if the following hold*

$$\mathcal{L}(x, \bar{y}) \ \geq \ \mathcal{L}(\bar{x}, \bar{y}) \ \geq \ \mathcal{L}(\bar{x}, y) \quad (x \in A,\ \ y \in B). \tag{10.11}$$

*The saddle point condition (10.11) is equivalent to*

$$0 \in f(\bar{x}, \bar{y}) + G(\bar{x}, \bar{y}), \tag{10.12}$$

*where $f$ and $G$ are defined on $A \times B$ by $f(x, y) = (\nabla_x \mathcal{L}(x, y), -\nabla_y \mathcal{L}(x, y))$ and $G(x, y) = N_A(x) \times N_B(y)$ with $N_A$ (resp. $N_B$) the normal cone to the set $A$ (resp. $B$). Hence, the variational problem (10.11) corresponds to generalized equation in formulation (10.1) and $(\bar{x}, \bar{y})$ can be approximated by the method (10.2).*

We need some basic definition and results.

**Definition 10.1 (Aubin–like continuity)** *A set–valued map $\Gamma : X \to 2^Y$ is said to be $M$–pseudo–Lipschitz around $(\alpha_0, \beta_0) \in gph\Gamma$, where $M > 0$ ($\Gamma$ is also called Aubin continuous at $(\alpha_0, \beta_0)$) if there exist constants $a$ and $b$ such that*

$$e(\Gamma(y) \cap \mathbb{B}_a(\beta_0), \Gamma(z)) \leq M\|y - z\| \quad (y, z \in \mathbb{B}_b(\alpha_0)),$$

*where the excess from the set $A$ to the set $C$ is defined by $e(C, A) = \sup_{x \in C} dist\,(x, A)$.*

**Remark 10.1** *The pseudo–lipschitzian property has been introduced by Aubin, see (cf. [218]). A basic characterization of this property of the inverse of a set–valued map is given by Graves theorem (cf. [335]). Other characterizations of Aubin continuity concept are obtained by Rockafellar (cf. [659]) using the Lipschitz continuity of the distance function $(x, y) \longrightarrow dist\,(y, \Gamma(x))$ around $(x_0, y_0)$, and by Mordukhovich (cf. [535, 660]) via the concept of coderivative of multiapplications $\mathcal{D}^*\Gamma(x/y)$, i.e.;*

$$v \in \mathcal{D}^*\Gamma(x/y)(u) \Longleftrightarrow (v, -u) \in N_{\mathrm{gph}\,\Gamma}(x, y). \tag{10.13}$$

*The Mordukhovich criterion says that $\Gamma$ of a closed graph is pseudo–Lipschitz around $(x_0, y_0)$ if and only if*

$$\| \mathcal{D}^*\Gamma(x_0/y_0) \|^+ = \sup_{u \in \mathbb{B}_1(0)} \quad \sup_{v \in \mathcal{D}^*\Gamma(x_0/y_0)(u)} \quad \| v \| < \infty. \tag{10.14}$$

*In fact, the Mordukhovich criterion plays a fundamental role in variational analysis and its applications. We refer the reader to (cf. [218, 219, 335, 336, 535, 536, 659, 660]) and the references given there for more details and applications of Aubin continuity concept.*

The first tool of our analysis is the fixed point theorem for set–valued map proved by Dontchev and Hager (cf. [335]). This theorem is a generalization of Picard fixed point theorem restricted to single–valued mappings.

**Lemma 10.1** *(cf. [335]) Let $\phi$ a set–valued map from $X$ into the closed subsets of $X$, let $\eta_0 \in X$ and let $r$ and $\lambda$ be such that $0 \leq \lambda < 1$ and the following conditions hold*

*(a)* dist $(\eta_0, \phi(\eta_0)) \leq r(1 - \lambda)$.

*(b)* $e(\phi(y) \cap \mathbb{B}_r(\eta_0), \phi(z)) \leq \lambda \parallel y - z \parallel$ forall $\mathrm{y, z} \in \mathbb{B}_{\mathrm{r}}(\eta_0)$.

*Then, $\phi$ has a fixed–point in $\mathbb{B}_r(\eta_0)$. That is, there exists $x \in \mathbb{B}_r(\eta_0)$ such that $x \in \phi(x)$. If $\phi$ is single–valued, then, $x$ is the unique fixed point of $\phi$ in $\mathbb{B}_r(\eta_0)$.*

By the second order Taylor expansion of $f$ at $y \in V$ with the remainder is given by integral form, the following lemmas are obtained directly.

**Lemma 10.2** *We suppose that the assumption (10.7) is satisfied on a convex neighborhood $V$. Then, for all $x$ and $y$ in $V$, we have the following*

$$\begin{aligned}&\parallel f(x) - f(y) - \nabla f(y)\,(x-y) - \frac{1}{2}\nabla^2 f(y)\,(x-y)^2 \parallel \\ &\leq \parallel x - y \parallel^2 \; \textstyle\int_0^1 (1-t)\,\omega(t \parallel x - y \parallel)\,dt.\end{aligned}$$

*In particular, if the assumption (10.9) is satisfied, then, for all $x$ in $V$, we have the following*

$$\begin{aligned}&\parallel f(x) - f(x^*) - \nabla f(x^*)\;(x-x^*) - \frac{1}{2}\nabla^2 f(x^*)\;(x-x^*)^2 \parallel \\ &\leq \parallel x - x^* \parallel^2 \; \textstyle\int_0^1 (1-t)\,\mu(t \parallel x - x^* \parallel)\,dt.\end{aligned}$$

**Lemma 10.3** *We suppose that the assumption (10.8) is satisfied on a convex neighborhood $V$. Then, for all $x$ and $y$ in $V$ we have the following*

$$\begin{aligned}&\parallel f(x) - f(y) - \nabla f(y)\,(x-y) - \frac{1}{2}\nabla^2 f(y)\,(x-y)^2 \parallel \\ &\leq \parallel x - y \parallel^{2+\theta} \; \textstyle\int_0^1 t^\theta\,(1-t)\,\sigma(t \parallel x - y \parallel)\,dt.\end{aligned}$$

**Remark 10.2** $\displaystyle\int_0^1 (1-t)\,\omega(t \parallel x - x^* \parallel)\,dt$ *and* $\displaystyle\int_0^1 t^\theta\,(1-t)\,\sigma(t \parallel x - x^* \parallel)\,dt$ *given in the previous lemmas are bounded by $\omega(diam\ (V))$ and $\sigma(diam\ (V))$, respectively, where $diam\ (V)$ is the diameter of neighborhood $V$.*

Before stating the main results on this section, we need to introduce some notations. First, for $k \in \mathbb{N}$ and $(x_k)$ defined in (10.2), let us define the set–valued mappings $Q : X \longrightarrow 2^Y$ and $\psi_k : X \longrightarrow 2^X$ by the following

$$Q(.) := A(., x^*) + G(.); \quad \phi_k(.) := Q^{-1}(Z_k(.)), \tag{10.15}$$

where $Z_k$ is defined from $X$ to $Y$ by

$$Z_k(x) := A(x, x^*) - A(x_k, x). \tag{10.16}$$

Let us mention that $x_1$ is a fixed point of $\phi_0$ if and only if $0 \in A(x_1, x_0) + G(x_1)$.

We will make the following assumptions in an open convex neighborhood $V$ of $x^*$

($\mathcal{H}0$) $\nabla f$ is $L$–Lipschitz on $V$ with $L > 0$ and there exists $L_0 > 0$ such that $\parallel \nabla^2 f(x^*) \parallel < L_0$;

($\mathcal{H}1$) The condition (10.7) is satisfied on $V$;

($\mathcal{H}1$) $^\star$ The condition (10.8) is satisfied on $V$;

($\mathcal{H}2$) The set–valued map $[A(.,x^*) + G(.)]^{-1}$ is pseudo–Lipschitz around $(0, x^*)$ with constants $M$, $a$ and $b$ (these constants are given by definition of Aubin continuity).

The main theorems of this section read as follows.

**Theorem 10.1** *Let $x^*$ be a solution of* (10.1). *We suppose that assumptions ($\mathcal{H}0$)–($\mathcal{H}2$) are satisfied and we denote by $\beta = M \displaystyle\int_0^1 (1-t)\,\omega(t\,a)\,dt$. Then, for every $C > \beta$, there exist $\delta > 0$ such that, for every starting point $x_0 \in \mathbb{B}_\delta(x^*)$, a sequence $(x_k)$ for (10.1), defined by* (10.2), *satisfies*

$$\| x_{k+1} - x^* \| \leq C \| x_k - x^* \|^2 . \tag{10.17}$$

*In other words,* (10.2) *generates $(x_k)$ with second order.*

**Theorem 10.2** *Let $x^*$ be a solution of* (10.1). *We suppose that assumptions ($\mathcal{H}0$), ($\mathcal{H}1$)$^\star$ and ($\mathcal{H}2$) are satisfied and we denote by $\beta' = M \displaystyle\int_0^1 t^\theta$ $(1-t)\,\sigma(t\,a)\,dt$. Then, for every $C' > \beta'$, there exist $\gamma > 0$ such that, for every starting point $x_0 \in \mathbb{B}_\gamma(x^*)$, a sequence $(x_k)$ for (10.1), defined by* (10.2), *satisfies*

$$\| x_{k+1} - x^* \| \leq C' \| x_k - x^* \|^{2+\theta} . \tag{10.18}$$

*In other words,* (10.2) *generates $(x_k)$ with superquadratic convergence. In particular, if $\theta = 1$, then, the convergence is cubic.*

Theorem 10.1 is showed as follows. Once $x_k$ is computed, we show that the function $\phi_k$ has a fixed point $x_{k+1}$ in $X$. This process allows us to prove the existence of a sequence $(x_k)$ satisfying (10.2). Now, we state a result which is the starting point of our algorithm. It will be very useful to prove theorem 10.1 and reads as follows.

**Proposition 10.1** *Under the hypotheses of theorem 10.1, there exists $\delta > 0$ such that, for all $x_0 \in \mathbb{B}_\delta(x^*)$ ($x_0 \neq x^*$), the map $\phi_0$ has a fixed point $x_1$ in $B_\delta(x^*)$ satisfying*

$$\| x_1 - x^* \| \leq C \| x_0 - x^* \|^2 .$$

**Proposition 10.2** *Under the hypotheses of theorem 10.2, there exists $\gamma > 0$ such that, for all $x_0 \in \mathbb{B}_\gamma(x^*)$ ($x_0 \neq x^*$), the map $\phi_0$ has a fixed point $x_1$ in $B_\gamma(x^*)$ satisfying*

$$\| x_1 - x^* \| \leq C' \| x_0 - x^* \|^{2+\theta} .$$

**Proof of Proposition 10.1.** By hypothesis ($\mathcal{H}2$), we have

$$e(Q^{-1}(y') \cap \mathbb{B}_a(x^*), Q^{-1}(y'')) \leq M \parallel y' - y'' \parallel \qquad (y', y'' \in \mathbb{B}_b(0)). \quad (10.19)$$

Fix $\delta > 0$ such that

$$\delta < \min\left\{a;\ \frac{1}{C};\ \sqrt{\frac{b}{5\,\beta}}\right\}. \quad (10.20)$$

To prove Proposition 10.1 we intend to show that both assertions (a) and (b) of Lemma 10.1 hold; where $\eta_0 := x^*$, $\phi$ is the function $\phi_0$ defined at the very beginning of this section and where $r$ and $\lambda$ are numbers to be set. According to the definition of the excess $e$, we have

$$\text{dist }(x^*, \phi_0(x^*)) \leq e\big(Q^{-1}(0) \cap \mathbb{B}_\delta(x^*), \phi_0(x^*)\big). \quad (10.21)$$

Moreover, for all $x_0 \in B_\delta(x^*)$ such that $x_0 \neq x^*$ we have by ($\mathcal{H}1$) and Lemma 10.2

$$\parallel Z_0(x^*) \parallel = \parallel A(x_0, x^*) \parallel \leq \frac{\beta}{M} \parallel x_0 - x^* \parallel^2 . \quad (10.22)$$

Then, (10.20) yields $\parallel Z_0(x^*) \parallel < b$. Hence, from (10.19), one has

$$e\big(Q^{-1}(0)\cap\mathbb{B}_\delta(x^*), \phi_0(x^*)\big) = e\big(Q^{-1}(0)\cap\mathbb{B}_\delta(x^*), Q^{-1}[Z_0(x^*)]\big) \leq \beta \parallel x^* - x_0 \parallel^2 .$$

By (10.21), we get

$$\text{dist }(x^*, \phi_0(x^*)) \leq \beta \parallel x^* - x_0 \parallel^2 . \quad (10.23)$$

Since $C > \beta$, there exists $\lambda \in ]0, 1[$ such that $C\ (1 - \lambda) \geq \beta$. Hence, we have

$$\text{dist }(x^*, \phi_0(x^*)) \leq C\,(1 - \lambda) \parallel x_0 - x^* \parallel^2 . \quad (10.24)$$

By setting $\eta_0 := x^*$ and $r := r_0 = C \parallel x^* - x_0 \parallel^2$, we can deduce from the last inequalities that assertion (a) in lemma 10.1 is satisfied.

Now, we show that condition (b) of Lemma 10.1 is satisfied. Since $\dfrac{1}{C} \geq \delta$ and $\parallel x^* - x_0 \parallel \leq \delta$, we have $r_0 \leq \delta \leq a$. Moreover by Lemma 10.2, we have, for $x \in \mathbb{B}_\delta(x^*)$,

$$\begin{aligned} \parallel Z_0(x) \parallel &= \parallel A(x, x^*) - A(x_0, x) \parallel \\ &\leq \parallel A(x, x^*) \parallel + \parallel A(x_0, x) \parallel \\ &\leq \beta \parallel x - x^* \parallel^2 + \beta \parallel x - x_0 \parallel^2 \leq 5\,\beta\,\delta^2. \end{aligned} \quad (10.25)$$

Then, by (10.20) we deduce that, for all $x \in \mathbb{B}_\delta(x^*)$, $Z_0(x) \in \mathbb{B}_b(0)$. Then, it follows that for all $x', x'' \in \mathbb{B}_{r_0}(x^*)$, we have

$$I = e(\phi_0(x') \cap \mathbb{B}_{r_0}(x^*), \phi_0(x'')) \leq e(\phi_0(x') \cap \mathbb{B}_\delta(x^*), \phi_0(x'')), \quad (10.26)$$

which yields, by (10.19),

$$
\begin{aligned}
I &\leq M \ \| Z_0(x') - Z_0(x'') \| \\
&\leq M \ \| \nabla f(x^*)(x' - x'') - \nabla f(x_0)(x' - x'') + \ \frac{1}{2}\nabla^2 f(x^*)(x' - x^*)^2 \\
&\quad - \frac{1}{2}\nabla^2 f(x^*)(x'' - x^*)^2 + \frac{1}{2}\nabla^2 f(x_0)(x'' - x_0)^2 - \frac{1}{2}\nabla^2 f(x_0)(x' - x_0)^2 \ \| \\
&\leq M \ \| \nabla f(x^*)(x' - x'') - \nabla f(x_0)(x' - x'') + \frac{1}{2}\nabla^2 f(x^*)(x' - x'' + x'' - x^*)^2 \\
&\quad - \frac{1}{2}\nabla^2 f(x^*)(x'' - x^*)^2 + \frac{1}{2}\nabla^2 f(x_0)(x'' - x_0)^2 - \frac{1}{2}\nabla^2 f(x_0)(x' - x'' + x'' - x_0)^2 \ \| \\
&= M \ \| \nabla f(x^*)(x' - x'') - \nabla f(x_0)(x' - x'') \\
&\quad + \frac{1}{2}(\nabla^2 f(x^*)(x' - x'')^2 - \nabla^2 f(x_0)(x' - x'')^2) \\
&\quad + \nabla^2 f(x^*)(x'' - x_0 + x_0 - x^*)(x' - x'') - \nabla^2 f(x_0)(x'' - x_0)(x' - x'') \ \| \\
&\leq M( \ \| \nabla f(x^*) - \nabla f(x_0) \| \| x' - x'' \| + \frac{1}{2} \| \nabla^2 f(x^*) - \nabla^2 f(x_0) \| \| x' - x'' \|^2 \\
&\quad + \| \nabla^2 f(x^*) - \nabla^2 f(x_0) \| \| x'' - x_0 \| \| x' - x'' \| \\
&\quad + \| \nabla^2 f(x^*) \| \| x_0 - x^* \| \| x' - x'' \| \ )
\end{aligned} \tag{10.27}
$$

By Assumptions $(\mathcal{H}0)$–$(\mathcal{H}1)$ and (10.20), we deduce that

$$
\begin{aligned}
I &\leq M(L\,\delta + \omega(a)\,\delta + 2\,\omega(a)\,\delta + L_0\,\delta) \ \| x' - x'' \| \\
&= M\,\delta\,(L + L_0 + 3\,\omega(a)) \ \| x' - x'' \| \,.
\end{aligned} \tag{10.28}
$$

Without loss of generality, we may assume that $\delta < \dfrac{\lambda}{M\,(L + L_0 + 3\,\omega(a))}$. Then, condition (b) of Lemma 10.1 is satisfied. Since both conditions of Lemma 10.1 are fulfilled, we can deduce the existence of a fixed point $x_1 \in \mathbb{B}_{r_0}(x^*)$ for the map $\phi_0$. Then, the proof of Proposition 10.1 is complete. ⊠

**Idea of the proof of Proposition 10.2.** The proof of Proposition 10.2 is the same one as that of the proof of Proposition 10.1. It is enough to make some modifications by choosing the constant $\gamma$ such that

$$
\gamma < \min\left\{a;\ \left(\frac{1}{C'}\right)^{\frac{1}{1+\theta}};\ \left(\frac{b}{(1 + 2^{2+\theta})\,\beta'}\right)^{\frac{1}{2+\theta}}\right\}. \tag{10.29}
$$

⊠

Now that we proved Proposition 10.1 and Proposition 10.2, the proof of Theorem 10.1 and Theorem 10.2 is straightforward as it is shown below.

**Proof of Theorem 10.1 and Theorem 10.2.** Proceeding by induction, keeping $\eta_0 = x^*$ and setting $r_k = C \ \| x_k - x^* \|^2$ and $r'_k = C' \ \| x_k - x^* \|^{2+\theta}$, the application of Proposition 10.1 and Proposition 10.2 to the map $\phi_k$, respectively, gives the desired results. ⊠

**Remark 10.3** *Theorem 10.1 and Theorem 10.2 remain true under (10.9) and (10.10).*

## 10.2 Convergence Under Slant Condition

In this section we present a new results for the local convergence of the Newton–like method to a unique solution of nondifferentiable variational inclusions in a Banach space setting. Using the Lipschitz–like property of set–valued mappings and the concept of slant differentiability hypothesis on the operator involved, as was inaugurated by Chen, Nashed and Qi (cf. [279]). The linear convergence of the Newton–like method is also established. Our results extend the applicability of Newton–like method (cf. [194], [279]) to variational inclusions.

We are concerned with the problem of approximating a locally unique solution $x^\star \in \mathcal{D}$ of the generalized equation (10.1), where $f$ is replaced by $F$. A large number of problems in applied mathematics and engineering are solved by finding the solutions of generalized equation (10.1), introduced by Robinson (cf. [646]). Consider an example of application to variational inequalities (cf. [660]).

Let $K$ be a convex set in $\mathbb{R}^n$ and $\varphi$ be a function from $K$ to $\mathbb{R}^n$. The variational inequality problem consists in seeking $k_0$ in $K$ such that, for each $k \in K$,

$$(\varphi(k_0), k - k_0) \ \geq \ 0, \tag{10.30}$$

where $(.,.)$ denotes the Euclidean scalar product on $\mathbb{R}^n$ defined by $(x, y) = \sum_{i=1}^{n} x_i \, y_i$ for all $x$, $y$ in $\mathbb{R}^n$. Let $\mathcal{I}_K$ denote the convex indicator function of $K$ and $\partial$ the subdifferential operator. Then, the problem (10.30) is equivalent to

$$0 \in \varphi(k_0) + \mathcal{H}(k_0) \tag{10.31}$$

with $\mathcal{H} = \partial \mathcal{I}_K$ (also called the normal cone of $K$). The variational inequality problem (10.30) is equivalent to (10.31) which is a generalized equation in formulation (10.1). In the particular case $G = \{0\}$, (10.1) is a nonlinear equation in the form (2.1) Most of the numerical approximation methods of $x^\star$ of (10.1) require the expensive computation of the Fréchet–derivative $F'(x)$ of operator $F$ at each step, for example Newton's method

$$0 \in F(x_n) + F'(x_n) \ (x_{n+1} - x_n) + G(x_{n+1}) \quad (x_0 \in \mathcal{D}, \ n \geq 0). \tag{10.32}$$

Alternative methods to (10.32) are given by

(i) The Secant method

$$0 \in F(x_n) + [x_n, x_{n-1}; F] \, (x_{n+1} - x_n) + G(x_{n+1}) \quad (x_{-1}, \, x_0 \in \mathcal{D}, \ n \geq 0), \tag{10.33}$$

(ii) Steffensen's method

$$0 \in F(x_n) + [g_1(x_n), g_2(x_n); F]\,(x_{n+1} - x_n) + G(x_{n+1}) \quad (x_0 \in \mathcal{D},\ n \geq 0), \tag{10.34}$$

where $g_i$ $(i = 1, 2)$ are continuous functions from a neighborhood $\mathcal{D}$ of $x^\star$ into $\mathcal{X}$ and $[x, y; F]$ is a first order divided difference of $F$ on the points $x$ and $y$.

A comprehensive bibliography on these methods and their variants is given in (cf. [139], [195]).

We are interested in a numerical method for solving generalized equation (10.1), when the involved function $F$ is slantly differentiable at $x^\star$. We proceed by replacing the term $F'(x_n)$ in method (10.32) by $A(x_n)$, where $A(x) \in \mathcal{L}(\mathcal{X}, \mathcal{Y})$. We consider the Newton–like method

$$0 \in F(x_n) + A(x_n)\,(x_{n+1} - x_n) + G(x_{n+1}) \quad (x_0 \in \mathcal{D},\ n \geq 0), \tag{10.35}$$

for approximating $x^\star$.

In the nonlinear equation case (i.e., $G = \{0\}$ in (10.1)), the method (10.35) becomes

$$x_{n+1} = x_n - A(x_n)^{-1}\,F(x_n) \quad (x_0 \in \mathcal{D},\ n > 0), \tag{10.36}$$

which was considered by Chen, Nashed and Qi (cf. [279]) for slantly differentiable operator. Consequently, we can approximate the solution $k_0$ of variational inequality (10.30) using our algorithm (10.35), when $\varphi$ is slantly differentiable. Here, we are motivated by the works in (cf. [194], [279], [280], [616]). Using the Lipschitz–like concept of set–valued mappings and slant differentiability hypothesis on the involved operator, we extend the applicability of Newton's method (cf. [279], [280]) to variational inequalities. We prove that Newton–like method (10.35) converges linearly.

First, we collect a number of basic definitions and properties on slant differentiability. We need the notion of slant differentiability (cf. [279]).

**Definition 10.2** *An operator $F : \mathcal{D} \subseteq \mathcal{X} \longrightarrow \mathcal{Y}$ is slantly differentiable at $x \in \mathcal{D}$ if there exists an operator $f^\circ : \mathcal{D} \longrightarrow \mathcal{L}(\mathcal{X}, \mathcal{Y})$ such that the family $\{f^\circ(x + h)\}$ of bounded linear mappings is uniformly bounded in the operator norm, for $h$ sufficiently small and*

$$\lim_{h \to 0} \frac{F(x + h) - F(x) - f^\circ(x + h)\,h}{\|\, h \,\|} = 0. \tag{10.37}$$

*The function $f^\circ$ is called a slanting function for $F$ at $x$.*

**Definition 10.3** *An operator $F : \mathcal{D} \subseteq \mathcal{X} \longrightarrow \mathcal{Y}$ is slantly differentiable in an open domain $\mathcal{D}_0 \subseteq \mathcal{D}$ if there exists an operator $f^\circ : \mathcal{D} \longrightarrow \mathcal{L}(\mathcal{X}, \mathcal{Y})$ such that $f^\circ$ is a slantly operator for $F$ at every point $x \in \mathcal{D}_0$. In this case, $f^\circ$ is called a slanting function for $F$ in $\mathcal{D}_0$.*

**Definition 10.4** *Assume $f^\circ : \mathcal{D} \longrightarrow \mathcal{L}(\mathcal{X}, \mathcal{Y})$ is a slantly operator for $F$ at $x \in \mathcal{D}$. The set*

$$\partial_S F(x) : = \{ \lim_{x_k \to x} f^\circ(x_k)\} \tag{10.38}$$

*is the slant derivative of $F$ associated with $f^\circ$ at $x \in \mathcal{D}$. Here, $\lim_{x_k \to x} f^\circ(x_k)$ means the limit of $f^\circ(x_k)$ for any sequence $\{x_k\} \subset \mathcal{D}$ such that $x_k \to x$ and $\lim_{x_k \to x} f^\circ(x_k)$ exists. Note that $\partial_S F(x) \neq \emptyset$ since $f^\circ(x) \in \partial_S F(x)$.*

We finally need the definitions of smoothing and semismoothing functions of $F$ (cf. [279], [675]).

**Definition 10.5** *An operator $f : \mathcal{D} \times (0, +\infty) \longrightarrow \mathcal{Y}$ is a smoothing mapping of $F$ if $f$ is continuously differentiable with respect to $x$ and, for any $x \in \mathcal{D}$ and $\epsilon > 0$,*

$$\| F(x) - f(x, \epsilon) \| \leq \mu \, \epsilon, \tag{10.39}$$

*where $\mu$ is a positive constant. The smoothing operator is said to satisfy the slant derivative consistency property at $\widehat{x}$ (in $\mathcal{D}$) if*

$$\lim_{\epsilon \to 0+} f'_x(x, \epsilon) = f^\circ(x) \in \mathcal{L}(\mathcal{X}, \mathcal{Y}) \tag{10.40}$$

*exists for $x$ in a neighborhood of $\widehat{x}$ (in $\mathcal{D}$) and $f^\circ$ serves as a slanting mapping for $F$ at $\widehat{x}$ (in $\mathcal{D}$).*

**Definition 10.6** *An operator $F : \mathcal{D} \subseteq \mathcal{X} \longrightarrow \mathcal{Y}$ is semismooth at $x$ if there is a slantly operator $f^\circ$ for $F$ in a neighborhood $N_x$ of $x$ such that $f^\circ$ and the associated slant derivative satisfy the conditions:*

*(i)* $\lim_{t \to 0+} f^\circ(x + t\, h)\, h$ *exists for every $h \in \mathcal{X}$,*

*and*

$$\lim_{h \to 0} \frac{\lim_{t \to 0+} f^\circ(x + t\, h)\, h - f^\circ(x + h)\, h}{\| h \|} = 0. \tag{10.41}$$

*(ii)*

$$f^\circ(x + h)\, h - V\, h = o(\| h \|) \qquad (V \in \partial_S F(x + h)). \tag{10.42}$$

**Definition 10.7** *1. A mapping $F : \mathcal{D} \subseteq \mathbb{R}^n \longrightarrow \mathbb{R}^n$ is said to be directionally differentiable at $x \in \mathcal{D}$ along direction $d$ if the following limit*

$$F'(x; d) : = \lim_{t \downarrow 0} \frac{F(x + t\, d) - F(x)}{t} \tag{10.43}$$

*exists. Note that a locally Lipschitz mapping $F$ admits directional derivative at $x \in \mathcal{D}$ if and only if (cf. [675])*

$$\lim_{d \to 0} \frac{F(x + d) - F(x) - F'(x; d)}{\| d \|} = 0.$$

2. *For $F : \mathcal{D} \subseteq \mathbb{R}^n \longrightarrow \mathbb{R}^n$ a locally Lipschitz continuous function, the limiting Jacobian of $F$ at $x \in \mathcal{D}$ is defined by*

$$\partial F(x) = \{\mathcal{M} \in \mathcal{L}(\mathbb{R}^n, \mathbb{R}^n) : \exists\, u^k \in \mathcal{D}, \lim_{k \longrightarrow \infty} F'(u^k) = \mathcal{M}\}. \tag{10.44}$$

3. *Let $F : \mathcal{D} \subseteq \mathbb{R}^n \longrightarrow \mathbb{R}^n$ be a locally Lipschitz continuous function. Clarke's Jacobian of $F$ at $x \in \mathcal{D}$ is defined by*

$$\partial^\circ F(x) = \overline{\text{co}}\ \partial F(x), \tag{10.45}$$

*where $\overline{\text{co}}\ \mathcal{A}$ is the closed convex envelope of $\mathcal{A} \subseteq \mathbb{R}^n$.*

Before presenting our main result of convergence of Newton–like method (10.35), we give the local convergence theorem restricted to the resolution of nonlinear equation (2.1) (cf. [194, 208]).

**Theorem 10.3** *Assume $F : \mathcal{D} \subseteq \mathcal{X} \longrightarrow \mathcal{Y}$ is a slantly differentiable operator at a solution $x^\star$ of equation $F(x) = 0$.*

*Let $f^\circ$ be a slanting operator for $F$ at $x^\star$, $A(x) \in \mathcal{L}(\mathcal{X}, \mathcal{Y})$ $(x \in \mathcal{D})$ with $A(x^\star)^{-1} \in \mathcal{L}(\mathcal{Y}, \mathcal{X})$,*

$$\| A(x^\star)^{-1}\ (A(x^\star + h) - A(x^\star))\ \| \leq L\ \| h \|, \tag{10.46}$$

$$\| A(x^\star)^{-1}\ (A(x^\star + h) - f^\circ(x^\star + h))\ \| \leq M\ \| h \| + M_0, \tag{10.47}$$

*as $\| h \| \rightarrow 0$. By the definition of slant differentiability, there exist $c > 0$ and $\delta = \delta(c) > 0$ such that, for all $\| h \| \leq \delta$, $\| f^\circ(x^\star + h) \| \leq c$ and*

$$\frac{\| F(x^\star + h) - F(x^\star) - f^\circ(x^\star + h)\, h \|}{\| h \|} \leq c. \tag{10.48}$$

*Assume that*

$$M_0 + c < 1. \tag{10.49}$$

*Set*

$$r = \min \left\{ \delta,\ \frac{1}{L},\ \frac{1 - c - M_0}{M + L} \right\} \tag{10.50}$$

*and*

$$U(x^\star, r) \subseteq \mathcal{D}. \tag{10.51}$$

*Then, the sequence $\{x_n\}$ generated by (10.36) is well defined, remains in $U(x^\star, r)$ for all $n \geq 0$ and converges to $x^\star$, provided that $x_0 \in U(x^\star, r)$. Moreover, the following estimates hold for all $k \geq 0$*

$$\| x_{k+1} - x^\star \| \leq q\ \| x_k - x^\star \|, \tag{10.52}$$

*where*

$$0 \leq q = \frac{c + M_0 + M\ \| x_0 - x^\star \|}{1 - L\ \| x_0 - x^\star \|} < 1. \tag{10.53}$$

We will consider the existence and the convergence of the sequence defined by (10.35) to the solution $x^\star$ of (10.1). The main result is as follows.

**Theorem 10.4** *Assume*

($C_1$) $F : \mathcal{D} \subseteq \mathcal{X} \longrightarrow \mathcal{Y}$ *is a slantly differentiable operator at a solution* $x^\star$ *of generalized equation* $0 \in F(x) + G(x)$, *where* $G$ *is a set–valued map from* $\mathcal{X}$ *to the subsets of* $\mathcal{Y}$ *with closed graph.*

($C_2$) *Let* $f^\circ$ *be a slanting operator for* $F$ *at* $x^\star$, $A(x) \in \mathcal{L}(\mathcal{X}, \mathcal{Y})$ $(x \in \mathcal{D})$ *and there exist a non–negative constants* $L$, $M$ *and* $M_0$ *such that*

$$\| A(x^\star + h) - A(x^\star) \| \leq L \ \| h \|, \tag{10.54}$$

$$\| A(x^\star + h) - f^\circ(x^\star + h) \| \leq M \ \| h \| + M_0 \quad \text{as} \quad \| h \| \rightarrow 0. \tag{10.55}$$

($C_3$) *Hypothesis (10.48) holds.*

($C_4$) *The set–valued map* $(G + A(x^\star)\ (. - x^\star))^{-1}$ *is pseudo–Lipschitz around* $(-F(x^\star), x^\star)$ *with constants* $M_1$, $a$ *and* $b$ *(These constants are given in Definition 10.1).*

($C_5$)

$$C_0 = M_1 \ (c + M \ \delta + M_0) < 1, \tag{10.56}$$

*where* $c$ *and* $\delta$ *are given by the definition of slant differentiability (see (10.48)).*

*Then, for every constant* $C$ *and* $\delta^\star$ *satisfying*

$$1 \geq C > C_0, \tag{10.57}$$

$$0 < \delta^\star < \delta_0^\star = \min\left\{a;\ \sqrt{\frac{b}{2\ L}};\ \frac{b}{2\ (c + M\ \delta + M_0)};\ \frac{C - C_0}{C\ M_1\ L}\right\}, \tag{10.58}$$

*respectively and for every starting point* $x_0$ *in* $\mathbb{B}_{\delta^\star}(x^\star)$ *(with* $x_0 \neq x^\star$*), there exists a sequence* $(x_k)$ *defined by (10.35) which is convergent to* $x^\star$ *and satisfies the following inequality for* $k \geq 0$

$$\| x_{k+1} - x^\star \| \leq C \ \| x_k - x^\star \| . \tag{10.59}$$

**Remark 10.4** *We need to introduce some notations (cf. [194], [195]). First, define the set–valued maps* $\Lambda : \mathcal{X} \rightrightarrows \mathcal{Y}$ *and* $\Theta_k : \mathcal{X} \rightrightarrows \mathcal{X}$ *by*

$$\Lambda(x) = F(x^\star) + A(x^\star)\ (x - x^\star) + G(x), \qquad \Theta_k(x) = \Lambda^{-1}(\Xi_k(x)) \quad (k \geq 0), \tag{10.60}$$

*where* $\Xi_k$ *is a mapping from* $\mathcal{X}$ *to* $\mathcal{Y}$ *defined by*

$$\Xi_k(x) = F(x^\star) - F(x_k) + A(x^\star)\ (x - x^\star) - A(x_k)\ (x - x_k) \qquad (k \geq 0). \tag{10.61}$$

*The proof of Theorem 10.4 is given by induction on $k$. We first state a result involving the starting point $x_0$. Let us note that the point $x_1$ is a fixed point of $\Theta_0$ if and only if $0 \in F(x_1) + A(x_0)\ (x_1 - x_0) + G(x_1)$. Once $x_k$ is computed, we show that the function $\Theta_k$ has a fixed point $x_{k+1}$ in $\mathcal{X}$. This process is useful to prove the existence of a sequence $(x_k)$ satisfying (10.35).*

**Proposition 10.3** *Under the assumptions of Theorem 10.4 and, for every distinct starting point $x_0$ in $\mathbb{B}_{\delta^\star}(x^\star)$ (with $x_0 \neq x^\star$), the set–valued map $\Theta_0$ has a fixed point $x_1$ in $\mathbb{B}_{\delta^\star}(x^\star)$ satisfying*

$$\| x_1 - x^\star \| \leq C \ \| x_0 - x^\star \|, \tag{10.62}$$

*where $C$ and $\delta^\star$ are given by Theorem 10.4.*

**Proof**

**Step 1** We prove that the first assumption in Lemma 10.1 is satisfied. By the pseudo–Lipschitzness hypothesis, we have

$$e(\Lambda^{-1}(y') \cap \mathbb{B}_a(x^\star), \Lambda^{-1}(y'')) \leq M_1 \ \| y' - y'' \| \quad (y', y'' \in \mathbb{B}_b(0)). \tag{10.63}$$

By (10.61), we have that

$$\begin{aligned} \| \Xi_0(x^\star) \| &= \| F(x^\star) - F(x_0) - A(x_0)\ (x^\star - x_0) \| \\ &= \| F(x_0) - F(x^\star) - f^\circ(x_0)(x_0 - x^\star) \\ &\quad + f^\circ(x_0)(x_0 - x^\star) - A(x_0)(x_0 - x^\star) \| \\ &\leq \| F(x_0) - F(x^\star) - f^\circ(x_0)(x_0 - x^\star) \| \\ &\quad + \| f^\circ(x_0)(x_0 - x^\star) - A(x_0)(x_0 - x^\star) \| \\ &\leq \| F(x_0) - F(x^\star) - f^\circ(x_0)(x_0 - x^\star) \| \\ &\quad + \| f^\circ(x_0) - A(x_0) \| \ \| x_0 - x^\star \| . \end{aligned} \tag{10.64}$$

Using (3.3) and (10.55), (10.64) becomes

$$\begin{aligned} \| \Xi_0(x^\star) \| &\leq c \ \| x_0 - x^\star \| + (M \ \| x_0 - x^\star \| + M_0) \ \| x_0 - x^\star \| \\ &\leq (c + M\ \delta + M_0) \ \| x_0 - x^\star \| . \end{aligned} \tag{10.65}$$

By (10.58) and (10.65), we obtain $\Xi_0(x^\star) \in \mathbb{B}_b(0)$. Hence, from (10.63) one gets

$$\begin{aligned} e\big(\Lambda^{-1}(0) \cap \mathbb{B}_{\delta^\star}(x^\star), \Theta_0(x^\star)\big) &= e\big(\Lambda^{-1}(0) \cap \mathbb{B}_{\delta^\star}(x^\star), \Lambda^{-1}[\Xi_0(x^\star)]\big) \\ &\leq M_1 \ (c + M\ \delta + M_0) \ \| x_0 - x^\star \| \\ &= C_0 \ \| x_0 - x^\star \| . \end{aligned} \tag{10.66}$$

According to the definition of excess $e$ and using (10.66), we have

$$\text{dist}\,(x^\star, \Theta_0(x^\star)) \leq e\big(\Lambda^{-1}(0) \cap \mathbb{B}_{\delta^\star}(x^\star), \Theta_0(x^\star)\big) \leq C_0 \;\; \| \, x_0 - x^\star \, \| \,. \quad (10.67)$$

Since $C > C_0$, there exists $\lambda \in [0, 1[$ such that

$$C\,(1 - \lambda) \geq C_0$$

and

$$\text{dist}\,(x^\star, \Theta_0(x^\star)) \leq C\,(1 - \lambda) \;\; \| \, x_0 - x^\star \, \| \,. \quad (10.68)$$

We can choose $\lambda = \dfrac{C - C_0}{C}$. Identifying $\eta_0$, $\phi$ and $r$ in Lemma 10.1 by

$$x^\star, \quad \Theta_0 \quad \text{and} \quad r_0 = C \;\; \| \, x_0 - x^\star \, \|,$$

respectively, we can deduce from the inequality (10.68) that the first assumption in Lemma 10.1 is satisfied.

**Step 2** We prove now that the second assumption of Lemma 10.1 is verified. Using (10.57) and (10.58), we have $r_0 \leq \delta^\star \leq a$. Moreover, for $x \in \mathbb{B}_{\delta^\star}(x^\star)$, we get in turn

$$\begin{aligned}
&\| \, \Xi_0(x) \, \| \\
&= \| \, F(x^\star) - F(x_0) + A(x^\star)\;(x - x^\star) - A(x_0)\;(x - x_0) \, \| \\
&\leq \| \, F(x_0) - F(x^\star) - f^\circ(x_0)(x_0 - x^\star) \, \| \qquad\qquad\qquad (10.69) \\
&\quad + \| \, f^\circ(x_0)(x_0 - x^\star) - A(x^\star)(x - x^\star) + A(x_0)(x - x^\star + x^\star - x_0) \, \| \\
&\leq \| \, F(x_0) - F(x^\star) - f^\circ(x_0)(x_0 - x^\star) \, \| \\
&\quad + \| \, f^\circ(x_0) - A(x_0) \, \| \; \| \, x_0 - x^\star \, \| + \| \, A(x_0) - A(x^\star) \, \| \; \| \, x - x^\star \, \| \,.
\end{aligned}$$

Using (10.48), (10.54) and (10.55), we obtain

$$\begin{aligned}
&\| \, \Xi_0(x) \, \| \\
&\quad \leq c \;\; \| \, x_0 - x^\star \, \| + (M\;\delta + M_0) \;\; \| \, x_0 - x^\star \, \| + L \;\; \| \, x_0 - x^\star \, \| \; \| \, x - x^\star \, \| \\
&\quad \leq c\;\delta^\star + (M\;\delta + M_0)\;\delta^\star + L\;\delta^{\star 2}. \qquad\qquad\qquad (10.70)
\end{aligned}$$

By (10.58), we deduce that, for all $x \in \mathbb{B}_{\delta^\star}(x^\star)$, we have $\Xi_0(x) \in \mathbb{B}_b(0)$. Then, it follows that for all $x', x'' \in \mathbb{B}_{r_0}(x^\star)$, we have

$$e(\Theta_0(x') \cap \mathbb{B}_{r_0}(x^\star), \Theta_0(x'')) \leq e(\Theta_0(x') \cap \mathbb{B}_{\delta^\star}(x^\star), \Theta_0(x'')),$$

which yields by (10.63) and the pseudo–Lipschitzness hypothesis to

$$\begin{aligned} & e(\Theta_0(x') \cap \mathbb{B}_{r_0}(x^\star), \Theta_0(x'')) \\ & \quad \leq M_1 \ \| \ \Xi_0(x') - \Xi_0(x'') \ \| \\ & \quad = M_1 \ \| \ A(x^\star)(x' - x'') - A(x_0) \ (x' - x'') \ \| \\ & \quad \leq M_1 \ \| \ A(x^\star) - A(x_0) \ \| \ \| \ x'' - x' \ \| \\ & \quad \leq M_1 \ L \ \| \ x_0 - x^\star \ \| \ \| \ x'' - x' \ \| \\ & \quad \leq M_1 \ L \ \delta^\star \ \| \ x'' - x' \ \| \leq \lambda \ \| \ x'' - x' \ \| \ . \end{aligned} \tag{10.71}$$

The second condition of Lemma 10.1 is satisfied. By Lemma 10.1, we can deduce the existence of a fixed point $x_1 \in \mathbb{B}_{r_0}(x^\star)$ for the map $\Theta_0$. This completes the proof of Proposition 10.3 is complete. ⊠

**Proof of Theorem 10.4.** Keep $\eta_0 = x^\star$ and for $k \geq 1$, $r := r_k = C \ \| \ x^\star - x_k \ \|$ . By Remark 10.4, the application of Proposition 10.3 to the map $\Theta_k$ gives the desired result. ⊠

**Remark 10.5** *We can cnlargc thc radius of convergence in Theorem 10.4 even further as follows: using inequalities (10.65), (10.70) and*

$$(c + M \ \delta + M_0) \ \eta - b \leq (c + M \ \delta + M_0) \ \eta + L \ \eta^2 - b,$$

*we can improve $\delta^\star$ given by (10.58) by considering the constant $\delta^{\star\prime}$*

$$\delta^{\star\prime} < \delta_0^{\star\prime} = \min \Big\{a; \ \delta_1 \Big\},$$

*where*

$$\delta_1 = \max \ \{\eta > 0 \ : \ (c + M \ \delta + M_0) \ \eta + L \ \eta^2 - b < 0\}.$$

In view of Theorem 10.4, we obtain an extension of (cf. [194, Theorem 1.3.2, page 56]).

**Proposition 10.4** *Assume that*

*(i) $F : \mathcal{D} \subseteq \mathcal{X} \longrightarrow \mathcal{Y}$ be slantly differentiable operator at a solution $x^\star$ of generalized equation $0 \in F(x) + G(x)$, where $G$ is a set–valued map from $\mathcal{X}$ to the subsets of $\mathcal{Y}$ with closed graph.*

*(ii) Let $f^\circ$ be a slanting operator for $F$ at $x^\star$, $A(x) \in \mathcal{L}(\mathcal{X}, \mathcal{Y})$ $(x \in \mathcal{D})$ and there exist a non–negative constants $L$, $\rho_1$ and $\rho_2$ such that, for any $x$ in a neighborhood of $x^\star$*

$$\| \ A(x^\star + h) - A(x^\star) \ \| \leq L \ \| \ h \ \|,$$

$$\| \ A(x^\star + h) - f^\circ(x^\star + h) \ \| \longrightarrow 0 \quad \text{as} \ \| \ h \ \| \to 0, \tag{10.72}$$

$$\| \ F(x) - F(x^\star) - f^\circ(x) \ (x - x^\star) \ \| \leq \rho_1 \ \| \ x - x^\star \ \| \tag{10.73}$$

*and*

$$\| A(x) - f^\circ(x) \| \leq \rho_2. \tag{10.74}$$

*(iii)* *The set–valued map* $(G + A(x^\star)\ (. - x^\star))^{-1}$ *is pseudo–Lipschitz around* $(-F(x^\star), x^\star)$ *with constants* $M_1$, $a$ *and* $b$.

*(iv)*

$$\overline{C_0} = M_1\ (\rho_1 + \rho_2) < 1. \tag{10.75}$$

*Then, for every constant* $\overline{C}$ *and* $\overline{\delta^\star}$ *such that*

$$1 \geq \overline{C} > \overline{C_0}, \tag{10.76}$$

$$0 < \overline{\delta^\star} < \overline{\delta_0^\star} = \min \left\{ a;\ \sqrt{\frac{b}{2\ L}};\ \frac{b}{2\ (\rho_1 + \rho_2)};\ \frac{\overline{C} - \overline{C_0}}{\overline{C}\ M_1\ L} \right\} \tag{10.77}$$

*respectively, and, for every starting point* $x_0$ *in* $\mathbb{B}_{\overline{\delta^\star}}(x^\star)$ *(with* $x_0 \neq x^\star$*), the conclusion of Theorem 10.4 holds with the constant* $C$ *is replaced by* $\overline{C}$ *in (10.59).*

**Idea of the proof** Proposition 10.4 can be proved in the same way as Theorem 10.4. We prove that the two assumption in Lemma 10.1 are satisfied. Using assumptions (i)–(iv), simply replace estimates (10.64) and (10.69)–(10.70) in the proof of Theorem 10.4 by

$$\| \Xi_0(x^\star) \| \quad \leq \quad (\rho_1 + \rho_2)\ \ \| x_0 - x^\star \|$$

and

$$\begin{aligned}\| \Xi_0(x) \| &\leq \| F(x_0) - F(x^\star) - f^\circ(x_0)(x_0 - x^\star) \| + \\ &\quad \| f^\circ(x_0) - A(x_0) \| \ \| x_0 - x^\star \| + \| A(x_0) - A(x^\star) \| \ \| x - x^\star \| \\ &\leq (\rho_1 + \rho_2)\ \ \| x_0 - x^\star \| + L\ \ \| x - x^\star \| \ \| x_0 - x^\star \| \\ &\leq (\rho_1 + \rho_2)\ \overline{\delta^\star} + L\ \overline{\delta^\star}^2,\end{aligned}$$

respectively. This completes the proof. ⊠

In view of Theorem 10.4 and Definitions 10.4–10.7, we obtain the special cases:

**Proposition 10.5** *Assume*

*(a)* $F : \mathcal{D} \subseteq \mathcal{X} \longrightarrow \mathcal{Y}$ *be slantly differentiable operator at a solution* $x^\star$ *of generalized equation* $0 \in F(x) + G(x)$, *where* $G$ *is a set–valued map from* $\mathcal{X}$ *to the subsets of* $\mathcal{Y}$ *with closed graph. Let* $f^\circ$ *be a slanting operator for* $F$.

*(b)* *The set–valued map* $(G + A(x^\star)\ (. - x^\star))^{-1}$ *is pseudo–Lipschitz around* $(-F(x^\star), x^\star)$.

*Then, Newton–like method*

$$0 \in F(x_n) + f^\circ(x_n)\,(x_{n+1} - x_n) + G(x_{n+1}) \quad (x_0 \in \mathcal{D},\ n \geq 0)$$

*converges to* $x^\star$ *with* $A = f^\circ$ *in condition (b).*

**Proof** The proof of Proposition 10.5 is the same one of the proof of Theorem 10.4. It is enough to make some modification by replacing the operator $A$ by $f^\circ$ in expressions (10.60) and (10.61). Conditions $(C_1)$, $(C_3)$ and $(C_4)$ of Theorem 10.4 are satisfied. The condition $(C_2)$ of Theorem 10.4 is also satisfied with $M = M_0 = 0$. Then, we can apply Theorem 10.4 by replacing $(C_5)$ by additional condition $M_1\ c < 1$. ⊠

**Remark 10.6** *We assume that both conditions (a) and (b) of Proposition 10.5 are verified. Our conjecture is given by the following two items:*

*(A) If* $f : \mathcal{D}\times(0,+\infty) \longrightarrow \mathcal{Y}$ *is smoothing function, which satisfies the slant consistency property (10.40) in a neighborhood of* $x^\star$*, then the smoothing Newton method*

$$0 \in F(x_n) + f'_x(x_n, \epsilon_n)\,(x_{n+1} - x_n) + G(x_{n+1}) \quad (x_0 \in \mathcal{D},\ n \geq 0)$$

*converges to* $x^\star$ *with* $A = f'_x(.,\varepsilon)$ *(*$\epsilon > 0$ *is a fixed parameter) in condition (b) of Proposition 10.5 and* $\epsilon_n > 0$ *is an appropriate real sequence converging to* $0$ *(cf. [276], [280]).*

*(B) If* $F$ *is semismooth at* $x^\star$*, then the semismoothing Newton method:*

$$0 \in F(x_n) + V(x_n)\,(x_{n+1} - x_n) + G(x_{n+1}) \quad (x_0 \in \mathcal{D},\ n \geq 0)$$

*converges to* $x^\star$ *with* $A = V$ *in condition (b) of Proposition 10.5 and* $V(x) \in \partial_S F(x)$ *for* $x \in \mathcal{D}$*.*

*Here are some points to solve this problem. However these points are insufficient. For (A) we replace the mappings in (10.60) and (10.61) by*

$$\Lambda(x) = F(x^\star) + f'_x(x^\star, \epsilon)\,(x - x^\star) + G(x), \quad \Theta_k(x) = \Lambda^{-1}(\Xi_k(x))$$

*and*

$$\Xi_k(x) = F(x^\star) - F(x_k) + f'_x(x^\star, \epsilon)\,(x - x^\star) - f'_x(x_k, \epsilon_k)\,(x - x_k)\ (k \geq 0),$$

*respectively. We suppose that* $\epsilon$ *and sequence* $\epsilon_k$ $(k \geq 0)$ *satisfy* $\epsilon \leq \parallel x_k - x^\star \parallel$ *and* $\epsilon_k \leq \parallel x_k - x^\star \parallel$ *for all* $k \geq 0$*. For (B) we replace the mappings in (10.60) and (10.61) by*

$$\Lambda(x) = F(x^\star) + V(x^\star)\,(x - x^\star) + G(x), \quad \Theta_k(x) = \Lambda^{-1}(\Xi_k(x))$$

*and*

$$\Xi_k(x) = F(x^\star) - F(x_k) + V(x^\star)\,(x - x^\star) - V(x_k)\,(x - x_k)\ (k \geq 0).$$

## 10.3 Newton–Josephy Method

In this section, we present a Kantorovich–type semilocal convergence analysis of the Newton–Josephy method for solving a certain class of variational inequalities. By using a combination of Lipschitz and center–Lipschitz conditions, and our new idea of recurrent functions, we provide an analysis with the following advantages over the earlier works (cf. [763], [761]) (under the same or less computational cost): weaker sufficient convergence conditions, larger convergence domain, finer error bounds on the distances involved and an at least as precise information on the location of the solution.

We are concerned with the problem of approximating a solution $x^\star \in \mathcal{D}$ of variational inequalities $VI(\mathcal{D}, F)$ such that

$$(y - x^\star)^T \, F(x^\star) \geq 0 \quad (y \in \mathcal{D}), \tag{10.78}$$

where $F$ is a Fréchet–differentiable operator defined on a non–empty, convex and closed subset $\mathcal{D}$ of $\mathbb{R}^n$, with values in $\mathbb{R}^n$. Moreover, the nonlinear complementarity problem is a special case of (10.78). In this case, we seek: $x^\star \in \mathcal{D} = \mathbb{R}^n_+$ such that

$$x^{\star T} \; F(x^\star) = 0, \quad F(x^\star) \in \mathcal{D}.$$

If $F(x) = \nabla G(x)$ $(x \in \mathcal{D})$ in (10.78), where $G \, : \, \mathcal{D} \longrightarrow \mathbb{R}$ is a real differentiable function, then, (10.78) corresponds to the first order necessary optimality condition of the following optimization–problem

$$\min_{x \in \mathcal{D}} G(x).$$

A particular case of the feasible set $\mathcal{D}$ in (10.78) is studied in (cf. [271]), where

$$\mathcal{D} = \{x \in \mathbb{R}^n \, : \, g_i(x) \geq 0, \; i = 1, \cdots, m; \;\; h_j(x) = 0, \; j = 1, \cdots, l\}$$

and $g_i$ $(1 \leq i \leq m)$, $h_j$ $(1 \leq j \leq l)$ are concave, affine real functions (cf. [271], [777], [778]). An elegant extensive study on the theory, applications of variational inequalities and their complete connections with other well–known problems in mathematical programming in finite dimensional spaces is presented in (cf. [410]).

Newton–Josephy method NJM generates a sequence $\{x_k\}$, so that $x_{k+1}$ solves the $F_k$ $k$–th linearized subproblem $VI(\mathcal{D}, F_k)$, where

$$F_k(x) = F(x_k) + F'(x_k) \, (x - x_k) \quad (k \geq 0) \tag{10.79}$$

and

$$(F'(x))_{ij} = \left( \frac{\partial F_i(x)}{\partial x_j} \right) \qquad (cf. \text{ [163], [469], [569]}). \tag{10.80}$$

NJM is used in (cf. [689]) for solving monotone variational inequalities. Moreover, the superlinear rate of convergence is established under standard conditions. A perturbed version of NJM is presented in (cf. [455]). Using D–gap function, the constrained variational inequality problem is solved by (10.79) in (cf. [585]). There is an extensive literature on the local as well as the semilocal convergence of Newton–type methods under various Lipschitz–type hypotheses on $F'$. A survey of such results can be found in (cf. [163], [194], [469], [569]).

In this section, we are motivated by the works of Wang–Shen in (cf. [761], Wang [760]–[763]) and optimization considerations. Wang and Shen provided a Newton–Kantorovich–type semilocal convergence result requiring computations only at the initial guess $x_0$. Their studies improve earlier works with more complicated hypotheses, which are difficult, or impossible to test (cf. [410], [462], [647], [647]). Using automatic differentiation, and interval computation (cf. [7]), techniques of hypotheses in (cf. [761]) can be tested. As far as we know the studies in (cf. [763], [761]) provide so far the weakest sufficient convergence conditions for NJM under certain hypotheses (see Theorem 10.6).

In this section, we expand the applicability of NJM. First, we introduce Lipschitz and center–Lipschitz conditions on $F'$. Using a combination of the two, and our new idea of recurrent functions, we provide under less or the same computational cost a new semilocal convergence analysis of NJM with the following advantages over the works in (cf. [761], [763])

(a) Weaker sufficient convergence conditions;

(b) Larger convergence domain;

(c) Finer error bounds on the distances $\| x_{k+1} - x_k \|$, $\| x_k - x^\star \|$ $(k \geq 0)$;

(d) An at least as precise information on the location of the solution $x^\star$.

Finally, we provide a numerical example, where we compare our error bounds with the corresponding ones in (cf. [761] and [763]). The Euclidean norm is used in this section. We shall use three lemmas.

**Lemma 10.4** *(cf. [761]) Let $B \in \mathbb{R}^{n\times n}$ be symmetric. Then, $B$ is positively definite if and only if there exists a symmetric positive definite matrix $A$ such that*

$$\| A - B \| \; \| A^{-1} \| < 1.$$

*If the above inequality holds, then*

$$\| B^{-1} \| \leq \frac{\| A^{-1} \|}{1 - \| I - A^{-1}\, B \|} \leq \frac{\| A^{-1} \|}{1 - \| A^{-1} \| \; \| A - B \|}.$$

**Lemma 10.5** *(cf. [242]) Let $\mathcal{D} \subseteq \mathbb{R}^n$ be nonempty, closed and convex. If $F : \mathcal{D} \subseteq \mathbb{R}^n \longrightarrow \mathbb{R}^n$ is strictly monotone in $\mathcal{D}$, then $VI(\mathcal{D}, F)$ has at most one solution. Moreover, if it is continuous and strongly monotone, then $VI(\mathcal{D}, F)$ is uniquely solvable.*

We finally need a lemma on majorizing sequences for NJM.

**Lemma 10.6** *Assume there exist constants $L_0 > 0$, $L > 0$ with $L_0 \le L$ and $\eta > 0$ such that*

$$h = \alpha\ \eta \le \frac{1}{2}, \tag{10.81}$$

*where*

$$\alpha = \frac{L_0\ (2\ L_0 + L + \sqrt{(2\ L_0 + L)^2 + 24\ L^2})}{6\ (2\ L_0 - 5\ L + \sqrt{(2\ L_0 + L)^2 + 24\ L^2})}. \tag{10.82}$$

*Then, the sequence $\{t_k\}$ $(k \ge 0)$ given by*

$$t_0 = 0, \quad t_1 = \eta, \quad t_{k+1} = t_k + \frac{L\ (t_k - t_{k-1})^2}{2\left(1 - \frac{1}{3}\ (L_0\ t_{k-1} + L\ (t_k - t_{k-1}))\right)} \quad (k \ge 1) \tag{10.83}$$

*is well defined, nondecreasing, bounded from above by $t^{\star\star}$ and converges to its unique least upper bound $t^\star \in [0, t^{\star\star}]$, where*

$$t^{\star\star} = \frac{2\ \eta}{2 - \delta}, \tag{10.84}$$

$$\delta = \frac{12\ L}{2\ L_0 + L + \sqrt{(2\ L_0 + L)^2 + 24\ L^2}}. \tag{10.85}$$

**Proof** The denominator in (10.82) is positive since $L_0 > 0$ and $L > 0$. We shall first show

$$0 \le t_{k+1} - t_k \le \frac{\delta}{2}\ (t_k - t_{k-1}) \le \cdots \le \left(\frac{\delta}{2}\right)^k \eta \quad (k \ge 1) \tag{10.86}$$

and

$$L_0\ t_{k-1} + L\ (t_k - t_{k-1}) < 3 \tag{10.87}$$

hold, using induction on $k$. For $k = 1$, (10.86) and (10.87) become

$$0 \le t_2 - t_1 = \frac{L\ \eta^2}{2\left(1 - \frac{1}{3}\ L\ \eta\right)} \le \frac{\delta}{2}\ \eta, \tag{10.88}$$

$$L\ \eta < 1, \tag{10.89}$$

respectively. It follows from (10.81)–(10.83) and (10.85) that estimates (10.86) and (10.87) hold for $k = 1$.
Let us assume (10.86) and (10.87) hold for all $m \le k$. Then, in view of (10.86), we have that

$$t_{k+1} - t_k \le \left(\frac{\delta}{2}\right)^k \eta \tag{10.90}$$

and

$$t_k \le \frac{1 - \left(\frac{\delta}{2}\right)^k}{1 - \frac{\delta}{2}}\ \eta < t^{\star\star}. \tag{10.91}$$

By the induction hypotheses to show (10.86) and (10.87), it suffices to prove that

$$L\ (t_{k+1}-t_k)+\frac{\delta}{3}\ (L_0\ t_k+L\ (t_{k+1}-t_k))\leq\delta \tag{10.92}$$

or

$$\left\{L\ \left(\frac{\delta}{2}\right)^k+\frac{\delta}{3}\ \left(L_0\ \frac{1-\left(\frac{\delta}{2}\right)^k}{1-\frac{\delta}{2}}+L\ \left(\frac{\delta}{2}\right)^k\right)\right\}\ \eta\leq\delta. \tag{10.93}$$

Estimate (10.93) motivates us (after dividing (10.93) by $\frac{\delta}{2}$) to set $s=\frac{\delta}{2}$ and introduce functions $f_k$ on $[0,1)$ by

$$f_k(s)=L\ s^{k-1}\ \eta+\frac{2}{3}\ (L_0\ (1+s+s^2+\cdots+s^{k-1})+L\ s^k)\ \eta-2\quad (k\geq 1). \tag{10.94}$$

We shall show instead of (10.93)

$$f_k(\frac{\delta}{2})\leq 0\qquad (k\geq 1). \tag{10.95}$$

We need a relationship between two consecutive functions $f_k$ as follows

$$\begin{aligned}
f_{k+1}(s)&=L\ s^k\ \eta+\frac{2}{3}\ (L_0\ (1+s+s^2+\cdots+s^{k-1}+s^k)+L\ s^{k+1})\ \eta-2\\
&=f_k(s)+L\ s^k\ \eta+\frac{2}{3}\ (L_0\ (1+s+\cdots+s^k)+L\ s^{k+1})\ \eta-2\\
&\quad -L\ s^{k-1}\ \eta-\frac{2}{3}\ (L_0\ (1+s+\cdots+s^{k-1})+L\ s^k)\ \eta+2\\
&=f_k(s)+g(s)\,s^{k-1}\ \eta,
\end{aligned} \tag{10.96}$$

where

$$g(s)=\frac{1}{3}\ (2\ L\ s^2+(2\ L_0+L)\ s-3\ L). \tag{10.97}$$

Note that $\frac{\delta}{2}$ given by (10.85) is the unique positive root of polynomial $g$. Define a function $f_\infty$ on $[0,1)$ by

$$f_\infty(s)=\lim_{k\longrightarrow\infty}f_k(s). \tag{10.98}$$

Then, we have by (10.94) and (10.98) that

$$f_\infty\left(\frac{\delta}{2}\right)=\lim_{k\longrightarrow\infty}f_k\left(\frac{\delta}{2}\right)=2\ \left\{\frac{L_0\ \eta}{3\ \left(1-\frac{\delta}{2}\right)}-1\right\}. \tag{10.99}$$

Using (10.96) and (10.97), we get that

$$f_\infty\left(\frac{\delta}{2}\right) = f_k\left(\frac{\delta}{2}\right) = f_{k-1}\left(\frac{\delta}{2}\right) = \cdots = f_1\left(\frac{\delta}{2}\right). \tag{10.100}$$

It follows from (10.95) and (10.100) that we only need to show

$$f_\infty\left(\frac{\delta}{2}\right) \leq 0, \tag{10.101}$$

which is true by (10.99) and (10.81). Hence, estimate (10.95) holds. It follows that sequence $\{t_n\}$ is nondecreasing, bounded from above by $t^{\star\star}$ and as such it converges to its unique least upper bound $t^\star \in [0, t^{\star\star}]$. This completes the proof of Lemma 10.6. ⊠

We can provide the main semilocal convergence result for NJM:

**Theorem 10.5** *Let $F : \mathcal{D} \subseteq \mathbb{R}^n \longrightarrow \mathbb{R}^n$ be a Fréchet–differentiable operator. Assume there exist $x_0 \in \mathcal{D}$, constants $\beta > 0$, $\gamma_0 > 0$, $\gamma > 0$, $\eta > 0$ with $\gamma_0 \leq \gamma$ such that for all $x, y \in \mathcal{D}$ $F'(x_0)$ is positive definite, $x_1$ is the unique solution of subproblem $VI(\mathcal{D}, F_0)$*

$$\| \widetilde{F'(x_0)}^{-1} \| \leq \beta, \qquad \| x_1 - x_0 \| \leq \eta, \tag{10.102}$$

$$\| F'(x) - F'(x_0) \| \leq \gamma_0 \ \| x - x_0 \|, \tag{10.103}$$

$$\| F'(x) - F'(y) \| \leq \gamma \ \| x - y \|, \tag{10.104}$$

$$h_{AH} = \overline{\alpha}\ \eta \leq \frac{1}{2}, \tag{10.105}$$

$$\overline{U}(x_0, r^\star) \subseteq \mathcal{D}, \tag{10.106}$$

*where*

$$\widetilde{F'(x)} = \frac{1}{2}\ (F'(x) + F'(x)^T) \qquad (x \in \mathcal{D}), \tag{10.107}$$

$$\overline{\alpha} = \frac{\beta\ \gamma_0}{2}\ \frac{2\ \gamma_0 + \gamma + \sqrt{(2\ \gamma_0 + \gamma)^2 + 24\ \gamma^2}}{2\ \gamma_0 - 5\ \gamma + \sqrt{(2\ \gamma_0 + \gamma)^2 + 24\ \gamma^2}}, \tag{10.108}$$

$$r^\star = \lim_{k\to\infty} r_k, \tag{10.109}$$

$$r_0 = 0, \quad r_1 = \eta,$$

$$r_{k+1} = r_k + \frac{3\ \beta\ \gamma\ (r_k - r_{k-1})^2}{2\ (1 - (\gamma_0\ \beta\ r_{k-1} + \gamma\ \beta\ (r_k - r_{k-1})))} \quad (k \geq 1). \tag{10.110}$$

*Then*

*(a) Scalar sequence $\{r_k\}$ is nondecreasing, bounded from above by $r^{\star\star}$, which is given by*

$$r^{\star\star} = \frac{2\eta}{2-\delta} \tag{10.111}$$

*and converges to its unique least upper bound $r^\star \in [0, r^{\star\star}]$. Moreover, estimate (10.86) holds with, $L_0$, $L$, $t^\star$, $t^{\star\star}$, $h$ given by*

$$L_0 = 3\ \gamma_0\ \beta,\ \ L = 3\ \gamma\ \beta,\ \ t^\star = r^\star,\ \ t^{\star\star} = r^{\star\star},\ \ h = h_{AH}. \tag{10.112}$$

(b) *Sequence* $\{x_k\}$ *generated by NJM is well defined, remains in* $\overline{U}(x_0, r^\star)$ *for all* $k \geq 0$ *and converges to a unique solution* $x^\star \in \overline{U}(x_0, r^\star)$ *of* $VI(\mathcal{D}, F)$.

*Moreover, the following estimates hold for all* $k \geq 0$

$$\| x_{k+1} - x_k \| \leq r_{k+1} - r_k \tag{10.113}$$

*and*

$$\| x_k - x^\star \| \leq r^\star - r_k. \tag{10.114}$$

**Proof**

(a) This part of the proof follows immediately from Lemma 10.6 by simply using (10.112).

(b) By using induction on $k$, we shall show that sequence $\{x_k\}$ is well defined,

$$\| x_{k+1} - x_k \| \leq r_{k+1} - r_k, \tag{10.115}$$

and

$$\overline{U}(x_{k+1}, r^\star - r_{k+1}) \subseteq \overline{U}(x_k, r^\star - r_k) \tag{10.116}$$

hold for all $k$.

Note that in view of (10.110)

$$r^\star \geq r_1. \tag{10.117}$$

Let $z \in \overline{U}(x_1, r^\star - r_1)$. Then, estimate

$$\| z - x_0 \| \leq \| z - x_1 \| + \| x_1 - x_0 \| \leq r^\star - r_1 + r_1 - r_0 = r^\star - r_0 \tag{10.118}$$

implies $z \in \overline{U}(x_0, r^\star - r_0)$. We also have

$$\| x_1 - x_0 \| \leq \eta = r_1 - r_0, \tag{10.119}$$

which together with (10.118) show (10.115) and (10.116) for $k = 0$. Let us assume $\{x_m\}$ is well defined and (10.115), (10.116) hold for all $m \leq k$.

Let $x \in \overline{U}(x_0, r^\star)$. Using (10.103) and (10.107), we get that

$$\begin{aligned}
\beta \ \| \widetilde{F}'(x) - \widetilde{F}'(x_0) \| &= \beta \ \| \frac{1}{2} \left(F'(x) + F'(x)^T\right) - \frac{1}{2} \left(F'(x_0) + F'(x_0)^T\right) \| \\
&\leq \beta \ \| F'(x) - F'(x_0) \| \\
&\leq \beta \ \gamma_0 \ \| x - x_0 \| \leq \beta \ \gamma_0 \ r^\star = L_0 \ t^\star < 1. \qquad (10.120)
\end{aligned}$$

By Lemma 10.4, we have $\widetilde{F}'(x)$, $F'(x)$ are positively definite and

$$\| \widetilde{F}'(x)^{-1} \| \leq \frac{\beta}{1 - \beta \ \gamma_0 \ \| x - x_0 \|}. \tag{10.121}$$

In particular, for $x = x_1$, Lemma 10.4 implies $VI(\mathcal{D}, F_1)$ is uniquely solvable and consequently $x_2$ is well defined. We have by the induction hypotheses

$$\| x_k - x_0 \| \leq \sum_{i=1}^{k} \| x_i - x_{i-1} \| \leq \sum_{i=1}^{k} (r_i - r_{i-1}) = r_k - r_0 = r_k \quad (10.122)$$

and

$$\| x_k + \theta (x_k - x_{k-1}) - x_0 \| \leq r_{k-1} + \theta (r_k - r_{k-1}) \leq r^\star \quad (\theta \in (0,1)). \quad (10.123)$$

Using (10.103), (10.104) and (10.115), we obtain the estimate:

$$\begin{aligned} &\|F(x_{k-1}) - F(x_k) - F'(x_k)\,(x_{k-1} - x_k)\,\| \\ &= \left\| \int_0^1 (F'(x_k + \theta\,(x_{k-1} - x_k)) - F'(x_k))\,(x_{k-1} - x_k)\,d\theta \right\| \\ &\leq \gamma \int_0^1 \| x_k + \theta\,(x_{k-1} - x_k) - x_k \| \; d\theta \; \| x_{k-1} - x_k \| \\ &= \frac{\gamma}{2} \; \| x_{k-1} - x_k \|^2 \leq \frac{\gamma}{2}\,(r_{k-1} - r_k)^2. \end{aligned} \quad (10.124)$$

The induction hypotheses are given by

$$\begin{aligned} &x_{k-1}, x_k \in \mathcal{D}, \\ &x_k, x_{k+1} \text{ solve } VI(\mathcal{D}, F_{k-1}),\; VI(\mathcal{D}, F_k),\; \textit{respectively}, \end{aligned} \quad (10.125)$$

$$(x_{k+1} - x_k)^T F_{k-1}(x_k) = (x_{k+1} - x_k)^T (F(x_{k-1}) + F'(x_{k-1})(x_k - x_{k-1})) \geq 0 \quad (10.126)$$

and

$$(x_k - x_{k+1})^T F_k(x_{k+1}) = (x_k - x_{k+1})^T (F(x_k) + F'(x_k)(x_{k+1} - x_k)) \geq 0. \quad (10.127)$$

It follows that estimates (10.126) and (10.127) give in turn that

$$\begin{aligned} &(x_k - x_{k+1})^T \; F'(x_{k-1})\,(x_k - x_{k+1}) \\ &\leq (x_k - x_{k+1})^T \; \big(F(x_k) - F(x_{k-1}) + F'(x_k)\,(x_{k+1} - x_k) \\ &\quad - F'(x_{k-1})\,(x_{k+1} - x_{k-1})\big) \\ &= (x_k - x_{k+1})^T \; \big((F'(x_k) - F'(x_{k-1}))\,(x_{k+1} - x_{k-1}) \\ &\quad - (F(x_{k-1}) - F(x_k) - F'(x_k)\,(x_{k-1} - x_k))\big) \\ &\leq \| x_k - x_{k+1} \| \; \big( \| F'(x_k) - F'(x_{k-1}) \| \, \| x_{k+1} - x_{k-1} \| \\ &\quad + \| F(x_{k-1}) - F(x_k) - F'(x_k)\,(x_{k-1} - x_k) \| \,\big) \\ &\leq \big(\gamma\,(\| x_{k+1} - x_k \| + \| x_k - x_{k-1} \|) \; \| x_k - x_{k-1} \| \\ &\quad + \frac{\gamma}{2} \; \| x_k - x_{k-1} \|^2 \,\big) \; \| x_{k+1} - x_k \|. \end{aligned} \quad (10.128)$$

We also have that

$$(x_k - x_{k+1})^T\, F'(x_{k-1})\,(x_k - x_{k+1}) = (x_k - x_{k+1})^T\, \widetilde{F}'(x_{k-1})\,(x_k - x_{k+1}) \geq \frac{\| x_{k+1} - x_k \|^2}{\| F'(x_{k-1})^{-1} \|} \geq \frac{\| x_{k+1} - x_k \|^2}{\beta_{k-1}}, \tag{10.129}$$

where

$$\| F'(x_{k-1})^{-1} \| \leq \beta_{k-1} \leq \frac{\beta}{1 - \beta\, \gamma_0\ \| x_{k-1} - x_0 \|} \quad (k > 1) \tag{10.130}$$

and

$$\beta_0 = \beta. \tag{10.131}$$

In view of (10.128) and (10.129), we obtain

$$\| x_{k+1} - x_k \| \leq \beta_{k-1}\, \big(\gamma\ \| x_k - x_{k-1} \|\ \ (\| x_{k+1} - x_k \| + \| x_k - x_{k-1} \|) + \frac{\gamma}{2}\ \| x_k - x_{k-1} \|^2 \,\big) \tag{10.132}$$

or

$$\begin{aligned} \| x_{k+1} - x_k \| &\leq \frac{3\,\gamma\,\beta_{k-1}\ \| x_k - x_{k-1} \|^2}{2\,(1 - \beta_{k-1}\,\gamma\ \| x_k - x_{k-1} \|)} \\ &\leq \frac{3\,\gamma\,\beta\,(r_k - r_{k-1})^2}{2\,(1 - \beta\,(\gamma_0\,(r_{k-1} - r_0) + \gamma\,(r_k - r_{k-1})))} \\ &= r_{k+1} - r_k, \end{aligned} \tag{10.133}$$

which implies (10.115) holds for all $k$.

Let $w \in \overline{U}(x_{k+1}, r^\star - r_{k+1})$. Then, we have that

$$\begin{aligned} \| w - x_k \| &\leq \| w - x_{k+1} \| + \| x_{k+1} - x_k \| \\ &\leq r^\star - r_{k+1} + r_{k+1} - r_k = r^\star - r_k. \end{aligned} \tag{10.134}$$

Hence, we deduce that

$$w \in \overline{U}(x_k, r^\star - r_k). \tag{10.135}$$

The induction for (10.115) and (10.116) is completed. Lemma 10.6 implies that $\{r_k\}$ is a complete sequence. In view of (10.115) and (10.116), $\{x_k\}$ ($k \geq 0$) is also a complete sequence, and as such it converges to some $x^\star \in \overline{U}(x_0, r^\star)$ ( since $\overline{U}(x_0, r^\star)$ is a closed set) with

$$(v - x_{k+1})^T\ (F(x_k) + F'(x_k)\ (x_{k+1} - x_k)) \geq 0 \quad (v \in \mathcal{D}). \tag{10.136}$$

Estimate (10.116) follows from (10.115) by using standard majorization techniques.

Finally, by letting $k \longrightarrow \infty$ in (10.136) and the continuity of $F$, $F'$, we get that

$$(x - x^\star)^T\ F(x^\star) \geq 0 \quad (x \in \mathcal{D}). \tag{10.137}$$

Hence, we deduce $x^\star$ is a solution of $VI\,(\mathcal{D}, F)$. This completes the proof of Theorem 10.5. ⊠

Note that the case $\gamma = 0$ is not included, since it is already covered in (cf. [761]).

**Remark 10.7** *Note that*

$$\gamma_0 \leq \gamma \tag{10.138}$$

*holds in general and* $\dfrac{\gamma}{\gamma_0}$ *can be arbitrarily large.*

We state the corresponding Theorem 1 given by Wang and Shen (cf. [761]), so we can compare it with our Theorem 10.5.

**Theorem 10.6** *Let $F\ :\ \mathcal{D} \subseteq \mathbb{R}^n \longrightarrow \mathbb{R}^n$ be a twice Fréchet–differentiable operator. Assume there exist $x_0 \in \mathcal{D}$, constants $\beta > 0$, $\gamma > 0$ and $\eta > 0$, such that, for all $x, y \in \mathcal{D}$*

*$F'(x_0)$ is positive definite, $x_1$ is the unique solution of subproblem $VI(\mathcal{D}, F_0)$*

$$\parallel \widetilde{F'(x_0)}^{\,-1} \parallel \leq \beta, \qquad \parallel x_1 - x_0 \parallel \leq \eta, \tag{10.139}$$

$$\parallel F''(x) \parallel \leq \gamma \quad (x \in \mathcal{D}), \tag{10.140}$$

$$h_{WS} = \overline{\overline{\alpha}}\ \eta \leq \frac{1}{2} \tag{10.141}$$

*and*

$$\overline{U}(x_0, s^\star) \subseteq \mathcal{D}, \tag{10.142}$$

*where $\widetilde{F'(x)}$ is defined by (10.107),*

$$\overline{\overline{\alpha}} = \frac{2\ \beta\ \gamma}{7 - \sqrt{33}}, \tag{10.143}$$

$$s^\star = \lim_{k \to \infty} s_k, \tag{10.144}$$

$$\begin{aligned} &s_0 = 0, \quad s_1 = \eta, \\ &s_{k+1} = s_k + \frac{3\ \beta\ \gamma\ (s_k - s_{k-1})^2}{2\ (1 - \gamma\ \beta\ s_{k-1})} \qquad (k \geq 1). \end{aligned} \tag{10.145}$$

*Then, the sequence $\{x_k\}$ generated by NJM is well defined, remains in $\overline{U}(x_0, s^\star)$ for all $k \geq 0$ and converges to a unique solution $x^\star \in \overline{U}(x_0, s^\star)$ of $VI(\mathcal{D}, F)$. Moreover, the following estimates hold for all $k \geq 0$*

$$\parallel x_{k+1} - x_k \parallel \leq s_{k+1} - s_k \tag{10.146}$$

*and*

$$\| x_k - x^\star \| \leq s^\star - s_k \leq \frac{1}{a} \left( \frac{3\,a}{2\,(1-a)} \right)^k \left( \frac{h_0}{a} \right)^{2^k - 1} \eta, \tag{10.147}$$

*where*

$$a = \frac{7-\sqrt{33}}{4} \quad \text{and} \quad h_0 = \beta\,\gamma\,\eta.$$

We can now compare Theorem 10.5 to Theorem 10.6

**Remark 10.8** *(a)* **Improvement 1: Case** $\gamma_0 = \gamma$**.** *Theorem 10.6 requires that operator $F$ is twice–Fréchet differentiable. However, according to Theorem 10.5, condition (10.140) can be replaced by the weaker (10.104).*

*(b)* **Improvement 2: Case** $\gamma_0 < \gamma$**.** *It follows from (10.105), (10.108), (10.141) and (10.143) that*

$$\overline{\alpha} < \overline{\overline{\alpha}}. \tag{10.148}$$

**Proof of estimate (10.148).** *In view of (10.108) and (10.143), we must have that*

$$\frac{\gamma_0}{2} \, \frac{2\,\gamma_0 + \gamma + \sqrt{(2\,\gamma_0+\gamma)^2 + 24\,\gamma^2}}{2\,\gamma_0 - 5\,\gamma + \sqrt{(2\,\gamma_0+\gamma)^2 + 24\,\gamma^2}} < \frac{2\,\gamma}{7-\sqrt{33}}. \tag{10.149}$$

*Set $\gamma_0 = \gamma\, t$, $(t \in (0,1))$. Then estimate (10.149) holds if*

$$f(t) < 0 \qquad (t \in (0,1)), \tag{10.150}$$

*where*

$$f(t) = t\, \frac{2\,t+1+\sqrt{(2\,t+1)^2+24}}{2\,t-5+\sqrt{(2\,t+1)^2+24}} - \frac{4}{7-\sqrt{33}}.$$

*The set of solutions for inequality (10.150) is $(0,\infty)$. For validating this result, see Figure 10.1. Using Maple to further validate (10.148), we derive $f$. Then, the derivative $f'$ is strictly positive (see Figure 10.1), and $f$ is strictly increasing. Hence, we have*

$$h_{WS} \leq \frac{1}{2} \Longrightarrow h_{AH} \leq \frac{1}{2}. \tag{10.151}$$

*But not necessarily vice versa, unless if $\gamma_0 = \gamma$. Indeed, we have that*

$$\overline{\alpha} = \overline{\overline{\alpha}}$$

*and*

$$h_{WS} = h_{AH}.$$

*Note that in this case, (10.81) or (10.151) coincide with the Kantorovich hypothesis (see also (10.145)). Indeed, let us define scalar iteration $\{w_n\}$ by*

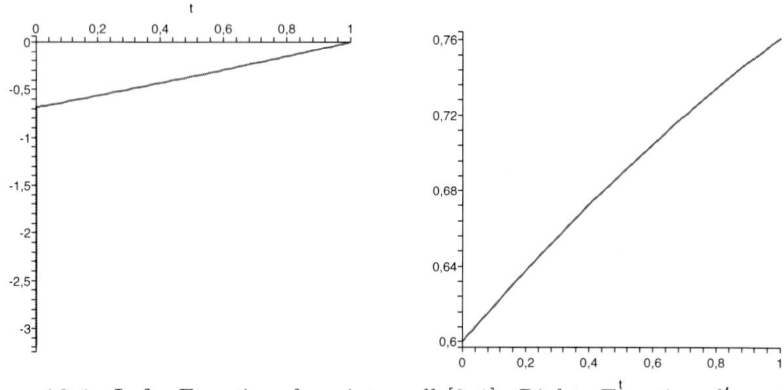


**Figure 10.1:** Left: Function $f$ on intervall $[0, 1]$. Right: Function $f'$ on intervall $[0, 1]$.

$$w_0 = 0, \quad w_1 = \eta, \quad w_{k+1} = w_k + \frac{\lambda\,(w_k - w_{k-1})^2}{2\,(1 - \lambda\, w_k)} \quad (n \geq 1).$$

*Then, the Kantorovich hypothesis for the convergence of sequence* $\{w_k\}$ *is given by*

$$h_K = \lambda\,\eta \leq \frac{1}{2}.$$

*Note that iteration* $\{w_k\}$ *is not exactly the same as* $\{s_k\}$ *given by (10.145). That is why (10.81) or (10.141) do not look exactly like the Kantorovich hypothesis. However, according to the proofs of Lemma 10.6, Theorem 1 in (cf. [761]), (10.81) and (10.141), respectively, are the correct analogs of the Kantorovich hypothesis. Moreover, using (10.110), (10.145) and a simple induction argument, we conclude:*

$$r_k < s_k \qquad (k > 1), \tag{10.152}$$

$$r_{k+1} - r_k < s_{k+1} - s_k \qquad (k > 1), \tag{10.153}$$

$$r^\star - r_k \leq s^\star - s_k \qquad (k \geq 1) \tag{10.154}$$

*and*

$$r^\star \leq s^\star. \tag{10.155}$$

*Note that under condition (10.105), the quadratic order of convergence for sequences* $\{x_n\}$ *and* $\{s_n\}$ *was established in (cf. [761]) (see (10.147)). In view of (10.151), (10.153) and (10.154), our estimates (10.112), (10.113) are even tighter than (10.146) and (10.147), respectively.*

(c) *The limit point* $r^\star$ *in (10.106) can be replaced by* $r^{\star\star}$ *given in closed form by (10.111). Hence, we do know the location of the solution of each subprogram.*

(d) *It follows from the proof of Theorem 10.5 that sequence* $\{v_k\}$ *given by*

$$v_0 = 0, \quad v_1 = \eta,$$

$$v_{k+1} = v_k + \frac{\beta\, \gamma_1\, (v_k - v_{k-1})^2}{2\,(1 - (\gamma_0\, \beta\, v_{k-1} + \overline{\gamma}_1\, \beta\, (v_k - v_{k-1})))} \qquad (k \geq 1) \qquad (10.156)$$

*is also majorizing sequence for* $\{x_k\}$.

*If* $\gamma_0 < \gamma$, *a simple inductive argument shows that*

$$v_k < r_k \qquad (k > 1), \qquad (10.157)$$

$$v_{k+1} - v_k < r_{k+1} - r_k \qquad (k \geq 1), \qquad (10.158)$$

$$v^\star - v_k \leq r^\star - s_k \qquad (k \geq 1) \qquad (10.159)$$

*and*

$$v^\star = \lim_{k \to \infty} v_k \leq r^\star, \qquad (10.160)$$

*where*

$$\gamma_1 = \begin{cases} 3\,\gamma_0, & \text{if} \quad k = 1, \\ 3\,\gamma, & \text{if} \quad k > 1, \end{cases} \qquad \overline{\gamma}_1 = \begin{cases} \gamma_0, & \text{if} \quad k = 1, \\ \gamma, & \text{if} \quad k > 1. \end{cases} \qquad (10.161)$$

*The motivation that leads to the definition of sequence* $\{v_k\}$ *is the following: Simply notice that in (10.128), we used (10.104) for the derivation of the upper bound*

$\| x_2 - x_1 \|$ *(given by (10.133)). However, the more precise than (10.104), condition (10.103) is actually needed.*

All the above (see, Remarks 10.7 and 10.8) justify the claims made in the Introduction of this section. Hence, the applicability of NJM has been extended under less or the same computational cost. In practice, the computation of $\gamma$ requires that of $\gamma_0$. Hence, (10.103) is not an additional hypothesis.

The proofs of the remaining results are omitted since they can be obtained from the corresponding ones in (cf. [761]) by simply replacing $\{s_k\}$, $s^\star$, $h_{WS}$ by $\{r_k\}$, $r^\star$, $h_{AH}$, respectively.

**Theorem 10.7** *Assume hypotheses of Theorem 10.5 hold. Then, for any* $y_0 \in \overline{U}(x_0, \eta) \cap \mathcal{D}$, *NJM* $\{y_k\}$ *is well defined, remains in* $\overline{U}(x_0, r^\star)$ *for all* $k \geq 0$ *and converges to a solution* $y^\star$ *of* $VI(\mathcal{D}, F)$, *which is contained in* $\overline{U}(x_0, \eta) \cap \mathcal{D}$.

**Theorem 10.8** *Assume hypotheses of Theorem 10.5 hold and the limit* $x^\star$ *of NJM* $\{x_k\}$ *is included in* $U(x_0, \eta)$, *then,* $x^\star$ *is the unique solution of* $VI(\mathcal{D}, F)$ *in* $U(x_0, \eta)$.

**Theorem 10.9** *Assume hypotheses of Theorem 10.5 hold and* $h_{AH} < \frac{1}{2}$, *then,* $x^\star$ *is the unique solution of* $VI(\mathcal{D}, F)$ *in* $U(x_0, \eta)$.

**Theorem 10.10** *Assume hypotheses of Theorem 10.5 hold and* $h_{AH} < \frac{1}{2}$, *then, starting from any* $y_0 \in \overline{U}(x_0, \eta)$, *NJM* $\{y_k\}$ *is quadratically convergent.*

Finally, the results obtained here can also be provided in affine invariant form, if we simply replace $F$ by $\widetilde{F'(x_0)}^{-1} F$ (cf. [163], [469], [569]). The advantages of this approach have already been explained in detail (cf. [343]). We further extend the applicability of NJM. In particular, we show Theorems 1.2 and 1.3 given by Wang (cf. [763]) can always be replaced by the following corresponding results.

The following two results correspond to Theorems 1.2 and 1.3 in (cf. [763]), respectively. The proofs are obtained by simply using (10.103) instead of (10.104) to obtain the upper bounds on the inverses of the operators involved and majorizing sequence $\{r_k\}$ instead of $\{s_k\}$ (see also (10.121)). The error bounds are obtained by letting $L_0 = \gamma_0\ \beta$ and $L = \gamma\ \beta$ in Lemma 2.11.

**Theorem 10.11** *Let* $F : \mathcal{D} \subseteq \mathbb{R}^n \longrightarrow \mathbb{R}^n$ *be a Fréchet–differentiable operator. Assume there exist* $x_0 \in \mathcal{D}$, *constants* $\beta > 0$, $\gamma_0 > 0$, $\gamma > 0$, $\eta > 0$ *with* $\gamma_0 \leq \gamma$ *such that, for all* $x, y \in \mathcal{D}$

$F'(x_0)$ *is positive definite,* $x_1$ *is the unique solution of subproblem* $VI(\mathcal{D}, F_0)$

$$
\begin{aligned}
& F_0(x) = F(x_0) + F'(x_0)\ (x - x_0), \\
& \| \widetilde{F'(x_0)}^{-1} \|_2 \leq \beta, \qquad \| x_1 - x_0 \|_2 \leq \eta, \\
& \| F'(x) - F'(x_0) \|_2 \leq \gamma_0\ \| x - x_0 \|_2, \\
& \| F'(x) - F'(y) \|_2 \leq \gamma\ \| x - y \|_2, \\
& h_{AH} = \overline{\alpha}\ \eta \leq \frac{1}{2}, \\
& \overline{U}(x_0, r^\star) \subseteq \mathcal{D},
\end{aligned}
\tag{10.162}
$$

*where*

$$\overline{\alpha} = \frac{\beta}{8}\ (\gamma + 4\ \gamma_0 + \sqrt{\gamma^2 + 8\ \gamma_0\ \gamma}),$$

$$r^\star = \lim_{k \to \infty} r_k,$$

$$r_0 = 0, \quad r_1 = \eta, \quad r_{k+1} = r_k + \frac{\beta\ \gamma\ (r_k - r_{k-1})^2}{2(1 - \beta\ \gamma_0\ r_{k-1})} \qquad (k \geq 1). \tag{10.163}$$

*Then*

(a) *Scalar sequence* $\{r_k\}$ *is nondecreasing, bounded from above by* $r^{\star\star}$, *which is given by*

$$r^{\star\star} = \frac{2\eta}{2 - \delta}$$

*and converges to its unique least upper bound* $r^\star \in [0, r^{\star\star}]$, *where*

$$\delta = \frac{4\ \gamma}{\gamma + \sqrt{\gamma^2 + 8\ \gamma_0\ \gamma}}.$$

(b) *Sequence $\{x_k\}$ generated by NJM is well defined, remains in $\overline{U}(x_0, r^\star)$ for all $k \geq 0$ and converges to a unique solution $x^\star \in \overline{U}(x_0, r^\star)$ of $VI(\mathcal{D}, F)$.*

*Moreover, the following estimates hold for all $k \geq 0$*

$$\| x_{k+1} - x_k \|_2 \leq r_{k+1} - r_k \leq \left(\frac{\delta}{2}\right)^k (2\ h_{AH})^{2^k - 1}\ \eta$$

*and*

$$\| x_k - x^\star \|_2 \leq r^\star - r_k \leq \left(\frac{\delta}{2}\right)^k \frac{(2\ h_{AH})^{2^k-1}\ \eta}{1 - (2\ h_{AH})^{2^k}} \quad (2\ h_{AH} < 1).$$

**Theorem 10.12** *Let $\ell = \{\ell_i\} \in \{\mathbb{R} \cup \{-\infty\}\}^n$, $u = \{u_i\} \in \{\mathbb{R} \cup \{+\infty\}\}^n$ be given with $\ell_i < u_i$, $i = 1, 2, \cdots, n$. Denote:*

$$[\ell, u] = \{x = \{x_i\} \in \mathbb{R}^n \,:\, \ell_i \leq x_i \leq u_i, \quad i = 1, 2, \cdots, n\}.$$

*$F \,:\, \mathcal{D} \supseteq [\ell, u] \longrightarrow \mathbb{R}^n$ is a Fréchet–differentiable operator defined on a convex set $\mathcal{D}$.*

*Assume there exist $x_0 \in \mathcal{D}$, constants $\beta > 0$, $\gamma_0 > 0$, $\gamma > 0$, $\eta > 0$ with $\gamma_0 \leq \gamma$ such that, for all $x, y \in \mathcal{D}$ $F'(x_0)$ is an $\mathcal{H}$–matrix whose diagonal elements are all positive,*

$$\begin{aligned}
&\| \overline{F'(x_0)^{-1}} \|_\infty \leq \beta, \qquad \| x_1 - x_0 \|_\infty \leq \eta,\\
&\| F'(x) - F'(x_0) \|_\infty \leq \gamma_0\ \| x - x_0 \|_\infty,\\
&\| F'(x) - F'(y) \|_\infty \leq \gamma\ \| x - y \|_\infty,\\
&h_{AH} = \overline{\alpha}\ \eta \leq \frac{1}{2},\\
&\overline{U}(x_0, r^\star) \subseteq \mathcal{D},
\end{aligned}$$

*where $h_{AH}$ is given in Theorem 10.11, $\overline{F'(x_0)^{-1}}$ is the comparison matrix of $F'(x_0)$ and $x_1$ is the unique solution of subproblem $VI([\ell, u], F_0)$, where $F_0$ is given Theorem 10.11. Then, the conclusions of Theorem 10.11 hold with $\| \,.\, \|_2$ replaced by the $\| \,.\, \|_\infty$ norm.*

**Remark 10.9** *The majorizing sequence $\{s_k\}$ used in (cf. [763]) is given by*

$$s_0 = 0, \quad s_1 = \eta, \quad s_{k+1} = s_k + \frac{\beta\ \gamma\ (s_k - s_{k-1})^2}{2\ (1 - \beta\ \gamma\ s_{k-1})} \qquad (k \geq 1). \tag{10.164}$$

*If $\gamma_0 = \gamma$, then, $r_k = s_k$ $(k \geq 0)$, and our results reduce to the ones in (cf. [763]). Otherwise, they constitute an improvement (see Remark 10.8). Note also that, if $\gamma_0 < \gamma$, then, we have that*

$$h_W = \beta\ \gamma\ \eta \leq \frac{1}{2} \Longrightarrow h_{AH} \leq \frac{1}{2},$$

*since $\overline{\alpha} < \gamma$ (but not vice versa). Moreover, according to Lemma 2.11, the ratio of quadratic convergence "$2\, h_{AH}$" (for $L = \gamma\, \beta$, $L_0 = \gamma_0\, \beta$) is smaller than "$2\, h_W$" given by Wang (cf. [763]). Furthermore, as noted in Remark 10.8, sequence $\{v_k\}$ given by*

$$v_0 = 0, \quad v_1 = \eta, \quad v_{k+1} = v_k + \frac{\beta\, \gamma_1\, (v_k - v_{k-1})^2}{2\, (1 - \beta\, \gamma_0\, v_{k-1})} \qquad (k \geq 1) \qquad (10.165)$$

*is also majorizing for $\{x_k\}$, where*

$$\gamma_1 = \begin{cases} \gamma_0 & \text{if} \quad k = 1, \\ \gamma & \text{if} \quad k > 1, \end{cases}$$

For example that $\gamma_0 < \gamma$, $h_{AH} \leq \frac{1}{2}$, but $h_{WS} > \frac{1}{2}$, the reader can refer to section 2.1, for even weaker sufficient convergence conditions.

**Remark 10.10** *We shall compare majorizing sequence $\{r_k\}$, $\{s_k\}$, $\{v_k\}$ with each other. Let us choose $\alpha := .7$ (here $\alpha$ is $\beta$ in (2.517)) in Example 2.11, then, we have*

$$\eta = .1, \quad \gamma_0 = 6.9, \quad \gamma = 7.8, \quad \beta = \frac{1}{3}.$$

*The hypotheses of our Theorem 10.12 and Theorem 1.3 in (cf. [763]) are satisfied. Then, we can provide the comparison table.*

**Comparison table**

| $k$ | (10.79) $\| x_{k+1} - x_k \|$ | (10.165) $v_{k+1} - v_k$ | (10.163) $r_{k+1} - r_k$ | (10.164) $s_{k+1} - s_k$ |
|---|---|---|---|---|
| 0 | .1414213562 | .1 | .1 | .1 |
| 1 | .1107329219e-1 | .1150000000e-1 | .1300000000e-1 | .1300000000e-1 |
| 2 | .1423335877e-3 | .2232792208e-3 | .2853246753e-3 | .2968918919e-3 |
| 3 | .2590350763e-7 | .8716252243e-7 | .1429985426e-6 | .1622603145e-6 |
| 4 | .3716428396e-10 | .1329207578e-13 | .3595020992e-13 | .4851937919e-13 |
| 5 | $\sim$ | .3091142050e-27 | .2272172307e-26 | .4338317565e-26 |
| 6 | $\sim$ | .1671749770e-54 | .9076552540e-53 | .3468434894e-52 |
| 7 | $\sim$ | .4889628797e-109 | .1448372442e-105 | .2216960533e-104 |

*The table shows that our error bounds (10.165) and (10.163) are tighter than (10.164) proposed in (cf. [763]).*

## Chapter 11

# Fast Two–Step Methods

We present in this chapter the convergence analysis of two classes of two–point Newton–type methods including Özban's and the midpoint methods.

## 11.1 Semilocal Convergence

The semilocal convergence of a Newton–type method used to solve nonlinear equations in a Banach space is studied. As two important applications, we also give the convergence analysis of two classes of two–point Newton–type methods including a method mentioned in (cf. [575]) and the midpoint method studied in (cf. [163, 171, 174]). Recently, an interest has been shown for this type of methods (cf. [440, 575]).

In this section, we are concerned with the problem of approximating a locally unique solution $x^*$ of (2.1). We use the two–point Newton–type method given for $x_0, y_0 \in \mathcal{D}$ by

$$\begin{aligned} z_n &= \frac{x_n + y_n}{2} \quad (n \geq 0), \\ x_{n+1} &= x_n - F'(z_n)^{-1}F(x_n), \qquad (11.1) \\ y_{n+1} &= x_{n+1} - F'(z_n)^{-1}F(x_{n+1}) \end{aligned}$$

to generate a sequence $\{x_n\}(n \geq 0)$ approximating $x^*$. The method uses two function evaluations and one inverse at each step.

Let us illustrate how this method is conceived. We start with the identity

$$F(x) - F(y) = \int_0^1 F'(y + \tau(x - y))\,\mathrm{d}\tau(x - y) \quad (x, y \in \mathcal{D}). \qquad (11.2)$$

If $x^*$ is a solution of equation (2.1), then identity (11.2) gives

$$F(x) = \int_0^1 F'(x + \tau(x^* - x))\,\mathrm{d}\tau(x^* - x). \qquad (11.3)$$

The linear operator in (11.3) can be approximated in different ways (cf. [163, 171, 440, 575, 575]). If for example

$$\int_0^1 F'(x + \tau(x^* - x))\,\mathrm{d}\tau = F'(x) \quad (x \in \mathcal{D}), \qquad (11.4)$$

then, (11.3) suggests the famous Newton's method (cf. [13, 163, 139, 171, 440, 469, 575, 765]). Another choice is given by

$$\int_0^1 F'(x+\tau(x^*-x))\,\mathrm{d}\tau = F'\left(\frac{x^*+x}{2}\right) \quad (x \in \mathcal{D}), \tag{11.5}$$

which leads to the implicit iteration:

$$x_{n+1} = x_n - F'\left(\frac{x_n + x_{n+1}}{2}\right)^{-1} F(x_n) \quad (n \geq 0). \tag{11.6}$$

Unfortunately, iterates in (11.6) can only be computed in very restrictive cases. Hence, we arrive at method (11.1). This method was first introduced in (cf. [765]), where the R–order $1+\sqrt{2}$, local convergence was established.

Recently, Gutiérrez (cf. [399]) used cubic instead of quadratic majorizing functions to establish new semilocal convergence theorem for Newton's method and explained the advantages and disadvantages of using cubic instead of quadratic majorizing functions (see also the monograph of Argyros (cf. [163])).

We will establish the R–order of the semilocal convergence of method (11.1) to be $1+\sqrt{2}$. Note that the results on the elegant work in (cf. [765]) rely on $x^*$ (local convergence), whereas the results obtained here rely on $x_0$ and $y_0$ (semilocal convergence). Especially, we first introduce the cubic majorizing functions of (11.1). Then, give the semilocal convergence analysis. Finally, as two applications of our results, we also give convergence analysis of two classes of two–point Newton–type methods including a method mentioned in (cf. [765]) and the midpoint method studied in (cf. [163, 171, 174]).

Let $\alpha \geq 0$, $\beta \geq 0$, and $\gamma \geq 0$ be fixed. Define parameter $m$ by

$$m = \frac{2}{\beta + \sqrt{\beta^2 + 4\alpha}} \quad (\alpha \neq 0 \text{ or } \beta \neq 0), \tag{11.7}$$

function $f$ on $[0, m]$ by

$$f(t) = \frac{\alpha}{3}t^3 + \frac{\beta}{2}t^2 - t + \gamma, \tag{11.8}$$

and scalar sequences $\{t_n\}$, $\{s_n\}$ $(n \geq 0)$ for $t_0 = 0$ and some $0 \leq s_0 \leq \gamma$ by

$$\begin{aligned} r_n &= \frac{t_n + s_n}{2} \quad (n \geq 0), \\ t_{n+1} &= t_n - f'(r_n)^{-1} f(t_n), \\ s_{n+1} &= t_{n+1} - f'(r_n)^{-1} f(t_{n+1}). \end{aligned} \tag{11.9}$$

It is well known (cf. [163, 399, 445]) that

$$f(m) \leq 0 \tag{11.10}$$

is a necessary and sufficient condition for $f$ to have positive roots $t^*$ and $t^{**}$ with

$$t^* \leq m \leq t^{**}. \tag{11.11}$$

Each one of the following conditions is equivalent to (11.10):

$$6\gamma\beta^3 + 9\gamma^2\alpha^2 + 18\gamma\beta\alpha \leq 3\beta^2 + 8\alpha \tag{11.12}$$

or

$$3\gamma\alpha^2 + 3\beta\alpha + \beta^3 \leq (\beta^2 + 2\alpha)^{3/2}. \tag{11.13}$$

Note also that using (11.8) we have

$$f'(t) < 0 \quad (t \in [0, m]). \tag{11.14}$$

We need the following proposition on majorizing sequences $\{t_n\}$, $\{s_n\}$ $(n \geq 0)$

**Proposition 11.1** *Under hypothesis* (11.12) *(or* (11.13)*), scalar sequences* $\{t_n\}$, $\{s_n\}$ *generated by* (11.9) *are well defined for all* $n \geq 0$ *and converge monotonically to* $t^*$ *with*

$$0 \leq t_n \leq s_n \leq t_{n+1} \leq t^*. \tag{11.15}$$

**Proof** We shall show using induction that, for all $k \geq 0$,

$$0 \leq t_k \leq s_k \leq t_{k+1} < t^*. \tag{11.16}$$

Estimate (11.16) holds true for $k = 0$ by the initial conditions and $t_1$ is well defined. Assume (11.16) holds for all $k = 0, 1, ..., n$. Then, there exists $u_n \in [t_{n+1}, t^*]$ such that

$$f(t_{n+1}) - f(t^*) = f'(u_n)(t_{n+1} - t^*). \tag{11.17}$$

In view of (11.9) and (11.17), we get

$$s_{n+1} - t^* = \frac{1}{2}(\alpha r_n + u_n + \beta)[(t_n - u_n) + (s_n - u_n)](t_{n+1} - t^*)f'(r_n)^{-1} < 0 \tag{11.18}$$

by the induction hypotheses, which implies $s_{n+1} < t^*$. Clearly, we have, $s_{n+1} \leq t_{n+2}$. As in (11.18), we show $t_{n+2} < t^*$. This completes the induction and the proof of the conclusions. ☒

We also need the following lemma on the R–order of convergence:

**Lemma 11.1** *(cf. [163, 765]) Let* $0 < a_0$, $a_1 < 1$, $p > 1$, $q \geq 0$ *and* $c \geq 0$ *be fixed. If a scalar sequence* $\{a_n\}$ *satisfies for all* $n \geq 0$

$$0 < a_{n+1} \leq c a_n^p a_{n-1}^q, \tag{11.19}$$

*then, it converges to zero with R–order*

$$\frac{p}{2} + \sqrt{\frac{p^2}{4} + q}. \tag{11.20}$$

Assume now that

$$y_0 \in \overline{U}(x_0, m) \subseteq \mathcal{D}, \tag{11.21}$$

$$F'(z_0)^{-1} \in L(\mathcal{Y}, \mathcal{X}), \tag{11.22}$$

$$\left\| F'(z_0)^{-1} F(x_0) \right\| \leq \gamma = s_0, \tag{11.23}$$

$$\left\| F'(z_0)^{-1} F''(x_0) \right\| \leq \beta \tag{11.24}$$

and

$$\left\| F'(z_0)^{-1} F'''(x) \right\| \leq \alpha \quad (x \in \overline{U}(x_0, m)). \tag{11.25}$$

Note that function $f$ defined by (11.8) satisfies the hypotheses (11.23)–(11.25). We need the following proposition on the upper bounds of $\|F'(z_0)^{-1} F''(x)\|$ and $\|F'(x)^{-1} F'(x_0)\|$.

**Proposition 11.2** *Suppose $F$ satisfies conditions* (11.21)–(11.25). *Then, the following hold*

$$\left\| F'(z_0)^{-1} F''(x) \right\| \leq f''(\|x - x_0\|) \tag{11.26}$$

*and*

$$F'(x)^{-1} \text{ exists and } \left\| F'(x)^{-1} F'(x_0) \right\| \leq -\frac{1}{f'(\|x - x_0\|)}. \tag{11.27}$$

**Proof** Using (11.24) and (11.25) we get

$$\begin{aligned}
&\left\| F'(z_0)^{-1} F''(x) \right\| \\
&\leq \left\| F'(z_0)^{-1} F''(x_0) \right\| + \left\| F'(z_0)^{-1} F''(x) - F'(z_0)^{-1} F''(x_0) \right\| \\
&= \left\| F'(z_0)^{-1} F''(x_0) \right\| + \left\| \int_0^1 F'(z_0)^{-1} F'''(x_0 + \tau(x - x_0))(x - x_0)\, \mathrm{d}\tau \right\| \\
&\leq \beta + \int_0^1 f'''(\tau \|x - x_0\|) \|x - x_0\| \, \mathrm{d}\tau \\
&= \beta + f''(\|x - x_0\|) - f''(0) = f''(\|x - x_0\|)
\end{aligned}$$

and

$$\begin{aligned}
\left\| F'(x_0)^{-1} F'(x) - I \right\| &= \left\| F'(x_0)^{-1} [F'(x) - F'(x_0)] \right\| \\
&= \left\| F'(x_0)^{-1} \int_0^1 F''(x_0 + \tau(x - x_0))(x - x_0)\, \mathrm{d}\tau \right\| \\
&\leq \int_0^1 f''(\tau \|x - x_0\|) \|x - x_0\| \, \mathrm{d}\tau \\
&= f'(\|x - x_0\|) - f'(0) = f'(\|x - x_0\|) + 1 < 1.
\end{aligned}$$

By the Banach Lemma on invertible operators, $F'(x)^{-1}$ exists and

$$\|F'(x)^{-1}F'(x_0)\| \leq \frac{1}{1-\|F'(x_0)^{-1}F'(x)-I\|} = -\frac{1}{f'(\|x-x_0\|)}.$$

This completes the proof of Proposition 11.2. ⊠

In the next proposition following our ideas in (cf. [171, 174]), we provide an Ostrowski-type representation for $F(x_n)$.

**Proposition 11.3** *Suppose $\mathcal{X}$ and $\mathcal{Y}$ are the Banach spaces, $\mathcal{D}$ is an open convex of the Banach space $\mathcal{X}$, $F : \mathcal{D} \subset \mathcal{X} \to \mathcal{Y}$ has thrice order continuous Fréchet derivatives. Moreover, assume sequences $\{x_n\}$, $\{y_n\}$ generated by* (11.1) *are well defined. Then, the following identity holds for all $n \geq 0$*

$$\begin{aligned}
&F(x_{n+1})\\
&= \int_0^1 F''\left(\frac{x_n+y_n}{2}+\tau\left(x_{n+1}-\frac{x_n+y_n}{2}\right)\right)(1-\tau)\,\mathrm{d}\tau\,(x_{n+1}-y_n)^2\\
&\quad +\frac{1}{2}\int_0^1 F''\left(\frac{x_n+y_n}{2}+\tau\left(x_{n+1}-\frac{x_n+y_n}{2}\right)\right)(1-\tau)\,\mathrm{d}\tau(x_{n+1}-y_n)(y_n-x_n)\\
&\quad +\frac{1}{2}\int_0^1 F''\left(\frac{x_n+y_n}{2}+\tau\left(x_{n+1}-\frac{x_n+y_n}{2}\right)\right)(1-\tau)\,\mathrm{d}\tau(y_n-x_n)(x_{n+1}-y_n)\\
&\quad +\frac{1}{4}\int_0^1\int_0^1 F'''\left(\frac{(1+\tau)x_n}{2}+\frac{(1-\tau)y_n}{2}+s\tau(x_{n+1}-x_n)\right)(1-\tau)\,\mathrm{d}s\mathrm{d}\tau\\
&\quad \times (x_{n+1}-x_n)(y_n-x_n)^2.
\end{aligned} \tag{11.28}$$

**Proof** Using (11.1), we have in turn that

$$\begin{aligned}
F(x_{n+1}) &= F(x_{n+1}) - F\left(\frac{x_n+y_n}{2}\right) - F'\left(\frac{x_n+y_n}{2}\right)\left(x_{n+1}-\frac{x_n+y_n}{2}\right)\\
&\quad + F\left(\frac{x_n+y_n}{2}\right) + F'\left(\frac{x_n+y_n}{2}\right)\left(x_{n+1}-\frac{x_n+y_n}{2}\right)\\
&= \int_0^1 F''\left(\frac{x_n+y_n}{2}+\tau\left(x_{n+1}-\frac{x_n+y_n}{2}\right)\right)(1-\tau)\,\mathrm{d}\tau\left(x_{n+1}-\frac{x_n+y_n}{2}\right)^2\\
&\quad + F\left(\frac{x_n+y_n}{2}\right) + F'\left(\frac{x_n+y_n}{2}\right)\left(x_{n+1}-\frac{x_n+y_n}{2}\right)
\end{aligned}$$

and

$$\begin{aligned}
&F\left(\frac{x_n+y_n}{2}\right) + F'\left(\frac{x_n+y_n}{2}\right)\left(x_{n+1}-\frac{x_n+y_n}{2}\right)\\
&\qquad = F\left(\frac{x_n+y_n}{2}\right) + F'\left(\frac{x_n+y_n}{2}\right)\left(x_{n+1}-x_n-\frac{y_n-x_n}{2}\right)\\
&\qquad = F\left(\frac{x_n+y_n}{2}\right) - F(x_n) - F'\left(\frac{x_n+y_n}{2}\right)\frac{y_n-x_n}{2}\\
&\qquad = -\frac{1}{4}\int_0^1 F''\left(\frac{x_n+y_n}{2}-\tau\left(\frac{y_n-x_n}{2}\right)\right)(1-\tau)\,\mathrm{d}\tau(y_n-x_n)^2.
\end{aligned}$$

Hence, we get that

$$
\begin{aligned}
&F(x_{n+1}) \\
&= \int_0^1 F''\left(\frac{x_n+y_n}{2}+\tau\left(x_{n+1}-\frac{x_n+y_n}{2}\right)\right)(1-\tau)\,\mathrm{d}\tau(x_{n+1}-y_n)^2 \\
&\quad+\frac{1}{2}\int_0^1 F''\left(\frac{x_n+y_n}{2}+\tau\left(x_{n+1}-\frac{x_n+y_n}{2}\right)\right)(1-\tau)\,\mathrm{d}\tau(x_{n+1}-y_n)(y_n-x_n) \\
&\quad+\frac{1}{2}\int_0^1 F''\left(\frac{x_n+y_n}{2}+\tau\left(x_{n+1}-\frac{x_n+y_n}{2}\right)\right)(1-\tau)\,\mathrm{d}\tau(y_n-x_n)(x_{n+1}-y_n) \\
&\quad+\frac{1}{4}\int_0^1 F''\left(\frac{x_n+y_n}{2}+\tau\left(x_{n+1}-\frac{x_n+y_n}{2}\right)\right)(1-\tau)\,\mathrm{d}\tau(y_n-x_n)^2 \\
&\quad-\frac{1}{4}\int_0^1 F''\left(\frac{x_n+y_n}{2}-\tau\left(\frac{y_n-x_n}{2}\right)\right)(1-\tau)\,\mathrm{d}\tau(y_n-x_n)^2 \\
&= \int_0^1 F''\left(\frac{x_n+y_n}{2}+\tau\left(x_{n+1}-\frac{x_n+y_n}{2}\right)\right)(1-\tau)\,\mathrm{d}\tau\,(x_{n+1}-y_n)^2 \\
&\quad+\frac{1}{2}\int_0^1 F''\left(\frac{x_n+y_n}{2}+\tau\left(x_{n+1}-\frac{x_n+y_n}{2}\right)\right)(1-\tau)\,\mathrm{d}\tau(x_{n+1}-y_n)(y_n-x_n) \\
&\quad+\frac{1}{2}\int_0^1 F''\left(\frac{x_n+y_n}{2}+\tau\left(x_{n+1}-\frac{x_n+y_n}{2}\right)\right)(1-\tau)\,\mathrm{d}\tau(y_n-x_n)(x_{n+1}-y_n) \\
&\quad+\frac{1}{4}\int_0^1\int_0^1 F'''\left(\frac{(1+\tau)x_n}{2}+\frac{(1-\tau)y_n}{2}+s\tau(x_{n+1}-x_n)\right)\tau(1-\tau)\,\mathrm{d}s\,\mathrm{d}\tau \\
&\quad\times(x_{n+1}-x_n)(y_n-x_n)^2.
\end{aligned}
$$

This completes the proof of Proposition 11.3. ⊠

In order for us to establish the order of convergence of scalar sequences $\{t_n\}$, $\{s_n\}$, we first present a local convergence conclusion for the methods (11.1) introduced by Werner (cf. [765]).

**Theorem 11.1** *Let $F : \mathcal{D} \subseteq \mathcal{X} \to \mathcal{Y}$ be a twice Fréchet-differentiable operator. Assume that*

$$
\begin{aligned}
F'(x)^{-1} \in L(\mathcal{Y},\mathcal{X}) \quad (x \in \mathcal{D}), \\
\sup_{x\in\mathcal{D}} \left\|F'(x)^{-1}\right\| \le \Gamma
\end{aligned}
\tag{11.29}
$$

*for all $x, y \in \mathcal{D}$ and $h \in \mathbb{R}^n$*

$$\|F'(x)-F'(y)\| \le L_1\|x-y\|, \tag{11.30}$$

$$\|F''(x)h-F''(y)h\| \le L_2\|x-y\|\,\|h\|. \tag{11.31}$$

*Set*

$$A = \frac{1}{2}\Gamma L_1 \quad \textit{and} \quad B = \frac{\Gamma L_2}{24}.$$

*Denote by* $r_0$, $r_1$ *the unique solutions of equations*

$$B\rho_0^2 + A\rho_1 = 1,$$
$$2A\rho_0^2 + A\rho_0\rho_1 = \rho_1$$

*on the interval* $\left(0, \frac{1}{A}\right)$. *Choose* $x_0 \in U(x^*, r_0)$, $y_0 \in U(x^*, r_1)$ *and assume*

$$U\left(x^*, \frac{1}{A}\right) \subseteq \mathcal{D} \text{ for } x^* \text{ such that } F(x^*) = 0.$$

*Set*

$$a_n = \|x_n - x^*\| \quad \text{and} \quad b_n = \|y_n - x^*\|. \tag{11.32}$$

*Then, sequences* $\{x_n\}$, $\{y_n\}$ *generated by method* (11.1) *are well defined for all* $n \geq 0$ *and converge to* $x^*$. *Moreover, the following estimates hold true*

$$b_{n+1} \leq A\left(2 + \frac{r_1}{r_0}\right) a_{n+1}a_n,$$
$$a_{n+1} \leq Ba_n^2 a_{n-1} + Aa_n b_n$$

*and*

$$a_{n+1} \leq Ca_n^2 a_{n-1} \text{ for some } C > 0.$$

*Furthermore, the R–order of convergence is* $1 + \sqrt{2}$.

By (11.9) we obtain the identities

$$t_{n+1} - t^* = f'(r_n)^{-1}\left[f'\left(\frac{t_n + t^*}{2}\right)(t_n - t^*) - f(t_n) + f(t^*)\right]$$
$$+ f'(r_n)^{-1}\left[f'(r_n) - f'\left(\frac{t_n + t^*}{2}\right)\right](t_n - t^*)$$

and

$$s_{n+1} - t^* = f'(r_n)^{-1}\left[f'(r_n)(t_{n+1} - t^*) - f(t_{n+1}) + f(t^*)\right].$$

In view of the above identities we arrive at the corollary:

**Corollary 11.1** *Let* $\delta > 0$ *be fixed and restrict function* $f$ *on* $\mathcal{D}_0 = [0, m_0]$, *where*

$$m_0 = \frac{2(1 + \delta^{-1})}{b + \sqrt{b^2 + 4a(1 + \delta^{-1})}}.$$

*Set* $\mathcal{X} = \mathcal{Y} = \mathbb{R}$, $\mathcal{D} = \mathcal{D}_0$, $F = f$, $\Gamma = \delta$, $L_1 = 2|a| m_0 + |b|$ *and* $L_2 = 2|a|$. *Assume* (11.12) *(or* (11.13)*) holds true. Then, scalar sequences* $\{t_n\}$, $\{s_n\}$ *generated by* (11.9) *are well defined and converge to* $t^*$ *with R–order of convergence of* $1 + \sqrt{2}$.

**Proof** Using the definition of scalar function $f$ on $\mathcal{D}_0$ given by (11.8), we can easily see that hypotheses (11.29) -(11.32) hold true for the above choices of $\Gamma$, $L_1$ and $L_2$. Hence, the conclusions follow according to Theorem 11.1and Lemma 11.1, since $p = 2$ and $q = 1$. This completes the proof of the corollary. ⊠

We can show the main semilocal convergence result for method (11.1):

**Theorem 11.2** *Assume hypotheses* (11.12) *(or* (11.13)*) and* (11.21)–(11.25) *hold true. Then, the sequences* $\{x_n\}$, $\{y_n\}$ *generated by* (11.1) *are defined* $x_n, y_n \in \overline{U}(x_0, t^*)$ *and converge to the unique solution* $x^*$ *in* $U(x_0, m)$ *with* $\|x_n - x^*\| \leq t^* - t_n$.

**Proof** We will prove by induction that the following estimates hold

$$\|x_n - x_0\| \leq t_n, \tag{11.33}$$

$$\|y_n - x_n\| \leq s_n - t_n, \tag{11.34}$$

$$\|y_n - x_0\| \leq s_n, \tag{11.35}$$

$$\left\| F'\left(\frac{x_n + y_n}{2}\right)^{-1} F'(x_0) \right\| \leq -f'\left(\frac{t_n + s_n}{2}\right)^{-1}, \tag{11.36}$$

$$\|x_{n+1} - x_n\| \leq t_{n+1} - t_n. \tag{11.37}$$

Estimate (11.33) holds true for $n = 0$, by the initial conditions. Assuming (11.33) holds true for $x_n$, we get

$$\|x_{n+1} - x_0\| \leq \|x_{n+1} - x_n\| + \|x_n - x_0\| \leq t_{n+1} - t_n + t_n = t_{n+1}.$$

By Proposition 11.3, we have in turn

$$\begin{aligned}
&\|F'(z_0)^{-1} F(x_{n+1})\| \\
&\leq \int_0^1 f''\left(\frac{t_n + s_n}{2} + \tau\left(t_{n+1} - \frac{t_n + s_n}{2}\right)\right)(1-\tau)\,\mathrm{d}\tau (t_{n+1} - s_n)^2 \\
&\quad + \int_0^1 f''\left(\frac{t_n + s_n}{2} + \tau\left(t_{n+1} - \frac{t_n + s_n}{2}\right)\right)(1-\tau)\,\mathrm{d}\tau (t_{n+1} - s_n)(s_n - t_n) \\
&\quad + \frac{1}{4}\int_0^1 \int_0^1 f'''\left(\frac{(1+\tau)t_n}{2} + \frac{(1-\tau)s_n}{2} + s\tau(t_{n+1} - t_n)\right)\tau(1-\tau)\,\mathrm{d}s\ \mathrm{d}\tau \\
&\qquad \times (t_{n+1} - t_n)(s_n - t_n)^2 \\
&= f(t_{n+1})
\end{aligned} \tag{11.38}$$

Hence, by (11.1), we get that

$$\begin{aligned}\|y_{n+1}-x_{n+1}\| &= \left\|-F'(x_{n+1})^{-1}F(x_{n+1})\right\| \\ &\leq \left\|-F'(x_{n+1})^{-1}F'(z_0)\right\| \leq \left\|F'(z_0)^{-1}F(x_{n+1})\right\| \\ &\leq -f'(t_{n+1})^{-1}f(t_{n+1}) = s_{n+1}-t_{n+1}\end{aligned} \tag{11.39}$$

and

$$\|y_{n+1}-x_0\| \leq \|y_{n+1}-x_{n+1}\| + \|x_{n+1}-x_0\| = s_{n+1}$$

since, by Proposition 11.2, we get

$$\left\|F'\left(\frac{x_{n+1}+y_{n+1}}{2}\right)^{-1}F'(x_0)\right\| \leq -f'\left(\frac{t_{n+1}+s_{n+1}}{2}\right)^{-1}. \tag{11.40}$$

By (11.1), (11.38) and (11.40), we get

$$\|x_{n+2}-x_{n+1}\| \leq \left\|F'(z_{n+1})^{-1}F'(x_0)\right\| \cdot \left\|F'(x_0)^{-1}F(x_{n+1})\right\| \leq t_{n+2}-t_{n+1}.$$

Hence, the sequences $\{x_n\}$, $\{y_n\}$ generated by (11.1) are well defined, $x_n, y_n \in \overline{U}(x_n, t^*)$ and $\{x_n\}$, $\{y_n\}$ converge to the solution $x^* \in \overline{U}(x_0, t^*)$ of the equation (2.1).

To show the uniqueness let us suppose $y^*$ is a solution of the equation (2.1) on $U(x_0, m)$, using (11.24), we get

$$\begin{aligned}&\left\|F'(x_0)^{-1}\int_0^1 F'(x^*+\tau(y^*-x^*))\,\mathrm{d}\tau - I\right\| \\ &\leq \left\|F'(x_0)^{-1}\int_0^1 F'(x^*+\tau(y^*-x^*)) - F'(x_0)\,\mathrm{d}\tau\right\| \\ &\leq \left\|F'(x_0)^{-1}\int_0^1\int_0^1 F''\left(x_0+s\left(x^*-x_0+\tau(y^*-x^*)\right)\right)\mathrm{d}s(x^*-x_0+\tau(y^*-x^*))\,\mathrm{d}\tau\right\| \\ &\leq \int_0^1\int_0^1 f''\left(s\left\|x^*-x_0+\tau(y^*-x^*)\right\|\right)\mathrm{d}s\,\mathrm{d}\tau\left\|x^*-x_0+\tau(y^*-x^*)\right\| \\ &= \int_0^1 f'\left(\left\|x^*-x_0+\tau(y^*-x^*)\right\|\right)\mathrm{d}\tau - f'(0) \\ &= \int_0^1 f'\left(\left\|(1-\tau)(x^*-x_0)+\tau(y^*-x_0)\right\|\right)\mathrm{d}\tau + 1 < 1.\end{aligned}$$

By the Banach Lemma, the inverse of $\int_0^1 F'(x^*+\tau(y^*-x^*))\,\mathrm{d}\tau$ exists. Because

$$F(y^*)-F(x^*) = \int_0^1 F'(x^*+\tau(y^*-x^*))\,\mathrm{d}\tau(y^*-x^*),$$

we deduce $y^* = x^*$. For $i > n$, we get

$$\|x_i-x_n\| \leq \|x_i-x_{i-1}\| + \|x_{i-1}-x_{i-2}\| + \ldots + \|x_{n+1}-x_n\| \leq t_i-t_n.$$

By letting $i \to \infty$, we get

$$\|x_n - x^*\| \le t^* - t_n.$$

This completes the proof of Theorem 11.2. ⊠

We generalize now Theorem 11.2 to a class of the two–point Newton–type methods given for $x_0, y_0 \in \mathcal{D}$ by

$$\begin{aligned} \xi_n &= \eta_n x_n + (1-\eta_n)y_n, \quad 0 \le \eta_n \le 1( \quad n \ge 0), \\ x_{n+1} &= x_n - F'(\xi_n)^{-1}F(x_n), \\ y_{n+1} &= x_{n+1} - F'(\xi_n)^{-1}F(x_{n+1}) \end{aligned} \tag{11.41}$$

to generate a sequence $\{x_n\}(n \ge 0)$ approximating $x^*$. The corresponding majorizing sequences of (11.41) are

$$\begin{aligned} r_n &= \eta_n t_{n+1} + (1-\eta_n)s_n, \quad 0 \le \eta_n \le 1 \quad (n \ge 0), \\ t_{n+1} &= t_n - f'(r_n)^{-1}f(t_n), \\ s_{n+1} &= t_{n+1} - f'(r_n)^{-1}f(t_{n+1}). \end{aligned} \tag{11.42}$$

Werner (cf. [765]) pointed out that Theorem 11.1 holds for two–point Newton–type methods (11.41) if choosing

$$\eta_n = \min\left\{1, \frac{1}{2} + \kappa\|F(x_n)\|\right\} \quad (\kappa > 0). \tag{11.43}$$

Thus, we can obtain

**Corollary 11.2** *Under the assumptions of Corollary 11.1, scalar sequences* $\{t_n\}$, $\{s_n\}$ *generated by* (11.42) *and* (11.43) *are well defined and converge to* $t^*$ *with R–order of convergence of* $1+\sqrt{2}$.

**Proof** The proof is a verbatim repetition of the proof of Corollary 11.1. We will omit the proof. ⊠

By Proposition 11.1, then, we have

**Theorem 11.3** *Under hypothesis* (11.12) *(or* (11.13)*), scalar sequences* $\{t_n\}$, $\{s_n\}$ *generated by* (11.42) *are well defined for all* $n \ge 0$ *and converge monotonically to* $t^*$ *with*

$$0 \le t_n \le s_n \le t_{n+1} \le t^*. \tag{11.44}$$

**Proof** We shall show using induction that, for all $k \ge 0$,

$$0 \le t_k \le s_k \le t_{k+1} < t^*. \tag{11.45}$$

Estimate (11.45) holds true for $k = 0$ by the initial conditions and $t_1$ is well defined. Assume (11.45) holds for all $k = 1, ..., n$. Then, there exists $u_n \in [t_{n+1}, t^*]$ such that

$$f(t_{n+1}) - f(t^*) = f'(u_n)(t_{n+1} - t^*). \tag{11.46}$$

In view of (11.55) and (11.46), we get that

$$\begin{aligned} s_{n+1} - t^* &= f'(r_n)^{-1}\left[f'(r_n)(t_{n+1} - t^*) - f(t_{n+1}) + f(t^*)\right] \\ &= f'(r_n)^{-1}\left[f'(r_n)(t_{n+1} - t^*) - f'(u_n)(t_{n+1} - t^*)\right] \\ &= f'(r_n)^{-1}(t_{n+1} - t^*)\left[\alpha(r_n^2 - u_n^2) + \beta(r_n - u_n)\right] \\ &= f'(r_n)^{-1}(t_{n+1} - t^*)(r_n - u_n)(\alpha(r_n + u_n) + \beta) \\ &= \eta_n f'(r_n)^{-1}(t_{n+1} - t^*)(t_n - u_n)(\alpha(r_{n+1} + u_n) + \beta) \\ &\quad + (1 - \eta_n) f'(r_n)^{-1}(t_{n+1} - t^*)(s_n - u_n)(\alpha(r_{n+1} + u_n) + \beta) < 0 \end{aligned} \tag{11.47}$$

by the induction hypotheses, which implies $s_{n+1} < t^*$. Clearly, we have $s_{n+1} \leq t_{n+2}$. As in (11.47), we show $t_{n+2} < t^*$. This completes the induction and the proof of the theorem. ⊠

An Ostrowski–type representation for $F(x_n)$ can also be provided for (11.42).

**Lemma 11.2** *Suppose $\mathcal{X}$ and $\mathcal{Y}$ are the Banach spaces, $\mathcal{D}$ is an open convex of the Banach space $\mathcal{X}$, $F : \mathcal{D} \subset \mathcal{X} \to \mathcal{Y}$ has thrice order continuous Fréchet derivatives. Moreover, assume sequences $\{x_n\}$, $\{y_n\}$ generated by (11.41) and (11.43) are well defined. Then, the following identity holds for all $n \geq 0$*

$$\begin{aligned} F(x_{n+1}) &= \int_0^1 F''\left(\xi_n + \tau(x_{n+1} - \xi_n)\right)(1 - \tau)\, \mathrm{d}\tau (x_{n+1} - y_n)^2 \\ &+ \eta_n \int_0^1 F''\left(\xi_n + \tau(x_{n+1} - \xi_n)\right)(1 - \tau)\, \mathrm{d}\tau (x_{n+1} - y_n)(y_n - x_n) \\ &+ \eta_n \int_0^1 F''\left(\xi_n + \tau(x_{n+1} - \xi_n)\right)(1 - \tau)\, \mathrm{d}\tau (y_n - x_n)(x_{n+1} - y_n) \\ &+ (\eta_n^2 - (1 - \eta_n)^2) \int_0^1 F''\left(\xi_n + \tau(x_{n+1} - \xi_n)\right)(1 - \tau)\, \mathrm{d}\tau (y_n - x_n)^2 \\ &+ (1 - \eta_n)^2 \int_0^1 \int_0^1 F'''\left(\xi_n - \tau(1 - \eta_n)(y_n - x_n) + s\tau(x_{n+1} - x_n)\right)\tau(1 - \tau)\, \mathrm{d}s\, \mathrm{d}\tau \\ &\times (x_{n+1} - x_n)(y_n - x_n)^2. \end{aligned} \tag{11.48}$$

**Proof** Using (11.41), we have in turn that

$$\begin{aligned} F(x_{n+1}) &= F(x_{n+1}) - F(\xi_n) - F'(\xi_n)(x_{n+1} - \xi_n) + F(\xi_n) + F'(\xi_n)(x_{n+1} - \xi_n) \\ &= \int_0^1 F''\left(\xi_n + \tau(x_{n+1} - \xi_n)\right)(1 - \tau)\, \mathrm{d}\tau\, (x_{n+1} - \xi_n)^2 + F(\xi_n) + F'(\xi_n)(x_{n+1} - \xi_n), \end{aligned}$$

and

$$\begin{aligned} F(\xi_n) + F'(\xi_n)(x_{n+1} - \xi_n) &= F(\xi_n) + F'(\xi_n)(x_{n+1} - x_n - (1-\eta_n)(y_n - x_n)) \\ &= F(\xi_n) - F(x_n) - F'(\xi_n)(1-\eta_n)(y_n - x_n) \\ &= -(1-\eta_n)^2 \int_0^1 F''(\xi_n - \tau(1-\eta_n)(y_n - x_n))(1-\tau)\,\mathrm{d}\tau (y_n - x_n)^2. \end{aligned}$$

This completes the proof of Lemma 11.2. ⊠

**Theorem 11.4** *Assume hypotheses* (11.12) *(or* (11.13)*) and* (11.21)–(11.25) *hold true. Then, the sequences* $\{x_n\}$, $\{y_n\}_{n\geq 0}$ *generated by* (11.41) *and* (11.43) *are defined* $x_n, y_n \in \overline{U}(x_0, t^*)$ *and converge to the unique solution* $x^*$ *in* $U(x_0, m)$ *with* $\|x_n - x^*\| \leq t^* - t_n$.

**Proof** The statement can be derived from Corollary 11.2, Theorem 11.3, Lemma 11.2 and the proof of Theorem 11.2. Especially, we need prove by induction that the following estimates hold

$$\|x_n - x_0\| \leq t_n, \tag{11.49}$$

$$\|y_n - x_n\| \leq s_n - t_n, \tag{11.50}$$

$$\|y_n - x_0\| \leq s_n, \tag{11.51}$$

$$\left\|F'(\xi_n)^{-1} F'(x_0)\right\| \leq -f'(r_n)^{-1}; \tag{11.52}$$

$$\|x_{n+1} - x_n\| \leq t_{n+1} - t_n. \tag{11.53}$$

The statement of the uniqueness of $x^*$ is a verbatim repetition of the corresponding proof of Theorem 11.2. We omit the details. ⊠

In what follows, we consider another class of two–point Newton–type methods stated below:

$$\begin{aligned} y_n &= x_n - F'(x_n)^{-1} F(x_n), \\ \xi_n &= \eta_n y_n + (1-\eta_n) x_n, \quad 0 \leq \eta_n \leq 1 \quad (n \geq 0), \\ x_{n+1} &= x_n - F'(\xi_n)^{-1} F(x_n) \end{aligned} \tag{11.54}$$

and its majorizing sequences are

$$\begin{aligned} s_n &= t_n - f'(t_n)^{-1} f(t_n), \\ r_n &= \eta_n s_n + (1-\eta_n) t_n, \quad 0 \leq \eta_n \leq 1 \quad (n \geq 0), \\ t_{n+1} &= t_n - f'(r_n)^{-1} f(t_n). \end{aligned} \tag{11.55}$$

Essentially, the same reasoning as in the above discussion about (11.41) and (11.43) can yield the semilocal convergence conclusion of (11.54). Note that the result corresponding to Corollary 11.1 for (11.54) uses the identities:

$$s_n - t^* = -f'(t_n)^{-1} \int_0^1 [f'(t^* + \tau(t_n - t^*)) - f'(t_n)](t_n - t^*)\,\mathrm{d}\tau$$

and

$$t_{n+1} - t^* = f'(r_n)^{-1} \left[ f(t^*) - f(t_n) - f'(r_n)(t^* - t_n) \right].$$

Especially, if choosing $\eta_n = \frac{1}{2}$, (11.54) is the midpoint method introduced by us in (cf. [171, 174]) by

$$\begin{aligned} y_n &= x_n - F'(x_n)^{-1} F(x_n) \quad (x_0 \in \mathcal{D}), \\ x_{n+1} &= x_n - F'\left(\frac{x_n + y_n}{2}\right)^{-1} F(x_n) \end{aligned} \tag{11.56}$$

and the corresponding majorizing sequences are

$$\begin{aligned} s_n &= t_n - f'(t_n)^{-1} f(t_n), \quad t_0 = 0, \\ t_{n+1} &= t_n - f'\left(\frac{t_n + s_n}{2}\right)^{-1} f(t_n). \end{aligned}$$

Using the same Ostrowski-type representation (11.28), under the hypotheses of Theorem 11.2 exactly as in the above proofs, we deduce that all the conclusions hold for $q = 0$, and $p = 3$. That is the R–order of convergence for the midpoint method is 3.

**Remark 11.1** *In (cf. [171, 174]), we used a quadratic majorizing function $f$ and hypotheses until the second Fréchet-derivative to get the R–order 3 of local convergence for the midpoint method.*

**Example 11.1** *We cannot compare our results for method* (11.1) *with others in the literature since we do not know of any semilocal results concerning this method. However, we can compare the results for the midpoint method obtained here with the ones in (cf. [171, 174]). In order for us to achieve this assume*

$$\left\| F'(x_0)^{-1} F''(x) \right\| \leq \beta_1 \tag{11.57}$$

*and*

$$\| F'(x_0)[F''(x) - F''(y)] \| \leq \beta_2 \| x - y \| \quad (x, y \in \mathcal{D}). \tag{11.58}$$

*Set*

$$\beta_3 = \sqrt{\frac{6\beta_1^2 + 7\beta_2}{6}} \tag{11.59}$$

*and*

$$h_0 = 2\beta_3 \gamma. \tag{11.60}$$

*The condition given in (cf. [171, 174]) corresponding to* (11.12) *is given by*

$$h_0 \leq 1. \tag{11.61}$$

*Let us consider quadratic operator equations of the form*

$$\Theta(x, x) + M(x) + v = 0, \tag{11.62}$$

*where $\Theta$ is a bounded bilinear operator on $\mathcal{X} \times \mathcal{X}$ and $M$ is a bounded linear operator on $\mathcal{X}$. Using* (11.40), (11.57) *and* (11.58), *we get*

$$\alpha = 0, \ \beta_1 = \beta_2 = 2\,\|\Theta\| \ \text{ and, for } x_0 \approx y_0, \quad \beta = 2\,\|\Theta\|\,.$$

*Then, condition* (11.12) *can be written as*

$$h_N = 2\beta\gamma \leq 1. \tag{11.63}$$

*By comparing* (11.61) *and* (11.63), *we deduce*

$$h_0 \leq 1 \Longrightarrow h_N \leq 1, \tag{11.64}$$

*but not vice versa unless if $\beta_2 = 0$. Hence, the improvement is given by*

$$\beta_4 = \frac{\beta}{\beta_3} < 1 \quad (\beta_2 \neq 0). \tag{11.65}$$

Finally, we complete this section with a numerical example.

**Example 11.2** *Let $\mathcal{X} = C[0,1]$ be the space of all functions continuous on the interval $[0,1]$ with norm:*

$$\|x\| = \max_{0 \leq s \leq 1} |x(s)|. \tag{11.66}$$

*Consider the Chandrasekhar-type equation (cf. [163, 171, 174, 469])*

$$P(x) = 0, \tag{11.67}$$

*where operator $P$ is defined on $\mathcal{X}$ by*

$$P(x)(s) = \lambda x(s) \int_0^1 \frac{s}{s+\tau} x(\tau)\ \mathrm{d}\tau - x(s) + 1 \tag{11.68}$$

*and $\lambda$ is a given real number. By* (11.62), *$M = -I$, $v = 1$ and $\Theta$ is such that*

$$\Theta(x,x)(s) = \lambda x(s) \int_0^1 \frac{s}{s+\tau} x(\tau)\ \mathrm{d}\tau. \tag{11.69}$$

*Set $x_0(s) \approx y_0(s) = 1$. Using the definition of the first and second Fréchet derivatives, we get*

$$\beta = \beta_1 = \beta_2 = 2\delta|\lambda| \ln 2,$$

*where $\delta = \|P'(x_0(s))^{-1}\|$ and $\gamma = \delta|\lambda|\ln 2$, which extend the applicability of our results in this case.*

# Chapter 12

# Fixed Point Methods

We present in this chapter the convergence analysis for successive methods to solve fixed points problems.

## 12.1 Successive Substitutions Methods

In this section, we are concerned with the problem of approximating a locally unique fixed point $x^\star$ of nonlinear equation

$$x = G(x), \tag{12.1}$$

where $G$ is a Fréchet–differentiable operator defined on a convex subset $\mathcal{D}$ of a Banach space $\mathcal{X}$ with values in $\mathcal{D}$. Recall that $x^\star$ is a fixed point of equation (12.1) if

$$x^\star = G(x^\star) \qquad (x^\star \in int\,\mathcal{D}). \tag{12.2}$$

The method of successive approximations

$$x_{k+1} = G(x_k) \qquad (k \geq 0,\ x_0 \in \mathcal{D}) \tag{12.3}$$

has been used by several authors to generate a sequence $\{x_k\}$ converging to $x^\star$ under various assumptions (cf. [139], [685]). The usual hypothesis is as follows

$$\| G'(x) \| \leq q < 1 \qquad (x \in \mathcal{D}), \tag{12.4}$$

which implies that operator is a contraction on $\mathcal{D}$ and as such the existence of $x^\star$ is guaranteed on $\mathcal{D}$. The contraction hypothesis (12.4) however is too strong in general.

As an example, consider $\mathcal{X} = \mathbb{R}$, $\mathcal{D} = \{x : \| x - x_0 \| \leq 1 - \alpha\}$, $\alpha \in [0, 1]$, $x_0 = 1$ and define function $G$ on $\mathcal{D}$ by

$$G(x) = x^3 + x - \alpha. \tag{12.5}$$

Using (12.5), we obtain that

$$\| G'(x) \| > 1 \quad (x \in \mathcal{D}). \tag{12.6}$$

Here, we avoid condition (12.4), by using instead of (12.4), the weaker condition:

$$\| G'(x_0) \| \leq q_0 < 1. \tag{12.7}$$

In particular, we are motivated by the work of Cătinaş (cf. [262], [266]). Cătinaş recently characterized the $q$–quadratic convergence of perturbed successive approximations and also provided sufficient conditions for the accelerated convergence of these sequences (cf. [263], [264], [266]).
We weaken the sufficient convergence conditions provided in (cf. [266]) under the same computational cost. This approach extends the applicability of the method of successive approximations. Moreover, the error estimates on the distances involved are finer and the information on the location of the fixed point at least as precise as in (cf. [266]).
We restate the relevant results obtained in (cf. [263], [264], [266]) to make the section as self contained as possible and for comparison reasons. The successive approximations converging quadratically to $x^\star$ are characterized by the following result.

**Theorem 12.1** *(cf. [264]) Let $G : \mathcal{D} \subseteq \mathcal{X} \longrightarrow \mathcal{D}$ be a differentiable operator on a neighborhood of $x^\star$. Assume*

$$\| G'(x) - G'(y) \| \leq L \ \| x - y \| \quad (x, y \in \mathcal{D}),$$

*there exists $x_0 \in \mathcal{D}$, such that the successive approximations (12.3) converge to $x^\star$ and $\mathcal{I} - G'(x_k)$ are invertible starting from a certain step. Then, the convergence is q–quadratic if and only if $G'$ has a zero eigenvalue and, starting from a certain step, the corrections $x_{k+1} - x_k$ are corresponding eigenvectors:*

$$G'(x^\star)\ (x_{k+1} - x_k) = 0 \quad (k \geq k_0).$$

*This condition holds equivalently if and only if the errors $x_k - x^\star$ are corresponding eigenvectors:*

$$G'(x^\star)\ (x_k - x^\star) = 0 \quad (k \geq k_0)$$

*or if and only if*

$$x_k \in x^\star + Ker\, G'(x^\star) \quad (k \geq k_0).$$

This result implies that, if $G'$ has no eigenvalue 0, there exists no sequence (12.3) converging to $x^\star$ with order 2. In such a case, one may choose to consider, for some $\{\delta_k\}_{k\geq 0} \subset \mathcal{X}$, the perturbed successive approximations:

$$x_{k+1} = G(x_k) + \delta_k \qquad (k \geq 0,\ x_0 \in \mathcal{D}). \tag{12.8}$$

Their quadratic convergence is characterized by the following result, which does not requires the existence of the eigenvalue 0.

**Theorem 12.2** *(cf. [264]) Let $G : \mathcal{D} \subseteq \mathcal{X} \longrightarrow \mathcal{D}$ be a differentiable operator. Assume*

*(a) conditions of Theorem 12.1 hold;*

*(b) the successive approximations (12.8) converge to $x^\star$.*

*Then, the convergence is q–quadratic, if and only if*

$$\| G'(x_k)\ (x_k - G(x_k)) + (\mathcal{I} - G'(x_k))\ \delta_k \| = \mathcal{O}(\| x_k - G(x_k) \|^2) \quad as\ k \longrightarrow \infty. \tag{12.9}$$

In (cf. [263]), it was shown that, if we write

$$\delta_k = (\mathcal{I} - G'(x_k))^{-1}\ (G'(x_k)\ (G(x_k) - x_k) + \gamma_k)$$

with $\{\gamma_k\}_{k\geq 0} \subset \mathcal{X}$, then, condition given by

$$\gamma_k = \mathcal{O}(\| x_k - G(x_k) \|^2) \quad as\ k \longrightarrow \infty$$

is equivalent to (12.9). Note that, in (cf. [263]), under the condition (12.4) in a neighborhood of $x^\star$, for a given $K > 0$, a natural choice (implied by the Banach lemma) for $\delta_k$ is:

$$\delta_k = (\mathcal{I} + \cdots + G'(x_k)^{i_k})\ G'(x_k)\ (G(x_k) - x_k)$$

with $i_k$ such that

$$\frac{q^{i_k+2}}{1-q} \leq K\ \| x_k - G(x_k) \|\ . \tag{12.10}$$

When applying Theorem 12.2 to characterize the quadratic convergence of the resulting sequence:

$$x_{k+1} = G(x_k) + (\mathcal{I} + \cdots + G'(x_k)^{i_k})\ G'(x_k)\ (G(x_k) - x_k) \quad (k \geq 0) \tag{12.11}$$

with $i_k$ given by (12.10), we must assume that this sequence converges to the fixed point $x^\star$. But is this assumption reasonable ? Cătinaş showed that under certain natural conditions the sequence converge to $x^\star$, so the answer is positive.

**Remark 12.1** *Solving equation (12.1) is equivalent to:*

$$F(x) = 0 \quad \text{with} \quad F(x) = x - G(x),$$

*for which the Newton method generates the iterates:*

$$s_k^N = -F'(x_k)^{-1}\ F(x_k),$$

$$x_{k+1} = x_k + s_k^N \quad (k \geq 0). \tag{12.12}$$

*In this setting, (12.11) may be rewritten as*

$$\begin{aligned} x_{k+1} &= x_k + (\mathcal{I} + \cdots + G'(x_k)^{i_k+1})\ (G(x_k) - x_k) \\ &= x_k + s_k \quad (k \geq 0) \end{aligned} \tag{12.13}$$

*with $i_k$ satisfying:*

$$\frac{q^{i_k+2}}{1-q} \leq K\ \| F(x_k) \|, \tag{12.14}$$

*i.e., as quasi–Newton method QNM (see also (cf. [569], [316], [265])).*

The following result gives sufficient Kantorovich–type conditions for the convergence of method (12.13).

**Theorem 12.3** *(cf. [266]) Let $G : \mathcal{D} \subseteq \mathcal{X} \longrightarrow \mathcal{D}$ be a differentiable operator. Assume*

*(a) condition (12.4) holds;*

*(b)*

$$\| G'(x) - G'(y) \| \leq L \ \| x - y \| \quad (x, y \in \mathcal{D}).$$

*Let $x_0 \in \mathcal{D}$, $K > 0$ and $r$ be chosen such that*

$$\nu = \left( \frac{L}{2\ (1-q)^2} + K\ (1+q) \right) \ \| F(x_0) \| < 1, \tag{12.15}$$

$$r = \frac{1}{(1-\nu)\ (1-q)} \ \| F(x_0) \| \tag{12.16}$$

*and*

$$\overline{U}(x_0, r) \subseteq \mathcal{D}.$$

*Then, the elements of the sequence defined by QNM remains in $\overline{U}(x_0, r)$ and converges to a fixed point $x^\star$ of $G$, which is unique in $\overline{U}(x_0, r)$. According to Theorem 12.2, the convergence is q–quadratic.*

**Remark 12.2** *Condition (12.15) is reasonable:*

*(a) $\| F(x_0) \|$ is sufficient small (which holds, e.g., when $x_0$ is sufficiently close to $x^\star$);*

*(b) $q$ is sufficiently small (see the result in (cf. [263]));*

*(c) $L$ is sufficiently small (the graph of $G$ is close to a constant in the case $\mathcal{X} = \mathbb{R}$);*

*(d) $K$ is sufficiently small (see the results of Dennis and Moré (cf. [316]): the linear systems are solved with increasingly precision, the iterates approaching to those given by Newton method).*

In the next, we present the extensions and improvements claimed in the introduction of this section.

**Theorem 12.4** *Let $G : \mathcal{D} \subseteq \mathcal{X} \longrightarrow \mathcal{D}$ be a differentiable operator. Assume there exist $x_0 \in \mathcal{D}$, $q_0 > 0$ such that*

$$\| G'(x_0) \| \leq q_0 < 1, \tag{12.17}$$

$$\| (\mathcal{I} - G'(x_0))^{-1} \| \leq a, \tag{12.18}$$

*operator $G'$ is $L$–Lipschitz continuous ($L > 0$) on $\mathcal{D}$*

$$\| G'(x) - G'(y) \| \leq L \ \| x - y \| \quad (x, y \in \mathcal{D}), \tag{12.19}$$

*and center $L_0$–Lipschitz continuous $(0 \leq L_0 \leq L)$ on $\mathcal{D}$*

$$\| G'(x) - G'(x_0) \| \leq L_0 \ \| x - x_0 \| \quad (x \in \mathcal{D}), \tag{12.20}$$

*$x_0 \in \mathcal{D}$, $K_0 > 0$, $K_1 > 0$, $a > 0$ and $r_1 > 0$ are such that*

$$K_0 \leq K_1, \tag{12.21}$$

$$\nu_1 = \left( \frac{L}{2\ (1 - q_1)^2} + K_1\ a\ (1 + q_1) \right) \ \| F(x_0) \| < 1, \tag{12.22}$$

$$q_1(r) = a\ L_0\ r, \quad q_1 = q_1(r_1), \tag{12.23}$$

$$a \leq \frac{1}{1 - q_0}, \tag{12.24}$$

$$r_1 \in \left(0, \frac{1}{a\ L_0}\right), \tag{12.25}$$

$$\left( \frac{\nu_0}{(1 - q_1)\ (1 - \nu_1)} + \frac{1}{1 - q_0} \right) \ \| F(x_0) \| \leq r_1, \tag{12.26}$$

$$\nu_0 = \left( \frac{L_0}{2\ (1 - q_0)^2} + K_0 \ \| F'(x_0) \| \ \right) \ \| F(x_0) \|,$$

$$a\ q_0^{i_0+2} \leq K_0 \ \| F(x_0) \|, \quad \frac{q_1^{i_k+2}}{1 - q_1} \leq K_1 \ \| F(x_k) \| \tag{12.27}$$

*and*

$$\overline{U}(x_0, r_1) \subseteq \mathcal{D}. \tag{12.28}$$

*Then, the sequence $\{x_n\}$ $(n \geq 0)$ generated by QNM is well defined, remains in $\overline{U}(x_0, r_1)$ for all $n \geq 0$ and converges to a fixed point $x^\star \in \overline{U}(x_0, r_1)$ of $G$, which is unique in $\overline{U}(x_0, r_1)$. The convergence is q–quadratic according to Theorem 12.2. Moreover, if*

$$r_1 \in \left(0, \frac{1 - q_0}{L_0}\right), \tag{12.29}$$

*$x^\star$ is the unique fixed point of $G$ in $\overline{U}(x_0, r_1)$.*

**Proof** Note that

$$q_0 \leq q_1, \tag{12.30}$$

$$L_0 \leq L \tag{12.31}$$

hold in general and $\frac{q_1}{q_0}$, $\frac{L}{L_0}$ can be arbitrarily large. Hence, by (12.21), (12.22), (12.24) and (12.27), we deduce $\nu_0 \in (0, \nu_1)$. Let $x \in \overline{U}(x_0, r_1)$. Using (12.20), (12.23) and (12.25), we obtain in turn that

$$\begin{aligned} \| (\mathcal{I} - G'(x_0))^{-1} \| \ \| (\mathcal{I} - G'(x)) - (\mathcal{I} - G'(x_0)) \| &\leq a \ \| G'(x) - G'(x_0) \| \\ &\leq a\ L_0 \ \| x - x_0 \| \\ &\leq a\ L_0\ r_1 < 1. \end{aligned} \tag{12.32}$$

It follows from (12.32) and the Banach lemma of invertible operators that

$$(\mathcal{I} - G'(x))^{-1} = \sum_{k=0}^{\infty} G'(x)^k \tag{12.33}$$

exists and

$$\| (\mathcal{I} - G'(x))^{-1} \| \leq \frac{a}{1 - a\, L_0 \ \| x - x_0 \|}. \tag{12.34}$$

Using (12.13) and (12.18), we get that

$$\| s_0 \| \leq \frac{\| F(x_0) \|}{1 - q_0}, \tag{12.35}$$

which implies $x_1 \in \overline{U}(x_0, r_1)$. In view of (12.17), (12.18), (12.20), (12.27) and (12.35), we obtain in turn that

$$\begin{aligned}
&\| F(x_1) \| \\
&= \| F(x_1) - F(x_0) - F'(x_0)\, s_0^N \| \\
&\leq \| F(x_1) - F(x_0) - F'(x_0)\, s_0 \| + \| F'(x_0)\, (s_0^N - s_0) \| \\
&\leq \| G(x_1) - G(x_0) - G'(x_0)\ s_0 \| + \| F'(x_0) \| \| s_0^N - s_0 \| \\
&\leq \int_0^1 \| (G'(x_0 + t\ (x_1 - x_0)) - G'(x_0))\, (x_1 - x_0)\ dt \| \\
&\quad + \| F'(x_0) \| \| s_0^N - s_0 \| \\
&\leq \frac{L_0}{2} \ \| s_0 \|^2 + \| F'(x_0) \| \| G'(x_0)^{i_0+2}\, (\mathcal{I} + G'(x_0) + \cdots)\, F(x_0) \| \\
&\leq \frac{L_0}{2}\, (1 + q_0 + \cdots + q_0^{i_0+1})^2 \ \| F(x_0) \|^2 + \frac{q_0^{i_0+2}}{1 - q_0} \ \| F'(x_0) \| \| F(x_0) \| \\
&\leq \frac{L_0}{2}\, \frac{(1 - q_0^{i_0+2})^2}{(1 - q_0)^2} \ \| F(x_0) \|^2 + K_0 \ \| F'(x_0) \| \| F(x_0) \|^2 \\
&= \nu_0 \ \| F(x_0) \| \leq \nu \ \| F(x_0) \| .
\end{aligned} \tag{12.36}$$

Similarly, for $k \geq 2$, we get that

$$\begin{aligned}
&\| F(x_k) \| \\
&= \| F(x_k) - F(x_{k-1}) - F'(x_{k-1})\ s_{k-1}^N \| \\
&\leq \| F(x_k) - F(x_{k-1}) - F'(x_{k-1})\ s_{k-1} \| + \| F'(x_{k-1})\ (s_{k-1}^N - s_{k-1}) \| \\
&\leq \| G(x_k) - G(x_{k-1}) - G'(x_{k-1})\ s_{k-1} \| \\
&\quad + \| \mathcal{I} - G'(x_{k-1}) \| \ \| s_{k-1}^N - s_{k-1} \|
\end{aligned}$$

$$\leq \int_0^1 \| (G'(x_{k-1} + t\ (x_k - x_{k-1})) - G'(x_{k-1}))\ s_{k-1}\ dt \|$$

$$+ (1 + \| G'(x_{k-1}) - G'(x_k) \| + \| G'(x_0) \|)\ \| s_{k-1}^N - s_{k-1} \| \quad (12.37)$$

$$\leq \frac{L}{2}\ \| s_{k-1} \|^2\ (1 + L_0\ \| x_{k-1} - x_0 \| + q_0)\ \| s_{k-1}^N - s_{k-1} \|$$

$$\leq \frac{L}{2}\ \| s_{k-1} \|^2 + (1 + q_0 + L_0\ r_1)\ \| s_{k-1}^N - s_{k-1} \|$$

$$\leq \frac{L}{2}\ (1 + q_1 + \cdots + q_1^{i_{k-1}+1})^2\ \| F(x_{k-1}) \|^2$$

$$+ (1 + q_1)\ \| G'(x_{k-1})^{i_{k-1}+2}\ (\mathcal{I} + G'(x_{k-1}) + \cdots)\ F(x_{k-1}) \|$$

$$\leq \frac{L}{2} \left( \frac{1 - q_1^{i_{k-1}+2}}{1 - q_1} \right)^2\ \| F(x_{k-1}) \|^2 + (1 + q_1)\ \frac{q_1^{i_{k-1}+2}}{1 - q_1}\ \| F(x_{k-1}) \|$$

$$\leq \frac{L}{2} \left( \frac{1 - q_1^{i_{k-1}+2}}{1 - q_1} \right)^2\ \| F(x_{k-1}) \|^2 + K_1\ (1 + q_1)\ \| F(x_{k-1}) \|^2$$

$$\leq \nu_1\ \| F(x_{k-1}) \|$$

$$\leq \nu_1^{k-1}\ \| F(x_1) \| \leq \nu_1^{k-1}\ \nu_0\ \| F(x_0) \| \leq \nu_1^k\ \| F(x_0) \| .$$

We also have that

$$\| x_k - x_{k-1} \| = \| s_{k-1} \| \leq \frac{1}{1 - q_1}\ \| F(x_{k-1}) \|$$

$$\leq \frac{\nu_1^{k-2}}{1 - q_1}\ \| F(x_1) \| \leq \frac{\nu_1^{k-2}\ \nu_0}{1 - q_1}\ \| F(x_0) \|, \quad (12.38)$$

$$\| x_k - x_0 \| \leq \| s_{k-1} \| + \| s_{k-2} \| + \cdots + \| s_0 \|$$

$$\leq \left( \frac{\nu_1^{k-2}\ \nu_0}{1 - q_1} + \frac{\nu_1^{k-3}\ \nu_0}{1 - q_1} + \cdots + \frac{\nu_1^0\ \nu_0}{1 - q_1} \right)\ \| F(x_0) \| + \| s_0 \|$$

$$= \frac{\nu_0}{1 - q_1}\ (\nu_1^{k-2} + \cdots + 1)\ \| F(x_0) \| + \| s_0 \|$$

$$\leq \left( \frac{\nu_0}{(1 - q_1)\ (1 - \nu_1)} + \frac{1}{1 - q_0} \right)\ \| F(x_0) \|$$

$$\leq r_0 \qquad \text{(by (12.26))} \quad (12.39)$$

and

$$\| x_{k+m} - x_k \| \leq \| x_{k+m} - x_{k+m-1} \| + \cdots + \| x_{k+1} - x_k \|$$

$$\leq \frac{\nu_1^{k+m-1} + \cdots + \nu^k}{1 - q_1} \parallel F(x_0) \parallel$$

$$\leq \frac{\nu_1^k}{(1-\nu_1)\ (1-q_1)} \parallel F(x_0) \parallel . \tag{12.40}$$

In view of (12.40), we deduce sequence $\{x_k\}$ is a Cauchy sequence in a Banach space $\mathcal{X}$ and as such it converges to some $x^\star \in \overline{U}(x_0, r_1)$ (since $\overline{U}(x_0, r_1)$ is a closed set). By letting $k \longrightarrow \infty$ in (12.37) and using the continuity of $F$, we deduce $F(x^\star) = 0$. Finally, to show the uniqueness, we can have for $x \in \overline{U}(x_0, r_1)$

$$\begin{aligned} \parallel G'(x) \parallel &\leq \parallel G'(x) - G'(x_0) \parallel + \parallel G'(x_0) \parallel \\ &\leq L_0 \ \parallel x - x_0 \parallel + q_0 \\ &\leq L\, r_1 + q_0 < 1 \qquad \text{(by (12.29))}. \end{aligned} \tag{12.41}$$

Hence, $G$ is a contraction on $\overline{U}(x_0, r_1)$. This completes the proof of Theorem 12.4. ⊠

It turnout that condition (12.19) can also dropped. However, in this case, the convergence is only linear. More precisely, we have the following semilocal convergence theorem for QNM.

**Theorem 12.5** *Let $G : \mathcal{D} \subseteq \mathcal{X} \longrightarrow \mathcal{D}$ be a differentiable operator. Assume there exist $x_0 \in \mathcal{D}$, $q_0 > 0$, such that*

$$\parallel G'(x_0) \parallel \leq q_0 < 1,$$

*operator $G'$ is center $L_0$–Lipschitz continuous ($L_0 > 0$) on $\mathcal{D}$*

$$\parallel G'(x) - G'(x_0) \parallel \leq L_0 \ \parallel x - x_0 \parallel,$$

*$x_0 \in \mathcal{D}$, $K_0 > 0$, $K_1 > 0$, $a > 0$ and $r_1 > 0$ are such that*

$$K_0 \leq K_1,$$

$$\nu_2 = \frac{2\ L_0\ r_1}{1 - q_1} + K_1\ a\ (1 + q_1) \ \parallel F(x_0) \parallel < 1, \tag{12.42}$$

$$q_1(r) = a\ L_0\ r, \quad q_1 = q_1(r_1),$$

$$a \leq \frac{1}{1 - q_0},$$

$$r_1 \in \left(0, \ \frac{1}{a\ L_0}\right),$$

$$\left(\frac{\nu_0}{(1 - q_1)\ (1 - \nu_1)} + \frac{1}{1 - q_0}\right) \parallel F(x_0) \parallel < r_1,$$

$$a\ q_0^{i_0+2} \le K_0\ \| F(x_0) \|, \quad \frac{q_1^{i_k+2}}{1-q_1} \le K_1\ \| F(x_k) \|,$$

$$\nu_0 \le \nu_2, \quad \text{or } \nu_0 < 1, \tag{12.43}$$

$$\nu_0 = \left( \frac{L_0}{2\ (1-q_0)^2} + K_0\ \| F'(x_0) \| \right)\ \| F(x_0) \|$$

*and*

$$\overline{U}(x_0, r_1) \subseteq \mathcal{D}.$$

*Then, the conclusions of Theorem 12.4 hold.*

**Proof** We only need to show the definition of $\nu_2$. The rest follows exactly as in Theorem 12.4. We have in turn for $k \ge 2$ that

$$\begin{aligned}
&\| G(x_k) - G(x_{k-1}) - G'(x_{k-1})\ s_{k-1} \| \\
&\le \int_0^1 \| (G'(x_{k-1} + t\ (x_k - x_{k-1})) - G'(x_0))\ s_{k-1}\ dt \| \\
&\quad + \| G'(x_0) - G'(x_{k-1}) \|\ \| s_{k-1} \| \\
&\le L_0 \left( \int_0^1 \| x_{k-1} + t\ (x_k - x_{k-1}) - x_0 \|\ dt + \| x_0 - x_{k-1} \| \right)\ \| s_{k-1} \| \\
&\le L_0 \left( \int_0^1 ((1-t)\ \| x_{k-1} - x_0 \| + t\ \| x_k - x_0 \|)\ dt + \| x_0 - x_{k-1} \| \right)\ \| s_{k-1} \| \\
&\le L_0 \left( \frac{1}{2}\ (r_1 + r_1) + r_1 \right)\ \| s_{k-1} \| \\
&\le 2\ L_0\ r_1\ (1 + q_1 + \cdots + q_1^{i_{k-1}+1}) \le \frac{2\ L_0\ r_1}{1-q_1}.
\end{aligned} \tag{12.44}$$

Then, using (12.44), estimate (12.37) becomes:

$$\| F(x_k) \| \le \nu_2\ \| F(x_{k-1}) \|,$$

where $\nu_2$ is given by (12.42). This completes the proof of Theorem 12.5. ⊠

**Remark 12.3** *Theorem 12.3 has been improved in several ways:*

**Case 1** *Under the hypotheses of Theorem 12.3, there exist $L_0$, $q_0$, $K_1$, $K_0$ satisfying (12.20), (12.31), (12.17), (12.24); $K_0 \le K$, $K_1 \le K$, respectively. Therefore, in this case, we obtain finer error bounds and an at least as precise information on the location of the solution.*

**Case 2** *Under hypotheses of Theorem 12.3 and Theorem 12.4, we see that our hypotheses are weaker.*

**Example 12.1** *Returning back to the example given in the introduction of the section, set*

$$F(x) = -\frac{1}{4}\ (x^3 - \alpha),$$

*then, we have that*

$$\| G'(x) \| \geq 1 \quad \text{on } \mathcal{D}.$$

*But*

$$\| G'(x_0) \| = \frac{1}{4} = q_0 < 1.$$

*We also have* $L_0 = 3-\alpha$, $L = 2\ (2-\alpha)$, $a = \frac{3}{4}$. *Hence, the results of Theorem 12.4 can apply, say for* $\alpha = .49$, *whereas Theorem 12.3 cannot. Sometimes condition (12.19) is violated but (12.20) holds. So Theorem 12.5 is useful in this case.*

In the rest of this section, we provide such examples.

**Example 12.2** *Let* $\mathcal{X} = \mathbb{R}$, $\mathcal{D} = [0, \infty)$, $x_0 = 1$ *and define a function* $F$ *on* $\mathcal{D}$ *by*

$$F(x) = \frac{x^{1+\frac{1}{i}}}{1+\frac{1}{i}} + c_1\ x + c_2, \tag{12.45}$$

*where* $c_1$, $c_2$ *are real parameters and* $i > 2$ *an integer. Then,* $F'(x) = x^{\frac{1}{i}} + c_1$ *is not Lipschitz on* $\mathcal{D}$. *However, center Lipschitz condition (12.45) holds for* $L_0 = (1+c_1)^{-1}$ $(c_1 \neq -1)$. *Indeed, we have*

$$\begin{aligned} \left\| F'(x_0)^{-1}\ [F'(x) - F'(x_0)] \right\| &= (1+c_1)^{-1}\ \left| x^{\frac{1}{i}} - x_0^{\frac{1}{i}} \right| \\ &= \frac{(1+c_1)^{-1}\ |x - x_0|}{x_0^{\frac{i-1}{i}} + \cdots + x^{\frac{i-1}{i}}} \leq L_0\ |x - x_0|\,. \end{aligned} \tag{12.46}$$

**Example 12.3** *We consider the integral equation*

$$u(s) = f(s) + \lambda \int_a^b G(s,t)\ u(t)^{1+\frac{1}{n}}\ dt \qquad (n \geq 1). \tag{12.47}$$

*Here,* $f$ *is a given continuous function satisfying* $f(s) > 0$, $s \in [a,b]$, $\lambda$ *is a real number and the kernel* $G$ *is continuous and positive in* $[a,b] \times [a,b]$. *For example, when* $G(s,t)$ *is the Green kernel, the corresponding integral equation is equivalent to the boundary value problem:*

$$u'' = \lambda\ u^{1+\frac{1}{n}},$$
$$u(a) = f(a), \quad u(b) = f(b)\,.$$

*Equations of the form (12.47) generalize equations of the form:*

$$u(s) = \int_a^b G(s,t)\ u(t)^n\, dt \tag{12.48}$$

*studied in (cf. [157]). Instead of (12.47) we can try to solve the equation* $F(u) = 0$, *where*

$$F : \Omega \subseteq \mathcal{C}[a,b] \to \mathcal{C}[a,b], \quad \Omega = \{u \in \mathcal{C}[a,b] : u(s) \geq 0, \ s \in [a,b]\}$$

*and*

$$F(u)(s) = u(s) - f(s) - \lambda \int_a^b G(s,t)\, u(t)^{1+\frac{1}{n}}\, dt.$$

*The norm we consider is the max–norm. The derivative* $F'$ *is given by*

$$F'(u)\ v(s) = v(s) - \lambda \left(1 + \frac{1}{n}\right) \int_a^b G(s,t)\ u(t)^{\frac{1}{n}}\ v(t)\, dt \quad (v \in \Omega).$$

*First of all, we notice that* $F'$ *does not satisfy a Lipschitz–type condition in* $\Omega$. *Let us consider, for instance,* $[a,b] = [0,1]$, $G(s,t) = 1$ *and* $y(t) = 0$. *Then* $F'(y)\ v(s) = v(s)$ *and*

$$\|F'(x) - F'(y)\| = |\lambda| \ \left(1 + \frac{1}{n}\right) \int_0^1 x(t)^{\frac{1}{n}}\ dt.$$

*If* $F'$ *were a Lipschitz function, then*

$$\|F'(x) - F'(y)\| \leq L_1 \ \|x - y\|$$

*or, equivalently, the inequality*

$$\int_0^1 x(t)^{\frac{1}{n}}\ dt \leq L_2 \ \max_{x \in [0,1]} x(s) \tag{12.49}$$

*would hold for all* $x \in \Omega$ *and for a constant* $L_2$. *But this is not true. Consider, for example, the functions*

$$x_j(t) = \frac{t}{j} \quad (j \geq 1, \quad t \in [0,1]).$$

*If these are substituted into (12.49),*

$$\frac{1}{j^{1/n}\ \left(1 + \frac{1}{n}\right)} \leq \frac{L_2}{j} \Longleftrightarrow j^{1-1/n} \leq L_2 \ \left(1 + \frac{1}{n}\right) \quad (j \geq 1).$$

*This inequality is not true when* $j \to \infty$. *(12.19) fails in this case. However, condition (12.20) holds. To show this, let* $x_0(t) = f(t)$ *and* $\alpha = \min_{s \in [a,b]} f(s)$, $\alpha > 0$. *Then, for* $v \in \Omega$,

$$\begin{aligned}
&\left\| \left[F'(x) - F'(x_0)\right]\ v \right\| \\
&= |\lambda| \ \left(1 + \frac{1}{n}\right) \ \max_{s \in [a,b]} \left| \int_a^b G(s,t) \ \left(x(t)^{\frac{1}{n}} - f(t)^{\frac{1}{n}}\right) \ v(t) \ dt \right| \\
&\leq |\lambda| \ \left(1 + \frac{1}{n}\right) \ \max_{s \in [a,b]} \frac{G(s,t) \ |x(t) - f(t)|}{x(t)^{(n-1)/n} + x(t)^{(n-2)/n} f(t)^{1/n} + \cdots + f(t)^{(n-1)/n}} \, dt \ \|v\|.
\end{aligned}$$

*Hence, we have*

$$\begin{aligned}\|F'(x) - F'(x_0)\| &\leq \frac{|\lambda|\ \left(1+\frac{1}{n}\right)}{\alpha^{(n-1)/n}}\ \max_{s\in[a,b]} \int_a^b G(s,t)\ dt\ \|x - x_0\| \\ &\leq K\ \|x - x_0\|,\end{aligned}$$

*where*

$$K = \frac{|\lambda|\ \left(1+\frac{1}{n}\right)}{\alpha^{(n-1)/n}}\ N, \qquad N = \max_{s\in[a,b]} \int_a^b G(s,t)\ dt.$$

*Set* $L_0 = \left\|F'(x_0)^{-1}\ K\right\|$. *Then, condition (12.20) holds for sufficiently small* $\lambda$.

# Bibliography


[1] Ahues, M., A note on perturbed fixed slope iterations, Appl. Math. Lett., 18 (2005), 375–380.

[2] Ahues, M., Newton Methods with Hölder derivative, Numer. Funct. Anal. Optim., (5–6) (2004), 379–395.

[3] Airapetyan, R.G., Continuous Newton method and its modification, Appl. Anal., 73 (2000), 463–484.

[4] Airapetyan, R.G., Ramm, A.G., Smimova, A.B., Continuous analog of Gauss-Newton method, Math. Models Methods Appl. Sci., 9 (1999), 463–474.

[5] Airapetyan, R.G., Ramm, A.G., Smimova, A.B., Continuous method for solving nonlinear ill–posed problems, Amer. Math. Soc., Providence RI, Fields. Inst. Commun., 25 (2000), 111–137.

[6] Airapetyan, R.G., Ramm, A.G., Dynamical systems and discrete methods for solving nonlinear ill–posed problems, Appl. Math. Reviews, vol. 1, World Sci. Publishers, River Edge, NJ, 2000, 491–536.

[7] Alefeld, G., Herzberger, J., Introduction to interval computations. Translated from the German by Jon Rokne, Computer Science and Applied Mathematics, Academic Press, New York, 1983.

[8] Allgower, E.L., Böhmer, K., Application of the mesh independence principle to mesh refinement strategies, SIAM J. Numer. Anal., 24 (1987), 1335–1351.

[9] Allgower, E.L., Böhmer, K., Potra, F.A., Rheinboldt, W.C., A mesh independence principle for operator equations and their discretizations, SIAM J. Numer. Anal., 23 (1986), 160–169.

[10] An, H–B., Bai, Z–Z., Directional secant method for nonlinear equations, J. Comput. Appl. Math., 175 (2005), 291–304.

[11] Amat, S., Bermúdez, C., Busquier, S., Gretay, J., Convergence by nondiscrete mathematical induction of a two step Secant's method, Rocky Mountain J. Math., 37 (2007), 359–369.

[12] Amat, S., Blanda, J., Busquier, S., A Steffensen's type method with modified functions, Riv. Mat. Univ. Parma, 7 (2007), 125–133.

[13] Amat, S., Busquier, S., Convergence and numerical analysis of a family of two–step Steffensen's method, Comput. Math. Appl., 49 (2005), 13–22.

[14] Amat, S., Busquier, S., On a Steffensen's type method and its behavior for semismooth equations, Appl. Math. Comput., 177 (2006), 819–823.

[15] Amat, S., Busquier, S., Third–order iterative methods under Kantorovich conditions, J. Math. Anal. Appl., 336 (2007), 243–261.

[16] Amat, S., Busquier, S., Candela, V.F., A class of quasi–Newton generalized Steffensen's methods on Banach spaces, J. Comput. Appl. Math., 149 (2002), 397–406.

[17] Amat, S., Busquier, S., Gutiérrez, J.M., Geometric constructions of iterative functions to solve nonlinear equations, J. Comput. Appl. Math., 157 (2003), 197–205.

[18] Amat, S., Busquier, S., Gutiérrez, J.M., On the local convergence of secant–type methods, Intern. J. Comput. Math., 81 (2004), 1153–1161.

[19] Amat, S., Busquier, S., Gutiérrez, J.M., Hernández, M.A., On the global convergence of Chebyshev's iterative method, J. Comput. Appl. Math., 220 (2008), 17–21.

[20] Amat, S., Busquier, S., Negra, M., Adaptive approximation of nonlinear operators, Numer. Funct. Anal. Optim., 25 (2004), 397–405.

[21] Amat, S., Busquier, S., Plaza, S., Dynamics of the King and Jarratt iterations, Aequat. Math., 69 (2005), 212–223.

[22] Amat, S., Busquier, S., Salanova, M.A., A fast Chebyshev's method for quadratic equations, Appl. Math. Comput., 148 (2004), 461–474.

[23] Amer, S.M., On solution of nonlinear singular integral equations with shift in generalized Hölder space, Chaos Solitions & Fractals, 12 (2001), 1323–1334.

[24] Amer, S.M., Dardery, S., About an approximation method for the solution of nonlinear singular integral equations with shift, Bull. Fac. Sci. Assiut Univ. C, 29 (2000), 1–15.

[25] Amer, S.M., Dardery, S., On the application of Newton–Kantorovich method to nonlinear singular integral equations, Proc. Pakistan Acad. Sci., 37 (2000), 79–86.

[26] Amer, S.M., Nagdy, A.S., On the modified Newton's approximation method for the solution of non–linear singular integral equations, Hokkaido Math. J., 29 (2000), 59–72.

[27] Anitescu, M., Coroian, D.I., Nashed, M.Z., Potra, F.A., Outer inverses and multi–body system simulation, Numer. Funct. Anal. Optim., 17 (1996), 661–678.

[28] Appell, J., The Newton–Kantorovich method for nonlinear integral operators, World Congress of Nonlinear Analysts, Vol. I–IV (Tampa, FL, 1992), de Gruyter, Berlin, 1996, 2419–2422.

[29] Appell, J., De Pascale, E., Evkuta, N.A., Zabrejko, P.P., On the two–step Newton method for the solution of nonlinear operator equations, Math. Nachr., 172 (1995), 5–14.

[30] Appell, J., De Pascale, E., Kalitvin, J.V., Zabrejko, P.P., On the application of the Newton–Kantorovich method to nonlinear partial integral equations, Z. Anal. Anwend., 15 (1996), 397–418.

[31] Appell, J., De Pascale, E., Lysenko, J.V., Zabrejko, P.P., New results on Newton–Kantorovich approximations with applications to nonlinear integral equations, Numer. Funct. Anal. Optim., 18 (1997), 1–17.

[32] Appell, J., De Pascale, E., Zabrejko, P.P., On the application of the Newton–Kantorovich method to nonlinear integral equations of Uryson type, Numer. Funct. Anal. Optim., 12 (1991), 271–283.

[33] Appell, J., De Pascale, E., Zabrejko, P.P., On the application of the method of successive approximations and the Newton–Kantorovich method to nonlinear functional-integral equations, Adv. Math. Sci. Appl., 2 (1993), 25–38.

[34] Argyros, I.K., Improving the rate of convergence of Newton methods on Banach spaces with a convergence structure and applications, Appl. Math. Lett., 10(1977), 21–28.

[35] Argyros, I.K., Quadratic equations and applications to Chandrasekhar's and related equations, Bull. Austral. Math. Soc., 32 (1985), 275–292.

[36] Argyros, I.K., On the cardinality of solutions of multilinear differential equations and applications, Int. J. Math. Math. Sci., 9 (1986), 757–766

[37] Argyros, I.K., On the approximation of some nonlinear equations, Aequat. Math., 32 (1987), 87–95.

[38] Argyros, I.K., On polynomial equations in Banach space, perturbation techniques and applications, Int. J. Math. Math. Sci., 10 (1987), 69–78.

[39] Argyros, I.K., Newton–like methods under mild differentiability conditions with error analysis, Bull. Austral. Math. Soc., 37 (1987), 131–147.

[40] Argyros, I.K., On Newton's method and nondiscrete mathematical induction, Bull. Austral. Math. Soc., 38 (1988), 131–140.

[41] Argyros, I.K., On a class of nonlinear integral equations arising in neutron transport, Aequat. Math., 36 (1988), 99–111.

[42] Argyros, I.K., The secant method and fixed points of nonlinear operators, Monatsh. Math., 106 (1988), 85–94.

[43] Argyros, I.K., On the number of solutions of some integral equations arising in radiative transfer, Int. J. Math. Math. Sci., 12 (1989), 297–304

[44] Argyros, I.K., Improved error bounds for a certain class of Newton–like methods, J. Approx. Theory Appl., 61 (1990), 80–98.

[45] Argyros, I.K., Error founds for the modified secant method, BIT, 20 (1990), 92–200.

[46] Argyros, I.K., On the solution of equations with nondifferentiable operators and the Ptăk error estimates, BIT, 30 (1990), 752–754.

[47] Argyros, I.K., On some projection methods for the approximation of implicit functions, Appl. Math. Lett., 32 (1990), 5–7.

[48] Argyros, I.K., The Newton–Kantorovich method under mild differentiability conditions and the Ptăk error estimates, Monatsh. Math., 101 (1990), 175–193.

[49] Argyros, I.K., The secant method in generalized Banach spaces, Appl. Math. Comp., 39 (1990), 111–121.

[50] Argyros, I.K., A mesh independence principle for operator equations and their discretizations under mild differentiability conditions, Comput., 45 (1990), 265–268.

[51] Argyros, I.K., On the convergence of some projection methods with perturbation, J. Comp. Appl. Math., 36 (1991), 255–258.

[52] Argyros, I.K., On an application of the Zincenko method to the approximation of implicit functions, Public. Math. Debrecen, 39 (1991), 1–7.

[53] Argyros, I.K., On an iterative algorithm for solving nonlinear equations, Beitrage zür Numerischen Math., 10 (1991), 83–92.

[54] Argyros, I.K., On a class of quadratic equations with perturbation, Funct. et. Approx. Comm. Math., XX, (1992), 51–63.

[55] Argyros, I.K., Improved error bounds for the modified secant method, Int. J. Computer Math., 43 (1992), 99–109.

[56] Argyros, I.K., Some generalized projection methods for solving operator equations, J. Comp. Appl. Math., 39 (1992), 1–6.

[57] Argyros, I.K., On the convergence of generalized Newton–methods and implicit functions, J. Comp. Appl. Math., 43 (1992), 335–342.

[58] Argyros, I.K., On the convergence of inexact Newton–like methods, Publ. Math. Debrecen, 42 (1992), 1–7.

[59] Argyros, I.K., On a mesh–independance principle for operator equations and the Secant method, Acta Math. Hung., 60 (1992), 7–19.

[60] Argyros, I.K., On the convergence of a Chebysheff–Halley–type method under Newton–Kantorovich hypothesis, Appl. Math. Lett., 5 (1993), 71–74.

[61] Argyros, I.K., Newton–like methods in partially ordered linear spaces, J. Approx. Theory Appl., 9 (1993), 1–10.

[62] Argyros, I.K., On the solution of undetermined systems of nonlinear equations in Euclidean spaces, Pure Math. Appl., 4 (1993), 199–209.

[63] Argyros, I.K., A convergence theorem for Newton–like methods under generalized Chen–Yamamato–type assumptions, Appl. Math. Comput., 61 (1994), 25–37.

[64] Argyros, I.K., On the discretization of Newton–like methods, Internat. J. Computer. Math., 52 (1994), 161–170.

[65] Argyros, I.K., On the midpoint method for solving nonlinear operator equations and applications to the solution of integral equations, Rev. Anal. Numer. Theory Approx., Tome 23 (1994), 139–152.

[66] Argyros, I.K., A multipoint Jarratt–Newton–type approximation algorithm for solving nonlinear operator equations in Banach spaces, Functiones et. approximatio, Commentarii Matematiki, XXIII, (1994), 97–108.

[67] Argyros, I.K., On Stirling's method, Tamkang J. Math., 27 (1995), 37–52.

[68] Argyros, I.K., Stirling's method and fixed points of nonlinear operator equations in Banach space, Publ. Inst. Math. Acad. Sin., 23 (1995), 13–20.

[69] Argyros, I.K., A unified approach for constructing fast two–step Newton–like methods, Monatsh. Math., 119 (1995), 1–22.

[70] Argyros, I.K., Results on controlling the residuals of perturbed Newton–like methods on Banach spaces with aconvergence structure, Southwest J. Pure Appl. Math., 1 (1995), 32–38.

[71] Argyros, I.K., On the method of tangent hyperbolas, J. Appr. Theory Appl., 12 (1996), 78–96.

[72] Argyros, I.K., On an extension of the mesh–independence principle for operator equations in Banach space, Appl. Math. Lett., 9 (1996), 1–7.

[73] Argyros, I.K., A generalization of Edelstein's theorem on fixed points and applications, Southwest J. Pure Appl. Math., 2 (1996), 60–64.

[74] Argyros, I.K., Chebysheff–Halley–like methods in Banach spaces, Korean J. Comp. Appl. Math., 4 (1997), 83–107.

[75] Argyros, I.K., Concerning the convergence of inexact Newton methods, J. Comp. Appl. Math., 79 (1997), 235–247.

[76] Argyros, I.K., General ways of constructing accelerating Newton–like iterations on partially ordered topological spaces, Southwest J. Pure Appl. Math., 2 (1997), 1–12.

[77] Argyros, I.K., On a new Newton–Mysovskii–type theorem with applications to inexact Newton–like methods and their discretizations, IMA J. Num. Anal., 18 (1997), 43–47.

[78] Argyros, I.K., On the convergence of two–step methods generated by point–to–point operators, Appl. Math. Comput., 82 (1997), 85–96.

[79] Argyros, I.K., Improved error bounds for Newton–like iterations under Chen–Yamamoto assumptions, Appl. Math. Lett., 10 (1997), 97–100.

[80] Argyros, I.K., Inexact Newton methods and nondifferentiable operator equations on Banach spaces with a convergence structure, Approx. Theory Appl., 13 (1997), 91–104.

[81] Argyros, I.K., A mesh independence principle for inexact Newton–like methods and their discretizations under generalized Lipschitz conditions, Appl. Math. Comp., 87 (1997), 15–48.

[82] Argyros, I.K., Concerning the convergence of inexact Newton methods, J. Comput. Appl. Math., 79 (1997), 235–247.

[83] Argyros, I.K., Smoothness and perturbed Newton–like methods, Pure Math. Appl., 8 (1997), 13–28.

[84] Argyros, I.K., The asymptotic mesh independence principle for inexact Newton–Galerkin–like methods, Pure Math. Appl., 8 (1997), 169–194.

[85] Argyros, I.K., A new convergence theorem for steffenson's method on Banach spaces and applications, Southwest J. Pure Appl. Math., 01 (1997), 23–29.

[86] Argyros, I.K., Polynomial operator equations in abstract spaces and applications, CRC Press LLC, Boca raton Florida USA, 1998.

[87] Argyros, I.K., On the convergence of a certain class of iterative procedures under relaxed conditions with applications, J. Comp. Appl. Math., 94 (1998), 13–21.

[88] Argyros, I.K., The theory and application of abstract polynomial equations, St.Lucie/CRC/Lewis Publ. Mathematics series, 1998, Boca Raton, Florida, U.S.A.

[89] Argyros, I.K., Sufficient conditions for constructing methods faster than Newton's, Appl. Math. Comp., 93 (1998), 169–181.

[90] Argyros, I.K., Improving the rate of convergence of some Newton–like methods for the solution of nonlinear equations containing a non–differentiable term, Revue d'Analyse Numérique et. de Théorie de l'approximation, 27 (1998), 191–202.

[91] Argyros, I.K., A new convergence theorem for the Jarratt method in Banach spaces, Comput. Math. Appl., 36 (1998), 13–18.

[92] Argyros, I.K., Improving the order and rates of convergence for the Super–Halley method in Banach spaces, Comp. Appl. Math., 5 (1998), 465–474.

[93] Argyros, I.K., Improved error bounds for a Chebysheff–Halley–type method, Acta Math. Hungarica, 84 (1999), 211–221.

[94] Argyros, I.K., On Newton's method under mild differentiability conditions and applications, Appl. Math. Comput., 102 (1999), 177–183.

[95] Argyros, I.K., Convergence domains for some iterative processes in Banach spaces using outer and generalized inverses, Comput. Anal. Appl., 1 (1999), 87–104.

[96] Argyros, I.K., Concerning the convergence of a modified Newton–like method, J. Anal. Appl. (ZAA), 18 (1999), 1–8.

[97] Argyros, I.K., Concerning the radius of convergence of Newton's method and applications, Korean J. Comp. Appl. Math., 6 (1999), 451–462.

[98] Argyros, I.K., Convergence rates for inexact Newton–like methods of singular points and applications, Appl. Math. Comp., 102 (1999), 185–201.

[99] Argyros, I.K., Relation between forcing sequences and inexact Newton iterates in Banach spaces, Comput., 63 (1999), 134–144.

[100] Argyros, I.K., An error analysis for the midpoint method, Tamkang J. Math., 30 (1999), 71–83.

[101] Argyros, I.K., A new convergence theorem for the inexact Newton method based on assumptions involving the second Fréchet–derivative, Comput. Appl. Math., 37 (1999), 109–115.

[102] Argyros, I.K., Forcing sequences and inexact Newton iterates in Banach space, Appl. Math. Lett., 13 (2000), 77–80.

[103] Argyros, I.K., Local convergence of inexact Newton–like iterative methods and applications, Comput. Math. Appl., 39 (2000), 69–75.

[104] Argyros, I.K., Advances in the Efficiency of Computational Methods and Applications, World Scientific Publ. Co., River Edge, NJ, 2000.

[105] Argyros, I.K., A mesh independence principle for perturbed Newton–like methods and their discretizations, Korean J. Comp. Appl. Math., 7 (2000), 139–159.

[106] Argyros, I.K., Newton methods on Banach spaces with a convergence structure and applications, Computers Math. with Appl., Intern. J., Pergamon Press, 40 (2000), 37–48.

[107] Argyros, I.K., Semilocal convergence theorems for a certain class of iterative procedures using outer or generalized inverses, Korean J. Comp. Appl. Math., 7 (2000), 29–40.

[108] Argyros, I.K., The effect of rounding errors on a certain class of iterative methods, Appl. Math., 27 (2000), 369–375.

[109] Argyros, I.K., Local convergence of Newton's method for nonlinear equations using outer or generalized inverses, Czecho. Math. J., 50 (2000), 603–614.

[110] Argyros, I.K., On a class of nonlinear implicit quasivariational inequalities, PanAmer. Math. J., 10 (2000), 101–109.

[111] Argyros, I.K., On the radius of convergence of Newton's method, Intern. J. Comput. Math., 77 (2001), 389–400.

[112] Argyros, I.K., A Newton–Kantorovich theorem for equations involving $m$-Fréchet-differentiable operators and applications in radiative transfer, J. Comp. Appl. Math., 131 (2001), 149–159.

[113] Argyros, I.K., A mesh independence principle for inexact Newton–type methods and their discretizations, Annales Univ. Sci. Budapest, Sect. Comp., 20 (2001), 31–53.

[114] Argyros, I.K., Error bounds for Newton's method under hypotheses on the $m$–th Fréchet derivative, Adv. Nonlinear Var. Inequal., 4 (2001), 23–33.

[115] Argyros, I.K., Error bounds for the midpoint method in Banach spaces, Comm. Appl. Nonlinear Anal., 08 (2001), 103–117.

[116] Argyros, I.K., A new semilocal convergence theorem for Newton's method in Banach space using hypotheses on the second Fréchet–derivative, J. Comput. Appl. Math., 139 (2001), 369–373.

[117] Argyros, I.K., On the radius of convergence of Newton's method, Int. J. Comput. Math., 77 (2001), 389–400.

[118] Argyros, I.K., Semilocal convergence theorems for Newton's method using outer inverses and hypotheses on the second Fréchet–derivative, Monatshefte fur Mathematik, 132 (2001), 183–195.

[119] Argyros, I.K., On general auxiliary problem principle and nonlinear mixed variational inequalities, Nonlinear Funct. Anal. Appl., 6 (2001), 247–256.

[120] Argyros, I.K., On an iterative procedure for approximating solutions of quasi variational inequalities, Adv. Nonlinear Var. Inequal., 4 (2001), 39–42.

[121] Argyros, I.K., On generalized variational inequalities, Adv. Nonlinear Var. Inequal., 4 (2001), 75–78.

[122] Argyros, I.K., On a semilocal convergence theorem for a class of quasi variational inequalities, Advances in Nonlinear Variational Inequalities, 4 (2001), 43–46.

[123] Argyros, I.K., On the convergence of a Newton–like method based on $m$–Fréchet–differentiable operators and applications in radiative transfer, J. Comput. Anal. Appl., 4 (2002), 141–154.

[124] Argyros, I.K., A unifying semilocal convergence theorem for Newton–like methods based on center Lipschitz conditions, Comput. Appl. Math., 21 (2002), 789–796.

[125] Argyros, I.K., A semilocal convergence analysis for the method of tangent hyperbolas, J. Concrete and Appl. Anal., 1 (2002), 135–144.

[126] Argyros, I.K., The asymptotic mesh independence principle for Newton–Galerkin methods using twice Fréchet differentiable operators without Lipschitz conditions, Comm. Appl. Nonlinear. Anal., 9 (2002), 67–75.

[127] Argyros, I.K., On the solution of generalized equations using $m$ ($m \geq 2$) Fréchet differential operators, Comm. Appl. Nonlinear Anal., 09 (2002), 85–89.

[128] Argyros, I.K., Results on the solution of generalized equations, Comm. Appl. Nonlinear Anal., 09 (2002), 103–107.

[129] Argyros, I.K., On the convergence and application of Newton's method under weak Hölder continuity assumptions, Intern. J. Comput. Math., 5 (2003), 767–780.

[130] Argyros, I.K., An improved error analysis for Newton–like methods under generalized conditions, J. Comput. Appl. Math., 157 (2003), 169–185.

[131] Argyros, I.K., On a Multistep Newton Method in Banach Spaces and the Ptăk Error Estimates, Adv. Nonlinear Var. Inequal., 6 (2003), 121–135.

[132] Argyros, I.K., New and generalized convergence conditions for the Newton–Kantorovich method, J. Appl. Anal., 9 (2003), 287–299.

[133] Argyros, I.K., On a theorem of L.V. Kantorovich concerning Newton's method, J. Comp. Appl. Math., 155 (2003), 223–230.

[134] Argyros, I.K., An improved convergence analysis and applications for Newton–like methods in Banach space, Numer. Funct. Anal. Optim., 24 (2003), 653–672.

[135] Argyros, I.K., A unified approach for constructing fast two–step methods in Banach space and their applications, PanAmer. Math. J., 13 (2003), 59–108.

[136] Argyros, I.K., On the Newton–Kantorovich hypothesis for solving equations, J. Comput. Appl. Math., 169 (2004), 315–332.

[137] Argyros, I.K., Convergence of Broyden's method, Comm. Appl. Nonlinear Anal., 11 (2004), 77–86.

[138] Argyros, I.K., On the comparison of a weak variant of the Newton–Kantorovich and Miranda theorems, J. Comput. Appl. Math., 166 (2004), 585–589.

[139] Argyros, I.K., A unifying local–semilocal convergence analysis and applications for two–point Newton–like methods in Banach space, J. Math. Anal. Appl., 298 (2004), 374–397.

[140] Argyros, I.K., Concerning the convergence and application of Newton's method under hypotheses on the first and second Fréchet derivative, Comm. Appl. Nonlinear Anal., 11 (2004), 103–119.

[141] Argyros, I.K., Newton Methods, Nova Science Publ. Corp., New York, 2005.

[142] Argyros, I. K., Concerning the "terra incognita" between convergence regions of two Newton methods, Nonlinear Anal., 62 (2005), 179–194.

[143] Argyros, I.K., An improved approach of obtaining good starting points for solving equations by Newton's method, Adv. Nonlinear Var. Ineq. 8 (2005), 133–142.

[144] Argyros, I.K., On a two–point Newton–like method of convergent order two, Int. J. Comput. Math., 88 (2005), 219–234.

[145] Argyros, I.K., Approximate solution of operator equations with applications, World Scientific Publ. Co., Hackensack, New Jersey, 2005, USA.

[146] Argyros, I.K., New sufficient convergence conditions for the Secant method, Czecho. Math. J., 55, (2005), 175–187.

[147] Argyros, I.K., On the Newton–Kantorovich method in Riemannian manifolds, Adv. Nonlinear Var. Ineq., 8 (2005), 81–85.

[148] Argyros, I.K., A convergence analysis for Newton–like methods for singular equations using outer or generalized inverses, Appl. Math., 32 (2005), 37–49.

[149] Argyros, I.K., Toward a unified convergence theory for Newton–like methods of "bounded deterioration", Adv. Nonlinear Var. Inequal. 8 (2005), 109–120.

[150] Argyros, I.K., Enlarging the convergence domain of Newton's method under regular smoothness conditions, Adv. Nonlinear Var. Inequal., 8 (2005), 121–129.

[151] Argyros, I.K., On the semilocal convergence of the Gauss–Newton method, Adv. Nonlinear Var. Inequal., 8 (2005), 93–99.

[152] Argyros, I.K., Relaxing the convergence conditions for Newton–like methods, J. Appl. Math. Comput., 21 (2006), 119–126.

[153] Argyros, I.K., An improved convergence analysis of a superquadratic method for solving generalized equations, Rev. Colombiana Math., 40 (2006), 65–73.

[154] Argyros, I.K., A Kantorovich–type analysis for a fast iterative method for solving equations, J. Math. Anal. Appl., 332 (2007), 97–108.

[155] Argyros, I.K., Approximating solutions of equations using Newton's method with a modified Newton's method iterate as a starting point, Revue Anal. Numér. Théor. Approx. 36 (2007), 123–137.

[156] Argyros, I.K., An improved unifying convergence analysis of Newton's method on Riemannian manifolds, J. Appl. Math. Comput., 25 (2007), 35–47.

[157] Argyros, I.K., Computational theory of iterative methods. Series: Studies in Computational Mathematics, 15, Editors: C.K. Chui, and L. Wuytack, 2007, Elsevier Publ. Co. New York, U.S.A.

[158] Argyros, I.K., On the gap between the semilocal convergence domain of two Newton methods, Appl. Math., 34 (2007), 193–204.

[159] Argyros, I.K., On the convergence of Broyden–like methods, Acta Math. Sin. (Engl. Ser.), 23 (2007), 2087–2096.

[160] Argyros, I.K., On the convergence of the Secant method under the gamma condition, Cent. Eur. J. Math., 5 (2007), 205–214.

[161] Argyros, I.K., A Kantorovich analysis of Newton's method on Lie groups, J. Concrete Appl. Anal., 6 (2008), 21–32.

[162] Argyros, I.K., Concerning the semilocal convergence of Newton's method and convex majorants, Rend. Circ. Mat. Palermo (2), 57 (2008), 331–341.

[163] Argyros, I.K., Convergence and applications of Newton–type iterations, Springer–Verlag, 2008, New York.

[164] Argyros, I.K., Approximating solutions of equations by combining Newton–like methods, J. Korea Soc. Educ. Ser. B: Pure and Appl. Math., 15 (2008), 35–45.

[165] Argyros, I.K., On a class of Newton–like methods for solving nonlinear equations, J. Comput. Appl. Math., 228 (2009), 115–122.

[166] Argyros, I.K., On the semilocal convergence of inexact Newton methods in Banach spaces, J. Comput. Appl. Math., 228 (2009), 434–443.

[167] Argyros, I.K., Concerning the convergence of Newton's method and quadratic majorants, J. Appl. Math. Comput., 29 (2009), 391–400.

[168] Argyros, I.K., Local convergence of Newton's method using Kantorovich's convex majorants, Revue Anal. Numér. Théor. Approx., 39 (2010), 97–106.

[169] Argyros, I.K., On the convergence region of Newton's method under Hölder continuity conditions, Intern. J. Comput. Math., 87 (2010), 317–326.

[170] Argyros, I.K., A semilocal convergence analysis for directional Newton methods, Math. Comput., AMS, 80 (2011), 327–343.

[171] Argyros, I.K., Chen, D., The midpoint method for solving equations in Banach spaces, Appl. Math. Lett., 5 (1992), 7–9.

[172] Argyros, I.K., Chen, D., Results on the Chebyshev method in Banach spaces, Proyecciones, 12 (1993), 119–128.

[173] Argyros, I.K., Chen, D., The midpoint method in Banach spaces and the Ptăk error estimates, Appl. Math. Comp., 62 (1994), 1–15.

[174] Argyros, I.K., Chen, D., On the midpoint iterative method for solving nonlinear operator equations and applications to the solution of integral equations, Revue D'analyse Numérique et. de Théorie de l'Approximation, 23 (1994), 139–152.

[175] Argyros, I.K., Chen, D., An inverse–free Jarratt type approximation in a Banach space, Approx. Theory Appl., 12 (1996), 19–30.

[176] Argyros, I.K., Chen, D., Qian, Q., The Jarratt method in a Banach space setting, J. Comput. Appl. Math., 51, (1994), 103–106.

[177] Argyros, I.K., Ezquerro, J.A., Gutiérrez, J.M., Hernández, M.A;, Hilout, S., On the semilocal convergence of efficient Chebyshev-Secant-type methods, J. Comput. Appl. Math., 235 (2011), 3195–3206.

[178] Argyros, I.K., Hilout, S., Newton's methods for variational inclusions under conditioned Fréchet derivative, Appl. Math., 34 (2007), 349–357.

[179] Argyros, I.K., Hilout, S., An improved local convergence analysis for Secant–like method, East Asian Math. J., 23 (2007), 261–270.

[180] Argyros, I.K., Hilout, S., Local convergence of Newton–like methods for generalized equations, Appl. Math. Comput., 197 (2008), 507–514.

[181] Argyros, I.K., Hilout, S., Multipoint method for generalized equations under mild differentiability conditions, Functiones et Approximatio, XXXVIII, (2008), 7–19.

[182] Argyros, I.K., Hilout, S., Hummel–Seebeck method for generalized equations under conditioned second Fréchet derivative, Nonlinear Funct. Anal. Appl., 14 (2009), 261–269.

[183] Argyros, I.K., Hilout, S., On the Local convergence of Gauss-Newton method, Punjab Univer. Math. J., 41 (2009), 23–33.

[184] Argyros, I.K., Hilout, S., A Fréchet derivative–free cubically convergent method for set–valued maps, Numerical Algorithms, 48 (2008), 361–371.

[185] Argyros, I.K., Hilout, S., Steffensen method for solving generalized equations, Serdica Math. J., 34 (2008), 455–466.

[186] Argyros, I.K., Hilout, S., On the midpoint method for solving generalized equations, Punjab Univ. J. Math., 40 (2008), 63–70.

[187] Argyros, I.K., Hilout, S., On the local convergence of Newton–type method in Banach spaces under a gamma–type condition, Proyecciones J. Math., 27 (2008), 1–14.

[188] Argyros, I.K., Hilout, S., Newton–like method for nonsmooth subanalytic variational inequalities, submitted for publication.

[189] Argyros, I.K., Hilout, S., On a Secant–like method for solving generalized equations, Math. Bohemica, 133 (2008), 313–320.

[190] Argyros, I.K., Hilout, S., Superquadratic method for generalized equations under relaxed conditions on the second derivative, submitted for publication.

[191] Argyros, I.K., Hilout, S., Newton–Steffensen–type method for perturbed nonsmooth subanalytic variational inequalities, submitted for publication.

[192] Argyros, I.K., Hilout, S., Traub–Potra–type method for set–valued maps, Aust. J. Math. Anal. Appl., accepted.

[193] Argyros, I.K., Hilout, S., A cubically convergent method without second order derivative for solving generalized equations, Internat. J. Modern Math., 3 (2008), 187–195.

[194] Argyros, I.K., Hilout, S., Efficient methods for solving equations and variational inequalities, Polimetrica Publisher, 2009.

[195] Argyros, I.K., Hilout, S., Aspects of the computational theory for certain iterative methods, Polimetrica Publisher, 2009.

[196] Argyros, I.K., Hilout, S., Enclosing roots of polynomial equations and their applications to iterative processes, Surveys Math. Appl., 4 (2009), 119–132.

[197] Argyros, I.K., Hilout, S., On the convergence of some iterative procedures under regular smoothness, PanAmer. Math. J., 19 (2009), 17–34.

[198] Argyros, I.K., Hilout, S., Extending the Newton–Kantorovich hypothesis for solving equations, J. Comput. Appl. Math., 234 (2010), 2993–3006.

[199] Argyros, I.K., Hilout, S., An improved local convergence analysis for Newton–Steffensen–type method, J. Appl. Math. Comput., 32 (2010), 111–118.

[200] Argyros, I.K., Hilout, S., A convergence analysis of Newton–like method for singular equations using recurrent functions, Numer. Funct. Anal. Optim., 31 (2010), 112–130.

[201] Argyros, I.K., Hilout, S., A Kantorovich–type analysis of Broyden's method using recurrent functions, J. Appl. Math. Comput., 32 (2010), 353–368.

[202] Argyros, I.K., Hilout, S., Convergence conditions for Secant–type methods, Czecho. Math. J., 60 (2010), 253–272.

[203] Argyros, I.K., Hilout, S., Improved generalized differentiability conditions for Newton–like methods, J. Complexity, 26 (2010), 316–333.

[204] Argyros, I.K., Hilout, S., Inexact Newton methods and recurrent functions, Appl. Math., 37 (2010), 113–126.

[205] Argyros, I.K., Hilout, S., On Newton–like methods of "bounded deterioration" using recurrent functions, Aequat. Math., 79 61–82, (2010).

[206] Argyros, I.K., Hilout, S., On the solution of systems of equations with constant rank derivatives, Numer. Algorithms, 57 (2011), no. 2, 235–253.

[207] Argyros, I.K., Hilout, S., A unifying theorem for Newtons method on spaces with a convergence structure, J. Complexity, 27 (2011), 39–54.

[208] Argyros, I.K., Hilout, S., On the convergence of Newton–like methods for solving equations using slantly differentiable operators, submitted.

[209] Argyros, I.K., Hilout, S., Tabatabai, M.A., Mathematical Modelling with Applications in Biosciences and Engineering, Nova Publishers, New York, 2011.

[210] Argyros, I.K., Hilout, S., Weaker conditions for the convergence of Newton's method, J. Complexity, 2012, in press.

[211] Argyros, I.K., Szidarovszky, F., Convergence of general iteration schemes, J. Math. Anal. Appl., 168 (1992), 42–52.

[212] Argyros, I.K., Szidarovszky, F., On the monotone of general Newton–like methods, Bull. Austral. Math. Soc., 45 (1992), 489–502.

[213] Argyros, I.K., Szidarovszky, F., The Theory and Applications of Iteration Methods, C.R.C. Press, Boca Raton, Florida, 1993.

[214] Argyros, I.K., Szidarovszky, F., On the convergence of modified contractions, J. Comput. Appl. Math., 55 (1994), 97–108.

[215] Argyros, I.K., Verma, R.U., Semilocal convergence theorems for a certain class of iterative procedures involving $m$–Fréchet differentiable operators, Math. Sci. Res. Hot-Line, 4 (2000), 1–12.

[216] Atkinson, K.E., The numerical evaluation of fixed points for completely continuous operators, SIAM J. Num. Anal., 10 (1973), 799–807.

[217] Atkinson, K.E., A Survey of Numerical Methods for the Solution of Fredholm Integral Equations of the Second Kind, SIAM, Philadelphia, 1976.

[218] Aubin, J.P., Lipschitz behavior of solutions to convex minimization problems, Math. Oper. Res., 9 (1984), 87–111.

[219] Aubin, J.P., Frankowska, H., Set–valued analysis, Birkhäuser, Boston, 1990.

[220] Avila, J.H., Continuation method for nonlinear equations, Technical Report TR-142, Computer Science Center, University of Maryland, January, 1971.

[221] Avila, J.H., The feasibility of continuation methods for nonlinear equations, SIAM J. Numer. Anal., 11 (1974), 102–122.

[222] Bai, Z.Z., A class of two–stage iterative methods for systems of weakly nonlinear equations, Numer. Algorithms, 14 (1997), 295–319.

[223] Bai, Z.Z., Wang, D.R., On the convergence of the factorization update algorithm, J. Comput. Math., 11 (1993), 236–249.

[224] Begehr, H., Efendiev, M.A., On the asymptotics of meromorphic solutions for nonlinear Riemann–Hilbert problems, Math. Proc. Cambridge Philos. Soc., 127(1999), 159–172.

[225] Belluce, L.P., Kirk, W.A., Fixed point theorems for a certain class of nonexpansive mappings, Proc. Amer. Math. Soc., 50 (1969), 144–146.

[226] Ben–Israel, A., A Newton–Raphson method for the solution of systems of operators, J. Math. Anal. Appl., 15 (1966), 243–252.

[227] Ben–Israel, A., Greville, T.N.E., Generalized Inverses: Theory and Applications, John Wiley and Sons, 1974.

[228] Ben–Israel, A., Levin, Y., Maple programs for directional Newton methods, are available at ftp://rutcor.rutgers.edu/pub/bisrael/Newton–Dir.mws

[229] Berinde, V., Remarks on the convergence of the Newton-Raphson method, Revue d'analyse Numér. Theory Approx., 24 (1995), 15–21.

[230] Bi, W., Wu, Q., Ren, H., Convergence ball and error analysis of Ostrowski–Traub's method, Appl. Math. J. Chinese Univ. Ser. B, 25 (2010), 374–378.

[231] Blum, L., Cucker, F., Shub, M., Smale, S., Complexity and real computation. With a foreword by Richard M. Karp., Springer–Verlag, New York, 1998.

[232] Bolte, J., Daniilidis, A., Lewis, A., The Lojasiewicz inequality for nonsmooth subanalytic functions with applications to subgradient dynamical systems, SIAM J. Optimization, 17 (2006), 1205–1223.

[233] Bolte, J., Daniilidis, A., Lewis, A., Tame functions are semismooth, Math. Program., Ser. B, 117 (2009), 5–19.

[234] Bonnans, J.F, Local study of Newton type algorithms for constrained problems, Optimization (Varetz, 1988), Lecture Notes in Math., 1405, Springer, Berlin, (1989), 13–24.

[235] Bonnans, J.F, An introduction to Newton type algorithms for nonlinearly constrained optimization problems, New methods in optimization and their industrial uses (Pau/Paris, 1987), 1–17, Internat. Schriftenreihe Numer. Math., 87, Birkhuser, Basel, 1989.

[236] Bonnans, J.F, Local analysis of Newton–type methods for variational inequalities and nonlinear programming, Appl. Math. Optim., 29 (1994), 161–186.

[237] Bonnans, J.F, Shapiro, A., Perturbation analysis of optimization problems, Springer Series in Operations Research, Springer–Verlag, New York, 2000.

[238] Bosarge, W.E., Falb, P.L., A multipoint method of third order, J. Optim. Theory Appl., 4 (1969), 156–166.

[239] Bosarge, W.E., Falb, P.L., Infinite dimensional multipoint methods and the solution of two point boundary value problems, Numer. Math., 14 (1970), 264–286.

[240] Brent, R.P., Algorithms for Minimization Without Derivatives, Prentice Hall, Englewood Cliffs, New Jersey, 1973.

[241] Brezinski, C., Dynamical systems and sequence transformations, Symmetries and integrability of difference equations, J. Phys. A, 34 (2001), 10659–10669.

[242] Brézis, H., Opérateurs maximaux monotones et. semi–groupes de contractions dans les espaces de Hilbert, North–Holland Mathematics Studies, No. 5, Notas de Matemática (50), North–Holland Publishing Co., Amsterdam–London; American Elsevier Publishing Co., Inc., New York, 1973.

[243] Browder, F.E., Petryshyn, W.V., The solution by iteration of linear functional equations in Banach spaces, Bull. Amer. Math. Soc., 72 (1996), 566–570.

[244] Browder, F.E., Nonexpansive nonlinear operators in Banach spaces, Proc. Nat. Acad. Sci. USA, 54 (1965), 1011–1014.

[245] Brown, P.N., A local convergence theory for combined inexact–Newton/finite–difference projection methods, SIAM J. Numer. Anal., 24 (1987), 407–434.

[246] Brown, P.N., Hindmarsh, A.C., Walker, H.F., Experiments with quasi–Newton methods in solving stiff ODE systems, SIAM J. Sci. Statist. Comput., 6 (1985), 297–313.

[247] Brown, P.N., Saad, Y., Convergence theory of nonlinear Newton–Krylov algorithms, SIAM J. Optim., 4 (1994), 297–330.

[248] Brown, P.N., Vassilevski, P.S., Woodward, C.S., On mesh–independent convergence of an inexact Newton–multigrid algorithm, SIAM J. Sci. Comput., 25 (2003), 570–590.

[249] Brown, K.M., Dennis Jr., J.M., Derivative free analogues of the Levenberg–Marquardt and Gauss algorithms for nonlinear least squares approximation, Numer. Math., 18 (1972) 289–297.

[250] Broyden, C.G., A class of methods for solving nonlinear simultaneous equations, Math. Comput., 19 (1965), 577–593.

[251] Broyden, C.G., Dennis, J.E., Moré, J., On the local and superlinear convergence of quasi–Newton methods, J. Inst. Math. Appl., 12 (1973), 223–246.

[252] Burmeister, W., Inversion freie verfahren zur lösung nichtlinearen operatorgleichungen, Zeit. Angew. Math. Mech., 52 (1972), 101–110.

[253] Byelostotskij, A.Ja.,, Some methods for the solution of functional equations (Russian), Uspekhi Matem. Nauk, 17 (1962), 192–193.

[254] Cabuzel, C., A multipoint iteration formula for solving a variational inclusions in the Hölder case, An. Stiint. Univ. Al. I. Cuza Iasi. Mat. (N.S.), 54 (2008), 147–160.

[255] Cabuzel, C., Piétrus, A., Solving variational inclusions by a method obtained using a multipoint iteration formula, Revista Math. Complut., 22 (2009), 63–74.

[256] Cabuzel, C., Piétrus, A., Solving variational inclusions by a multipoint iteration method under center–Hölder continuity conditions, Appl. Math., 34 (2007), 493–503.

[257] Cabuzel, C., Piétrus, A., Local convergence of Newton's method for subanalytic variational inclusions, Positivity, 12 (2008), 525–533.

[258] Candela, V., Marquina, A., Recurrence relations for rational cubic methods I: The Halley method, Comput., 44 (1990), 169–184.

[259] Candela, S., Marquina, A., Recurrence relations for rational cubic methods. II. The Chebyshev method, Comput., 45 (1990), 355–367.

[260] Caponetti, D., De Pascale, E., Zabrejko, P.P., On the Newton–Kantorovich method in $K$–normed spaces, Rend. Circ. Mat. Palermo, 49 (2000), 545–560.

[261] Cătinaş, E., On some iterative methods for solving nonlinear equations, Revue Anal. Numér. Theory Approx., 23 (1994), 47–53.

[262] Cătinaş, E., Inexact perturbed Newton methods, and applications to a class of Krylov solvers, J. Optim. Theory Appl., 108 (2001), 543–570.

[263] Cătinaş, E., On accelerating the convergence of the successive approximations method, Rev. Anal. Numér. Theory Approx., 30 (2001), 3–8.

[264] Cătinaş, E., On the superlinear convergence of the successive approximations method, J. Optim. Theory Appl., 113 (2002), 473–485.

[265] Cătinaş, E., The inexact, inexact perturbed, and quasi–Newton methods are equivalent models, Math. Comput., 74 (2005), 291–301.

[266] Cătinaş, E., Sufficient convergence conditions for certain accelerated successive approximations, Trends and applications in constructive approximation, 71–75, Internat. Ser. Numer. Math., 151, Birkhäuser, Basel, 2005.

[267] Cătinaş, E., Estimating the radius of the attraction balls, Appl. Math. Lett., 22 (2009), 712–714.

[268] Chan, R.H., Chung, H.L., Xu, S.F., The inexact Newton–like method for inverse eigenvalue problem, BIT, 43 (2003) 7–20.

[269] Chan, R.H., Xu, S.F., Zhou, H.M., On the convergence of a quasi–Newton method for inverse eigenvalue problem, SIAM J. Numer. Anal., 36 (1999), 436–441.

[270] Chandrasekhar, S., Radiative transfer, Dover Publ., New York, 1960.

[271] Chen, B., Harker, P.T., A continuation method for monotone variational inequalities, Math. Program., 69 (1995), 237–253.

[272] Chen, D., Kantorovich–Ostrowski convergence theorems and optimal error bounds for Jarratt's iterative method, Intern. J. Comput. Math., 31 (1990), 221–235.

[273] Chen, D., On the convergence of a class of generalized Steffensen's iterative procedures and error analysis, Int. J. Comput. Math., 31 (1989), 195–203.

[274] Chen, J., The convergence analysis of inexact Gauss–Newton methods for nonlinear problems, Comput. Optim. Appl., 40 (2008), 97–118.

[275] Chen, J., Li, W., Convergence behaviour of inexact Newton methods under weak Lipschitz condition, J. Comput. Appl. Math., 191 (2006), 143–164.

[276] Chen, X., On the convergence of Broyden–like methods for nonlinear equations with nondifferentiable terms, Ann. Inst. Statist. Math., 42 (1990), 387–401.

[277] Chen, X., Nashed, M.Z., Convergence of Newton–like methods for singular operator equations using outer inverses, Numer. Math., 66 (1993), 235–257.

[278] Chen, X., Nashed, Z., Qi, L., Convergence of Newton's method for singular smooth and nonsmooth equations using adaptive outer inverses, SIAM J. Optim., 7 (1997), 445–462.

[279] Chen, X., Nashed, M.Z., Qi, L., Smoothing methods and semismooth methods for nondifferentiable operator equations, SIAM J. Numer. Anal., 38 (2000), 1200–1216.

[280] Chen, X., Qi, L., Sun, D., Global and superlinear convergence of the smoothing Newton method and its application to general box constrained variational inequalities, Math. Comput., 67 (1998), 519–540.

[281] Chen, X., Yamamoto, T., Convergence domains of certain iterative methods for solving nonlinear equations, Numer. Funct. Anal. Optim., 10 (1989), 37–48.

[282] Chen, Y., Cai, D., Inexact overlapped block Broyden methods for solving nonlinear equations, Appl. Math. Comput., 136 (2003), 215–228.

[283] Courant, R., Variational methods for the solution of problems of equilibrium and vibrations, Bull. Amer. Math. Soc., 49 (1943), 1–23.

[284] Chui, C.K., Quak, F., Wavelets on a bounded interval. In: Numerical Methods of Approximation Theory, Vol. 9, (eds: Braess D. and Larry L. Schumaker, Intern. Ser. Num. Math., Vol. 105), Basel: Birkhäuser Verlag, 53–75.

[285] Chow, S.N., Hale, J.K., Methods of Bifurcation Theory, Springer–Verlag, New York, 1962.

[286] Cianciaruso, F., Convergence theorems for Newton–like methods for operators with generalized Hölder derivative, Fixed Point Theory, 5 (2004), 21–35.

[287] Cianciaruso, F., Convergence of Newton–Kantorovich approximations to an approximate zero, Numer. Funct. Anal. Optim., 28 (2007), 631–645.

[288] Cianciaruso, F., A further journey in the "terra incognita" of the Newton–Kantorovich method, Nonlinear Funct. Anal. Appl., 15 (2010), 173–183.

[289] Cianciaruso, F., De Pascale, E., The Newton–Kantorovich approximations for the nonlinear integro–differential equations of mixed type, Ricerche Mat., 51, (2002), 249–260.

[290] Cianciaruso, F., De Pascale, E., The Newton–Kantorovich approximations for nonlinear singular integral equations with shift, J. Integral Equat. Appl., 14 (2002), 223–237.

[291] Cianciaruso, F., De Pascale, E., Newton–Kantorovich aproximations when the derivative is Hölderian: Old and new results, Numer. Funct. Anal. Optim., 24 (2003), 713–723.

[292] Cianciaruso, F., De Pascale, E., Estimates of majorizing sequences in the Newton–Kantorovich method, Numer. Funct. Anal. Optim., 27 (2006), 529–538.

[293] Cianciaruso, F., De Pascale, E., Estimates of majorizing sequences in the Newton–Kantorovich method: A further improvement, J. Math. Anal. Appl., 322 (2006), 329–335.

[294] Cianciaruso, F., De Pascale, E., Zabrejko, P.P., Some remarks on the Newton-Kantorovič approximations, Atti Sem. Mat. Fis. Univ. Modena, 48 (2000), 207–215.

[295] Cohen, G., Auxiliary problem principle and decomposition of optimization problems, J. Optim. Theory Appl., 32, 3, (1980), 277–305.

[296] Cohen, G., Auxiliary problem principle extended to variational inequalities, J. Optim. Theory Appl., 59 (1988), 325–333.

[297] Collatz, L., Functional Analysis and Numerisch Mathematik, Springer–Verlag, New York, 1964.

[298] Cordero, A., Torregrosa, J.R., A class of multi–point iterative methods for nonlinear equations, Appl. Math. Comput., 197 (2008), 337–344.

[299] Craven, B.D., Nashed, M Z., Generalized implicit function theorems when the derivative has no bounded inverse, Nonlinear Anal., 6 (1982), 375–387.

[300] Danes, J., Fixed point theorems, Nemyckii and Uryson operators, and continuity of nonlinear mappings, Comment. Math. Univ. Carolinae, 11 (1970), 481–500.

[301] Darbo, G., Punti uniti in trasformationa codominio non compatto, Rend. Sem. Mat. Univ. Padova, 24 (1955), 84–92.

[302] Daubechies, I., Ten Lectures in Wavelets, (Conf. Board Math. Sci. (CBMS) Vol. 61), Society for Industrial and Applied Mathematics (SIAM), Philadelphia, PA, 1992.

[303] Davis, H.T., Introduction to nonlinear differential and integral equations, Dover Publications, Inc., New York, 1962.

[304] Decker, D.W., Keller, H.B., Kelley, C.T., Convergence rates of Newton's method at singular points, SIAM J. Numer. Anal., 20 (1983), 296–314.

[305] Dedieu, J.P., Penality functions in subanalytic optimization, Optim., 26 (1992), 27–32.

[306] Dedieu, J.P., Estimations for the separation number of a polynomial system, J. Symbolic Comput., 24 (1997), 683–693.

[307] Dedieu, J.P., Priouret, P., Malajovich, G., Newton's method on Riemannian mnifolds: Convariant $\alpha$ theory, IMA J. Numer. Anal., 23 (2003), 395–419.

[308] Dembo, R.S., Eisenstat, S.C., Steihaug, T., Inexact Newton methods, SIAM J. Numer. Anal., 19 (1982), 400–408.

[309] Demidovich, N.T., Zabrejko, P.P., Lysenko, Ju.V., Some remarks on the Newton–Kantorovich mehtod for nonlinear equations with Hölder continuous linearizations (in Russian), Izv. Akad. Nauk Belorus, 3 (1993), 22–26.

[310] Dennis, J.E., Jr., On Newton–like methods, Numer. Math., 11 (1968), 324–330.

[311] Dennis, J.E., On the Kantorovich hypothesis for Newton's method, SIAM J. Numer. Anal., 6 (1969), 493–507.

[312] Dennis, J.E., On the convergence of Newton–like methods, Numerical methods for nonlinear algebraic equations, (Proc. Conf., Univ. Essex, Colchester, 1969), Gordon and Breach, London, (1970), 163–181.

[313] Dennis, J.E., Toward a unified convergence theory for Newton–like methods, in Nonlinear Functional Analysis and Applications (L.B. Rall, ed.), Academic Press, New York, (1971), 425–472.

[314] Dennis, J.E., A brief survey of convergence results for quasi–Newton methods, Nonlinear programming (Proc. Sympos., New York, 1975), SIAM–AMS Proc., Vol. IX, Amer. Math. Soc., Providence, R. I., (1976), 185–199.

[315] Dennis, J.E., A brief introduction to quasi–Newton methods, Numerical analysis (Proc. Sympos. Appl. Math., Atlanta, Ga., 1978), Proc. Sympos. Appl. Math., XXII, Amer. Math. Soc., Providence, R.I., (1978), 19–52.

[316] Dennis, J.E., Moré, J.J., A characterization of superlinear convergence and its application to quasi–Newton methods, Math. Comput., 28 (1974), 549–560.

[317] Dennis, J.E., Moré, J.J., Quasi–Newton methods, motivation and theory, SIAM Rev., 19 (1977), 46–89.

[318] Dennis, J.E., Moré, J.J., Quasi–Newton methods, motivation and theory, (Chinese) Translated from the English by Wen Yu Sun., Appl. Math. Math. Comput., 1 1983, 1–30.

[319] Dennis, J.E., Schnabel, R.B., Least change secant updates for quasi–Newton methods, SIAM Rev., 21 (1979), 443–459.

[320] Dennis, J.E., Sheng, S.B., Vu, P.A., A memoryless augmented Gauss–Newton method for nonlinear least–squares problems, J. Comput. Math., 6 (1988), 355–374.

[321] Dennis, J.E., Walker, H.F., Inaccuracy in quasi–Newton methods: local improvement theorems, Mathematical programming at Oberwolfach, II (Oberwolfach, 1983), Math. Program. Stud., 22 (1984), 70–85.

[322] De Pascale, E., Zabrejko, P.P, New convergence criteria for Newton–Kantorovich method and some applications to nonlinear integral equations, Rend. Sem. Mat. Univ. Padova, 100 (1998), 211–230.

[323] De Pascale, E., Zabrejko, P.P., Convergence of the Newton–Kantorovich method under Vertgeim conditions: a new improvement, Z. Anal. Anwend., 17 (1998), 271–280.

[324] De Pascale, E., Zabrejko, P.P., Fixed point theorems for operators in spaces of continuous functions, Fixed Point Theory, 5 (2004), 117–129.

[325] Deuflhard, P., Newton Methods for Nonlinear Problems, Springer–Verlag, New–York, 2004.

[326] Deuflhard, P., Heindl, G., Affine invariant convergence theorems for Newton's method and extensions to related methods, SIAM J. Numer. Anal., 16 (1979), 1–10.

[327] Deuflhard, P., Potra, F.A., Asymptotic mesh independence of Newton–Galerkin methods via a refined Mysovskii theorem, SIAM J. Numer. Anal., 29 (1992), 1395–1412.

[328] Deuflhard, P., Schiela, A., Weiser, M., Asymptotic mesh independence of Newton's method, revisited, SIAM J. Numer. Anal., 42 (2005), 1830–1845.

[329] Diaconu, A., On the approximation of solutions of equations in Banach spaces using approximant sequences, Conference on Analysis, Functional Equations, Approximation and Convexity, in Honor of E. Popoviciv, Cluj–Napora, October 15–16, (1999), 62–72.

[330] Dingguo, P., Weiwen, T., Globally convergent inexact generalized Newton's methods for nonsmooth equations, J. Comput. Appl. Math., 138 (2002), 37–49.

[331] Do Carmo, M.P., Riemannian Geometry, Birkhauser, 1992, Boston, USA.

[332] Dontchev, A.L., Local analysis of a Newton–type method based on partial linearization. Renegar, James (ed.) et. al., The mathematics of numerical analysis. 1995 AMS-SIAM summer seminar in applied mathematics, Providence, RI: AMS. Lect. Appl. Math., 32 (1996), 295–306.

[333] Dontchev, A.L., Local convergence of the Newton method for generalized equation, C.R.A.S., 322, Serie I, (1996), 327–331.

[334] Dontchev, A.L., Uniform convergence of the Newton method for Aubin continuous maps, Serdica Math. J., 22 (1996), 385–398.

[335] Dontchev, A.L., Hager, W.W., An inverse function theorem for set–valued maps, Proc. Amer. Math. Soc., 121 (1994), 481–489.

[336] Dontchev, A.L., Quincampoix, M., Zlateva, N., Aubin criterion for metric regularity, J. Convex Anal., 13 (2006), 281–297.

[337] Dontchev, A.L., Rockafellar, R.T., Characterizations of strong regularity for variational inequalities over polyhedral convex sets, SIAM J. Optim. 6 (1996), 1087–1105.

[338] Dontchev, A.L., Rockafellar, R.T., Ample parameterization of variational inclusions, SIAM J. Optim. 12 (2001), 170–187.

[339] Dontchev, A.L., Rockafellar, R.T., Regularity and conditioning of solution mappings in variational analysis, Set–Valued Anal., 12 (2004), 79–109.

[340] Döring, B., Iterative lösung gewisser randwertprobleme und integralgleichungen, Appl. Mat., 24, (1976), 1–31.

[341] Dunford, N., Schwartz, J.T., Linear operators. Part I, Int. Publ. Leyden, (1963).

[342] Duvuat, G., Lions, J.L., Inequalities in Physics and Mechanics, Springer–Verlag, Berlin, 1976.

[343] Eaves, B.C., A locally quadratically convergent algorithm for computing stationary point, Dpt of Operations Rearch, Stanford University, 1978.

[344] Edelstein, M., On fixed and periodic points under contractive mappings., J. London Math. Soc., 37 (1962), 74–79.

[345] Edelstein, M., A remark on a theorem of M.A. Krasnoselskii, Amer. Math. Monthly, 73 (1966), 509–510.

[346] Eisenstat, S.C., Walker, H.F., Globally convergent inexact Newton methods, SIAM J. Optim., 4 (1994), 393–422.

[347] Eisenstat, S.C., Walker, H.F., Choosing the forcing terms in an inexact Newton method, Special issue on iterative methods in numerical linear algebra, (Breckenridge, CO, 1994), SIAM J. Sci. Comput., 17 (1996), 16–32.

[348] Ezquerro, J.A., Gutiérrez, J.M., Hernández, M.A., Romero, N., Rubio, M.J., The Newton method: from Newton to Kantorovich. (Spanish), Gac. R. Soc. Mat. Esp., 13 (2010), 53–76.

[349] Ezquerro, J.A., Gutiérrez, J.M., Hernández, M.A., Salanova, M. A., Solving nonlinear integral equations arising in radiative transfer, Numer. Funct. Anal. Optim., 20 (1999), 661–673.

[350] Ezquerro, J.A., Hernández, M.A., Avoiding the computation of the second Fréchet–derivative in the convex acceleration of Newton's method, J. Comput. Appl. Math., 96 (1998), 1–12.

[351] Ezquerro, J.A., Hernández, M.A., Multipoint super–Halley type approximation algorithms in Banach spaces, Numer. Funct. Anal. Optim., 21 (2000), 845–858.

[352] Ezquerro, J.A., Hernández, M.A., Recurrence relations for Chebyshev–type methods, Appl. Math. Optim., 41 (2000), 227–236.

[353] Ezquerro, J.A., Hernández, M.A., A special type of Hammerstein integral equations, Int. Math. J., 1 (2002), 557–566.

[354] Ezquerro, J.A., Hernández, M.A., On an application of Newton's method to nonlinear operators with $\omega$–conditioned second derivative, BIT, 42 (2002), 519–530.

[355] Ezquerro, J.A., Hernández, M.A., Generalized differentiability conditions for Newton's method, IMA J. Numer. Anal., 22 (2002), 187–205.

[356] Ezquerro, J.A., Hernández, M.A., On the $R$–order of convergence of Newton's method under mild differentiability conditions, J. Comput. Appl. Math., 197(1) (2006), 53–61.

[357] Ezquerro, J.A., Hernánadez, M.A., The Ulm method under mild differentiability conditions, Numer. Math., 109 (2008), 193–207.

[358] Ezquerro, J.A., Hernández, M.A., An improvement of the region of accessibility of Chebyshev's method from Newton's method, Math. Comput., 78 (2009), 1613–1627.

[359] Ezquerro, J.A., Hernández, M.A., New iterations of $R$–order four with reduced computational cost, BIT Numer. Math., 49 (2009), 325–342.

[360] Ezquerro, J.A., Hernández, M.A., An optimization of Chebyshev's method, J. Complexity, 25 (2009), 343–361.

[361] Ezquerro, J.A., Hernández, M.A., Romero, N., Newton–type methods of high order and domains of semilocal and global convergence, Appl. Math. Comput., 214 (2009), 142–154.

[362] Ezquerro, J.A., Hernández, M.A., Salanova, M.A., Remark on the convergence of the midpoint method under mild differentiability conditions, J. Comp. Appl. Math., 98 (1998), 305–309.

[363] Feinstauer, M., Zernicek, A., Finite element solution on nonlinear elliptic problems, Numer. Math., 50 (1987), 471–475.

[364] Ferreira, O.P., Local convergence of Newton's method from the view point of the majorant principle, IMA. J. Numer. Anal., 29 (2009), 746–759.

[365] Ferreira, O.P., Convergence of Newton's method in Banach space from the view of the majorant principle, IMA J. Numer. Anal. 29 (2009), 746–759.

[366] Ferreira, O.P., Svaiter, B.F., Kantorovich's theorem on Newton's method in Riemannian manifolds, J. Complexity, 18 (2002), 304–329.

[367] Ferreira, O.P., Svaiter, B.F., Kantorovich's majorants principle for Newton's method, Comput. Optim. Appl., 42 (2009), 213–229.

[368] Ferreira, O.P., Svaiter, B.F., Kantorovich's theorem on Newton's method, submitted for publication.

[369] Ferreira, O.P., Goncalves, M.L.N, Local convergence analysis of inexact Newton–like methods under majorant condition, preprint, Comput. Optim. Appl., 48 (2011), 1–21.

[370] Ferris, M.C., Pang, J.S., Engineering and economic applications of complementarity problems, SIAM Rev., 39 (1997), 669–713.

[371] Foerster, H., Frommer, A., Mayer, G., Inexact Newton methods on a vector supercomputer, J. Comp. Appl. Math, 58 (1995), 237–253.

[372] Frontini, M., Sormani, E., Modified Newton's method with third–order convergence and multiple roots, J. Comput. Appl. Math., 156 (2003), 345–354.

[373] Frontini, M., Sormani, E., Third–order methods from quadrature formulae for solving systems of nonlinear equations, Appl. Math. Comput., 149 (2004), 771–782.

[374] Fujimoto, T., Global asymptotic stability of nonlinear difference equations I, Econ. Lett., 22 (1987), 247–250.

[375] Fujimoto, T., Global asymptotic stability of nonlinear difference equations II, Econ. Lett., 23 (1987), 275–277.

[376] Galantai, A., The theory of Newton's method, J. Comput. Appl. Math., 124 (2000) 25–44.

[377] Galperin, A., Kantorovich majorization and functional equations, Numer. Funct. Anal. Optim., 24 (2003), 783–811.

[378] Galperin, A., On convergence domains of Newton's and modified Newton methods, Numer. Funct. Anal. Optim., 26 (2005), 385–405.

[379] Galperin, A., Secant method with regularity continuous divided differences, J. Comput. Appl. Math., 193 (2006), 574–575.

[380] Galperin, A., Optimal iterative methods for nonlinear equations, Numer. Funct. Anal. Optim., 30 (2009), 499–522.

[381] Galperin, A., Ulm's method without derivatives, Nonlinear Anal., 71 (2009), 2094–2113.

[382] Galperin, A., The second Broyden's method for operators with regularly continuous divided differences, preprint.

[383] Galperin, A., Waksman, Z., Regular smoothness and Newton's method, Numer. Funct. Anal. Optim., 15 (1994), 813–858.

[384] Galperin, A., Waksman, Z., Newton–type methods under regular smoothness, Numer. Funct. Anal. Optim., 17 (1996), 259–291.

[385] Gander, W., On Halley's iteration method, Amer. Math. Monthly, 92 (1985), 131–134.

[386] Gaskparo, M.G., Morini, B., Inexact Newton–like methods: Forcing terms and conditioning, J. Optim. Theory. Appl., 107 (2000), 573–589.

[387] Gavurin, M.K., Nonlinear functional equations and continuous analogs of iterative methods, Izv. Vuzov. Ser. Math., 5 (1958), 18–31.

[388] Geoffroy, M.H., A secant type method for variational inclusions, Preprint.

[389] Geoffroy, M.H., Hilout, S., Piétrus, A., Acceleration of convergence in Dontchev's iterative method for solving variational inclusions, Serdica Math. J., 29 (2003), 45–54.

[390] Geoffroy, M.H., Hilout, S., Piétrus, A., Stability of a cubically convergent method for generalized equations, Set–Valued Anal., 14 (2006), 41–54.

[391] Geoffroy, M.H., Piétrus, A., Superquadratic method for solving generalized equations in the Hölder case, Ricerche di Math. LII, fasc. 2 (2003), 231–240.

[392] Geoffroy, M.H., Piétrus, A., Local convergence of some iterative methods for solving generalized equations, J. Math. Anal. Appl., 290 (2004), 497–505.

[393] Gill, P.E., Murray, W., Wright, M.A., Practical Optimization, Academic Press, 1991, London.

[394] Glowinski, R., Lions, J.L., Trémolières, R., Numerical analysis of variational inequalities, North–Holland, Amsterdam, 1982.

[395] Gragg, W.B., Tapia, R.A., Optimal error bounds for the Newton–Kantorovich theorem, SIAM J. Numer. Anal., 11 (1974), 10–13.

[396] Graves, L.M., Riemann integration and Taylor's theorem in general analysis, Trans. Amer. Math. Soc., 29 (1927), 163–177.

[397] Grigat, E., Sachs, G., Predictor–corrector continuation method for optimal control problems, in: Variational calculus, optimal control and applications (Trassenheide, 1996), Volume 124, Int. Ser. Numer. Math., Birkhäuser, Basel, (1998), 223–232.

[398] Guo, X., On semilocal convergence of inexact Newton methods, J. Comput. Math., 25 (2007), 231–242.

[399] Gutiérrez, J.M., A new semilocal convergence theorem for Newton's method, J. Comput. Appl. Math., 79 (1997), 131–145.

[400] Gutiérrez, J.M., Hernández, M.A., Recurrence relations for the super-Halley method, Comput. Math. Appl., 36 (1998), 1–8.

[401] Gutiérrez, J.M., Hernández, M.A., An acceleration of Newton's method: Super–Halley method, Appl. Math. Comp., 117 (2001), 223–239.

[402] Gutiérrez, J.M., Hernández, M.A., Salanova, M.A., Accessibility of solutions by Newton's method, Internat. J. Comput. Math., 57 (1995), 239–247.

[403] Gutiérrez, T.M., Hernández, M.A., Salanova, M.A., A family of the Chebyshev–Halley type methods in Banach spaces, Bull. Austral. Math. Soc., 55 (1997), 113–130.

[404] Gwinner, J., Generalized Stirling–Newton methods, In W. Oettli, K. Ritter (eds), Optimization and Operations Research, Oberwolfach, 1975, Lecture Notes Economics and Mathematical Systems, 11 (1976), 99–135.

[405] Hadeller, K.P., Shadowing orbits and Kantorovich's theorem, Numer. Math., 73 (1996), 65–73.

[406] Hald, O.H., On a Newton–Moser type method, Numer. Math., 23 (1975), 411–425.

[407] Han, D., Wang, X., The error estimates of Halley's method, Numer. Math. J. Chinese Univ. (English Ser.), 6 (1997), 231–240.

[408] Han, S., Pang, J., Rangaraj, N., Globally convergent Newton methods for nonsmooth equations, Math. Oper. Res., 17 (1992), 586–607.

[409] Han, D.F., Wang, X., Convergence of a deformed Newton method, Appl. Math. Comput., 94 (1998), 65–72.

[410] Harker, P.T., Pang, J.S., Finite dimensional variational inequality and nonlinear complementarity problems: a survey of theory, algorithms and applications, Math. Program., 48 (1990), 161–220.

[411] Hartman, P., Ordinary differential equations, John Wiley & Sons, Inc., New–York, London, Sydney, 1964.

[412] Häubler, W.M., A Kantorovich–type convergence analysis for the Gauss–Newton method, Numer. Math., 48 (1986), 119–125.

[413] Heinkenschloss, M., Kelley, C.T., Tran, H.T., Fast algorithms for nonsmooth compact fixed–point problems, SIAM J. Numer. Anal., 29 (1992), 1769–1792.

[414] Helgason, S., Differential Geometry, Lie Groups and Symmetric Spaces, Pergamon Press, Oxford, 1982.

[415] Hellinger, E., Toeplitz, O., Integralgleichungen und Gleichungen mit unendlichvielen Unbekannten, (German), Chelsea Publishing Company, New–York, (1953), 1335–1616.

[416] Hernández, M.A., Second–derivative–free variant of the Chebyshev method for nonlinear equations, J. Optim. Theory Appl., 104 (2000), 501–515.

[417] Hernández, M.A., The Newton method for operators with Hölder continuous first derivatives, J. Optim. Theory Appl., 109 (2001), 631–648.

[418] Hernández, M.A., A modification of the classical Kantorovich conditions for Newton's method, J. Comp. Appl. Math., 137 (2001), 201–205.

[419] Hernández, M.A., Chebyshev's approximation algorithms and applications, Comput. Math. Appl., 41 (2001), 433–445.

[420] Hernández, M.A., Rubio, M.J., The Secant method and divided differences Hölder continuous, Appl. Math. Comp., 124 (2001), 139–149.

[421] Hernández, M.A., Rubio, M.J., Semilocal convergence of the Secant method under mild convergence conditions of differentiability, Comput. Math. Appl., 44 (2002), 277–285.

[422] Hernández, M.A., Rubio, M.J., The Secant method for nondifferentiable operators, Appl. Math. Lett., 15 (2002), 395–399.

[423] Hernández, M.A., Rubio, M.J., An uniparametric family of iterative processes for solving nondifferentiable equations, J. Math. Anal. Appl., 275 (2002), 821–834.

[424] Hernández, M.A., Rubio, M.J., $\omega$–conditioned divided differences to solve nonlinear equations, Monografías del Semin. Matem. García de Galdeano, 27 (2003), 409–417.

[425] Hernández, M.A., Rubio, M.J., A modification of Newton's method for nondifferentiable equations, J. Comp. Appl. Math., 164/165 (2004), 323–330.

[426] Hernández, M.A., Rubio, M.J., Ezquerro, J.A., Secant–like methods for slving nonlinear integral equations of the Hammerstein type, J. Comput. Appl. Math., 115 (2000), 245–254.

[427] Hernández, M.A., Rubio, M.J., Ezquerro, J.A., Solving a special case of conservative problems by Secant–like method, Appl. Math. Cmput., 169 (2005), 926–942.

[428] Hernández, M.A., Salanova, M.A., Modification of the Kantorovich assumptions for semilocal convergence of the Chebyshev methods, J. Comput. Appl. Math., 126 (2000), 131–143.

[429] Hernández, M.A., Salanova, M.A., A new third–order iterative process for solving nonlinear equations, Monat. Math., 133 (2001), 131–142.

[430] Higle, J.L., Sen, S., On the convergence of algorithms with applications to stochastic and nondifferentiable optimization, SIE Working Paper #89-027, University of Arizona, 1989.

[431] Hille, E., Philips, R.S., Functional Analysis and Semigroups, Amer. Math. Soc. Coll. Publ., New–York, 1957.

[432] Hilout, S., Steffensen–type methods on Banach spaces for solving generalized equations, Adv. Nonlinear Var. Inequal., 10 (2007), 105–113.

[433] Hilout, S., An uniparametric Newton–Steffensen–type methods for perturbed generalized equations, Adv. Nonlinear Var. Inequa., 10 (2007), 115–124.

[434] Hilout, S., Superlinear convergence of a family of two–step Steffensen–type methods for generalized equations, Internat. J. Pure Appl. Math., 40 (2007), 1–10.

[435] Hilout, S., A two–step Steffensen–type method for nonsmooth variational inclusions, Comm. Appl. Nonlinear Anal., 14 (2007), 27–34.

[436] Hilout, S., Convergence analysis of a family of Steffensen–type methods for generalized equations, J. Math. Anal. Appl., 339 (2008), 753–761.

[437] Hilout, S., An uniparametric Secant–type method for nonsmooth generalized equations, Positivity, 12 (2008), 281–287.

[438] Hilout, S., Piétrus, A., A semilocal convergence of a secant–type method for solving generalized equations, Positivity, 10 (2006), 673–700.

[439] Hiriart–Urruty, J.B., Lemaréchal, C., Convex analysis and minimization algorithms (two volumes): I. Fundamentals II. Advanced theory and bundle methods, Grundlehren der Mathematischen Wissenschaften, Vol. 305 and 306, Springer–Verlag, Berlin, 1993.

[440] Homeir, H., A modified method for root finding with cubic convergence, J. Comput. Appl. Math., 157 (2003), 227–230.

[441] Homeier, H., A modified Newton method with cubic convergence: the multivariable case, J. Comput. Appl. Math., 169 (2004), 161–169.

[442] Hoppe, R.H.W., Numerical methods for large–scale nonlinear systems, Handouts published on–line on the web, (2005), 1–108.

[443] Hu, N., Shen, W., Li, C., Kantorovich's type theorems for systems of equations with constant rank derivatives, J. Comput. Appl. Math., 219 (2008), 110–122.

[444] Hu, Z., A new semilocal convergence theorem for Newton's method involving twice Fréchet–differentiability at only one point, J. Comput. Appl. Math., 181 (2005), 321–325.

[445] Huang, Z.D., A note of Kantorovich theorem for Newton iteration, J. Comput. Appl. Math., 47 (1993), 211–217.

[446] Huang, Z.D., On Newton's method under Hölder continuous derivative, J. Math. Anal. Appl., 270 (2002), 332–339.

[447] Huang, Z.D., On the convergence of inexact Newton method, J. Zheijiang University, Natur. Sci. Edition, 30 (2003), 393–396.

[448] Huang, Z.D., The convergence ball of Newton's method and the uniqueness ball of equations under Hölder–type continuous derivatives, Comput. Math. Appl., 25 (2004), 247–251.

[449] Hummel, P.M., Seebeck, C.L.Jr., A generalized of Taylor's expansion, Amer. Math. Monthly, 56 (1949), 243–247.

[450] Hwang, D.M., Kelley, C.T., Convergence of Broyden's method in Banach spaces, SIAM J. Optim., 2 (1992), 505–532.

[451] Ioffe, A.D., Tikhomirov, V.M., Theory of extremal problems, North Holland, Amsterdam, 1979.

[452] Ip, Chi–M., Kyparisis, J., Local convergence for quasi–Newton methods for $B$–differentiable operators, Math. Program., 56 (1992), 71–89.

[453] Isac, G., Leray–Schauder type alternative, Complementarity problems and Variational inequalities, Volume 87 (Nonconvex optimization and its applications), Springer, 2006.

[454] Ishikawa, S., Fixed points by a new iteration method, Proc. Amer. Math. Soc., 44 (1974), 147–150.

[455] Izmailov, A.F, Solodov, M.V., Inexact Josephy–Newton framework for generalized equations and its applications to local analysis of Newtonian methods for constrained optimization, Comput. Optim. Appl., 46 (2010), 347–368.

[456] Jarratt, P., Multipoint iterative methods for solving certain equations, Comput. J., 8 (1965/1966), 398–400.

[457] Jarratt, P., Some efficient fourth order multipoint methods for solving equations, BIT, 9 (1969), 119–124.

[458] Jean–Alexis, C., Piétrus, A., A variant of Newton's method for generalized equations, Rev. Colombiana Mat., 39 (2005), 97–112.

[459] Jean–Alexis, C., A cubic method without second order derivative for solving generalized equations, Compt. Rend. Acad. Bulg. Sci., 59 (2006), 1213–1218.

[460] Jerome, J.W., An adaptive Newton algorithm based on numerical inversion: regularization as postconditioner., Numer. Math., 47 (1985), 123–138.

[461] Jin, Q., A convergence analysis of the iteratively regularized Gauss–Newton method under the Lipschitz condition, Inverse Problems, 24 (2008), 1–24.

[462] Josephy, N.H., Newton's method for generalized equations, Technical Report No. 1965, Mathematics Research Center, University of Wisconsin, Madison, WI, 1979.

[463] Josephy, N.H., Quasi–Newton method for generalized equations, Technical Summary Report No. 1966, Mathematics Research Center, University of Wisconsin–Madison, June 1979, available from National Technical Information Service, Springfield, VA 22161, under Accession No. A077 097.

[464] Josephy, N.H., A Newton method for the PIES energy model, Technical Summary Report No. 1971, Mathematics Research Center, University of Wisconsin–Madison, June 1979, available from National Technical Information Service, Springfield, VA 22161, under Accession No. A077 102.

[465] Josephy, N.H., Hogan's PIES example and Lemke's algorithm, Technical Summary Report No. 1972, Mathematics Research Center, University of Wisconsin–Madison, June 1979, available from National Technical Information Service, Springfield, VA 22161, under Accession No. A077 103.

[466] Kaltenbacher, B., On Broyden's method for the regularization of nonlinear ill–posed problems, Numer. Funct. Anal. Optim., 19 (1998), 807–833.

[467] Kanno, S., Convergence theorems for the method of tangent hyperbolas, Math. Japon., 37 (1992), 711–722.

[468] Kantorovich, L.V., On Newton's method for functional equations (Russian), Dokl. Akad. Nauk. SSSR, 59 (1948), 1237–1240.

[469] Kantorovich, L.V., Akilov, G.P., Functional Analysis, Pergamon Press, Oxford, 1982.

[470] Kelley, C.T., Identification of the support of nonsmoothness, in Large scale optimization (Gainesville, FL, 1993), Kluwer Acad. Publ., Dordrecht, 1994, 192–205.

[471] Kelley, C.T., Iterative methods for linear and nonlinear equations, (with separately available software), Frontiers in Applied Mathematics, 16,

Society for Industrial and Applied Mathematics (SIAM), Philadelphia, PA, 1995.

[472] Kelley, C.T., Sachs, E.W., Multilevel algorithms for constrained compact fixed point problems, Iterative methods in numerical linear algebra (Copper Mountain Resort, CO, 1992), SIAM J. Sci. Comput., 15 (1994), 645–667.

[473] Kikuchi, F., Finite element analysis of a nondifferentiable nonlinear problem related to MHD equilibria, J. Fac. Sci. Univ. Tokyo Sect. IA Math., 35 (1988), 77–101.

[474] Kinderlehrer, D., Stampacchia, G., An introduction to variational inequalities and their applications, Academic Press, Editors: S. Eilenberg and H. Bass, 1980.

[475] King, A.J., Rockafellar, R.T., Sensitivity analysis for nonsmooth generalized equations, Math. Programming, 55 (1992), 193–212.

[476] King, R.F., Tangent methods for nonlinear equations, Numer. Math., 18 (1972), 298–304.

[477] Kneisl, K., Julia sets for the super–Newton method, Cauchy's method and Halley's method, Chaos, 11 (2001), 359–370.

[478] Krishnan, S., Manocha, D., An efficient surface intersection algorithm based on lower–dimensional formulation, ACM Trans. on Graphics, 16 (1997), 74–106.

[479] Kummer, B., Newton's method for non-differentiable functions, in J. Guddat et/ al. (eds.), Advances in Mathematical Optimization, Akademie-Verlag, Berlin, (1988), 171–194.

[480] Kummer, B., Newton's method based on generalized derivatives for nonsmooth functions: convergence analysis, in Oettli, W. and Pallaschke, D. (eds.): Advances in Optimization, Springer, Berlin, (1991), 171–194.

[481] Klatte, D., Kummer, B., Nonsmooth equations in optimization. Regularity, calculus, methods and applications. Nonconvex Optimization and its Applications, 60, Kluwer Academic Publishers, Dordrecht, 2002.

[482] Klatte, D., Kummer, B., Newton methods for stationary points: an elementary view of regularity conditions and solution schemes, Optim., 56 (2007), 441–462.

[483] Klatte, D., Kummer, B., Optimization methods and stability of inclusions in Banach spaces, Math. Program., Ser. B, 117 (2009), 305–330.

[484] Kojima, M., Shindo, S., Extensions of Newton and quasi–Newton methods to systems of $PC^1$ equations, J. Oper. Res. Soc. Japan, 29 (1986), 352–374.

[485] H. J. Kornstaedt, Funktionallongleichungen und iterations verfahren, Aequat. Math., 13 (1975), 21–45.

[486] Kornstaedt, H.J., Ein allgemeiner Konvergenzstaz für verschärfte Newton–Verfahrem, ISNM 28, Birkhaüser Verlag, Basel and Stuttgart, 1975, 53–69.

[487] Krasnosel'skii, M.A., Positive solutions of operator equations, Goz. Isdat. Fiz. Mat. Moscow 1962; Transl. by R. Flaherty and L. Boron, P. Noordhoff, Groningen, 1964.

[488] Krasnosel'skii, M.A., Topological Methods in the Theory of Nonlinear Integral Equations, Pergamon Press, London, 1966.

[489] Krasnosel'skii, M.A., Rutitskii, Ya.B., On the theory of equations with concave operators, SMZH, 10, 3, (1969).

[490] Krasnosel'skii, M.A., Vainikko, G.M., Zabrejko, P.P., Rutiskii, Ya.B.,Stetsenko, V.Ya., Approximate Solution of Operator Equations, Wolters–Noordhoff Publishing, Groningen, 1972.

[491] Krasnosel'skii, M.A., Zabrejko, P.P., Geometrical Methods of Nonlinear Analysis, Springer–Verlag, New York, 1984.

[492] Kung, H.T., The complexity of obtaining starting points for solving operator equations by Newton's method, Technical Report NR 044–422, Department of Computer Science, Carnegie–Mellon University, Pittsburgh, PA, 15213, October, 1975.

[493] Kuratowski, C., Sur les espaces complets, Fund. Math., 15 (1930), 301–309.

[494] Kurchatov, V.A., Optimization with respect to the number of arithmetic operations and entropy of difference methods of linearization (Russian), Izv. Vyssh. Uchebn. Zaved., 4 (1990), 33–37.

[495] Kwon, U.K., Redheffer, R.M., Remarks on linear equations in Banach space, Arch. Rational Mech. Anal., 32 (1969), 247–254.

[496] Laasonen, P., Ein überquadratisch konvergenter iterativer algorithmus, Ann. Acad. Sci. Fenn. Ser I, 450 (1969), 1–10.

[497] Lancaster, P., Error analysis for the Newton–Raphson method, Numer. Math., 9 (1968), 55–68.

[498] Laumen, M., A Kantorovich theorem for the structured PSB update in a Hilbert space, J. Optim. Theory Appl., 105 (2000), 391–415.

[499] Levin, Y., Ben–Israel, A., Directional Newton methods in $n$ variables, Mathematics of Computation, AMS, 71 (2002), 251–262.

[500] Lewy, H., Stampacchia, G., On the regularity of the solution of a variational inequality, Comm. Pure Appl. Math., 22 (1969), 153–188.

[501] Li, C., Shen, W.P., Local convergence of inexact methods under the Hölder condition, J. Comput. Appl. Math., 222 (2008), 544–560.

[502] Li, C., Zhang, W–H., Jin, X–Q., Convergence and uniqueness properties of Gauss–Newton's method, Comput. Math. Appl., 47 (2004), 1057–1067.

[503] Li, C., Wang, J., Convergence of the Newton method and uniqueness of zeros of vector fields on Riemannian manifolds, Sci. China Ser. A, 48 (2005), 1465–1478.

[504] Li, C., Hu, N., Wang, J., Convergence bahavior of Gauss–Newton's method and extensions to the Smale point estimate theory, J. Complexity, 26(3), (2010), 268–295.

[505] Li, D., Fukushima, M., Globally convergent Broyden–like methods for semismooth equations and applications to VIP, NCP and MCP, Optimization and numerical algebra (Nanjing, 1999), Ann. Oper. Res., 103 (2001), 71–97.

[506] Li, D., Zeng, J., Zhou, S., Convergence of Broyden–like matrix, Appl. Math. Lett., 11 (1998), 35–37.

[507] Lions, J.–L., Stampacchia, G., Variational inequalities, Comm. Pure Appl. Math., 20 (1967), 493–519.

[508] Liskovetz, O.A., Variational methods for solving nonstable problems, Nauka Technika, Minsk, (1981), (in Russian).

[509] Liusternik, L.A., Sobolev, V.J., Elements of functional analysis, Ungar Publ., 1961.

[510] Lukács, G., The generalized inverse matrix and the surface–surface intersection problem. Theory and practice of geometric modeling (Blaubeuren, 1988), 167–185, Springer, Berlin, 1989.

[511] Lysenko, J.V., Conditions for the convergence of the Newton–Kantorovich method for nonlinear equations with Hölder linearizations (in Russian), Dokl. Akad. Nauk BSSR, 38 (1994), 20–24.

[512] Ma, C., A smoothing Broyden–like method for the mixed complementarity problems, Math. Comput. Model., 41 (2005), 523–538.

[513] Magaril–Il'yaev, G.G., Tikhomirov, V.M., Convex analysis: theory and applications, Translated from the 2000 Russian edition by Dmitry Chibisov and revised by the authors, Translations of Mathematical Monographs, 222, AMS, Providence, RI, 2003.

[514] Maĭstrovskiĭ, G.D., The optimality of the Newton method, (Russian) Dokl. Akad. Nauk SSSR, 204 (1972), 1313–1315.

[515] Matveev, A.F., Yunganns, P., On the construction of an approximate solution of a nonlinear integral equation of permeable profile (Russian), Differ. Uravn., 33 (1997), 1242–1252, 1295. Translation in Differential Equations, 33, (1997), 1249–1259, (1998).

[516] Mann, W.R., Mean value methods in iteration, Proc. Amer. Math. Soc., 4 (1953), 506–510.

[517] Marcotte, P., Wu, J.H., On the convergence of projection methods, J. Optim. Theory Appl., 85 (1995), 347–362.

[518] Măruşter, S., Quasi–nonexpansivity and two classical methods for solving nonlinear equations, Proc. Amer. Math. Soc., 62 (1976), 119–123.

[519] McCormick, S.F., A revised mesh refinement strategy for Newton's method applied to nonlinear two–point boundary value problems, Numer. Treat. Diff. Equ. Appl. Proc., 679, Lecture Notes Math., Springer, (1978), 15–23.

[520] Meyer, P.W., Die Anwendung Verallgemeinerter Normen zer Fehlerabschätzung Bei Iteration Sverfahren, Dissertation, Düsseldort, 1980.

[521] Meyer, P.W., Das modifizierte Newton-Verfahren in verallgemeinerten Banach-Räumen, Numer. Math., 43 (1984), 91–104.

[522] Meyer, P.W., Newton's method in generalized Banach spaces, Numer. Funct. Anal. Optim., 9 (1987), 244–259.

[523] Meyer, P.W., A unifying theorem on Newton's method, Numer. Funct. Anal. Optim., 13 (1992), 463–473.

[524] Miel, G.J., Unified error analysis for Newton–type methods, Numer. Math., 33 (1979), 391–396.

[525] Miel, G.J., Majorizing sequences and error bounds for iterative methods, Math. Comput., 34 (1980), 185–202.

[526] Mifflin, R.,Semismooth and semiconvex functions in constrained optimization, SIAM J. Control Optim., 15 (1977), 959–972.

[527] Migovich, F.M., On the convergence of projection–iterative methods for solving nonlinear operator equations, Dopov. Akad. Nauk. Ukr. RSR, Ser. A, 1 (1970), 20–23.

[528] Minty, G.J., Monotone (nonlinear) operators in Hilbert space, Duke Math. J., 29 (1973), 341–346.

[529] Minty, G.J., On the monotonicity of the gradient of a convex function, Pacific J. Math., 14 (1964), 243–247.

[530] Miranda, C., Un osservatione su un teorema d, Brouwer, Ball. Unione Mat. Ital., Serr. 11, 3 (1940), 5–7.

[531] Mirsky, L., An Introduction to linear algebra, Clarendon Press, Oxford, England, 1955.

[532] Moore, R.H., Approximations to nonlinear operator equations and Newton's method, Numer. Math., 12 (1968), 23–34.

[533] Moore, R.E., A test for existence of solutions to nonlinear systems, SIAM J. Numer. Anal., 14 (1977), 611–615.

[534] Moore, R.E., Methods and applications of interval analysis, SIAM Publications, Philadelphia, Pa., 1979.

[535] Mordukhovich, B.S., Complete characterization of openness metric regularity and Lipschitzian properties of multifunctions, Trans. Amer. Math. Soc., 340 (1993), 1–36.

[536] Mordukhovich, B.S., Stability theory for parametric generalized equations and variational inequalities via nonsmooth analysis, Trans. Amer. Math. Soc., 343 (1994), 609–657.

[537] Mordukhovich, B.S., Coderivatives of set–valued mappings: calculs and applications, Nonlinear Anal., 30 (1997), 3059–3070.

[538] Mordukhovich, B.S., Variational analysis and generalized differentiation, I. Basic theory, II. Applications (two volumes), Springer, Grundlehren Series, Vol. 330 and 331, 2006.

[539] Moret, I., A note on Newton–type iterative methods, Comput. 33 (1984), 65–73.

[540] Moret, I., On the behaviour of approximate Newton methods, Comput., 37 (1986), 185–193.

[541] Moret, I., On a general iterative scheme for Newton–type methods, Numer. Funct. Anal. Optim., 9 (1987/1988), 1115–1137.

[542] Moret, I., A Kantorovich–type theorem for inexact Newton methods, Numer. Funct. Anal. Optim., 10 (1989), 351–365.

[543] Morini, B., Convergence behaviour of inexact Newton methods, Math. Comp., 68 (1999), 1605–1613.

[544] Moser, J., Stable and random motions in dynamical systems with special emphasis on celestial mechanics, Herman Weil Lectures, Annals of Mathematics Studies 77, Princeton Univ. Press, Princeton, NJ, 1973.

[545] Moudafi, A., Proximal point algorithm extended to equilibrium problems, J. Nat. Geom., 15 (1999), 91–100.

[546] Moudafi, A., Second–order differential proximal methods for equilibrium problems, JIPAM. J. Inequal. Pure Appl. Math., 4 (2003), Article 18, 7 pp., (electronic).

[547] Moudafi, A., A perturbed inertial proximal method for maximal monotone operators, Comm. Appl. Nonlinear Anal., 11 (2004), 101–107.

[548] Moudafi, A., A hybrid inertial projection–proximal method for variational inequalities, JIPAM. J. Inequal. Pure Appl. Math., 5 (2004), Article 63, 5 pp., (electronic).

[549] Moudafi, A., On finite and strong convergence of a proximal method for equilibrium problems, Numer. Funct. Anal. Optim., 28 (2007), 1347–1354.

[550] Mühlig, H., Lösung praktischer Approximationsaufgaben durch parameteridentifikation, ZAMM 73 (1993), 837–839.

[551] Mukaidani, H., Shimomura, T., Mizukami, K., Asymptotic expansions and a new numerical algorithm of the algebraic Riccati equation for multiparameter singularly perturbed systems, J. Math. Anal. Appl., 267 (2002), 209–234.

[552] Muroya, Y., Practical monotonous iterations for nonlinear equations, Mem. Fac. Sci. Kyushu Univ., Ser. A, 22 (1968), 56–73.

[553] Muroya, Y., Left subinverses of matrices and monotonous iterations for nonlinear equations, Memoirs of the Faculty of Science and Engineering, Waseda University, 34 (1970), 157–171.

[554] Mysovskii, I., On the convergence of Newton's method (in Russian), Trudy Mat. Inst. Steklov, 28 (1949), 145–147.

[555] Nagatou, K., Yamamoto, N., Nakao, M.T., An approach to the numerical verification of solutions for nonlinear elliptic problems with local uniqueness, Numer. Funct. Anal. Optim., 20 (1999), 543–565.

[556] Nashed, M.Z., Generalized inverse mapping theorems and related applications of generalized inverses in nonlinear analysis. Nonlinear equations in abstract spaces (Proc. Internat. Sympos., Univ. Texas, Arlington, Tex., 1977), 217–252, Academic Press, New York, 1978.

[557] Nayakkankuppam, M.V., Overton, M.L., Conditioning for semidefinite programs, Math. Program. Ser. A., 85 (1999), 525–540.

[558] Necepurenko, M.T., On Chebysheff's method for functional equations (Russian), Usephi, Mat. Nauk, 9 (1954), 163–170.

[559] Nerekenov, T.K., Necessary and sufficient conditions for uryson and nemytskii operators to satisfy a Lipschitz condition (Russian), VINITI 1459, 81, Alma–Ata, 1981.

[560] Nesterov, Y., Nemirovskii, A., Interior–point polynomial algorithms in convex programming, SIAM Studies in Appl. Math., 13, Philadelphia, PA, (1994).

[561] Nesterov, Y., Nemirovsky, A., Interior Point Polynomial Methods in Convex Programming, SIAM, Philadelphia, 1994.

[562] Neumaier, A., Shen, Z., The Krawczyk operator and Kantorovich's theorem, J. Math. Anal. Appl., 149 (1990), 437–443.

[563] Noble, B., The numerical solution of nonlinear integral equations and related topics, University Press, Madison, WI, (1964).

[564] Noor, K.I., Noor, M.A., Iterative methods for a class of variational inequalities, in Numerical Analysis of singular perturbation problems, Hemker and Miller, eds., Academic Press, New York, (1985), 441–448.

[565] Noor, M.A., An iterative scheme for a class of quasivariational inequalities, J. Math. Anal. Appl., 110 (1985), 463–468.

[566] Noor, M.A., Generalized variational inequalities, Appl. Math. Lett., 1 (1988), 119–122.

[567] Ojnarov, R., Otel'baev, M., A criterion for a Uryson operator to be a contraction (Russian), Dokl. Akad. Nauk. SSSR, 255 (1980), 1316–1318.

[568] Okuguchi, K., Expectations and stability in oligopoly models, Springer–Verlag, New York, (1976).

[569] Ortega, L.M., Rheinboldt, W.C., Iterative Solution of Nonlinear Equations in Several Variables, Academic press, New York, 1970.

[570] Ostrowski, A.M., Sur la convergence et l'estimation des erreurs dans quelques procédés de résolution des équations numériques. (French), Memorial volume dedicated to D. A. Grave [Sbornik posvjaščenii pamjati D. A. Grave], 213–234. publisher unknown, Moscow, 1940.

[571] Ostrowski, A.M., Solution of Equations and Systems of Equations, 2nd ed., Academic Press, New York, (1966).

[572] Ostrowski, A.M., La méthode de Newton dans les espaces de Banach, C. R. Acad. Sci. Paris Sér. A–B, 272 (1971), 1251–1253.

[573] Ostrowski, A.M., Solution of Equations in Euclidean and Banach Spaces, Academic press, New York, 1973.

[574] Owren, B., Welfert, B., The Newton Iteration on Lie Groups, BIT, 40 (2000), 121–145.

[575] Özban, A.Y., Some new variants of Newton's method, Appl. Math. Lett., 17 (2004), 677–682.

[576] Paardekooper, M.H.C., An upper and a lower bound for the distance of a manifeld to a nearby point, J. Math. Anal. Appl., 150 (1990), 237–245.

[577] Pandian, M.C., A convergence test and componentwise error estimates for Newton–type methods, SIAM J. Num. Anal., 22 (1985), 779–791.

[578] Pang, Jong–Shi, Newton's method for $B$–differentiable equations, Math. Oper. Res., 15 (1990), 311–341.

[579] Parida, P.K., Gupta, D.K., Recurrence relations for semilocal convergence of a Newton–like method in Banach spaces, J. Math. Anal. Appl., 345 (2008), 350–361.

[580] Patrikalakis, N.M., Surface–to–surface intersections, IEEE Computer graphics and applications, 13 (1993), 89–95.

[581] Păvăloiu, I., Sur la méthode de Steffensen pour la résolution des équations opérationnelles non linéaires, Rev. Roum. Math. Pures Appl., 13 (1968), 857–861.

[582] Păvăloiu, I., Introduction in the theory of approximation of equations solutions, Dacia Ed., Cluj–Napoca, 1976.

[583] Păvăloiu, I., Rezolvarea equaţiilor prin interpolare. Dacia Publ. Cluj–Napoca, Romania, (1981).

[584] Păvăloiu, I., A convergence theorem concerning the method of chord, Revue d'Analyse Numérique et. de Théorie de l'Approximation, 21 (1992), 59–65.

[585] Peng, Ji–M., Kanzow, C., Fukushima, M., A hybrid Josephy–Newton method for solving box constrained variational inequality problems via the D–gap function, Optim. Methods Softw, 10 (1999), 687–710.

[586] Penrose, R., A generalized inverse for matrices, Proc. Cambridge Philos. Soc., 51 (1955), 406–413.

[587] Pereyra, V., Iterative methods for solving nonlinear least square problems, SIAM J. Numer. Anal., 4 (1967), 27–36.

[588] Petryshin, W.V., Williamson, T.E., Strong and weak convergence of the sequence of successive approximation for quasi–nonexpansive mappings, J. Math. Anal. Appl., 43 (1973), 459–497.

[589] Petzeltova, H., Remark on a Newton–Moser type method, Comment. Math. Univ. Carolinae, 21 (1980), 719–725.

[590] Piétrus, A., Generalized equations under mild differentiability conditions, Rev. Real. Acad. Ciencias de Madrid, 94 (2000), 15–18.

[591] Piétrus, A., Does Newton's method converges uniformly in mild differentiability context?, Rev. Colombiana Math., 32 (2000), 49–56.

[592] Polyak, B.T., Introduction to optimization. Translated from the Russian. With a foreword by D.P. Bertsekas. Translations Series

in Mathematics and Engineering. Optimization Software, Inc., Publications Division, New York, 1987.

[593] Potra, F.A., On a modified secant method, Revue Anal. Numér. Théor. Approx., 8 (1979), 203–214.

[594] Potra, F.A., The rate of convergence of a modified Newton's process. With a loose Russian summary., Aplikace Matematiky, 26(1) (1981), 13–17.

[595] Potra. F.A., An application of the induction method of V. Ptăk to the study of Regula Falsi, Aplikace Matematiky, 26 (1981), 111–120.

[596] Potra, F.A., On the convergence of a class of Newton–like methods, Iterative Solution of Nonlinear Systems of Equations, (Oberwolfach, 1982), Lecture Notes in Math., 953, Springer, Berlin–New York, 1982, 125–137.

[597] Potra, F.A., An error analysis for the secant method, Numer. Math., 38 (1981/82), 427–445.

[598] Potra, F.A., On the a posteriori error estimates for Newton's method, Beiträge Zur Numer. Math., 12 (1984), 125–138.

[599] Potra, F.A., On an iterative algorithm of order 1.839... for solving nonlinear equations, Numer. Funct. Anal. Optim., 7 (1984–85), 75–106.

[600] Potra, F.A., Sharp error bounds for a class of Newton–like methods, Libertas Math., 5 (1985), 71–84.

[601] Potra, F.A., On $Q$–order and $R$–order of convergence, J. Optim. Theory Appl., 63 (1989), 415–431.

[602] Potra, F.A., The Kantorovich theorem and interior point methods, Math. Program., 102 (2005), 47–70.

[603] Potra, F.A., Ptăk, V., Nondiscrete induction and double step secant method, Math. Scand., 46 (1980), 236–250.

[604] Potra, F.A., Ptăk, V., On a class of modified Newton processes, Numer. Func. Anal. Optim., 2 (1980), 107–120.

[605] Potra, F.A., Ptăk, V., Sharp error bounds for Newton's process, Numer. Math., 34 (1980), 63–72.

[606] Potra, F.A., Pták, V., A generalization of regula falsi, Numer. Math., 36 (1980/81), 333–346.

[607] Potra, F.A., Ptăk, V., Nondiscrete induction and an inversion–free modification of Newton's method, Casopis pro pestováni matematiky, 108 (1983), 333–341.

[608] Potra, F.A., Ptăk, V., Nondiscrete induction and iterative processes, Research Notes in Mathematics 103, Pitman Advanced Publishing program, Boston, 1984.

[609] Potra, F.A., Wright, S.J., Interior–point methods, J. Comp. Appl. Math., 124 (2000), 281–302.

[610] Ptăk, V., The rate of convergence of Newton's process., Numer. Math., 25 (1976), 279–285.

[611] Pták, V., Nondiscrete mathematical induction and iterative existence proofs, Linear Algebra Appl, 13 (1976), 223–238.

[612] Pousin, J., Rappaz, J., Consistency, stability, a priori and a posteriori errors for Petrov-Galerkin's method applied to nonlinear problems, Numer. Math., 69 (1994), 213–231.

[613] Proinov, P.D., General local convergence theory for a class of iterative processes and its applications to Newton's method, J. Complexity, 25 (2009), 38–62.

[614] Proinov, P.D., New general convergence theory for iterative processes and its applications to Newton–Kantorovich type theorems, J. Complexity, 26 (2010), 3–42.

[615] Qi, L., Convergence analysis of some algorithms for solving nonsmooth equations, Math. Oper. Res., 18 (1993), 227–244.

[616] Qi, L., Sun, J., A nonsmooth version of Newton's method, Math. Program., 58 (1993), 353–367.

[617] Rall, L.B., Computational solution of nonlinear operator equations, Wiley, New York, (1968).

[618] Rall, L.B., Nonlinear functional analysis and applications, Academic Press, New York, (1971).

[619] Rall, L.B., A comparison of the existence theorems of Kantorovich and Moore, SIAM J. Numer. Anal., 17 (1980), 148–161.

[620] Ralph, D., On branching numbers of normal manifolds, Technical Report 92–1283, Department of Computer Science, Cornell University, Ithaca, NY, May 1992.

[621] Ralph, D., Global convergence of damped Newton's method for nonsmooth equations via the path search, Math. Oper. Res., 19 (1994), 352–389.

[622] Ralston, A., Rabinowitz, P., A first course in numerical analysis, 2nd Edition, Mc Graw–Hill, 1978.

[623] Ramm, A.G., Linear ill–posed problems and dynamical systems, J. Math. Anal. Appl., 258 (2001), 448–456.

[624] Ramm, A.G., Acceleration of convergence of a continuous analog of the Newton method, Appl. Anal., 81 (2002), 1001–1004.

[625] Ramm, A.G., Dynamical systems method for solving operator equations, Commun. Nonlinear. Sci. Numer. Simul., 9 (2004), 383–402.

[626] Ramm, A.G., Smimova, A.B., Continuous regularized Gauss–Newton–type algorithm for nonlinear ill–posed equations with simultaneous updates of inverse derivative, Int. J. Pure. Appl. Math., 2 (2002), 23–34.

[627] Ramm, A.G., Smimova, A.B., On stable numerical differentiation, Math. Comput., 70 (2001), 1131–1153.

[628] Ramm, A.G., Smimova, A.B., Favini, A., Continuous modified Newton's–type method for nonlinear operator equations, Annali di Math., 182 (2003), 37–52.

[629] Rappaz, J., Approximation of a nondifferentiable nonlinear problem related to MHD equilibria., Numer. Math., 45 (1984), 117–133.

[630] Redheffer, R., Walter, W., A comparison theorem for difference inequalities, J. Differ. Equat., 44 (1982), 111–117.

[631] Ren, H., On the local convergence of a deformed Newton's method under Argyros–type condition, J. Math. Anal. Appl., 321 (2006), 396–404.

[632] Ren, H., Wu, Q.B., The convergence ball of the Secant method under Hölder continuous divided differences, J. Comput. Appl. Math., 194 (2006), 284–293.

[633] Ren, H., Wu, Q.B., Mysovskii–type theorem for the Secant method under Hölder continuous Fréchet derivative, J. Math. Anal. Appl., 320 (2006), 415–424.

[634] Ren, H., Wu, Q., Convergence ball of a modified secant method with convergence order 1.839..., Appl. Math. Comput., 188 (2007), 281–285.

[635] Renegar, J., A polynomial–time algorithm, based on Newton's method, for linear programming, Math. Progr., 40, 1, Ser. A, (1988), 59–93.

[636] Renegar, J., Shub, M., Unified complexity analysis for Newton LP methods, Math. Program. Ser. A., 53 (1992), 1–16.

[637] Rheinboldt, W.C., A unified convergence theory for a class of iterative processes, SIAM J. Numer. Anal., 5 (1968), 42–63.

[638] Rheinboldt, W.C., An adaptive continuation process for solving systems of nonlinear equations, Polish Academy of Science, Banach Ctr. Publ., 3 (1977), 129–142.

[639] Rheinboldt, W.C., Solution fields of nonlinear equations and continuation methods, SIAM J. Numer. Anal., 17 (1980), 221–237.

[640] Rheinboldt, W.C., Error estimation for nonlinear parametrized problems, Proceedings of the international conference on innovative methods for nonlinear problems, (New Orleans, La., 1984), Pineridge, Swansea, (1984), 295–311.

[641] Rheinboldt, W.C., On a theorem of S. Smale about Newton's method for analytic mappings, Appl. Math. Lett., 1 (1988), 69–72.

[642] Rheinboldt, W.C., On the sensitivity of solutions of parameterized equations, SIAM J. Numer. Anal., 30 (1993), 305–320.

[643] Rheinboldt, W.C., On the theory and error estimation of the reduced basis method for multi–parameter problems, Nonlinear Anal., 21 (1993), 849–858.

[644] Rheinboldt, W.C., Methods for solving systems of nonlinear equations, Second edition, CBMS–NSF, Regional Conference Series in Applied Mathematics, 70, Society for Industrial and Applied Mathematics (SIAM), Philadelphia, PA, 1998.

[645] Rheinboldt, W.C., Numerical continuation methods: a perspective, Numerical analysis, 2000, Vol. IV, Optimization and nonlinear equations, J. Comput. Appl. Math., 124 (2000), 229–244.

[646] Robinson, S.M., Generalized equations and their solutions, part I: basic theory, Math. Program. Study, 10 (1979), 128–141.

[647] Robinson, S.M., Strongly regular generalized equations, Math. Oper. Res., 5 (1980), 43–62.

[648] Robinson, S.M., Generalized equations and their solutions, part II: applications to nonlinear programming, Math. Program. Study, 19 (1982), 200–221.

[649] Robinson, S.M., Generalized equations, in: A. Bachem, M. Grŏtschel and B. Korte, eds., Mathematical programming: the state of the art (Springer, Berlin), (1982), 346–367.

[650] Robinson, S.M., Generalized equations. Mathematical programming: the state of the art (Bonn, 1982), 346–367, Springer, Berlin, 1983.

[651] Robinson, S.M., Newton's method for a class of nonsmooth functions, Set–Valued Anal., 2 (1994), 291–305.

[652] Robinson, S.M., Nonsmooth continuation for generalized equations, Recent advances in optimization (Trier, 1996), 282–291, Lecture Notes in Econom. and Math. Systems, 452, Springer, Berlin, 1997.

[653] Robinson, S.M., A reduction method for variational inequalities, Math. Program., 80 (1998), 161–169.

[654] Robinson, S.M., Structural methods in the solution of variational inequalities, Nonlinear optimization and related topics (Erice, 1998), 369–380, Appl. Optim., 36, Kluwer Acad. Publ., Dordrecht, 2000.

[655] Robinson, S.M., Variational conditions with smooth constraints: structure and analysis, Math. Program., Ser. B, 97 (2003), 245–265.

[656] Robinson, S.M., Solution continuity in monotone affine variational inequalities, SIAM J. Optim., 18 (2007), 1046–1060.

[657] Robinson, S.M., Lu, S., Solution continuity in variational conditions, J. Glob. Optim., 40 (2008), 405–415.

[658] Rockafellar, R.T., Convex analysis, Princeton University Press, Princeton, 1967.

[659] Rockafellar, R.T., Lipschitzian properties of multifunctions, Nonlinear Anal., 9 (1984), 867–885.

[660] Rockafellar, R.T., Wets, R. J–B., Variational analysis, A Series of Comprehensives Studies in Mathematics, Springer, 317, 1998.

[661] Rokne, J., Newton's method under mild differentiability conditions with error analysis, Numer. Math., 18 (1972), 401–412.

[662] Roos, C., Vial, J.–Ph., Terlaky, T., Theory and algorithms for linear optimization: an interior point approach, Wiley Interscience Series in Discrete Mathematics and Optimization, John Wiley and Sons, 1997.

[663] Schlenkrich, S., Walther, A., Global convergence of quasi–Newton methods based on adjoint Nroyden updates, Appl. Numer. Math., 59 (2009), 1120–1136.

[664] Schomber, H., Monotonically convergent iterative methods for nonlinear systems of equations, Numer. Math., 32 (1979), 97–104.

[665] Schmidt, J.W., Eine übertragung der Regula Falsi and gleichungen in Banachräumen, I, Z. Angew. Math. Mech., 43 (1963), 97–110.

[666] Schmidt, J.W., Eine Übertragung der Regula Falsi auf Gleichungen in Banachräumen, II, Nichtlineare Gleichungssysteme, (German), Z. Angew. Math. Mech., 43 (1963), 97–110.

[667] Schmidt, J.W., Überlinear konvergente Mehrschrittverfahren vom Regula falsi– und Newton–Typ, (German), Z. Angew. Math. Mech., 53 (1973), 103–114.

[668] Schmidt, J.W., Untere Fehlerschranken fur Regula–Falsi Verhafren, Period. Hungar., 9 (1978), 241–247.

[669] Schmidt, J.W., Leonhardt, H., Eingrenzung von Losungen mit der Regala Falsi, Computing, 6 (1970), 318–329.

[670] Schmidt, J.W., Schwetlick, H., Ableitungsfreie Verfahren mit hohere Konvergenzgeschwindigkeit, Comput., 3 (1968), 215–226.

[671] Schwetlick, H., Numerische Lösung nichtlinearer Gleichungen, R. Oldenburg Verlag, Munchen Wien, 1979.

[672] Schröder, J., Nichtlineare majoranten beim verfahren der schrittweisen näherung, Arch. Math., 7 (1956), 471–484.

[673] Sergeev, A.S., The method of chords, (Russian), Sibirsk. Mat. Z., 2 (1961), 282–289.

[674] Shakhno, S.M., Gnatyshyn, O.P., On an iterative algorithm of order 1.839... for solving the nonlinear least squares problems, Appl. Math. Comput., 161 (2005) 253–264.

[675] Shapiro, A., On concepts of directional differentiability, J. Optim. Theory Appl., 66, (1990), 477–487.

[676] Shapiro, A., Sensitivity analysis of parametrized programs via generalized equations, SIAM J. Control Optim., 32 (1994), 553–571.

[677] Shapiro, A., A variational principle and its applications, Parametric optimization and related topics, IV (Enschede, 1995), Approx. Optim., 9, Lang, Frankfurt am Main, (1997), 341–357.

[678] Shapiro, A., Sensitivity analysis of generalized equations, Optimization and related topics, 1, J. Math. Sci. (N.Y.), 115 (2003), 2554–2565.

[679] Shapiro, A., Sensitivity analysis of parameterized variational inequalities, Math. Oper. Res., 30 (2005), 109–126.

[680] Shub, M., Smale, S., Complexity of Bezout's theorem 1: geometric aspects, J. Amer. Math. Soc., 6 (1993), 459–501.

[681] Slugin, S.N., Approximate solution of operator equations on the basis of Caplygin method, (Russian), Dokl. Nauk SSSR, 103 (1955), 565–568.

[682] Slugin, S.N., Monotonic processes of bilateral approximation in a partially ordered convergence group, Soviet. Math., 3 (1962), 1547–1551.

[683] Smale, S., Newton method estimates from data at one point. The merging of disciplines: new directions in pure, applied and computational mathematics (R. Ewing, K. Gross, C. Martin, eds), Springer–Verlag, New York, (1986), 185–196.

[684] Smale, S., Newton's contribution and the computer revolution, Math. Medley, 17 (1989), 51–57.

[685] Smith, D.A., Ford, W.F., Sidi, A., Extrapolation methods for vector sequences, SIAM Rev., 29 (1987), 199–233.

[686] Smith, S.T., Optimization techniques on Riemannian manifolds, Hamiltonian and gradient flows, Algorithms and control, Fields Inst. Commun., AMS, 3 (1994), 113–146.

[687] Solodov, M.V., Svaiter, B.F., A globally convergent inexact Newton method for systems of monotone equations, Reformulation: nonsmooth, piecewise smooth, semismooth and smoothing methods, (Lausanne, 1997), Appl. Optim., 22, Kluwer Acad. Publ., Dordrecht, (1999), 355–369.

[688] Solodov, M.V., Svaiter, B.F., A truly globally convergent Newton–type method for the monotone nonlinear complementarity problem, SIAM J. Optim., 10 (2000), 605–625.

[689] Solodov, M.V., Svaiter, B.F., A new proximal–based globalization strategy for the Josephy–Newton method for variational inequalities, Optim. Methods Softw., 17 (2002), 965–983.

[690] Stakgold, I., Green's Functions and Boundary Value Problems, Wiley, New York, 1998.

[691] Stampacchia, G., Formes bilinéaires coercitives sur les ensembles convexes, Comptes Rendus de L'Académie des Science de Paris, 258 (1964), 4413–4416.

[692] Stampacchia, G., Variational inequalities, Theory and Applications of Monotone Operators, (Proc. NATO Advanced Study Inst., Venice, 1968), Edizioni "Oderisi", Gubbio, (1969), 101–192.

[693] Stampacchia, G., Regularity of solutions of some variational inequalities, Nonlinear Functional Analysis, (Proc. Sympos. Pure Math., Vol. XVIII, Part 1, Chicago, Ill., 1968), Amer. Math. Soc., Providence, R.I., (1970), 271–281.

[694] Stampacchia, G., On the regularity of solutions of variational inequalities, Proc. Internat. Conf. on Functional Analysis and Related Topics, (Tokyo, 1969), Univ. of Tokyo Press, Tokyo, (1970), 285–289.

[695] Stirling, J., Methodus differentialis: sive tractatus de summatione et. interpolations serierum infiniterum, W. Boyer, London, 1730.

[696] Stoer, J., Bulirsch, K., Introduction to numerical analysis, Springer–Verlag, 1976.

[697] Stoffer, D., Palmer, K.J., Rigorous verification of chaotic behaviour of maps using validated shadowing, Nonlinearity, 12 (1999), 1683–1698.

[698] Szidarovszky, F., Yakowitz, S., Principles and procedures in numerical analysis, Plenum, New York, 1978.

[699] Szidarovszky, F., Bahill, T., Linear systems theory, CRC Press, Boca Raton, FL, 1992.

[700] Tan, K.K., Xu, H.K., Approximating fixed points of nonexpansive mappings by the Ishikawa iteration process, J. Math. Anal. Appl., 178 (1993), 301–308.

[701] Tapia, R.A., The Kantorovich theorem for Newton's method, Amer. Math. Monthly, 11 (1971), 10–13.

[702] Tapia, R.A., Newton's method for optimization problems with equality constraints, SIAM J. Numer. Anal., 11 (1974), 874–886.

[703] Tapia, R.A., A general approach to Newton's method for Banach space problems with equality constraints, Bull. Amer. Math. Soc., 80 (1974), 355–360.

[704] Tapia, R.A., A stable approach to Newton's method for general mathematical programming problems in $\mathbb{R}^n$, Collection of articles dedicated to Magnus R. Hestenes., J. Optim. Theory Appl., 14 (1974), 453–476.

[705] Taylor, A.E., Introduction to functional analysis, Wiley, New York, (1957).

[706] Tishyadhigama, S., Polak, E., Klessig, R., A comparative study of several convergence conditions for algorithms modeled by point–to–set maps, Math. Program. Stud., 10 (1979), 172–190.

[707] Törnig, W., Monoton konvergente Iterationsverfahren zür Lösung michtlinearer differenzen–randwertprobleme, Beiträge zür Numer. Math., 4 (1975), 245–257.

[708] Traub, J.F., Iterative Methods for the Solution of Equations, Englewood Cliffs, New Jersey: Prentice Hall, 1964.

[709] Traub, J.F. (editor), Analytic Computational Complexity, Academic Press, 1975.

[710] Traub, J.F., Woźniakowsi, H., Convergence and complexity of Newton iteration for operator equations, J. Assoc. Comput. Mach., 26 (1979), 250–258.

[711] Traub, J.F., Woźniakowsi, H., A general theory of optimal algorithms, ACM Monograph Series, Academic Press, Inc., Harcourt Brace Jovanovich, Publishers, New York, London, 1980.

[712] Traub, J.F., Woźniakowsi, H., Optimal radius of convergence of interpolatory iterations for operator equations, Aequat. Math., 21 (1980), 159–172.

[713] Traub, J.F., Woźniakowsi, H., Convergence and complexity of interpolatory–Newton iteration in a Banach space, Comput. Math. Appl., 6 (1980), 385–400.

[714] Tsuchiya, T., An application of the Kantorovich theorem to nonlinear finite element analysis, Numer. Math., 84 (1999), 121–141.

[715] Tsuchiya, T., Babuska, I., A priori error estimates of finite element solutions of parametrized strongly nonlinear boundary value problems, J. Comp. Appl. Math., 79 (1997), 41–66.

[716] Ulbrich, M., Semismooth Newton methods for operator equations in function spaces, SIAM J. Optim., 13 (2003), 805–841.

[717] Uko, L.U., On a class of general strongly nonlinear quasivariational inequalities, Revista di Math. Pura ed Appl., 11 (1992), 47–55.

[718] Uko, L.U., Generalized equations and the generalized Newton method, Math. Program., 73 (1996), 251–268.

[719] Uko, L.U., Adeyeye, J.O., Generalized Newton–iterative methods for nonlinear operator equations, Nonlinear Studies, 8 (2001), 465–477.

[720] Uko, L.U., Argyros, I.K., A weak Kantorovich existence theorem for the solution of nonlinear equations, J. Math. Anal. Appl., 2 (2008), 909–914.

[721] Uko, L.U., Argyros, I.K., A generalized Kantorovich theorem for nonlinear equations based on function splitting, Rend. Circ. Mat. Palermo (2), 58 (2009), 441–451.

[722] Uko, L.U., Argyros, I.K., An extension of Argyros' Kantorovich-type solvability theorem for nonlinear equations, submitted for publication.

[723] Uko, L.U., Argyros, I.K., Generalized equations, variational inequalities, and a weak Kantorovich theorem, Numerical Algorithms, 52 (2009), 321–333.

[724] Uko, L.U., Orozco, J.C., Some $p$–norm convergence results for Jacobi and Gauss–Seidel iterations, Revista Colombiana de Math., 38 (2004), 65–71.

[725] Uko, L.U., Velásquez Ossa, R.E., Convergence Analysis of a one–step intermediate Newton iterative scheme, Revista Colombiana de Math., 35 (2001), 21–27.

[726] Ulm, S., On iterative methods with successive approximation of the inverse operator (in Russian), Tzv. Akad. Nauk Est. SSR, 16 (1967), 403–411.

[727] Urabe, M., Convergence of numerical iteration in solution of equations, J. Sci. Hiroshima Univ., Ser. A, 19 (1976), 479–489.

[728] Uzawa, H., The stability of dynamic processes, Econometrica, 29 (1961), 617–631.

[729] Vainberg, M.M., Variational method and method of monotone operators (in Russian), Nauka, M. (1972).

[730] Vainikko, G.M., Galerkin's perturbation method and the general theory for approximate methods for nonlinear equations, USSR Comp. Math. Math. Phys., 4 (1967), 723–751.

[731] Van Den Dries, L., Miller, C., Geometric categories and $O$–minimal structures, Duke Math. J., 84 (1996), 497–540.

[732] Vandergraft, J.S., Newton's method for convex operators in partially ordered spaces, SIAM J. Numer. Anal., 4 (1967), 406–432.

[733] Varga, R.S., Matrix iterative analysis, Prentice–Hall, Englewood Cliffs, NJ, 1962.

[734] Varona, J.L., Graphic and numerical comparison between iterative methods, Math. Intelligencer, 24 (2002), 37–46.

[735] Verma, R.U., A class of projection–contraction methods applied to monotone variational inequalities, Appl. Math. Lett., 13 (2000), 55–62.

[736] Verma, R.U., General class of relaxed pseudococoercive nonlinear variational inequalities and relaxed projection methods, Adv. Nonlinear Var. Inequal., 8 (2005), 131–140.

[737] Verma, R.U., Nonlinear pseudococoercive variational problems and projection methods, Nonlinear Funct. Anal. Appl., 10 (2005), 641–649.

[738] Verma, R.U., General convergence analysis for two-step projection methods and applications to variational problems, Appl. Math. Lett., 18 (2005), 1286–1292.

[739] Verma, R.U., General projection systems and relaxed cocoercive nonlinear variational inequalities, ANZIAM J., 49 (2007), 205–212.

[740] Verma, R.U., Existence theorems for a class of generalized vector variational inequality problems, Fixed Point Theory and Applications, Ed. by Y. J. Cho, J. K. Kim, S. M.Kang, Vol. 6, Nova Sci. Publ., New York, (2007), 165–170.

[741] Verma, R.U., Two–step models for projection methods and their applications, Math. Sci. Res. J., 11 (2007), 444–453.

[742] Vertgeim, B.A., On conditions for the applicability of Newton's method, (Russian) Dokl. Aknd. N., SSSR, 110 (1956), 719–722.

[743] Vertgeim, B.A., On some methods for the approximate solution of nonlinear functional equations in Banach spaces, Uspekhi Mat. Nauk, 12 (1957), 166–169 (in Russian); English transl.: Amer. Math. Soc. Transl., 16 (1960), 378–382.

[744] Vertgeim, B.A., Optimal alternation of the fundamental and the modified Newton–Kantorovich process, (Russian), Optimal. Planirovanie, 17 (1970), 10–31.

[745] Vertgeim, B.A., Certain generalizations of the Newton–Kantorovich method of approximate solution of nonlinear operator equations, (Russian), Optimizacija Vyp., 16 (1975), 84–90.

[746] Walker, H.F., Watson, L.T., Least–change Secant update methods for underdetermined systems, SIAM J. Numer. Anal., 27 (1990), 1227–1262.

[747] Wang, D.R., Zhao, F.G., The theory of Smale's point estimation and its applications, Linear/nonlinear iterative methods and verification of solution (Matsuyama, 1993), J. Comput. Appl. Math., 60 (1995), 253–269.

[748] Wang, G.R., Wei, Y.M., Qiao, S.Z., Generalized inverse: theory and computations, Science Press, Beijing, New York, 2004.

[749] Wang, J.H., Li, C., Kantorovich's theorem for Newton's method on Lie groups, J. Zhejiang Univ. Sci. A, 8 (2007), 978–986.

[750] Wang, X.H., The convergence ball on Newton's method, (in Chinese), Chinese Science Bulletin, A Special Issue of Mathematics, Physics, Chemistry, 25 (1980), 36–37.

[751] Wang, X., Convergence of Newton's method and inverse function theorem in Banach space, Math. Comput., 68 (1999), 169–186.

[752] Wang, X.H., Convergence on the iteration of Halley family in weak condition, Chinese Sci. Bull., 42 (1997), 552–555.

[753] Wang, X.H., Convergence of Newton's method and uniqueness of the solution of equations in Banach space, IMA J. Numer. Anal., 20 (2000), 123–134.

[754] Wang, X.H., Han, D.F., On dominating sequence method in the point estimate and Smale theorem, Sci. China Ser. A, 33 (1990), 135–144.

[755] Wang, X.H., Han, D.F., Criterion $\alpha$ and Newton's method under weak conditions, (Chinese) Math. Numer. Sin., 19(1) (1997), 103–112; translation in Chinese J. Numer. Math. Appl., 19(2) (1997), 96–105.

[756] Wang, X.H., Li, C., The local and global behaviors of methods for solving equations, (Chinese) Kexue Tongbao (Chinese), 46(6) (2001), 444–451.

[757] Wang, X.H., Li, C., Local and global behavior for algorithms of solving equations, Chinese Sci. Bull., 46(6) (2001), 441–448.

[758] Wang, X.H., Li, C., Convergence of Newton's method and uniqueness of the solution of equations in Banach spaces, II, Acta Math. Sin. (Engl. Ser.), 19 (2003), 405–412.

[759] Wilkinson, J.H., The Algebraic Eigenvalue Problem, Clarendon, Oxford, 1965.

[760] Wang, Z., Semilocal convergence of Newton's method for finite–dimensional variational inequalities and nonlinear complementarity problems, Doctoral thesis, Fakultăt fűr Mathematik, Universităt Karlsruhe, Germany, June 2005.

[761] Wang, Z., Shen, Z., Kantorovich theorem for variational inequalities, Appl. Math. Mech., 25 (2004), 1291–1297.

[762] Wang, X.H., Xuan, X.H., Random polynomial space and computational complexity theory, Sci. Sinica Ser. A, 30 (1987), 673–684.

[763] Wang, Z.Y., Extensions of the Newton–Kantorovich theorem to variational inequality problems, preprint, 2009 http://math.nju.edu.cn/.

[764] Weerakoon, S., Fernando, T.G.I., A variant of Newton's method with accelerated third–order convergence, Appl. Math. Lett., 13 (2000), 87–93.

[765] Werner, W., Ŭber ein Verfahren der Ordung $1+\sqrt{2}$ zur Nullstellenbestmmung, Numer. Math., 32, (1979), 333–342.

[766] Werner, W., Some supplementary results on the $1+\sqrt{2}$ order method for the solution of equations, Numer. Math., 38 (1982), 383–392.

[767] Werner, W., Newton–like method for the computation of fixed points, Comput. Math. Appl., 10 (1984), 77–84.

[768] Wolfe, M.A., A quasi–Newton method with memory for unconstrained function minimization, J. Inst. Math. Appl., 15 (1975), 85–94.

[769] Wolfe, M.A., Extended iterative methods for the solution of operator equations, Numer. Math., 31 (1978), 153–174.

[770] Wolfe, M.A., An existence–convergence theorem for a class of iterative methods, J. Optim. Theory Appl., 31 (1980), 153–174.

[771] Wolfe, M.A., A modification of Krawczyk's algorithm, SIAM J. Numer. Anal., 17 (1980), 376–379.

[772] Wolfe, M.A., On the convergence of some methods for determining zeros of order–convex operators, Comput., 26 (1981), 45–56.

[773] Wright, S.J., Primal–Dual Interior–Point Methods, SIAM, Philadelphia, 1997.

[774] Wu, J.W., Brown, D.P., Global asymptotic stability in discrete systems, J. Math. Anal. Appl., 140 (1989), 224–227.

[775] Wu, Q., Ren, H., A note on some new iterative methods with third–order convergence, Appl. Math. Comput., 188 (2007), 1790–1793.

[776] Wu, Q., Zhao, Y., Third–order convergence theorem by using majorizing function for a modified Newton method in Banach space, Appl. Math. Comput., 175 (2006), 1515–1524.

[777] Xiao, B., Harker, P.T., A nonsmooth Newton method for variational inequalities, I: theory, Math. Programming, 65 (1994), 151–194.

[778] Xiao, B., Harker, P.T., A nonsmooth Newton method for variational inequalities: II: numerical results, Math. Program., 65 (1994), 195–216.

[779] Xu, H., Glover, B. M., New version of the Newton method for nonsmooth equations, J. Optim. Theory and Appl., 93 (1997), 395–415.

[780] Xu, H., Chang, X. W., Approximate Newton methods for nonsmooth equations, J. Optim. Theory Appl., 93 (1997), 373–394.

[781] Xu, X.B., Li, C., Convergence of Newton's method for systems of equations with constant rank derivatives, J. Comput. Math., 25 (2007), 705–718.

[782] Xu, X.B., Li, C., Convergence criterion of Newton's method for singular systems with constant rank derivatives, J. Math. Anal. Appl., 345 (2008), 689–701.

[783] Xue, Z.Q., Zhou, H.Y., Cho, Y.J., Iterative solution of nonlinear equations for $m$–accretive operators in Banach spaces, J. Non. Con. Anal., 1 (2000), 313–320.

[784] Yakoubsohn, J.C., Finding zeros of analytic functions : $\alpha$–theory for Secant type method, J. of Complexity, 15, (1999), 239–281.

[785] Yamamoto, T., Error bounds for Newton's process derived from the Kantorovich conditions, Japan J. Appl. Math., 2 (1985), 285–292.

[786] Yamamoto, T., A method for finding sharp error bounds for Newton's method under the Kantorovich assumptions, Numer. Math., 49 (1987), 203–230.

[787] Yamamoto, T., Newton's method and its applications, (Japanese) Sugaku Expositions, 1, (1988), no. 2, 219–238, Sūgaku, 37 (1985), 1–15.

[788] Yamamoto, T., A unified derivation of several error bounds for Newton's process, Proceedings of the international conference on computational and applied mathematics, (Leuven, 1984), J. Comput. Appl. Math., 12/13, (1985), 179–191.

[789] Yamamoto, T., Error bounds for Newton–like methods under Kantorovich type assumptions, Japan J. Appl. Math., 3 (1986), 295–313.

[790] Yamamoto, T., A convergence theorem for Newton's method in Banach spaces, Japan J. Appl. Math., 3 (1986), 37–52.

[791] Yamamoto, T., Error bounds for Newton's method under the Kantorovich assumptions, The merging of disciplines: new directions in pure, applied, and computational mathematics (Laramie, Wyo., 1985), Springer, New York, (1986), 197–208.

[792] Yamamoto, T., A convergence theorem for Newton–like methods in Banach spaces, Numer. Math., 51, (1987), 545–557.

[793] Yamamoto, T., Uniqueness of the solution in a Kantorovich–type theorem of Häussler for the Gauss–Newton method, Japan J. Appl. Math., 6 (1989), 77–81.

[794] Yamamoto, T., Historical developments in convergence analysis for Newton's and Newton–like methods, Numerical analysis 2000, Vol. IV, Optimization and nonlinear equations, J. Comput. Appl. Math., 124, (1–2), (2000), 1–23.

[795] Ye, Y., Interior Point Algorithms: Theory and Analysis, Wiley–Interscience Series in Discrete Mathematics and Optimization, John Wiley and Sons, 1997.

[796] Ypma, T.J., Numerical solution of systems of nonlinear algebraic equations, Ph. D. thesis, Oxford, 1982.

[797] Ypma, T.J., Affine invariant convergence results for Newton's methods, BIT, 22 (1982), 108–118.

[798] Ypma, T.J., Local convergence of difference Newton–like methods, Math. Comput., 41 (1983), 527–536.

[799] Ypma, T.J., The effect of rounding error on Newton–like methods, IMA J. Numer. Anal., 3 (1983), 109–118.

[800] Ypma, T.J., Convergence of Newton–like iterative methods, Numer. Math., 45 (1984), 241–251.

[801] Ypma, T.J., Local convergence of inexact Newton methods, SIAM J. Numer. Anal., 21 (1984), 583–590.

[802] Yau, L., Ben–Israel, A., The Newton and Halley methods for complex roots, Amer. Math. Monthly, 105 (1998), 806–818.

[803] Zaanen, A.C., Linear analysis, North–Holland Publ., Amsterdam, 1953.

[804] Zabrejko, P.P., $K$–metric and $K$–normed linear spaces: A survey, Collect. Math., 48 (1997), 825–859.

[805] Zabrejko, P.P., Majorova, N.L., On the solvability of nonlinear Uryson integral equations (Russian), Kach. Pribl. Metody Issled. Oper. Uravn., 3 (1978), 61–73.

[806] Zabrejko, P.P., Nguen, D.F., The majorant method in the theory of Newton–Kantorovich approximations and the Ptăk error estimates, Numer. Funct. Anal. Optim., 9 (1987), 671–684.

[807] Zabrejko, P.P., Zlepko, P.P., On a generalization of the Newton-Kantorovich method for an equation with a non-differentiable operator (Russian), Ukr. Mat. Zhurn. 34 (1982), 365–369.

[808] Zabrejko, P.P., Zlepko, P.P., On majorants of Uryson integral operators (Russian), Kach. Pribl. Metody Issled. Oper. Uravn., 8 (1983), 67–76.

[809] Zehnder, E.J., A remark about Newton's method, Comm. Pure Appl. Math., 27 (1974), 361–366.

[810] Zeidler, E., Nonlinear Functional Analysis and Its Applications, I: Fixed Point Theorems, Springer–Verlag, New York, 1986.

[811] Zhao, F., Wang, D., The theory of Smale's point estimation, and its applications, J. Comput. Appl. Math., 60 (1995), 253–259.

[812] Zinčenko, A.I., Some approximate methods of solving equations with non–differentiable operators, (Ukrainian), Dopovidi Akad. Nauk Ukraïn. RSR, (1963), 156–161.

[813] Zinčenko, A.I., A class of approximate methods for solving operation equations with nondifferentiable operators, Dopovidi Akad. Nauk Ukrain. RSR, (1963), 156–161.

[814] Zlepko, P.P., Migovich, F.M., an application of a modification of the Newton–Kantorovich method to the approximate construction of implicit functions (Ukrainian), Ukrainskii Math. Zhurnal, 30 (1978), 222–226.

[815] Zuhe, S., Wolfe, M.A., A note on the comparison of the Kantorovich and Moore theorems, Nonlinear Anal., 15 (1990), 229–232.

# Index